S0-CAH-207

Lecture Notes in Chemistry 76

Springer
Berlin
Heidelberg
New York
Barcelona
Hong Kong
London
Milan
Paris
Singapore
Tokyo

Pekka Pyykkö

Relativistic Theory of Atoms and Molecules III

A Bibliography 1993–1999

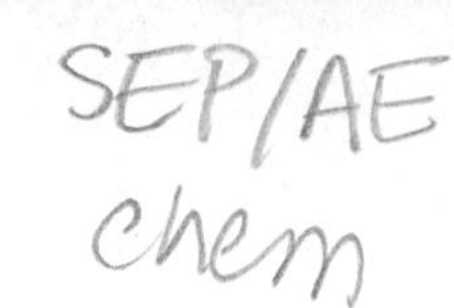

Author

Prof. Pekka Pyykkö
Department of Chemistry
University of Helsinki
POB 55 (A.I. Virtasen aukio 1)
FIN-00014 Helsinki
Finland

E-Mail: *Pekka.Pyykko@helsinki.fi*

Cataloging-in-Publication Data applied for

Pyykkö, Pekka:
Relativistic theory of atoms and molecules : a bibliography / P. Pyykkö. - Berlin ; Heidelberg ; New York ; London ; Paris ; Tokyo ; Hong Kong ; Barcelona ; Budapest : Springer

3. 1993 - 1999. - 2000
(Lecture notes in chemistry ; Vol. 76)
ISBN 3-540-41398-7

ISSN 0342-4901
ISBN 3-540-41398-7 Springer-Verlag Berlin Heidelberg New York

Springer-Verlag Berlin Heidelberg New York
a member of BertelsmannSpringer Science+Business Media GmbH

Printed in Germany

Typesetting: Camera ready by author
Printed on acid-free paper SPIN: 10790835 51/3142 - 543210

Preface

Seven years have elapsed since "RTAM II" was published in 1993. Together with "RTAM I", published in 1986, it contained 6577 references. The present volume covers the years 1993-99. The running numbers in the present volume begin from 6578 and end at 10369. Gradually, the present volumes seem to be forming a *Köchel-Verzeichnis* of relativistic calculations on atoms and molecules. In the present volume, a certain degree of completeness on molecular production work was still attempted, although certainly not reached. In the future that would be both more and more difficult and less and less meaningful.

The molecular production work using all methods is now collected to the single Table 7.10 and ordered by the element, or group of elements, considered characteristic. This will facilitate the comparisons between different methods.

The data base corresponding to the latter part of the book will be made available in the World Wide Web at *http://www.csc.fi/lul/rtam*. The analysis in the tables is not available in the Web.

I thank the *Royal Society of Chemistry* and my coauthor *Herman Stoll* for permission to include in the present Table 7.10 material from my review on pseudopotential calculations in *Royal Society Periodical Reports, Chemical Modelling, Applications and Theory (2000).*

I apologise for any inadvertent omissions or characterisations that the author would not find fitting. No legal responsibility is accepted in such cases.

I thank *Jonas Jusélius* for technical advice and The Academy of Finland for a Research Professorship in August 1995–July 2000. I also thank the Springer-Verlag for constructive cooperation during the past years.

Helsinki, 31 October, 2000

Pekka Pyykkö

Contents

List of Tables

Chapter 1

Introduction

Table 1.1. gives a list of treatises and other general references. For a list of symbols and acronyms, see the end of the book.

Table 1.1: Monographs, reviews and other general references.

Reference	Comments
Kessler (1985)	Polarized electrons.
Lindgren and Morrison (1986)	Atomic many-body theory.
Benn and Tucker (1988)	Introduction to spinors.
Ryder (1988)	Text on quantum field theory. New derivation of Dirac equation.
Akai et al. (1990)	Theory of hyperfine interactions in metals.
Lindgren (1992,1994)	Reviews on the atomic many-body problem.
Yarkony (1992)	Spin-forbidden chemistry within BP approximation. Review.
T.-N. Chang (1993)	MBPT of atomic structure and photoionization.
Drake (1993ab)	High-precision calculations and QED effects for 2- and 3-electron atoms.
Drake (1993c)	He Rydberg levels.
Fanchi (1993)	Parametrized relativistic quantum theory. Book.
F. Gross (1993)	Relativistic quantum mechanics and field theory. Book.
Johansson and Brooks (1993)	Theory of cohesion in lanthanides and actinides.
Johnson (1993)	Relativistic MBPT on highly-charged ions.
Y.-K. Kim (1993ab)	Recent and unsettled questions in atomic structure theory.
Labzowsky et al. (1993b)	Treatise on relativistic and QED effects in atoms.
Levin and Micha (1993)	Long-range Casimir forces. Book.
Lindgren et al. (1993a)	Heavy-ion spectroscopy and QED in atomic systems.
Mårtensson-Pendrill (1993a)	MBPT in atomic structure calculations.
Mårtensson-Pendrill (1993b)	PNC effects in atomic systems.
Minaev and Lunell (1993); Minaev (1996); Minaev and Ågren (1995,1996)	SO effects in organic reactions.
Mohr (1993,1994)	QED in few-electron systems.
Norman and Koelling (1993)	Electronic structure of f electron metals.
Sandars (1993)	P and/or T violation. Review.
Sapirstein (1993)	Theory of many-electron atoms. Review on QED and many-body aspects.
Adler (1994)	Quaternionic QM and quantum fields.
Altmann and Herzig (1994)	Comprehensive point-group theory tables.
Balasubramanian (1994)	Lanthanide and actinide molecules.
Dyall (1994c)	Review on DF/GTO for polyatomic molecules.
Ermler and Marino (1994,1996)	Reviews on PP methods.
Froese Fischer and Jönsson (1994)	Review on MCHF atomic calculations.
Grant (1994ab,1996)	Reviews on relativistic atomic calculations.
Greiner (1994,1997)	Relativistic quantum mechanics. Wave equations. Book.
Malli (1994ab)	Relativistic and electron correlation effects in molecules and solids.
Marian (1994)	Relativistic calculations on transition metal compounds.
Schweber (1994)	History of QED.
Bauschlicher et al. (1995)	Transition-metal compounds.
Borovskii et al. (1995)	Plasma of multicharged ions.
Eichler and Meyerhof (1995)	Relativistic atomic collisions.

Reference	Comments
Engel (1995); Engel and and Dreizler (1996)	Relativistic density functional theory.
Hess et al. (1995)	Review on SO effects and methods.
Kaldor (1995,1997)	Reviews on the Fock-space CC approach.
Lindgren et al. (1995b)	Few-body problems in atomic physics.
Porteous (1995)	Clifford algebras and classical groups.
Robinson (1995)	Textbook on special relativity.
Schädel (1995)	Chemistry of transactinides.
Ågren et al. (1996)	SO phenomena in molecules. Handled with response theory.
Aksela et al. (1996a)	Resonant and nonresonant Auger recombination.
Almlöf and Gropen (1996)	Relativistic effects in chemistry.
Cundari et al. (1996)	Review on PP methods.
Dolg and Stoll (1996)	Review on lanthanide chemistry.
Ebert (1996)	Magneto-optical effects in TM systems.
Ebert and Schütz (1996)	SO-influenced spectroscopies of magnetic solids.
Frenking et al. (1996,1999); Frenking and Pidun (1997)	Transition-metal compounds. Especially PP work.
Gordon and Cundari (1996)	PP studies on transition-metal compounds. Review.
E. K. U. Gross et al. (1996)	DFT of time-dependent phenomena.
Labzowsky (1996)	Treatise on QED of atoms.
Lindgren (1996)	Relativistic many-body and QED calculations on atomic systems.
Mohr (1996)	Tests of fundamental physics.
Musaev and Morokuma (1996)	Potential-energy surfaces of TM-catalyzed chemical reactions.
Pachucki et al. (1996)	Theory of energy levels and precise two-photon spectroscopy of H and D.
Pershina (1996)	Electronic structure and properties of transactinides and their compounds.
Roos et al. (1996)	Review multiconfiguration PT for molecules. Includes PT relativity.
Rösch et al. (1996)	Review on DK DFT molecular calculations.
Sapirstein (1996a)	Quantum electrodynamics.
Siegbahn (1996)	Electronic structure calculations for molecules containing transition metals.
van der Lugt (1996)	Polyanions, especially Zintl anions, in ionic alloys A review.
Yndurain (1996)	Relativistic QM. Book.
Andersen et al. (1997)	Negative alkaline-earth ions.
Baerends and Gritsenko (1997)	Quantum chemical view on density-functional theory.
Balasubramanian (1997a)	Relativistic quantum chemistry: methods.
Balasubramanian (1997b)	Relativistic quantum chemistry: applications.
Bouchiat and Bouchiat (1997)	Review on parity violation in atoms.
Edmonds (1997)	Book on relativity.
Ehlotzky et al. (1997)	Review on electron-atom collisions in laser fields.
Fricke et al. (1997)	Review on superheavy elements as treated by DFT-DVM.
Hess (1997)	Review on relativistic effects in heavy-element chemistry.
Kaltsoyannis (1997a)	Relativistic effects in inorganic and organometallic chemistry.
Y.-K. Kim (1997)	Review on relativistic atomic structure calculations.

Reference	Comments
Mohr (1997)	QED corrections in heavy atoms.
Persson et al. (1997b)	Theoretical survey of QED tests in highly charged ions.
Pyykkö (1997)	Strong closed-shell interactions in inorganic chemistry.
Rudzikas (1997)	Theoretical atomic spectroscopy.
Schädel et al. (1997ab)	Experimental chemistry of Sg (E106).
Schwabl (1997)	Advanced QM. Book. Dirac equation, QED.
Snygg (1997)	Book on Clifford algebra.
Stöcker et al. (1997)	Structure of vacuum and elementary matter. Conference proceedings.
Tomonaga (1997)	History of spin.
Dolg (1998)	Lanthanides and actinides.
Dzuba et al. (1998)	Atomic MBPT.
Godefroid et al. (1998)	Review on atomic structure calculations.
Grabo et al. (1998)	Optimized effective potential DFT. Review.
Hess (1998)	Review on relativistic theory and applications in chemistry.
Lindgren (1998)	Electron correlation and QED.
I. Martin (1998)	Relativistic quantum defect orbital method.
Nemoshkalenko and Antonov (1998)	Computational methods in solid-state physics.
Sapirstein (1998)	Theory of many-electron atoms. Review on QED and many-body aspects.
Schwerdtfeger and Seth (1998)	Relativistic effects in superheavy elements.
Shabaev et al. (1998c)	QED and nuclear effects in highly charged ions.
Soff et al. (1998)	QED in strong fields. Review.
Strange (1998)	Relativistic QM with condensed-matter and atomic physics applications. Book.
Datz et al. (1999)	Atomic physics. Review.
Ellis and Guenzburger (1999)	Review on the Discrete Variational Method in DFT.
Fröhlich and Frenking (1999)	Bonding in transition-metal complexes. Carbonyls, carbenes, π-bonded systems, dihydrogen complexes.
Greiner and Gupta (1999)	Edited book on superheavy elements.
Kaldor and Eliav (1999)	High-accuracy calculations for heavy and superheavy elements.
Kleppner (1999)	Atomic physics in the 20th century.
Lindgren (1999)	QED effects in strong nuclear fields.
Pershina and Fricke (1999)	Electronic structure and chemistry of the heaviest elements.
Pettersson et al. (1999b)	New rare-gas-containing neutral molecules. Review.
Quiney et al. (1999)	Ab initio relativistic quantum chemistry.

Chapter 2

One-Particle Problems

Table 2.1: The Dirac equation: historical notes, interpretative studies, symmetry properties and non-relativistic limits.

Reference	Comments
Takabayasi (1955)	Structure of Dirac wave function.
Takabayasi and Vigier (1957)	Relativistic hydrodynamics of Dirac matter.
Gorshkov (1962)	Scattering of Dirac particle from Coulomb field with finite-nucleus and screening corrections.
Barut (1973)	Extensions of Dirac Hamiltonian using Lie algebra.
Triebel (1980)	Dirac operators. Textbook of higher analysis.
Ryder (1988)	Alternative derivation of Dirac equation starting from plane waves.
Briegel et al. (1991)	Square root of the KG equation.
Aldaya et al. (1993)	Poincaré group and the position operator.
Anandan and Mazur (1993)	Geometric phase for the KG equation.
Avery et al. (1993)	Four-currents in Dirac theory.
Barut (1993)	On 'new constants of motion' for free Dirac electron.
Barut and Cruz (1993)	Classical spinning particle with anomalous magnetic moment.
Barut and Unal (1993)	Poisson brackets and symplectic structures for classical and quantum Zitterbewegung.
Bechler (1993)	Summation formulae for spherical spinors.
Bednar and Kolar (1993)	Relativistic spin projection operators.
Bohm (1993)	R and NR dynamical groups.
Comay (1993)	Interpretation of the Lorentz-Dirac equation.
Corben (1993)	Factors of 2 in magnetic moment, SO coupling, Thomas precession.
Faisal and Radożycki (1993)	Bound KG particle in an intense laser field. 3D model.
Frisk and Guhr (1993)	SO coupling in the semiclassical (WKB) approximation.
Frochaux (1993abc)	Relativistic corrections to Schrödinger equation from quantum field theory.
Hestenes (1993)	Zitterbewegung modelling.
Ito (1993)	Renormalization of relativistic wave function at origin.
Kowalski (1993)	Relativistic nature of the Schrödinger equation. Only one free-particle phase velocity is meaningful.
Lämmerzahl (1993,1994)	Pseudodifferential operator square root of the KG equation.
Lounesto (1993)	Clifford algebras and Hestenes spinors.
Mocanu et al. (1993)	On Thomas rotation paradox.
Okolowski and Slomiana (1993)	R aspects of NR QM. Comment on Dieks and Nienhuis (1990)[4035].
Ord and McKeon (1993)	Feynman chessboard model and the (3+1)D Dirac equation.
Pauri (1993)	Constants of motion for free Dirac particle.
Pavšič et al. (1993)	Classical models for spin.
Pilkuhn and Staudner (1993)	Double Dirac equation and decay rates of bound particles.
Rodrigues et al. (1993)	Zitterbewegung and electron structure.
Shin and Rafelski (1993)	Relativistic classical ($\hbar \to 0$) limit of quantum theory.
Vaz and Rodriguez (1993)	Zitterbewegung and the EM field of the electron.
Y. X. Zheng (1993)	Regularity of weak solutions to a 2D modified Dirac-KG equation.
Aliev et al. (1994)	Path-integral representation for Dirac propagator.
Barut (1994)	Localized mass-less wavelets with half-integer spin.
Barut and Cruz (1994)	The Zitterbewegung of the relativistic electron. Comments on Sherry (1988).

Reference	Comments
Barut and Pavsic (1994)	Dirac's shell model of the electron. Moving relativistic charged membranes.
Z. Chen et al. (1994)	Upper and lower bounds for 1-electron Dirac Hamiltonian. Tests on Coulomb potential.
Forte (1994)	Relativistic quantum theory with fractional spin and statistics.
Greiner (1994,1997)	Relativistic quantum mechanics. Wave equations. Book.
Grosse et al. (1994)	Order and spacings of KG energy levels.
Hill and Krauthauser (1994)	Variational collapse avoided by using $1/h_{\mathrm{D}}$ instead of h_{D}. Application on the H atom.
Land et al. (1994)	Selection rules for dipole radiation from a relativistic bound state.
Pillin (1994)	q-deformed relativistic wave equations.
Rivas (1994)	Dirac equation obtained by quantizing a classical spinning particle.
Shankar and Mathur (1994)	Thomas precession, Berry potential and the 'meron'.
Vrbik (1994)	Dirac equation and Clifford algebra.
Waxman (1994)	Fredholm determinant for Dirac operator. 1D, some $V(x)$. Periodic boundary conditions.
Aldaya and Guerrero (1995)	A finite enlargement of the (1+1)D Poincaré group. Boost $i\frac{\partial}{\partial\pi}$.
Aparicio et al. (1995)	Feynman parametrization of the Dirac equation.
Augenstein (1995)	Quotes Y. Frenkel [Z. Phys. 47 (1928) 786] and D. Iwanenko and L. Landau [Z. Phys. 48 (1928) 340] as precursors for the Dirac equation.
Baylis and Sienkiewicz (1995)	Represent electron-scattering polarization data by 'polarization trajectories' on a Poincaré sphere.
Blancarte et al. (1995)	High- and low-energy estimates for the Dirac equation. Levinson theorem for the zero-mass limit.
Caspersson (1995)	Dirac equation at semiclassical limit.
Cooper et al. (1995)	Supersymmetry and the Dirac equation. Coulomb and Lorentz scalar potentials. Magnetic field.
Gaioli and Garcia Alvarez (1995)	Intrinsic parity in Ryder's (1988) derivation of the Dirac equation.
Jefimenko (1995)	Retardation and relativity for a moving line charge.
Land and Horwitz (1995)	Zeeman effect for a relativistic bound state. Zeeman splitting in manifestly covariant form.
Ochs and Sorg (1995)	Relativistic Schrödinger equations and the particle-wave duality.
Petráš (1995)	The $SO(3,3)$ group as common basis for the Dirac and Proca equations.
Phatak et al. (1995)	Semiclassical features of Dirac particle in cavity.
Pilkuhn (1995)	An M1 hyperfine operator between two Dirac particles, both having anomalous magnetic moments.
Boeyens and Kassman (1996)	Introduction of spin into the Schrödinger equation.
Crisp (1996)	Relativistic neoclassical radiation theory.
De Leo (1996)	Quaternions and special relativity.
Goodmanson (1996)	A graphical representation of the Dirac algebra.
Hamilton (1996)	Relativistic precession.
Jefimenko (1996)	Relativistic force transformation from Lorentz force law.
Lerner (1996)	Derive Dirac equation from relativistic representation of spin.
Ord (1996)	Relationships between the Schrödinger and Dirac equations and the random walk Brownian motion formalism of diffusion.

Reference	Comments
Oudet (1996)	Discusses interpretation of g-factors and spin.
Philpott (1996)	Thomas precession and the Liénard-Wiechert field.
Robson and Staudte (1996)	An eight-component relativistic wave equation for spin-$\frac{1}{2}$ particles. Solved for the H atom.
Salcedo and Ruiz Arriola (1996)	Wigner transformation for the determinant of Dirac operators.
Zhuang and Heinz (1996)	Kinetic equations for Dirac particle with EM, scalar, and pseudoscalar interactions.
Booth and Radford (1997)	Dirac-Maxwell equations with cylindrical symmetry.
Dahl (1997)	Physical origin of Runge-Lenz vector.
Gitman and Shelepin (1997)	Poincaré group and relativistic wave equations in 2+1 dimensions. Anyons, with any spin and statistics.
Gomes et al. (1997)	Relativistic corrections to Aharonov-Bohm scattering.
Y. S. Huang et al. (1997)	Alternative to KG for spin-0 particles.
Isozaki (1997)	Inverse scattering theory for Dirac operators.
Jung (1997)	Geometrical approach to inverse scattering for Dirac equation.
Nakamura (1997)	Path space measures for Dirac and Schrödinger equations.
Omnès (1997)	Localization of relativistic particles.
Pelc and Horwitz (1997)	Complete sets of states in relativistic scattering theory.
Zakrzewski (1997)	Localization of relativistic systems.
Bechouche et al. (1998)	NR limit of the Dirac equation for time-dependent EM fields.
Bolte and Keppeler (1998, 1999)	Semiclassical time evolution of R spin-$\frac{1}{2}$ particles. Dirac equation at the limit $\hbar \to 0$.
Breuer and Petruccione (1998)	Quantum-state diffusion for a Dirac electron.
Cohen and Leung (1998)	Relativistic corrections to atomic sum rules.
S.-H. Dong et al. (1998)	Relativistic Levinson theorem in 2D.
Finster (1998)	Local $U(2,2)$ symmetry in relativistic quantum mechanics.
Gosselin and Polonyi (1998)	Path integral for relativistic equations of motion. KG and Dirac.
Popov et al. (1998)	Relativistic version of the imaginary time method. Applications on ionization.
Reginatto (1998)	Derives the Pauli equation using the principle of minimum Fisher information.
Ruijgrok (1998)	General requirements for a relativistic quantum theory.
Ryder (1998)	Relativistic spin operator for Dirac particles.
Sharma et al. (1998)	NR limit for particle of arbitrary spin in external field.
Unterberger (1998)	A calculus of observables on a Dirac particle.
K. Ziegler (1998)	Delocalization of 2D Dirac fermions with random mass. Role of broken supersymmetry.
Avetissian et al. (1999)	Eikonal wave function for Dirac particle in arbitrary potential and radiation fields.
Baylis and Yao (1999)	Relativistic dynamics of charges in EM fields using eigenspinors.
Bohm and Kaldass (1999)	Relativistic partial-wave analysis using the velocity basis of the Poincaré group.
Boya and Byrd (1999)	Clifford periodicity from finite groups.
Bracken and Melloy (1999)	Localizing the relativistic electron. Free electron 4-vector density $(\rho(x,t), j(x,t)/c)$ can be localized using only positive-energy states.
des Cloizeaux (1999)	Dirac equation formulated in terms of local observables.

Reference	Comments
S.-H. Dong et al. (1999)	Levinson theorem for the KG equation in 2D.
Edmonds (1999)	Dirac's equation in half of his algebra.
García-Calderón et al. (1999)	Low-energy relativistic effects in discussing time-dependent tunneling.
Jaekel and Reynaud (1999)	Observable Dirac electron in accelerated frames. Localization observables.
Mattes and Sorg (1999a)	Second-order mixtures in relativistic Schrödinger theory.
Mattes and Sorg (1999b)	Two-particle systems with EM interactions in relativistic Schrödinger theory.
Pissondes (1999)	Recovers the KG equation in 'scale-relativity theory'.
Şimşek (1999)	Negative-energy levels for a ND Dirac equation. Spherically symmetrical, purely imaginary, linear potential.
Tkachuk and Roy (1999)	Supersymmetry of a spin-$\frac{1}{2}$ particle on the real line. Rotating magnetic field + a scalar potential.
B.-X. Wang and Lange (1999)	Attractors for the KG equation.
Z.-C. Wang and Li (1999)	Geometric phase in Dirac theory. (Cp. Berry). Lorentz invariant.

The Table 2.2 below discusses transformations of the Dirac equation. For further examples from quantum chemical calculations, see also the Table 7.3.

Table 2.2: The Dirac equation: further transformations.

Reference	Comments
Fraga and Karwowski (1974)	Two nuclear-mass-dependent terms in the BP equation. Of same order as the hyperfine interaction.
Arriola and Salcedo (1993)	Semiclassical expansion of Dirac Hamiltonians.
Mustafa and Sever (1993)	Shifted $1/N$ expansion for a relativistic spin-$\frac{1}{2}$ particle.
Q.-S. Wang and Stedman (1993)	Derives from FW an E1 term of form $q\mathbf{s}\cdot\dot{\mathbf{A}}\times\mathbf{p}/2m^2c^2$. Applications to lanthanides.
Faulkner (1994)	Scattering matrices for non-spherical SR potential.
Fearing et al. (1994)	Compare the FW and Pauli approximations in nuclear physics. FW found better.
Aucar et al (1995b)	Bethe sum rule for the no-pair Hamiltonian.
Costella and McKellar (1995)	The FW transformation.
Grigoryan and Grigoryan (1995)	Quasiclassical FW transformation and canonical quantization of $D = 2n$-dimensional relativistic particles in external EM field.
Lucha and Schöberl (1995)	Semirelativistic Hamiltonians of apparently nonrelativistic form. Effective mass
van Lenthe et al. (1995)	Solves the 2nd-order Dirac equation for the large component in a Slater basis. Tests on U atom.
Amore et al. (1996)	Effective potentials for nuclear physics via FW.
Ebert et al. (1996);	SO effects for Dirac equation and spin-dependent potentials
Guil and Mañas (1996)	The 2D heat equation yields the KG equation, which is here expressed in terms of solutions of the Dirac equation. KD = Kadomtsev-Petiashvili equations.
Lévai and Del Sol Mesa (1996)	Transform the Dirac equation into a Schrödinger-like one for minimal and non-minimal couplings. Polynomial potentials.
Ebert et al. (1996);	SO effects for Dirac equation and spin-dependent potentials

Reference	Comments
Ebert et al. (1997b)	divided into two parts, diagonal and off-diagonal.
Staudte (1996)	An eight-component wave equation for spin-$\frac{1}{2}$ particles. Part II.
van Lenthe et al. (1996a)	Solve the Dirac equation using the FW transformation and large components only. Tests on U atom.
Dyall (1997)	'Normalized elimination of small components'.
Lagmago Kamta et al. (1998)	Semiclassical KG dipole matrix elements.
Nikitin (1998)	New classes of external fields having an exact FW transformation.
Kocinski (1999)	A 5-dimensional form of the Dirac equation. Pauli limit derived.
Nowakowski (1999)	QM current of Pauli equation.

Table 2.3: The Dirac equation: solutions for hydrogen-like systems.

Reference	Comments
Gorshkov (1962)	Scattering of Dirac particle from Coulomb field with finite-nucleus and screening corrections.
Barut and Bornzin (1971)	$SO(4,2)$-Formulation for Dirac and KG equations. With or without magnetic charges.
Tzara (1985)	KG Coulomb problem in momentum space.
Moss and Sadler (1986)	Electric quadrupole moments of one-electron atoms. In $j = \frac{1}{2}$ states due to hfs. Resolves earlier inconsistencies.
Hofstetter et al. (1992)	Ionization of hydrogen by relativistic heavy projectile.
Biedenharn et al. (1993)	Dirac-Coulomb problem and κ-Poincaré group.
Bleyer (1993)	Coulomb problem for generalized Dirac equation.
Bondarev and Kuten (1993)	Polarizability tensors for H-like $ns_{1/2}$ and $np_{1/2}$ states. Due to M1 hfs.
Choe et al. (1993)	Tests of B-spline basis sets on H-like O and Hg.
Cohen and Kuharetz (1993)	DC wave function in Whittaker form. Further solutions to the Darwin-Gordon (1928) ones are presented.
Guiasu and Koniuk (1993)	Particle interpretation of the DC solutions. Effective 'no-pair' one-electron solution found, starting from Fock space.
C. Hofmann et al. (1993)	Distorted-wave scattering solution of Dirac electron from a Coulomb potential.
Horwitz (1993)	Dynamical group of the relativistic Kepler problem. $SO(4,1)$ for the bound states.
Kwato Njock et al.(1993)	E1 radial integrals in semiclassical Coulomb approximation.
Papp (1993)	Dirac-Coulomb and linear potentials using 'β functions'.
Pisani et al. (1993)	Discuss relativistic effects in atoms and molecules using LCAO calculations on H-like systems as example.
Tangerman and Tjon (1993)	Exact supersymmetry in NR H-atom. Pauli particle in Coulomb field.
Yao and Chu (1993)	Pseudospectral methods for bound and resonance state problems with the Dirac equation. H-like atom as example.
Anthony and Sebastian (1994)	Rel. corrections to g-factors of H-like atoms and positronium.
Z. Chen et al. (1994)	Upper and lower bounds for 1-electron Dirac Hamiltonian. Tests on Coulomb potential.
Hill and Krauthauser (1994)	Variational collapse avoided by using $1/h_{\mathrm{D}}$ instead of h_{D}. Application on the H atom.
Kwato Njock et al. (1994a)	Supersymmetry-based QDT of Dirac equation for central potentials.
Le et al. (1994)	DC Green function. Polarizability of H-like atoms.

Reference	Comments
Lucha and Schöberl (1994)	Variational approach to spinless relativistic Coulomb problem.
Melić (1994)	Hydrogenic atom with spinless nucleus in magnetic field.
Norbury et al. (1994)	Exact numerical solution of spinless Salpeter equation in momentum space for Coulomb potential.
Salamin (1994)	Expectation values $\langle nl\|r^{\beta}\|nl\rangle$ with DC wave functions.
Shabaev (1994a)	M1 and E2 hfs of H-like ions. Finite-nucleus corrections.
Talman (1994)	Spurious solutions in matrix approximations to the DC problem.
Tong et al. (1994a)	Relativistic effect on atomic radiative processes. A 'zero' of the H-like E1 transition matrix element found.
van Leeuwen et al. (1994)	ZORA solutions for H-like atoms are scaled Dirac ones. Also ZORA KG.
Baltz (1995)	Coulomb potential from a particle in uniform ultrarelativistic motion.
Baluja (1995)	Dipole polarizability of H-like ions.
Dahl and Jørgensen (1995)	Johnson-Lippmann operator, supersymmetry, normal-mode representations.
Deco et al. (1995)	Scattering involving Coulomb potentials. Ion-atom collisions.
Horbatsch and Shapoval (1995a)	KG Coulomb problem in the Feshbach-Villars representation.
Horbatsch and Shapoval (1995b)	Relativistic two-particle scattering resonances in the Tamm-Dancoff approximation.
Horbatsch and Shapoval (1995c)	Analysis of the DC problem using momentum-space free-particle functions.
Indelicato and Mohr (1995)	Asymptotic expansion of the DC radial Green function.
Le Yaouanc et al. (1995)	High-order expansion of eigenvalues of a relativistic Coulomb equation.
Nag and Roychoudhury (1995)	$1/N$ solution of the DC equation.
Rosenberg (1995)	Electron scattering by heavy hydrogenic ion.
Shabaev et al. (1995)	Hfs of H-like ions.
Kołakowska (1996)	Minimax principle for the DC problem. Spurious roots eliminated, no upper bounds obtained.
Le et al. (1996)	Dynamic polarizability of H-like atoms.
Lucha and Schöberl (1997)	Analytical upper bounds on energy levels of the spinless relativistic Coulomb problem.
Molzberger and Schwarz (1996)	Effects of different order in α^2 on energies of H-like atoms. Compare Dirac with 1st- and 2nd-order DK.
Robson and Staudte (1996)	An eight-component relativistic wave equation for spin-$\frac{1}{2}$ particles. Solved for the H atom.
Villalba (1996)	The DC problem. Solution of Cohen and Kuharetz (1993) for $119 < Z < 137$ unphysical.
Andrae (1997)	Recursive formulae for $\langle r^k\rangle$, -6 $\leq k \leq$ 5. Numerical values for Z = 1, 80, 137.
Barysz (1997)	Compares Pauli PT, DPT and Dirac levels for H-like systems.
Goodman and Ignjatović (1997)	A simpler solution for the DC problem.
Hylton and Snyderman (1997)	Analytic basis set for high-Z atomic QED. He-like ions. Sturmian DC functions.
Manakov and Zapriagaev (1997)	DC problem by 2nd-order Dirac equation approach. Green functions given.

Reference	Comments
Moiseiwitsch (1997)	Virial theorem for electron capture from H-like atom.
Pyykkö and Seth (1997)	H-like relativistic correction factors for nuclear quadrupole coupling. Analytical solution.
Sergheyev (1997)	Relativistic Coulomb problem for modified Stueckelberg equation.
Shabaev et al. (1997b)	Ground-state M1 hfs splitting of H-like ions. Z=49-83.
Simulik (1997)	Hydrogen spectrum in classical electrodynamics.
Stahlhofen (1997)	Algebraic solutions of DC problems. Symmetries. Relations with Dirac oscillators.
Staruszkiewicz (1997)	The DC problem. The role of the classical Coulomb field.
Szmytkowski (1997)	DC Sturmians and the DC Green function. Relativistic polarizability of the H-like atom, Z=1-137.
Szymanowski et al. (1997)	Two-photon bound-bound amplitudes in H-like atoms. Uses DC Green function in a Sturmian basis.
Zolotorev and Budker (1997,1999)	PNC in relativistic hydrogenic ions. Circular dichroism on the $1s \rightarrow 2s$ transition due to interference between the M1 and PNC-induced E1 amplitudes.
Al-Jaber (1998)	N-dimensional hydrogen atom at BP level.
J.-J. Chang (1998)	Similarities between Dirac-Coulomb and Schrödinger-Coulomb radial functions. Photoionization of H.
Dragić et al. (1998)	H atom with added magnetic SO and spin-spin interactions. No deep-lying extra levels found.
Eichler et al. (1998)	Alignment caused by photoionization and in radiative electron capture into excited states of high-Z H-like ions.
Indelicato and Mohr (1998a)	Coordinate-space approach to bound-electron SE. Coulomb-field calculation.
Martínez-y-Romero et al. (1998,1999)	Non-unitary $SU(2)$ representations in a Dirac H atom.
Matrasulov (1998,1999)	Chaotic ionization of an H-like atom in intensive monochromatic field.
Pivovarov (1998)	Coherent excitation of H-like heavy ions penetrating through a crystals.
Prosser (1998)	Energy spectrum of H atom in photon field.
Santos et al. (1998b)	Two-photon decay rates of $2s$-level in H-like ions. Z=1-100.
Shabaev (1998b)	TP between M1 hfs levels of H-like ions related to the bound-electron g-factor.
Shabaev et al. (1998a)	Recoil correction to H-like ground-state energy.
Shabaev et al. (1998b,1999)	Recoil correction to H-like energy levels.
Szmytkowski (1998a)	Continuum DC Sturmian functions.
van Lenthe et al. (1998)	Relation between Dirac and ZORA levels for M1 hfs, g-factors, of H-like systems.
Benvegnù (1999)	1D $-1/\|x\|$.
Holzscheiter and Charlton (1999)	Ultra-low energy antihydrogen. Review on experiments and possibilities.
Hua et al. (1999)	Exact, analytical solution of time-dependent Dirac equation.
Kónya and Papp (1999)	Coulomb Sturmian matrix elements of Coulomb Green's operators for KG and 2nd-order Dirac equations.
Moore (1999)	Magnetic shielding (NMR shift) for H-like Dirac atom. Closed-form expressions.
Vrejoiu et al. (1999)	Retardation corrections to angular distributions

Reference	Comments
	of free-bound transitions in H-like atoms.
Yamanaka and Ichimura (1999)	Nuclear polarizability contributions to $1s$ energies of H-like ^{208}Pb and ^{238}U. Transverse polarization is even more important than longitudinal one.
Yerokhin and Shabaev (1999)	First-order SE in H-like systems. $\|\kappa\| \leq 5$, $n \leq 5$. Uses DC Green function.

Table 2.4: The Dirac equation: Solutions for various non-Coulomb fields.

Reference	Comments
Gorshkov (1962)	Scattering of Dirac particle from Coulomb field with finite-nucleus and screening corrections.
Acker et al. (1966)	Muon in a two-parameter Fermi nuclear potential.
Gorshkov et al. (1967)	Screening effects in K-shell photoionization.
Mikhailov and Polikanov (1968)	Central potential. PT by reducing the Dirac equation to a Riccati equation.
Cook (1971)	Relativistic harmonic oscillator with intrinsic spin structure.
Coudray and Coz (1971)	Construction of V for fixed energy. KG and Dirac. The inversion problem.
Coulter and Adler (1971)	The 1D square-well potential.
Skyrme (1971)	Kinks and the Dirac equation.
Cusson et al. (1985)	Time-dependent 3D Dirac equation in heavy-ion scattering.
Bär (1990)	The spectrum of the Dirac operator.
Arai (1993)	Dirac-Weyl operator with a strongly singular gauge potential.
Artimovich and Ritus (1993)	Deep well potential.
Atanasov and Bankova (1993)	Relativistic Schrödinger equation in a linear +Coulomb potential.
Bautista (1993)	Dirac particle with anomalous magnetic moment μ' in a homogeneous magnetic field. Uniformly accelerated Rindler coordinates.
S. J. Blundell (1993)	The Dirac comb and the Kronig-Penney model.
Bosanac (1993)	Wave packets of spin-$\frac{1}{2}$ particles in an EM field.
Burgess and Jensen (1993)	Fermions near 2D surfaces. (3+1)D versus (2+1)D.
Caliceti and Cherubini (1993)	Relativistically deformed harmonic oscillator.
Centelles et al. (1993)	Dirac oscillator, $V(r) = \frac{1}{2}Kr^2$.
J.-J. Chang (1993,1998)	Dirac quantum-defect theory.
Z. Chen and Goldman (1993)	H-like atom with a finite nucleus in magnetic field.
Clerk and McKellar (1993)	Disordered 1D δ-function array.
Coutinho and Perez (1993)	Aharonov-Bohm scattering of Dirac particles by magnetic flux.
Crawford (1993)	Dirac oscillator and local automorphism invariance.
Droz-Vincent (1993)	Relativistic two-body problem in time-dependent external potentials.
Fujimoto and Kawakami (1993)	SO effects on persistent currents in mesoscopic Hubbard rings, threaded by an Aharonov-Bohm flux. Interplay of SO and $e-e$ interaction crucial.
Gambhir and Ring (1993)	Solution of Dirac equation using GTO basis for nuclei. Recipe for avoiding spurious states.
Gitman and Saa (1993)	Dirac-Pauli particle with anomalous magnetic momentum in external EM field.
J. Gonzalez et al. (1993)	Electronic spectrum of fullerenes from the Dirac equation.

Reference	Comments
Grigoryan and Grigoryan (1993)	Canonical quantization of Dirac particle in external magnetic field.
Hachem (1993)	Zeeman effect for Dirac electron.
Hatsugai and Lee (1993)	Localization of Dirac fermions in a 2D lattice.
Hersbach (1993a)	Relativistic calculations in momentum space. 'From positronium to quarkonium'.
Hersbach (1993b)	Relativistic linear potential in momentum space.
L. M. Jones (1993)	Another Dirac oscillator.
S. H. Kim (1993)	Thomson scattering by a relativistic electron.
Koutroulos and Papadopoulos (1993)	Dirac equation for rectangular spherical well. Model for hypernucleus.
Ktitarev and Yegikian (1993)	Feynman path integral for Dirac system with analytic potential.
Le and Nguyen (1993)	H-like atom in Dirac monopole + Aharonov-Bohm field.
Maier et al. (1993)	Time-dependent 1D Dirac square well: Pair production.
Matrasulov et al. (1993)	Near-continuum states of relativistic electron in an electric dipole field.
Maung et al. (1993)	Two-body system with confining linear+Coulomb interactions.
McQuarrie and Vrscay (1993)	KG equation with Coulomb + perturbing vector or scalar λr^k potentials. Rayleigh-Schrödinger PT of arbitrary order.
Mohamed et al. (1993)	Dirac operator with periodic potential. Asymptotic band spectrum.
Moshinsky and Loyola (1993)	Barut equation for particle-antiparticle system with Dirac oscillator interaction.
Nogami and Toyama (1993)	Supersymmetry aspects of 1D Dirac equation with a Lorentz scalar potential.
Ovcharov and Fedosov (1993)	Correlated coherent states of electron in field of plane EM wave.
Papp (1993a)	Dirac-Coulomb and linear potentials using 'β functions'.
Piekarewicz (1993)	Levinson theorem for Dirac particles.
Poliatzky (1993ab,1994)	Levinson theorem for Dirac equation.
Rosenberg and Zhou (1993)	Dirac and KG particles in a multimode laser field: Generalized Volkov solutions.
Roy (1993ac)	Boundary conditions at a δ-function in a 1D Dirac equation.
Roy (1993b)	Tunneling through a double δ-function barrier.
Roy and Khan (1993a)	Relativistic effects on impurity states.
Roy and Khan (1993b)	Relativistic effects on tunneling through multi-barrier systems.
Roychoudhury and Panchanan (1993)	Modified $1/N$ expansion for the Dirac equation for a screened Coulomb potential.
Safonov (1993)	Crystal-field anisotropies handled via the metric tensor of the space-time, for Dirac electrons.
Sakamoto (1993)	$N-$body bound state solution of Dirac particles in (1+1)D.
Salamin (1993)	Dirac particle with anomalous magnetic moment in a plane EM field.
Schmidt (1993)	Nonvanishing instability intervals in periodic Dirac systems.
Semay and Ceuleneer (1993)	Two-body Dirac equation and Regge trajectories.
Semay et al. (1993)	Two-body Dirac equation with diagonal central potentials.
Shabaev (1993a)	Finite-nuclear-size corrections to energy levels of multicharged ions. Expressed in terms of nuclear $\langle r^n \rangle$; n=2,4. Z=1-100.

Reference	Comments
Shishkin and Villalba (1993)	Electrically neutral Dirac particles in external fields. Exact solutions. Neutron in magnetic field.
Suzuki and Nogami (1993)	Variable-phase approach to the Dirac bound states.
Tashkova and Donev (1993)	Relativistic electron tunneling through structured 1D barrier.
Toyama et al. (1993)	Construct a 1D transparent potential of Lorentz scalar type. An infinite family exists.
Vonsovskii and Svirskii (1993)	Klein paradox and zitterbewegung for electron in a field with constant scalar potential.
Vonsovskii et al. (1993)	Electron zitterbewegung in linear crystal with alternating parity.
Willis (1993)	"Classical description of the absence of bound states for strong Coulomb fields."
Yao and Chu (1993)	Pseudospectral methods for bound and resonance state problems with the Dirac equation. H-like atom as example.
Zarzo and Martínez (1993)	Quantum relativistic harmonic oscillator. Spectrum of zeros in its wave functions from WKB.
Arvieu and Rozmej (1994, 1995)	"Spin-orbit pendulum": transitions from pure to mixed spin states.
Bagrov et al. (1994ab)	Maslov's 'complex germ method for a Dirac particle at the $\hbar \to 0$ limit.
Barut and Duru (1994)	Path-integral quantization for confined two-body problem.
Benvegnù and Dąbrowski (1994)	Relativistic point interaction.
Bergerhoff and Soff (1994)	Scalar potentials in the Dirac equation.
Bortman and Ron (1994)	Hysteresis in an atomic system (trapped ion, driven by a laser).
Clarkson (1994)	Spheres, pseudospheres.
Clemence (1994a)	Periodic Dirac system.
Clemence (1994b)	Levinson theorem for perturbed Coulomb potential.
Friedberg et al. (1994)	A new way of removing fermion doubling for solutions of the Dirac equation on a lattice.
A. Gonzalez et al. (1994)	Particle-antiparticle system with a Dirac oscillator interaction.
Grypeos et al. (1994,1995)	Dirac equation for neutrons in nuclei or Λ-particles in hypernuclei solved. Spherical, rectangular scalar and vector wells.
Helffer and Parisse (1994)	Decay of eigenfunctions for Dirac and KG. Tunneling.
Hersbach (1994)	Relativistic meson spectroscopy in momentum space.
Idlis et al. (1994)	Application of supersymmetry and factorization in solution of the Dirac equation.
Mijatović et al. (1994)	Scattering and bound states of a relativistic, neutral spin-$\frac{1}{2}$ particle in a magnetic field. System of N magnetic barriers or magnetic δ functions.
Khasanov (1994)	Constructs scalar and vector potentials for the Dirac equation with discrete eigenvalues in the continuum part.
Kwato Njock et al. (1994)	Supersymmetry-based QDT of Dirac equation for central potentials.
Z.-F. Luo et al. (1994)	Energy-dependent potential and hyperfine mass splitting of quarkonium.
Mattes and Sorg (1994)	Relativistic Schrödinger equation and the Bohm-Aharonov effect. Scalar localization field + extended vector field.

Reference	Comments
Melić (1994)	Hydrogenic atom with spinless nucleus in magnetic field.
W. Moreau et al. (1994)	Relativistic 1D harmonic and anharmonic oscillator.
C. Müller (1994)	Finite-element solution of time-dependent Dirac equations.
Müller-Nehler and Soff (1994)	Electron excitations in superheavy quasimolecules.
Nag and Roychoudhury (1994)	Exact solutions of two body Dirac equations.
Nagel (1994)	Relativistic Hermite polynomial is a Gegenbauer polynomial. Eigenfunctions of the 1D harmonic oscillator.
Newton (1994)	Comment on Poliatzky (1993ab).
Qian and Su (1994)	SO interaction and Aharonov-Anandan phase in mesoscopic rings.
Rosenberg (1994a)	Infrared radiative corrections to potential scattering of a Dirac electron.
Rosenberg (1994b)	Minimum principle for potential scattering of Dirac electron.
Roy (1994)	Tunneling through a multiple δ-function barrier.
Roy et al. (1994)	Relativistic equation for a slowly varying potential.
D. K. Roy and Singh (1994)	Relativistic effects on the tunnel effect.
Scherer et al. (1994)	Low-energy Compton scattering by a Dirac proton with anomalous magnetic moment. FW used.
Shabaev (1994a)	M1 and E2 hfs of H-like ions. Finite-nucleus corrections.
Sørensen and Belkacem (1994)	DC wave functions in momentum space.
Teychenné et al. (1994)	Oscillatory relativistic motion in a power-law or sinusoidal potential well. Classical (non-quantum).
Villalba (1994a)	Exact solution of the 2D Dirac oscillator.
Villalba (1994b)	Dirac equation for central fields. Separation in standard or rotating coordinate systems.
Villalba (1994c)	KG H-atom in magnetic monopole and Aharonov-Bohm potentials.
Waxman (1994)	Fredholm determinant for Dirac operator. 1D, some $V(x)$. Periodic boundary conditions.
Young and Norrington (1994)	Solution of relativistic asymptotic equations in electron-ion scattering. Program.
Zakout and Sever (1994)	Relativistic, heavy $q\bar{q}$ bound states. Parametrized effective potentials.
Baltz (1995)	Coulomb potential from a particle in uniform ultrarelativistic motion.
Cooper et al. (1995)	Supersymmetry and the Dirac equation. Coulomb and Lorentz scalar potentials. Magnetic field.
Danilov (1995)	Periodic potential. Resolvent estimates and spectrum.
Delbourgo (1995)	Square root of the harmonic oscillator.
Dominguez-Adame and Rodriguez (1995)	A 1D screened Coulomb potential.
Grigoryan and Grigoryan (1995)	Quasiclassical FW transformation and canonical quantization of $D = 2n$-dimensional relativistic particles in external EM field.
Halasz and Verbaarschot (1995)	Kalkreuter's (1995) Dirac spectrum for $SU(2)$ gauge theory. Statistical analysis of eigenvalues.
Kalkreuter (1995); Kalkreuter and Simma (1996)	Spectrum of squared Dirac operator using multigrid or conjugate gradient algorithms. $SU(2)$ gauge fields.
Kholomaj (1995,1998)	Dirac electron in arbitrary periodic electrostatic field.

Reference	Comments
Le Yaouanc et al. (1995)	High-order expansion of eigenvalues of a relativistic Coulomb equation.
A. Martin and Stubbe (1995)	Dirac equation for spherically symmetrical vector potential. Bargmann- and Calogero-type bounds for number of eigenvalues. Dito for KG.
Momberger et al. (1995)	Numerical, momentum-space solution of time-dependent Dirac equation for Au^{79+} or U^{92+} impinging on U^{91+}.
Moshinsky et al. (1995)	Two-body system with Dirac oscillator interaction. Symmetry Lie algebra.
Nana Engo et al. (1995)	Relativistic semiclassical dipole matrix elements for $nlj \to n'l'j'$ transitions in non-hydrogenic ions.
Panchanan et al. (1995)	$V(r) = -\frac{1}{r} + kr + gr^2$, $V(r) = -(1+\beta)/r$ solved. Reviews analytical solutions.
Rosenberg (1995)	Electron scattering by heavy hydrogenic ion.
Sucher (1995)	Confinement in relativistic potential models.
Tezuka (1995)	Dirac equation with attractive monomial (ar^n) or polynomial vector potential cannot confine a particle. A scalar potential, larger than vector potential, is needed.
Villalba (1995)	Dirac H-atom in magnetic monopole and Aharonov-Bohm potentials. Exact solution.
Alberto et al. (1996)	Relativistic particle in a 1D box. Avoid Klein paradox by a Lorentz scalar potential.
Avetissian and Movsissian (1996)	Scattering of Dirac electrons in arbitrary static potentials. Short- and long-range.
Camarda (1996)	Eigenvalues for quaternion matrix with a band structure.
Falomir et al. (1996)	Determinants of Dirac operators with local boundary conditions. 2D disk under baglike conditions.
Funakubo et al. (1996)	Numerical approach to CP-violating Dirac equation.
D.-S. Guo et al. (1996)	Dirac electron in multimode quantized radiation field.
Hagen and Park (1996)	Relativistic, (2+1)D Aharonov-Bohm-Coulomb problem. $1/r$, scalar and vector.
Kegley et al. (1996)	Deformed spherical oscillator with strong SO. Schrödinger equation solved in a lattice. Cylindrical coordinates.
J.-H. Kim et al. (1996)	Construct a potential for a constant-period classical (non-quantum) oscillator.
Lévai and Del Sol Mesa (1996)	Transform the Dirac equation into a Schrödinger-like one for minimal and non-minimal couplings. Polynomial potentials.
Maksudov and Allakh-verdiev (1996)	Spectral theory of non-self-adjoint Dirac operators.
T. A. Marian (1996)	Higher-order multipole expansions for Dirac and Pauli Hamiltonians. (Atomic electron in radiation field). Includes retardation up to 4th order.
Nenciu and Purice (1996)	1D periodic Dirac Hamiltonians: semiclassical and high-energy asymptotics for the gaps.
Nogami and Toyama (1996)	Coherent state of the Dirac oscillator.
Ogurisu (1996)	Neutral Dirac particle with anomalous magnetic moment in asymptotically constant magnetic field.
Protopapas et al. (1996,1997)	Atomic physics with super-high intensity lasers. Important mass-shift effects.

Reference	Comments
Roy (1996)	'Fibonacci lattice' of rectangular potential wells.
Savchenko (1996)	Dirac particle with anomalous magnetic moment in a circularly polarized wave.
Szmytkowski and Hinze (1996)	NR and R R-matrix expansions at the reaction volume boundary.
Tkachuk and Vakarchuk (1996)	The N=4 supersymmetry of electron in magnetic field.
Valiev and Pazdzerskij (1996)	Dirac equation for a singular point potential.
Znojil (1996)	Harmonic oscillations in quasirelativistic regime. $V(r) = \sqrt{A + Br^2}$.
Alonso and De Vincenzo (1997);	Dirac particle in a 1D box.
Alonso et al. (1997)	1D or spherical box. Boundary conditions discussed.
Bose (1997)	Relativistic Schrödinger equation, $V(r) = -\alpha/r + \beta/r^{1/2}$.
Cotaescu and Draganescu (1997)	Operator algebra of the relativistic oscillator.
Falsaperla et al. (1997)	Two methods for solving the Dirac equation without variational collapse (Lehmann-Maehly and Kato methods). H atom in magnetic field.
Karat and Shulz (1997)	Self-adjoint extensions of the Pauli equation in presence of magnetic monopole.
Khelashvili and Kiknadze (1997)	Bound states in the continuum for quasipotentials, arising from QED.
Kylstra et al. (1997)	1D Dirac model atom in an intense laser field. Time-dependent problem solved in momentum space using B-splines.
R. T. Lewis et al. (1997)	Essential spectrum of relativistic multi-particle operators.
Lieb et al. (1997ab)	Stability of matter in classical EM fields. (Dirac electrons).
D.-H. Lin (1997)	Relativistic fixed-energy amplitudes of step and square well potentials.
Lobanov (1997)	Electron polarization in a pulsed EM field.
Lun and Buckman (1997); Lun et al. (1998)	Extract SO interactions from scattering phase shifts via inversion. Obtain effective $V_{SO}(r)$.
Olsen and Kunashenko (1997)	Channeling of relativistic electrons in crystals. Electron-positron production. Dirac equation in cylindrical coordinates.
Qian et al. (1997)	Persistent currents from competition between Zeeman coupling and SO interaction.
Rathe et al. (1997)	Intense laser-atom dynamics with 2D Dirac equation.
Rozmej et al. (1997)	'Collapse and revival' in wave-packet dynamics due to SO interaction.
Rutkowski and Kozlowski (1997)	Relativistic H atom in static, uniform $\mathbf{B}$. DPT.
Skarzhinsky and Audretsch (1997)	Scattering of scalar and Dirac particles by a magnetic tube of finite radius.
Stahlhofen (1997)	Algebraic solutions of DC problems. Symmetries. Relations with Dirac oscillators.
Torres del Castillo and Cortés-Cuautli (1997)	Dirac equation in the field of a magnetic monopole.
Toyama et al. (1997)	The Dirac harmonic oscillator under a FW transformation.
Villalba (1997)	Exact solution of Dirac equation for pseudoscalar potentials: 1D $1/x$, cylindrical $1/\rho$, spherical $1/r$. No bound states.
Vonsovskii and Svirskii (1997)	Zitterbewegung and uncertainties of velocity and acceleration in Dirac theory.

Reference	Comments
Wachter et al. (1997)	Relativistic corrections to the central $q\bar{q}$ potential in meson theory.
Alberto et al. (1998)	Infinite spherical well. Uses Lorentz scalar potential. Discuss LS coupling.
Arbatsky and Braun (1998)	Quadratic Zeeman effect for highly excited H.
Asaga et al. (1998)	g-factor of a tightly bound electron.
Atakishiyev et al.(1998)	Meixner oscillators. Have equidistant levels.
Barakat et al. (1998)	Perturbed Coulomb potentials in the KG equation.
Berbenni-Bitsch et al. (1998)	Microscopic universality in spectrum of lattice Dirac operator.
Bruce and Roa (1998)	Dirac particle in a vectorlike Coulomb potential.
Dattoli et al. (1998)	'Relativistic Hermite polynomials'.
Dragić et al. (1998)	H atom with added magnetic SO and spin-spin interactions. No deep-lying extra levels found.
Ermolaev (1998)	3D potentials for atoms in superintense laser fields.
X.-C. Gao et al. (1998)	KG particle in time-dependent, homogeneous $\mathbf{E}$.
Huber and Leeb (1998); Huber (1995)	SO potentials from inversion of radial Dirac equations.
Lichnerowicz (1998)	First eigenvalue of Dirac operator for Kähler manifold of even complex dimension.
Q.-G. Lin (1998)	Levinson theorem for a 2D Dirac particle in a central field.
Mallampalli and Sapirstein (1998c)	Solves $H = \sqrt{\mathbf{p}^2 + m^2} - m - \frac{Z\alpha}{r}$ in momentum space using B-splines. No singularity problems in $\mathbf{p}$-space
Milhorat (1998)	Spectrum of the Dirac equation on Gr_2 (C^{m+2}).
Mudry et al. (1998)	2D Dirac particle in strong imaginary vector potential + random impurity potential.
C. Müller et al. (1998)	Finite-element solution of 1D time-dependent Dirac equations. A solution for 'fermion doubling'.
Nedjadi et al. (1998)	Extended relativistic oscillator for $S = 1$ particles.
Nikitin (1998)	New classes of external fields having an exact FW transformation.
Nogami and Toyama (1998)	Reflectionless potentials for 1D Dirac equation. Pseudoscalar potentials.
Pal'chikov (1998)	En and Mn TP of arbitrary n for H-like atoms. Z=1-92.
Schön and Köppel (1998)	Geometric phases and quantum dynamics in SO-coupled systems. Molecules with vibronic interactions.
Segev and Wells (1998)	Time-dependent Dirac equation solved at ultrarelativistic limit for pair production in heavy-ion collisions. Light-front variables.
Sugawara-Tanabe and Arima (1998);Sugawara-Tanabe et al. (1999)	Hidden pseudospin symmetry in the Dirac equation. Spherical or axially deformed potentials.
Szmytkowski (1998b)	Variational R-matrix methods for the Dirac equation.
Taïeb et al. (1998)	Signature of relativistic effects in atom-laser interactions at ultrahigh intensities. 1D 'soft-core Coulomb' model: $V(x) = -q/\sqrt{2 + v^2}$. KG.
Villalba and Pino (1998)	2D hydrogen atom in magnetic field. KG and Pauli.
Wen et al. (1998)	High harmonics of H atom in ultrastrong laser field. MC simulation.
Yu and Takahashi (1998)	Dirac electron in two-frequency circularly polarized EM wave. Radiation and pair production.

Reference	Comments
Benvegnù (1999)	1D -1/|x|.
Bohun and Copperstock (1999)	Dirac-Maxwell solitons. Potential of uniformly charged sphere.
Brau (1999)	1D spinless Salpeter eq. for Coulomb + hard-core potential. Results wrong.
Braun et al. (1999)	Numerical solution of time-dependent Dirac equation. Split-operator method on a (3+1)D grid. Discuss the Klein paradox and the *Zitterbewegung*.
J. Chen et al. (1999)	High-order harmonics for H-like atom in ultrastrong laser field.
Darewych (1999)	Integral identities and bounds for scattering solutions.
Hall (1999)	Spectral comparison theorem for Dirac equation with $V(r)$.
Y. S. Huang et al. (1999)	Relativistic H-atom in spherical cavity. Applications to spherical quantum dots.
Kapshai and Alferova (1999)	1D scattering problem from superposition of δ-potentials.
D.-H. Lin (1999)	Path integral for the 3D Aharonov-Bohm-Coulomb system.
Q.-G. Lin (1999)	Levinson theorem for a 1D Dirac particle in a symmetrical field.
Matrasulov et al. (1999)	Dirac electron in the field of two opposite charges. $Z < 137$ and $Z > 137$ considered.
Mustafa and Odeh (1999)	KG for harmonic oscillator. Cp. Znojil (1996).
Narayan Vaidya and Barbosa da Silva Filho (1999)	Green function for charged spin-$\frac{1}{2}$ particle with anomalous magnetic moment in a plane wave EM field.
Nitta et al. (1999)	Motion of wave packet in Klein paradox.
Ouyang et al. (1999)	Dirac particles in twisted tubes.
Panek et al. (1999)	Electron scattering in powerful laser field. Angular and polarization effects. Electron handled as laser-dressed KG particle.
Rozmej and Arvieu (1999)	The Dirac oscillator. Permits exact FW.
Şimşek (1999)	Negative-energy levels for a ND Dirac equation. Spherically symmetrical, purely imaginary, linear potential.
Soliman and Abelraheem (1999)	Modification of band structure in intense laser fields. 1D model. New gaps introduced.
Toyama and Nogami (1999)	Search a 1D Dirac harmonic oscillator potential. (Only bound states, equally spaced.) Combination of scalar ($\beta S(x)$) and pseudoscalar ($\alpha\beta f(x)$) terms works.
Wessels et al. (1999)	Discretizing the 1D Dirac equation. Use time symmetry to conserve probability density.

Table 2.5: Relativistic virial theorems.

Reference	Comments
Rutkowski et al. (1993)	Relativistic virial theorem for diatomic molecules. H_2^+. Large-R and small-R limits.
Semay (1993)	Virial theorem for two-body Dirac equation.
Moiseiwitsch (1997)	Virial theorem for electron capture from H-like atom.
Cohen and Leung (1998)	Relativistic corrections to atomic sum rules using the FW approximation.
Cohen and Leung (1999)	Comment on Romero and Aucar (1998).

Chapter 3

Quantum Electrodynamical Effects

Table 3.1: QED and other higher-order effects: Methods.

Reference	Comments
Håkansson (1950)	Bethe logarithm for helium-like atoms.
Baranger et al. (1953)	Relativistic correction to Lamb shift. Hydrogen $2s$, 7.13 MHz.
Sommerfield (1958)	Electron g-factor to 4th order.
Todorov (1971)	Quasipotential equation in relativistic eikonal approximation.
Sapirstein and Yennie (1990)	Theory of hydrogenic bound states. Review.
Compagno and Salamone (1991)	Radiative corrections to charge density of an electron in an oscillator potential.
Shabad (1991)	VP and an electron gas in external field.
Crater et al. (1992); Crater and Van Alstine (1994)	Two-body interactions for spinning particles.
Milonni and Shih (1992b)	Casimir forces. Review.
K. T. Cheng et al. (1993)	Lamb shifts for non-Coulomb potentials. Finite nuclear size, core-Hartree potential.
Compagno et al. (1993)	Radiative corrections to charge density of H-like atoms. Main effect eigenvalue-induced exponential at large r. Related to Welton potential.
Devoto et al. (1993)	High-Z Lamb shifts in stripped atoms can be obtained from H-like values for a given electron density.
Frochaux (1993abc)	Relativistic corrections to Schrödinger equation from quantum field theory.
Hostler (1993,1994)	SE for an electron in an external potential. Commutator expansion. Reduced Green function.
Ionescu et al. (1993)	Collective excitations of the QED vacuum.
Ito and Gross (1993)	Compton scattering from relativistic composite systems.
Khriplovich et al. (1993)	Logarithmic corrections in the two-body QED problem.
Labzowsky (1993)	Adiabatic S-matrix approach in QED of large-Z He-like ions.
Labzowsky et al. (1993a)	Review on higher-order QED corrections for high-Z ions.
Labzowsky et al. (1993b)	Treatise on relativistic and QED effects in atoms.
Lindgren et al. (1993b)	2nd-order QED corrections for few-electron atoms. Reducible Breit-Coulomb correction and mixed SE-VP correction.
Lindgren et al. (1993c)	Bound-state SE using partial-wave renormalization.
Mohr (1993,1994)	QED in few-electron systems.
Pachucki (1993a)	QED corrections to charge density in H-like atoms.
Pachucki (1993b)	Two-loop corrections to Lamb shift of s-states.
Pachucki (1993c)	Higher-order binding corrections to Lamb shift.
Pachucki et al. (1993)	Nuclear-structure correction to the Lamb shift are -60 Hz and -19 kHz for H and D $1s$, respectively.
Persson (1993)	QED effects in highly charged ions. Thesis.
Persson et al. (1993a, 1998)	New approach to SE calculations. Convergent sums for each partial wave. H-like ions, Li-like U.
Persson et al. (1993b)	Accurate calculation of higher-order VP terms (beyond Uehling potential). Can be used with arbitrary atomic potential.
Quiney and Grant (1993,1994)	Partial-wave mass-renormalization in atomic QED calculations.
Rosenberg (1993)	Derives effective Hamiltonian for a two-electron atom. Extremum principles.

Reference	Comments
Sapirstein (1993)	Theory of many-electron atoms. Review on QED and many-body aspects.
Schneider et al. (1993a)	Källén-Sabry energy shift for H-like atoms with finite nuclei. (The VP shift of order $\alpha^2(Z\alpha)$). Z=1-100.
Seke (1993)	Gauge independence of NR Lamb shift including retardation effects.
Shabaev (1993b)	A Schrödinger-like equation derived from QED for the relativistic few-electron atom.
Soff (1993)	Radiative corrections in strong Coulomb fields. Emphasizes higher-order VP, nuclear-size effects on SE.
Sommerer (1993)	Relativistic two-body wave equations.
Yakhontov and Grant (1993)	Parameter-free renormalization in self-mass correction.
Barut and Saradzhev (1994)	Two-body system in (1+1)D QED.
Cavicchi and Vairo (1994)	New method for Lamb-shift calculation.
Hnizdo (1994)	VP potentials for extended nuclear charges.
Khriplovich et al. (1994)	Infrared divergence, Thomson scattering and the Lamb shift.
Kijowski (1994)	On electrodynamical self-interaction.
Labelle et al. (1994)	$O(m\alpha^8)$ contributions to decay of orthopositronium.
Labzowsky et al. (1994)	Non-resonant corrections (deviation of line profiles from Lorentz form) in Lamb-shift measurements for high-Z H-like ions.
Mil'shtein and Khriplovich (1994)	Large relativistic corrections to positronium decay.
Padden (1994)	Coulomb correction to VP tensor in superstrong **B**.
Rosenberg (1994a)	Infrared radiative corrections to potential scattering of a Dirac electron.
Schweber (1994)	History of QED.
Seke (1994)	Spontaneous decay of excited atomic states in NR QED.
Shabaev (1994b)	QED theory of electron recombination with highly charged ions.
Tang and Finkelstein (1994)	Relativistically covariant symmetry in QED.
Aucar et al. (1995b)	Bethe sum rule for the no-pair Hamiltonian.
Acikgoz et al. (1995)	Calculation of VP using self-field QED without infinities.
Hartemann and Luhmann (1995)	Radiation damping force on accelerated charged particle.
Kinoshita (1995)	New value for α^3 electron anomalous magnetic moment.
Koures and Harris (1995)	Gaussian-basis QED.
Labzowsky and Tokman (1995)	Reference state two-photon corrections for high-Z few-electron ions. Applications on Li-like ones.
Lindgren et al. (1995a)	Full QED calculation of two-photon exchange in He-like systems.
Mitrushenkov et al. (1995)	2nd-order loop-after-loop SE correction for few-electron multicharged ions.
Pachucki (1995)	Radiative recoil correction to Lamb shift.
Pachucki and Karshenboim (1995)	Nuclear-spin-dependent recoil correction to Lamb shift.
Rivelles (1995)	Comments on nonlocal and covariant symmetry in QED.
Rocchi and Sacchetti (1995)	Radiative corrections to Compton cross section.

Reference	Comments
Schäfer and Reinhardt (1995)	VP as test of C and CPT invariance. Muonic $4f_{7/2}$ and $5g_{9/2}$ levels in ^{209}Pb.
Eides (1996)	Weak-interaction contributions to hfs and the Lamb shift.
Jentschura and Pachucki (1996)	Higher-order Lamb shifts of $2P$ states.
Khriplovich et al. (1996)	Nature of the Darwin term and $(Z\alpha)^4 m^3/M^2$ contribution to the Lamb shift for arbitrary spin of the nucleus.
Kinoshita (1996)	The fine structure constant, α.
Labzowsky (1996)	Treatise on QED of atoms.
Labzowsky and Mitrushenkov (1996)	2nd-order SE for tightly bound atomic electron.
Labzowsky et al. (1996)	Vacuum polarization-nuclear polarization corrections to the Lamb shift.
Mallampalli and Sapirstein (1996)	4th-order VP contribution to Lamb shift. Z=5-100.
Mohr (1996)	Tests of fundamental physics.
Pachucki (1996a)	Lamb shift in muonic hydrogen.
Pachucki et al. (1996)	Theory of energy levels of H and D.
Pak et al. (1996)	D-line of Na-like ions. Includes a phenomenological Lamb-shift model.
Persson et al. (1996a)	Second-order SE-VP contributions to Lamb shift. H-like and Li-like systems, Z=70...92.
Persson et al. (1996b)	Two-electron Lamb shifts for He-like ions.
Sapirstein (1996a)	Quantum electrodynamics.
Seke (1996)	Complete Lamb shift to order α^5 in NR QED.
Shelyuto (1996)	Suggests a new gauge for the photon propagator. One-loop radiative corrections.
Beier et al. (1997a)	Review on Lamb shifts of H-like ions.
Beier et al. (1997b)	Two-loop VP contribution for H-like ions.
Beier et al. (1997c,1998)	Influence of nuclear size on QED corrections of H-like ions.
Blundell et al. (1997a)	Effects of non-Coulomb fields on radiative corrections. Electron screening, Zeeman field, M1 hyperfine fields as example.
Gonzalo and Santos (1997)	Radiative corrections to Zeeman effect of $2\,^3P$ states of He.
Hylton and Snyderman (1997)	Analytic basis set for high-Z atomic QED. He-like ions. Sturmian DC functions.
Jaekel and Reynaud (1997)	Movement and fluctuations of the vacuum.
Jallouli and Sazdjian (1997a)	Two-body potentials for two spin-0 or spin-$\frac{1}{2}$ particles from summation of Feynman diagrams.
Jallouli and Sazdjian (1997b)	Incorporation of anomalous magnetic moments.
Jentschura et al. (1997a)	Bound $\mu^+\mu^-$ system.
Jentschura et al. (1997b)	Lamb shift of $3P$ and $4P$ states and the determination of α.
B. D. Jones and Perry (1997)	Lamb shift in a light-front Hamiltonian approach.
Karshenboim (1997c)	Radiative corrections to light muonic atom decay.
Khelashvili and Kiknadze (1997)	Bound states in the continuum for quasipotentials, arising from QED.
Labzowsky and Goidenko (1997)	Multiple commutator expansion for the Lamb shift in a strong Coulomb field.

Reference	Comments
Labzowsky et al. (1997b)	Non-resonant QED corrections to radiative electron capture of highly charged ions.
Mohr (1997)	QED corrections in heavy atoms.
Persson et al. (1997b)	Theoretical survey of QED tests in highly charged ions.
Quiney et al. (1997)	Relativistic calculation of EM properties of molecules.
Stöcker et al. (1997)	Structure of vacuum and elementary matter. Conference proceedings.
Tulub et al. (1997)	Simulate the QED effects in many-electron atoms by a renormalized nucleus (right radius, changed charge inside).
Yerokhin et al. (1997a)	Two-electron SE contribution to the ground-state energy of He-like ions.
Au and Chu (1998)	H-like atoms with finite mass: two-photon processes lead to effective scalar photon interaction.
Dmitriev and Fedorova (1998)	New adiabatic QED method for bound states. Applied on two-photon exchange diagrams.
Dmitriev et al. (1998)	New, direct renormalization of bound electron self-energy.
Greiner (1998)	Correlations in the vacuum.
Jallouli and Sazdjian (1998)	Relativistic effects in pionium ($\pi^+\pi^-$) lifetime. Corrections of order $O(\alpha)$.
Karshenboim et al. (1998b)	One-loop VP correction to energy and wave function at origin in exotic atoms.
Khriplovich et al. (1998)	$O(\alpha^7(\ln\alpha)mc^2)$ corrections to FS splittings and $O(\alpha^6(\ln\alpha)mc^2)$ corrections to He energy levels. (Comment on T. Zhang (1996)).
Khriplovich and Sen'kov (1998)	Nucleon polarizability contribution to H-atom Lamb shift and H-D isotope shift.
Labzowsky et al. (1998a)	SE with partial-wave renormalization using B-splines. Results for H-like ns ions, Z=10-100, n=1-8.
Labzowsky et al. (1998bc)	2nd-order SE counterterms in bound-state QED.
Labzowsky and Tokman (1998)	Reference state two-photon Coulomb-Breit corrections for high-Z few-electron ions. Applications on Li-like ones.
Lindgren et al. (1998)	Analysis of SE for tightly bound electrons.
Low (1998)	Run-away electrons in relativistic spin-$\frac{1}{2}$ QED.
Mallampalli and Sapirstein (1998a)	4th-order SE contribution to Lamb shift.
Mallampalli and Sapirstein (1998b)	Perturbed-orbital two-loop Lamb shift. Z=0.5-5.
Mallampalli and Sapirstein (1998c)	Bethe logarithm for hydrogen using B-spline basis.
Mickelsson (1998)	VP and the geometric phase: Gauge invariance. Quantized fermions in external vector potentials.
Moussa and Baseia (1998)	Nonlocality of single photon in cavity QED.
Pachucki (1998a)	Energy of He $n\ ^3S_1$ states in order $m\alpha^6$. Effective Hamiltonian approach. Regularized Coulomb, $V(R) = -\frac{Z\alpha}{r}(1-\mathrm{e}^{-\lambda m\alpha r})$.
Pachucki (1998b)	Singlet S states of He to order $m\alpha^6$.
Pachucki (1998c)	Simple derivation of He Lamb shift.
Pachucki and Karshenboim (1998)	Positronium to order $m\alpha^6$.

Reference	Comments
Passante (1998)	Level shifts of accelerated H atom. Contribution from vacuum fluctuations affected.
Persson et al. (1998)	Renormalization corrections to partial-wave procedure.
Plunien et al. (1998)	Exact two-loop VP correction to Lamb shift in H-like ions.
Pyykkö et al. (1998)	Estimated valence-electron Lamb shifts for Li-E119, Cu-E111. About -1% of kinetic Dirac shifts for large Z.
Sapirstein (1998)	Theory of many-electron atoms. Review on QED and many-body aspects.
Schaden et al. (1998)	Unified treatment of some Casimir energies and Lamb shifts. Lamb shift of an atom in a dielectric medium, like a dilute gas.
Seke (1998)	Spontaneous decay of excited atomic states in R/NR QED.
Shabaev (1998a)	QED theory of nuclear recoil in atoms to all orders in αZ.
Soff et al. (1998)	QED in strong fields. Review.
Sucher (1998)	What is the force between the electrons? Searches an effective electron-electron potential.
Sunnergren (1998)	Complete one-loop QED calculations for few-electron ions. Applications on energies, g-factors and hfs. Thesis.
Yerokhin et al. (1998)	Two-electron SE corrections to ground-state energy of Li-like ions, $Z = 20 - 100$.
Bach et al. (1999)	Stability of the relativistic electron-positron field.
Bednyakov et al. (1999)	Electroweak radiative corrections.
Beier et al. (1999)	QED effects on radiative electron capture.
Coutinho et al. (1999)	Two definitions for electric polarizability, with or without the vacuum background.
Hughes and Kinoshita (1999)	Anomalous g values of the electron and muon. A review.
Ionescu and Belkacem (1999)	Relativistic collisions of highly-charged ions. Time evolution of QED vacuum solved numerically.
Ionescu et al. (1999)	Inner-shell photoionization at relativistic energies. Vacuum-assisted processes most probable at high E. 'Vacuum spark'.
Jentschura et al. (1999)	Nonperturbative numerical evaluation of 1-photon SE for H-like ions with low Z of 1-5.
Labelle and Zebarjad (1999)	Derivation of Lamb shift using effective field theory.
Labzowsky et al. (1999a)	QED corrections to ns electron g factors of neutral K-Fr, Ba^+.
Milonni et al. (1999)	Lamb shift of an atom in a dielectric medium.
Nefiodov et al. (1999a)	New approach to electron SE. Based on multiple commutator expansion and partial-wave renormalization.
Nefiodov et al. (1999b)	QED effects of radiative interference in recombination of electrons with heavy multicharged ions. Dielectronic recombination of He-like U.
Pachucki (1999a)	QED effects on He FS.
Pachucki (1999b)	Proton structure effects in muonic hydrogen.
Pachucki and Karshenboim (1999)	Higher-order recoil corrections to energy levels of two-body systems.
Panat and Paranjape (1999)	Dipolar and quadrupolar contributions to SE of H-like atom between two metallic slabs.
Ritchie and Weatherford (1999)	Quantum classical correspondence in NR QED. Lamb shift of a harmonically bound electron.
Sapirstein et al. (1999)	Potential-independence in relativistic MBPT achieved

Reference	Comments
	if negative-energy states are included. He-like ions. 'No-pair' not enough.
Vrejoiu et al. (1999)	Retardation corrections to angular distributions of free-bound transitions in H-like atoms.
Webb et al. (1999)	Search for time variation of the fine-structure constant α.
Yerokhin et al. (1999a)	Two-electron SE correction to $2p_{1/2} - 2s$ transition energy in Li-like ions, $Z = 18 - 100$.
Yerokhin et al. (1999b)	Screened SE and VP in high-Z Li-like ions.
Yerokhin and Shabaev (1999)	First-order SE in H-like systems. $\lvert\kappa\rvert \leq 5$, $n \leq 5$, Z=74,83,90,92.

Table 3.2: QED: Hyperfine interactions.

Reference	Comments
Bodwin et al. (1985)	Reviews recoil effects in QED of hfs.
Bodwin and Yennie (1988)	Recoil corrections to hydrogen hfs.
G. Li et al. (1993)	One-logarithmic recoil in muonium hfs.
Kinoshita and Nio (1994); Nio and Kinoshita (1997)	Muonium hfs.
Schneider et al. (1994a)	VP contribution to M1 hfs of H-like atoms.
T. Zhang and Xiao (1994)	Positronium hfs, α^6 corrections.
Eides and Shelyuto (1995)	M1 hfs contributions of order $\alpha^2(Z\alpha)^5$.
Schneider (1995)	M1 hfs of one-electron atoms. Thesis.
Shabaeva and Shabaev (1995)	Interelectronic contribution to hfs of Li-like ions. $1/Z$ expansion, including QED effects. Z=5-100.
Labzowsky et al. (1995)	Introduce the 'dynamic proton model' for hfs of H-like ^{209}Bi. Interaction between valence electron and valence proton at QED level.
Karshenboim (1996b)	Leading logarithmic corrections to muonium hfs.
Pachucki (1996b)	$\alpha(Z\alpha)^2 E_F$ correction to M1 hfs in H-like atoms.
Persson et al. (1996c)	SE correction to hfs of H-like atoms.
Yan et al. (1996)	FC term of M1 hfs in Li $2s, 2p, 3s$, Be$^+$ $2s$. Includes QED.
Blundell and Cheng (1997b)	Muonium hfs splitting.
Karshenboim (1997a)	Nuclear-structure-dependent radiative corrections to hydrogen hfs.
Labzowsky et al. (1997ac)	M1,E2,M3 hfs of the $2p_{3/2}$ state of H-, Li-, B- and N-like $^{209}_{83}$Bi. Interaction between valence electron and valence proton treated at QED level.
Pachucki (1997)	Positronium hyperfine structure.
Persson et al. (1997a)	Radiative corrections to electron g-factor in H-like ions.
Shabaev et al. (1997b)	Ground-state M1 hfs splitting of H-like ions. Z=49-83.
Tulub et al. (1997)	Simulate the QED effects in many-electron atoms by a renormalized nucleus (right radius, changed charge inside). Application on M1 hfs of H-like Bi.
Yerokhin et al. (1997b)	SE correction to $1s$ and $2s$ hfs splitting in H-like ions.
Beiersdorfer (1998b)	M1 hfs in Li-like U.
Karshenboim et al. (1998ab)	Analytic Uehling correction to hfs of muonic atoms.
Shabaev et al. (1998c)	QED and nuclear effects in highly charged ions.
Shabaev et al. (1998e)	Ground-state M1 hfs splitting and lifetime for Li-like

Reference	Comments
	ions, Z=49-83. Includes QED.
Sunnergren et al. (1998)	Radiative corrections to hfs of H-like systems.
Czarnecki et al. (1999a)	Positronium hyperfine splitting. Analytical $O(m\alpha^6)$.
Czarnecki et al. (1999b)	α^2 Corrections to parapositronium decay.
Faustov et al. (1999)	Hadronic VP and muonium M1 hfs.
Shabaev (1999)	Hfs of highly charged (H- and Li-like) ions. By combining them, nuclear factors can be eliminated. TP. QED included.

Table 3.3: QED: Energy levels.

Reference	Comments
Acker et al. (1966)	VP for muons in a Fermi nuclear potential.
Fricke (1969)	VP in muonic atoms. Quadrupole part of VP, muonic or pionic pairs included.
Geersten (1969)	Lamb shift and vibrational spectra of H_2^+, HD^+.
Jeziorski and Kołos (1969)	H_2^+ Lamb shift and the IP of H_2.
Blomqvist (1972)	VP in exotic atoms. Muonic Pb.
Feldman and Fulton (1988)	Lamb shifts for many-electron atoms.
Klarsfeld (1977)	Analytical expressions for VP in muonic atoms.
Kim et al. (1991) [=Ref. 4829 in RTAM II]	Lamb-shifts for resonance transitions of Li-, Na- and Cu-like ions.
Bukowski and Jeziorski (1993)	Lamb shift for muonic H_2^+ (ppμ etc.).
Beiersdorfer et al. (1993,1998b)	The $2s_{1/2} - 2p_{3/2}$ Lamb shift in Li-like to Ne-like U.
Berry et al. (1993)	QED and $1s2s$ - $1s2p$ triplet state energies of He-like systems, Z=2-92.
Berseth and Darewych (1993)	QED corrections to Ps^-.
Blundell (1993ab);	QED effects in Li-like, Na-like and Cu-like ions.
Blundell (1994)	Screening effects included.
Blundell (1993c)	Two-photon graphs for He-like ions.
Drake et al. (1993)	Energy corrections of order $mc^2\alpha^6 \ln \alpha$ for He.
Fell (1993)	Single-transverse-photon contributions of order $\alpha^6 \ln \alpha$ to energy levels of positronium.
Babb and Spruch (1994)	Retardation (Casimir) potential for Rydberg H_2 from QED.
Eides et al. (1994)	Light-by-light-scattering $\alpha^2(Z\alpha)^5 m$ contribution to H-atom Lamb shift.
Haftel and Mandelzweig (1994a)	Relativistic, finite-size and QED corrections for the 2 ^1S state of He.
Shabaev and Fokeeva (1994)	Reducible part of two-photon diagrams in the QED of multicharged ions.
Artemyev et al. (1995ab)	Nuclear recoil corrections for H-like and Li-like systems to all orders of αZ.
Adkins and Shiferaw (1995);	Positronium hyperfine levels
Eides and Grotch (1995ab)	α^6 corrections to S levels.
Eides and Shelyuto (1995)	Lamb-shift contributions of order $\alpha^2(Z\alpha)^5$.
Johnson et al. (1995)	$2s_{1/2} - 2p_{3/2}$ transitions in Li-like to Ne-like U. Includes one-loop QED corrections.
Kukla et al. (1995)	He-like Ar^{16+}, $1s2s\ ^3S$ – $1s2p\ ^3P$. Uncalculated QED terms estimated as $0.15(Z\alpha)^4$ a.u.
Eides (1996)	Weak-interaction contributions to hfs and the Lamb shift.
Erokhin and Shabaev (1996)	Screened SE diagrams to ground-state Lamb shift of

Reference	Comments
	two-electron atoms.
Nefiodov et al. (1996)	Nuclear polarization effects on spectra of multicharged ions. H-like ^{208}Pb, ^{238}U. Effects small.
Pak et al. (1996)	D-line of Na-like ions. Includes a phenomenological Lamb-shift model.
M. S. Safronova et al. (1996a)	$n = 2$ levels of Be-like systems, Z=4-100. Includes Lamb shift.
M. S. Safronova et al. (1996b)	$n = 2$ levels of B-like systems, Z=5-100. Includes Lamb shift.
T. Zhang (1996ab, 1997); T. Zhang and Drake (1994ab,1996)	QED corrections to FS of He.
Adkins et al. (1997); Adkins and Sapirstein (1998)	Positronium hyperfine interval.
Artemyev et al. (1997)	VP screening corrections for He-like ground states.
Beier et al. (1997a)	Review on Lamb shifts of H-like ions.
Beier et al. (1997c, 1998)	Influence of nuclear size on QED corrections of H-like ions.
Bhatia and Drachman (1997)	Rydberg states of Li.
Eides and Grotch (1997a)	Recoil corrections of order $(Z\alpha)^6(m/M)m$ to H-like s levels.
Eides and Grotch (1997b)	Radiative correction to nuclear-size effects and isotope shifts.
Eides et al. (1997)	Lamb-shift contribution of order $\alpha^2(Z\alpha)^5 m$
Elander and Yarevsky (1997)	QED corrections for antiprotonic helium.
Friar and Payne (1997a)	Higher-order nuclear-polarizability corrections for hydrogen (H or D).
Friar and Payne (1997b)	Higher-order nuclear-size corrections for hydrogen.
V. G. Ivanov and Karshenbojm (1997a)	Radiative corrections to level widths of light mesonic atoms.
V. G. Ivanov and Karshenbojm (1997b)	VP contribution of order $\alpha(Z\alpha)^6 mc^2$ to H-like levels
Karshenboim (1997a)	Two-loop logarithmic corrections in hydrogen Lamb shift.
Karshenboim (1997b)	Lamb shift of excited s levels of H and D.
King (1997,1999)	High-precision calculations on the Li atom. Review.
Korobov and Bakalov (1997)	Metastable states of antiprotonic He.
Martynenko and Faustov (1997)	FS of positronium.
Pal'chikov et al. (1997)	Assesses accuracy of Lamb-shift measurements in H.
M. S. Safronova et al. (1997ab)	$n = 3$ levels of Be-like systems, Z=4-30,54. Includes Lamb shift.
Yan and Drake (1997)	FS of Li $2p$ state. Includes QED.
Yan and Drake (1998)	Relativistic and QED energies in Li $2s$, $2p$ states and in Li$^+$.
Beiersdorfer et al. (1998)	The $2p_{1/2} - 2p_{3/2}$ transition of Be-like to F-like U.
Bhatia and Drachman (1998a)	Lamb shift in two-electron systems. He-like ground-states, $Z \leq 10$.
Indelicato et al. (1998)	K-, L-, and M-shell IP for elements with Z=10-100. The $1s$ Lamb shift of heavy elements.
Indelicato and Mohr (1998a)	Coordinate-space approach to bound-electron SE. Coulomb-field calculation.
Indelicato and Mohr (1998b)	H-like $6s$ and $8d$ state SE. New results for SE screening.
Labzowsky et al. (1998a)	SE with partial-wave renormalization using B-splines. Results for H-like ns ions, Z=10-100, n=1-8.

Reference	Comments
Marrocco et al. (1998)	QED shifts of Rydberg levels between parallel metal plates.
Pyykkö et al. (1998)	Lamb shifts of ns valence electrons of neutral alkali metals Li-E119 and coinage metals Cu-E111 estimated.
Santos et al. (1998a)	$2s_{1/2} - 2p_{3/2}$ transitions of Li-like to Ne-like Bi, Th and U. Includes QED via Welton potential.
Shabaev et al. (1998a)	Recoil correction to H-like ground-state energy.
Shabaev et al. (1998b,1999)	Recoil correction to H-like energy levels.
Shabaev et al. (1998c)	QED and nuclear effects in highly charged ions.
Artemyev et al. (1999)	VP screening corrections for Li-like ions, Z=20-100.
Drake and Goldman (1999)	Bethe logarithms for Ps^-, H^-, He-like atoms.
Friar et al. (1999)	Hadronic VP and the Lamb shift.
Goidenko et al. (1999)	2nd-order SE in H-like ions. Z=3-92.
Jentschura et al. (1999)	Nonperturbative numerical evaluation of 1-photon SE for H-like ions with low Z of 1-5.
Karshenboim et al. (1999)	Analytic VP contribution to energy of H-like muonic or electronic atoms.
Kinoshita and Nio (1999)	6th-order VP contribution to Lamb shift of muonic hydrogen.
Labzowsky et al. (1999b)	Lamb shifts of ns valence electrons of neutral Li-E119, Cu-E111, Hg^+, Tl^{2+} calculated.

Table 3.4: QED: Interatomic and intermolecular interactions.

Reference	Comments
Milonni and Shih (1992b)	Casimir forces. Review.
Ford (1993)	Spectrum of the Casimir effect and the Lifshitz theory.
Iacopini (1993)	Casimir effect at macroscopic distances.
Levin and Micha (1993)	Long-range Casimir forces. Book.
F. Luo et al. (1993)	Effect of retardation on binding of He_2. R^{-7} instead of R^{-6} important.
Power and Thirunamachandran (1993)	Derive the retarded van der Waals forces between neutral molecules from vacuum-fluctuation-induced dipole moments of the monomers.
Babb and Spruch (1994)	Retardation (Casimir) potential for Rydberg H_2 from QED.
Marinescu et al. (1994a)	Interaction of two alkali atoms. Includes retardation.
Power and Thirunamachandran (1994)	Derive the fully retarded dispersion potentials, including N-body terms, N=1-4.
Spruch et al. (1994)	Asymptotic Casimir interaction of a pair of finite systems.
Jamieson et al. (1995)	Retarded dipole-dipole dispersion for He.
F. Zhou and Spruch (1995)	van der Waals and Casimir interactions of an electron or of an atom with multilayered walls.

Table 3.5: QED: High-field ionization processes.

Reference	Comments
Åberg (1993)	QED of multiphoton ionization.

Chapter 4

Multielectron Atoms: Methods

Table 4.1: General methods and basic theory for multielectron atoms.

Reference	Comments
Foldy and Krajcik (1975)	Separable solutions for directly interacting particle systems. Expansion in α^2.
Jáuregui et al. (1991)	Relativistic CI for atoms.
Crater et al. (1992); Crater and Van Alstine (1994)	Two-body interactions for spinning particles.
Lindgren (1992,1994)	Reviews on the atomic many-body problem.
Barut et al. (1993)	Relativistic two-fermion equation.
Ilyabaev and Kaldor (1993)	Relativistic coupled-cluster approach for open-shell atoms. Tested on Li, C, O, F, Na.
Indelicato and Desclaux (1993)	Projection operators needed in MCDF.
Ishikawa and Quiney (1993)	Relativistic MBPT based on DFB wave functions.
Ito and Gross (1993)	Compton scattering from relativistic composite systems.
Johnson (1993)	Relativistic MBPT on highly-charged ions.
Y.-K. Kim (1993ab)	Recent and unsettled questions in atomic structure theory.
Lindgren et al. (1993a)	Heavy-ion spectroscopy and QED in atomic systems. Symposium proceedings.
Mårtensson-Pendrill (1993a)	MBPT in atomic structure calculations.
Piekarewicz (1993b)	Salpeter's approach to the relativistic two-body problem.
Pilkuhn and Staudner (1993)	Double Dirac equation and decay rates of bound particles.
Rabinowitch (1993)	Generalization of the Dirac equation for N electrons.
Rosenberg (1993)	Derives effective Hamiltonian for a two-electron atom. Extremum principles.
Sakamoto (1993)	$N-$body bound state solution of Dirac particles in (1+1)D.
Shabaev (1993b)	A Schrödinger-like equation derived from QED for the relativistic few-electron atom.
Sommerer (1993)	Relativistic two-body wave equations.
Yokojima et al. (1993)	Derivation of Bethe-Salpeter type N-body bound state equation.
Bieroń et al. (1994)	Ground-state energy of He, He-like Ho, U, with and without negative-energy projection operators.
Boero and Cortona (1994)	Transverse exchange in DFT.
Chandra and Hess (1994)	Finite-nucleus model for DK calculations.
Darevich (1994)	Hamiltonian variational method for few-particle systems.
Froese Fischer and Jönsson (1994)	Review on MCHF atomic calculations.
Grant (1994ab,1996)	Reviews on relativistic atomic calculations.
Jáuregui et al. (1994,1996)	Relativistic atomic CI: a variational principle.
Malvetti and Pilkuhn (1994)	Equal-time relativistic two-body equations.
Mourad and Sazdjian (1994)	Two-fermion relativistic wave equations. Constraint theory. Pauli-schrödinger form.
Nagy (1994)	Relativistic DFT for ensembles of of excited states.
Applebaum (1995)	Fermion stochastic calculus in Dirac-Fock space.
Broyles (1995)	Derives the Dirac-Breit Hamiltonian.
Devine and Wallace (1995)	Instant two-body equation in Breit frame.
Indelicato (1995)	Projection operators in MCDF calculations. Applied on ground states of He-like atoms.
Kenny et al. (1995)	QMC correlation and Breit energies for He-like systems, Be and Ne. Mass-polarization included.

Reference	Comments
Lindgren et al. (1995b)	Few-body problems in atomic physics.
Pilkuhn (1995)	An M1 hyperfine operator between two Dirac particles, both having anomalous magnetic moments.
Dzuba et al. (1996bc)	MBPT method for more than one valence electron.
Essén (1996)	Review derivation of the Darwin Lagrangian. Macroscopic consequences of the magnetic part.
Evans et al. (1996)	The spectrum of the electron equation by Bethe and Salpeter.
Häckl and Pilkuhn (1996)	Transformation of Breit operators into hyperfine-like operators.
Indelicato (1996)	Correlation and negative continuum effects for the M1 transition in two-electron ions. MCDF.
Kolakowska et al. (1996)	Minimax variational approach to relativistic two-electron problem.
Lindgren (1996)	Relativistic many-body and QED calculations on atomic systems.
Parpia et al. (1996)	The GRASP92 atomic MCDF package.
Phillips and Wallace (1996)	Derive a 3D bound-state equation from the 4D Bethe-Salpeter one for two-body boson or fermion systems.
L.-J. Wu (1996)	R MBPT including 'pseudoconfigurations'. Be ground state.
Badnell (1997)	Two-body, non-fine-structure operators (contact spin-spin, two-body Darwin, orbit-orbit) incorporated into AUTOSTRUCTURE code.
Bijtebier (1997); Bijtebier and Broekaert (1997)	Solutions of the Bethe-Salpeter equation.
Cea (1997)	Vacuum stability for 3D fermions problem for large m.
Ishikawa et al. (1997)	Finite-nucleus models for GTO spinors.
Jáuregui et al. (1997ab)	Eigenvalue spectra of the 'no pair' Hamiltonian. Tests on 1- and 2-electron systems.
Jáuregui et al. (1997c)	Buildup of many-electron atoms from QED.
Y.-K- Kim (1997)	Review on relativistic atomic structure calculations.
King (1997,1999)	High-precision calculations on the Li atom. Review.
Kołakowska (1997)	Explicitly correlated (r_{12}), Hylleraas-type relativistic wave functions for He-like systems.
Bunge et al. (1998ab)	Decoupling of positive- and negative energy states.
Entralgo et al. (1998)	The problem of two point particles with spin.
Feldmann et al. (1998)	S states of 3-electron atoms. BP Hamiltonian and explicitly correlated wave functions lead to individual divergencies which, however, cancel.
Y.-K. Kim et al. (1998)	Failure of MCDF at NR limit. The wave function for specific J may not reduce to the LS limit. Possible remedies discussed.
Klink (1998ab)	Relativistic QM for particles, each of which is a bound state of a mass operator.
Lindgren (1998)	Review on electron correlation and QED.
Long and Crater (1998)	Two-body Dirac equations for general covariant interactions. Their coupled Schrödinger-like forms.
Lucenti et al. (1998)	Dirac observables and spin bases for N particles.
I. Martin (1998)	Relativistic quantum defect orbital method.
Monahan and McMillan (1998)	Faddeev equation for relativistic two- and three-body system.

Reference	Comments
Reiher (1998)	Numerical algorithms for MCDF. Thesis.
Ruijgrok (1998)	General requirements for a relativistic quantum theory for two or more particles. Quasipotential theory recommended.
Sigg and Sorg (1998)	Nature of exchange terms in a two-particle relativistic Schrödinger theory. A potential.
Sucher (1998)	What is the force between the electrons? Searches an effective electron-electron potential.
Sunnergren (1998)	Symmetrical spectrum method to avoid fermion doubling.
Dürr et al. (1999)	'Hypersurface Bohm-Dirac models.' N entangled but noninteracting Dirac particles.
Esteban and Séré (1999)	Claim, from a mathematical point of view, that the DF equations have infinitely many solutions.
Lindgren (1999)	QED effects in strong nuclear fields. Review.
Noyes and Jones (1999)	Relativistic three-body problem. Only the masses m_a, m_b, m_c defined.
Reiher and Hinze (1999)	Self-consistent treatment of the Breit term in MCDF.
Sapirstein et al. (1999)	Potential-independence in relativistic MBPT achieved if negative-energy states are included. He-like ions. 'No-pair' not enough.

Table 4.2: Published programs for atoms.

Reference	Comments
Eissner (1991b)	SUPERSTRUCTURE. A BP-level atomic code.
Parpia et al. (1993)	Complete active spaces (CAS) for atomic calculations.
Perger et al. (1993)	Continuum solver for the GRASP code.
Salvat and Mayol (1993)	Partial wave analysis for electron and positron elastic scattering.
Young and Norrington (1994)	Solution of relativistic asymptotic equations in electron-ion scattering.
Berrington et al. (1995)	RMATRIX I. The Belfast atomic R-matrix code.
Fritzsche and Grant (1995); Fritzsche (1997)	Expansion of jj-coupled symmetry functions into determinants.
Kroger and Kroger (1995)	Program for angular coefficients of relativistic one-electron hfs parameters.
Szmytkowski (1995)	Relativistic multi-channel variable phase program for asymptotic equations of electron-atom and electron-ion scattering.
Ankudinov et al. (1996)	Single-configuration version of the DF code of Desclaux (1975).
Jönsson et al. (1996)	HFS92: A program for atomic hfs calculations. MCDF.
Parpia et al. (1996)	The GRASP92 atomic MCDF package.
Pöschl et al. (1996); Pöschl (1997,1998)	B-spline finite-element codes. Primarily for nuclear problems.
Jönsson and Froese Fischer (1997)	SMS92: Relativistic isotope shift calculations.
Fritzsche et al. (1999)	Expansion package RATIP for GRASP92 for various transition and ionization properties.
Chernysheva and Yakhontov (1999)	A DF code. Continuum states and muons can be included.
Fuchs and Scheffler (1999)	PP for polyatomic systems from SR DFT theory.

Table 4.3: Numerical four-component methods.

Reference	Comments
Froese Fischer and Parpia (1993)	Accurate spline solution of radial Dirac equations.
Sapirstein and Johnson (1996)	Use of splines in atomic physics.
Sapirstein et al. (1999)	Potential-independence in relativistic MBPT achieved if negative-energy states are included. He-like ions. 'No-pair' not enough.

Table 4.4: Four-component approaches for many-electron atoms..

Reference	Comments
Da Silva et al. (1993ab)	Universal Gaussian basis set for DF calculations.
Gambhir and Ring (1993)	Solution of Dirac equation using GTO basis for nuclei. Recipe for avoiding spurious states.
Ley-Koo et al. (1993)	Two-body integrals for relativistic atomic calculations. Both Coulomb and Breit.
Malli et al. (1993ab, 1994)	Universal Gaussian basis for DF calculations.
Koc and Ishikawa (1994)	Single-Fock-operator method for DF on open-shell atoms. Li-K, B-In.
Koc et al. (1994)	Relativistic CI calculations on open-shell atoms.
S. N. Datta (1995)	DFB tests on He-Ne.
Minami and Matsuoka (1995)	GTO basis for $_{86}$Rn-$_{94}$Pu. DF.
Deineka (1996)	B-spline methods.
Dzuba et al. (1996bc)	MBPT method for more than one valence electron. Combination of MBPT and CI.
Jorge et al. (1996,1997ab); Jorge and da Silva (1998)	GTO-basis sets.
Jorge and da Silva (1996ab,1997)	Generator coordinate DFB.
Koc et al. (1996)	Relativistic modification of asymptotic CI. C-like systems, $Z \leq 106$.
Ley-Koo et al. (1997)	Relativistic atomic r_{12}^{-1} integrals using perimetric coordinates.
Vilkas et al. (1997)	2nd-order MRCI based on MCDF functions.
Visscher and Dyall (1997)	DF atomic calculations using different nuclear charge distributions. Z=1-109.
Malli and Ishikawa (1998)	Geometric series of Gaussian exponents from 'generator coordinate DF method' for open-shell atoms.
Vilkas et al. (1998b)	Relativistic MR MP method based on MCDF reference functions.
Chaudhuri et al. (1999a)	GTO-basis MBPT.
Vilkas et al. (1999)	Relativistic MR MBPT for quasidegenerate systems.

Table 4.5: Relativistic density functional theory.

Reference	Comments
Dreizler (1993)	Relativistic DFT.
Holas and March (1993)	Relativistic DFT reduced to the Dirac equation.
March (1993)	Completely local, relativistic DFT. Role of the virial. Applied to Thomas-Fermi theory and heavy atoms in magnetic fields.
Nagy (1994)	Relativistic DFT for ensembles of of excited states.
Rajagopal (1994)	DFT including EM fields in condensed matter.
Engel (1995); Engel and Dreizler (1996)	Reviews on relativistic density functional theory.
Engel et al. (1995)	Local and non-local relativistic exchange-correlation functionals.
Rieger and Vogl (1995)	Relativistic self-interaction-free DFT.
Vijayakumar and Gopinathan (1995)	Atomic tests of the authors' Ξ functional. Relaxation effects, Koopmans' theorem discussed.
Engel et al. (1996)	GGA for relativistic exchange-only energy functional.
E. K. U. Gross et al. (1996)	DFT of time-dependent phenomena.
Kenny et al. (1996)	QMC for relativistic homogeneous electron gas. PT relativity.
Rajagopal and Buot (1996)	Fundamentals of time-dependent DFT, including EM fields.
Engel et al. (1997)	Relativistic corrections to exchange-correlation functional.
Higuchi and Hasegawa (1997, 1998)	Relativistic current- and spin-density functional theory. Applications on Ln^{3+} ions; Ln=Ce-Yb.
Engel et al. (1998)	Relativistic optimized potential method for DFT. Exact transverse exchange and MPn-based correlation potential. No-pair gives gauge independence.
Facco Bonetti et al. (1998)	Relativistic exchange-correlation energy functional. Gauge dependence of no-pair correlation energy.
Grabo et al. (1998)	Optimized effective potential DFT. Review.
Kreibich et al. (1998)	Approximate relativistic optimized potential method.
W.-J. Liu and Dolg (1998)	IPn and $d-f$ transitions of La-Yb; n=1-4. Several density functionals compared with QR PP.
W.-J. Liu et al. (1998c)	IPn; n=1-4, $5f-6d$ excitation energies of Ac-No. DFT and PP methods compared.
Tong and Chu (1998)	Relativistic DFT with self-interaction correction. Right $(-1/r)$ long-range behaviour. Atomic ground states, Z=2-106.
Engel and Dreizler (1999)	DFT that claims to include van der Waals effects.
Eschrig and Servedio (1999)	DFT approach to open-shell atoms (C-Pb + ions).
Schmid et al. (1999)	Relativistic GGA tested on 5d TM. Band structure and cohesive properties. Relativity changes exchange at 1% level, is negligible for correlation functional.

Table 4.6: Thomas-Fermi calculations.

Reference	Comments
March (1993)	Completely local, relativistic DFT. Role of the virial. Applied to TF theory and heavy atoms in magnetic fields.
Shivamoggi and Mulser (1993)	Relativistic TF model for atoms in a strong magnetic field.
Shivamoggi (1995)	Relativistic TF theory with thermal effects, with or without a very strong magnetic field.
Lieb et al. (1996)	Stability of relativistic matter via TF theory. Finds stability for $Z \leq 59$ if $\alpha = 1/137$.
March (1997)	TF theory and $1/Z$ expansions for large Z.

Table 4.7: Independent-particle models.

Reference	Comments
J.-J. Chang (1993,1998)	Dirac quantum-defect theory.
U. I. Safronova et al. (1993)	Screening theory for transition energies of highly charged ions.
Stein (1993)	A local, 1-VE PP for Li-Cs atoms. Tested on $s - f$ states.
Bielinska-Waz et al. (1994)	Core-polarization effects in the relativistic quantum-defect-orbital theory.
Nana Engo et al. (1997)	Supersymmetry-inspired R and QR quantum-defect theory. Applications on transition matrix elements.
Schweizer et al. (1999)	Local model potential for Li-Cs and Li-like ions, Z=3-10.

The following Table 4.8 contains the methodological developments, including the new pseudopotentials, since RTAM II. No distinction has been made between between the pseudopotentials with nodeless pseudo wavefunctions, and the 'model potentials' whose wave functions contain the nodes. For a steady update of the Stuttgart energy-consistent PP, see *http://www.theochem.uni-stuttgart.de/*.

Table 4.8: Pseudopotentials: Methodological work.

Reference	Comments
Stevens et al. (1984)	PP for Li-Ne, Na-Ar. (NR.)
Igel-Mann (1987)	PP for main-group and d^{10} TM elements.
Vanderbilt (1990)	Soft self-consistent PP.
Gonze et al. (1991)	Analysis of separable pseudopotentials.
Troullier and Martins (1991)	PP for plane-wave calculations. R/NR for $Z<72$, some $Z>72$.
Kaupp (1992)	PP applications. Thesis.
Bergner et al. (1993)	Energy-consistent PP for Groups 13-17 (B-In, .. F-I).
Cundari and Stevens (1993)	PP for Ce-Lu. Pd-like (46-e) core.
Czuchaj et al. (1993)	Ne-Xe, Ba PP.
Dolg et al. (1993c)	Energy-consistent PP for Ha (element 105). SO.
Ehlers et al. (1993)	An f-polarization function for $3d, 4d, 5d$ TM PP. Optimized for the CISD energy of the lowest s^1d^n state.
Häussermann et al. (1993)	Energy-consistent R/NR PP for Hg.
Hemstreet et al. (1993)	SO PP for solids. III-V semiconductors.
Höllwarth et al. (1993)	A d-type polarization function for main-group B-Bi. An f-polarization function for Zn-Hg.
Leininger et al. (1993)	MH^+, M=Fe-Os. Compare several PP.
Marino and Ermler (1993)	Core-valence correlation potential operator. Cs.
Stein (1993)	A local, 1-VE PP for Li-Cs atoms. Tested on $s-f$ states.
Ermler and Marino (1994,1996)	Review on PP methods.
Fernandez Pacios and Gomez (1994)	Triple-zeta GTO basis for PP.
Kresse and Hafner (1994)	Norm-conserving and ultrasoft PP for first-row and transition elements.
Küchle et al. (1994)	PP for Ac-Lr. SO potential included.
Mitas (1994)	PP QMC calculation of IP and EA of Fe.
Ross et al. (1994)	PP for Ln, Ce→Lu. 54-e core. SO.
Shukla and Banerjee (1994)	A four-component relativistic DF method for valence electrons only. An effective core-valence potential constructed. Tested on atoms, Li-S.
Casarrubios and Seijo (1995)	Test of WB MP on Pt.
Huzinaga (1995)	Review on 'active electrons in chemistry'.
Koseki et al. (1995)	Effective Z for SO splitting using Stevens' pseudo wave functions.
Nicklass et al. (1995)	R/NR PP for Ne-Xe. SO.
Schwerdtfeger et al. (1995a)	Accuracy of the PP approach tested on InCl, $InCl_3$.
Seijo (1995)	Derive PP including SO part from the WB Hamiltonian. F-At, Tl→Rn.
Titov and Mosyagin (1995)	New PP for Cu-Au.
Tuan and Pitzer (1995a)	PP for Hf.
Tupitsyn et al. (1995)	'Generalized ECP:s' for Hg→Bi. Atomic tests only.
Wittborn and Wahlgren (1995)	New PP for $5d$ metals.
Charpentier et al. (1996)	PP including semicore for Ba, Ce, Th. Plane wave basis.

Reference	Comments
Cundari et al. (1996)	Review on pseudopotential methods.
Dolg (1996a)	Energy-consistent 1-VE PP for Group 1 (Li-Cs).
Dolg (1996bc)	Accuracy of the 7-VE energy-consistent PP for Group 17 tested. Differential effects of the nodelessness are small.
Frenking et al. (1996,1999); Frenking and Pidun (1997)	Reviews on transition-metal compounds.
Gordon and Cundari (1996)	Pseudopotential studies on transition-metal compounds. Review.
Kaupp (1996e)	PP applications on TM systems. Habilitation.
Leininger et al. (1996a)	9-VE PP for K-Cs.
Leininger et al. (1996b)	13-VE and 21-VE PP for In.
F. Nogueira et al. (1996)	Transferability of *local* PP, based on solid-state electron density.
Blaudeau and Curtiss (1997)	Basis sets for the PP of K, Ca, Ga-Kr. PP of Hurley et al. (1986) used.
Eichkorn et al. (1997)	Auxiliary basis sets for Z = 1-56, 72-85. RI method.
Flad and Dolg (1997)	Accuracy of PP for TM in MC calculations. Sc→Cr. IP within 0.1 eV.
Flad et al. (1997)	SO coupling in variational MC calculations. Pb, Bi^+, Po^{2+}.
Leininger et al. (1997)	R/NR 13-VE and 21 VE PP for Ga, Tl.
Menchi and Bosin (1997)	DFT PP in QMC.
Mosyagin et al. (1997)	Generalized PP for Hg-Tl. Claims higher accuracy. Has many parameters.
Nash et al. (1997)	PP for $_{95}$Am-(E118). SO also given.
Pollack et al. (1997)	Density-based local PP for 16 simple metals.
Sakai et al. (1997)	PP for main-group elements, Li-Rn.
Seth et al. (1997)	PP for E111 and E112.
Wildman et al. (1997)	New PP for Tl→Rn. $5d6s6p$ valence shells. Special consideration on f shells.
Buenker et al. (1998)	SO splittings and PP. Tl ground state as example.
Casarrubios and Seijo (1998)	WB-based MP for B-Ba (Z = 5-56).
Decker et al. (1998)	Calibration of model potentials for main-group elements. N_2-Sb_2, F_2-I_2, XY (X=C-Sn, Y=O-Te), OH_2-TeH_2, NH_3-SbH_3, ...
Delley (1998)	Scattering-theory approach to scalar relativistic corrections to bonding. Local pseudopotential can include them.
Féret and Pascale (1998)	Two electrons in a PP. Ba.
Hartwigsen et al. (1998)	'Separable dual-space Gaussian PP' for H-Rn. LDA.
Y. S. Kim et al. (1998)	Kramers-unrestricted molecular PP calculations at HF or MP2 level. SO PP used.
W.-J. Liu and Dolg (1998)	IPn and $d-f$ transitions of La-Yb; n=1-4. Several density functionals compared with QR PP.
W.-J. Liu et al. (1998c)	IPn; n=1-4, $5f-6d$ excitation energies of Ac-No. DFT and PP methods compared.
Miyoshi et al. (1998)	PP for main-group elements Ga-Kr, In-Xe, Tl-Rn.
Rakowitz et al. (1998)	Ir^+. A new SO operator tested.
Sakai et al. (1998)	PP for Ln (La-Lu). 46-e core.
Sanoyama et al. (1998)	Low-lying levels of Ln^{3+}; Ln=Ce-Eu, Tb-Yb. Compare different PP. Semi-core correlation $(4d, 5s, 5p)$ important.
Schautz et al. (1998)	2-VE PP for Zn-Hg.
Schimmelpfennig et al.	The 'atomic-mean-field integral (AMFI)'

Reference	Comments
(1998ab)	approach to SO CI. Tested on the Pt atom.
Seth (1998)	PP for E111-E120.
Y. X. Wang and Dolg (1998)	10-VE ($5s5p6s$) PP for Yb.
Alatalo et al. (1999)	Truncated pseudopotentials for alloy calculations. Cut-offs in momentum space introduced. Pd-Al alloys as example.
Casarrubios and Seijo (1999)	WB-based MP for La, Hf-Hg (Z = 57,72-80). V_{SO} included.
Fuchs and Scheffler (1999)	PP for polyatomic systems from SR DFT theory.
Rakowitz (1999)	Development of PP methods, especially for SO. Thesis.
Rakowitz et al. (1999ab)	PP for TM, Sc to Hg.
Ramer and Rappe (1999)	Nonlocal PP with improved transferability. H,Si,Ca,Zr,Pb.
Titov and Mosyagin (1999)	'Generalized Relativistic ECP'. Many further parameters give moderate increase in accuracy in atomic tests.
Trail and Bird (1999a)	Core reconstruction in PP calculations. Full Dirac.
Trail and Bird (1999b)	Uses this method for calculating structure factors.

The Table 4.9 contains the available references on atoms in strong laser fields. No complete coverage is attempted. Some of these references also appear in other tables.

Table 4.9: Atoms in strong laser fields.

Reference	Comments
Faisal and Radozycki (1993)	3D model for bound particle in laser pulse.
Ovcharov and Fedosov (1993)	Correlated coherent states of electron in field of plane EM wave.
Radozycki and Faisal (1993)	Multiphoton ejection of strongly bound relativistic electrons.
Rosenberg and Zhou (1993)	Dirac and KG particles in a multimode laser field: Generalized Volkov solutions.
Keitel (1996)	Ultra-energetic electron ejection in relativistic atom. Laser field interaction.
Meyerhofer et al. (1996)	Relativistic mass-shift effects during high-intensity laser–electron interactions. Multiphoton scattering.
Ehlotzky et al. (1997)	Electron–atom collisions in a laser field.
Kylstra et al. (1997)	1D model atom in intense laser field.
Protopapas et al. (1997)	Atomic physics with super-high intensity lasers. Review.
Rathe et al. (1997)	Intense laser-atom dynamics with 2D Dirac equation.
Szymanowski et al. (1997)	Relativistic calculation of two-photon bound-bound transition amplitudes in hydrogenic atoms.
Crawford and Reiss (1998)	Relativistic ionization of hydrogen by linearly polarized light.
Goreslavsky and Popru-shenko (1998)	Relativistic deflection of photoelectron trajectories in elliptically polarized laser fields.
Krainov (1998)	Energy distribution of relativistic electrons in the tunneling ionization of atoms by super-intense laser radiation.
Matrasulov (1998,1999)	Chaotic ionization of an H-like atom in intensive mono-chromatic field.
Reiss (1998)	Introduction to special issue on 'Relativistic effects in strong electromagnetic fields. After low-field E1, next comes 'figure-8' pattern due to magnetic component. Full relativistic effects when the kinetic electron energy approaches mc^2.
Q. Su et al. (1998)	Relativistic suppression of wave packet spreading.
Szymanowski and Maquet (1998)	Relativistic signatures in laser-assisted scattering at high field intensities.
Taïeb et al. (1998)	Signature of relativistic effects in atom-laser interactions at ultrahigh intensities. 1D 'soft-core Coulomb' model: $V(x) = -q/\sqrt{2+v^2}$. KG.
J.-H. Wen et al. (1998)	High harmonics of H atom in ultrastrong laser field. MC.
Yu and Takahashi (1998)	Dirac electron in two-frequency circularly polarized EM wave. Radiation and pair production.
J. Chen et al. (1999)	High-order harmonics for H-like atom in ultrastrong laser field.
Krainov (1999)	Energy and angular distribution of relativistic electrons in the tunnelling ionization of atoms by circularly polarized light.
Panek et al. (1999)	Angular and polarization effects in relativistic potential scattering of electrons in a powerful laser field.
Soliman and Abelraheem (1999)	Modification of band structure in intense laser fields. 1D model. New gaps introduced.

Table 4.10: $1/Z$ and other similar expansions for many-electron atoms.

Reference	Comments
Shabaeva and Shabaev (1995)	Interelectronic contribution to hfs of Li-like ions. $1/Z$ expansion, including QED effects. Z=5-100.
Weiss and Kim (1995)	Relativistic modifications of charge expansion theory. Introduce shells with same (nj).
March (1997)	Thomas-Fermi theory and $1/Z$ expansions for large Z.
Yan et al. (1998)	$1/Z$ extrapolations for $2s$ and $2p$ states of Li-like systems.
Safronova and Shlyaptseva (1999)	Autoionization rates and energies for C-, N-, O-, and F-like ions with $Z = 6 - 54$. $1/Z$ expansion.

Table 4.11: Related nuclear calculations.

Reference	Comments
Centelles et al. (1993)	Semiclassical nuclear mean-field theory.
Fritz et al. (1993)	Dirac effects in HF calculations on finite nuclei employing realistic forces. Increase the binding energy *and* the charge radius for ^{16}O and ^{40}Ca.
Gambhir and Ring (1993)	Solution of Dirac equation using GTO basis for nuclei. Recipe for avoiding spurious states.
Yokojima et al. (1993)	Derivation of Bethe-Salpeter type N-body bound state equation.
Fearing et al. (1994)	Compare the FW and Pauli approximations in nuclear physics. FW found better.
Warrier and Gambhir (1994)	SO splitting in relativistic nuclear mean field theory.
Zamick and Zheng (1994)	Nuclear structure with Dirac phenomenology.
Forest et al. (1995)	Relativistic nuclear Hamiltonians. Combines two- and three-body potentials in rest frame with boosts.
Nefiodov et al. (1996)	Nuclear polarization effects on spectra of multicharged ions. H-like ^{208}Pb, ^{238}U. Effects small.
Phillips and Wallace (1996)	Derive a 3D bound-state equation from the 4D Bethe-Salpeter one for two-body boson or fermion systems.
Pöschl et al. (1996); Pöschl (1997,1998)	B-spline finite-element codes. Primarily for nuclear problems.
Bijtebier (1997); Bijtebier and Broekaert (1997)	Solutions of the Bethe-Salpeter equation.
Ring (1997)	Relativistic mean-field program for even-even nuclei.
Stoitsov et al. (1997)	Relativistic nuclear calculations using a transformed harmonic-oscillator basis.
Klink (1998ab)	Relativistic QM for particles, each of which is a bound state of a mass operator.
Tomaselli et al. (1998)	M1 $1s$ hfs in H-like atoms with one-hole nuclei. ^{165}Ho...^{207}Pb.
Blum and Brockmann (1999)	Dirac sea effects and the stability of nuclear matter.

Chapter 5

Multielectron Atoms: Results

Table 5.1: Tabulations of atomic ground-state properties.

Reference	Comments
Chakravorty et al. (1993)	Atomic ground-state energies for 3-18 electrons, Z=3-36, .. Includes Dirac-Breit +QED.
García de la Vega and Miguel (1994)	Momentum expectation values, $\langle p^n \rangle$, n = -2 to +4 for atomic ground states, Cs, U. DF/HF.
Ishikawa and Koc (1994)	Ground states of Ne-Rn, Zn-Hg, Ne-like ions.
Indelicato (1995)	Projection operators in MCDF calculations. Applied on ground states of He-like atoms, Z=10-100.
Kenny et al. (1995)	QMC relativistic and Breit corrections for He-like systems, Be and Ne. Much smaller Breit contributions than at DF level.
Ishikawa and Koc (1996)	Li-like systems, $Z \leq 20$.
Kotochigova et al. (1996,1997)	LDA ground-state energies for atoms 1-92.
Cioslowski et al. (1997)	MCDF core-electron densities, Z=3-118.
Forstreuter et al. (1997)	Ln atoms and ions. Ln=Ce→Lu. DFT tested.
King (1997,1999)	High-precision calculations on the Li atom. Review.
King et al. (1998)	Hylleraas-type calculation of BP relativistic corrections for the Li ground state.
Frolov (1999)	Ground-state properties of the positronium anion, Ps^-. BP.

Table 5.2: Energy levels.

Reference	Comments
Y. Zhang et al. (1992b)	Na-like Nb.
Arp et al. (1993)	K-shell spectra of Ca, Cr, Mn, Cu. $1s \rightarrow 4p$ important.
Berry et al. (1993)	QED and $1s2s$ - $1s2p$ triplet state energies of He-like systems, Z=2-92.
Berseth and Darewych (1993)	Relativistic corrections to Ps^-.
Birkett et al. (1993)	$2\ ^3P_0$–$^3P_1$ FS splitting of He-like Ag. Hyperfine quenching.
M.H. Chen et al. (1993)	$n = 2$ triplet states of He-like ions, Z=5-100.
X.-L. Cheng et al. (1993)	Ni-like ions, $47 \leq Z \leq 54$.
Chevary (1993)	Ground-state DF configurations for M^-, M=Sr-Ra,Yb,La,Lu.
Chou et al. (1993ad)	Hg-like ions,
Chou et al. (1993bd)	Cd-like ions.
Chou et al. (1993c)	Mg-like ions.
K. T. Chung and Zhu (1993)	$2snl$ states of the Be atom. FS, isotope shifts.
Connerade et al. (1993)	Rydberg levels for alkali metals.
Drake (1993c)	Review on He Rydberg levels.
Harra et al. (1993)	He-like Ni. $1s2l$, $1s3l$. MCDF.
Ilyabaev and Kaldor (1993)	Relativistic coupled-cluster approach for open-shell atoms. Li, C, O, F, Na.
Kronfeldt et al. (1993b)	Fine and hyperfine structure of Er I, $4f^{11}5d6s6p$.
Liaw (1993b)	Alkali metal atoms in the Brueckner approximation.
C. D. Lin (1993)	Classification and properties of doubly excited states.
Lindroth and Indelicato (1993,1994)	Inner-shell transitions in heavy atoms.
Marketos et al. (1993)	O III excitation energies.
Marques et al. (1993a)	Hyperfine quenching of Be-like $1s^22s2p\ ^3P_0$.
Marques et al. (1993b)	Hyperfine quenching of Mg-like $3s3p\ ^3P_0$.

Reference	Comments
I. Martin et al. (1993)	Li-like sequence. QDT.
Migdalek and Stanek (1993)	$5s^2 - 5s5p$ transitions in the Sr sequence.
Rath and Patnaik (1993)	$2p$ SO splitting of Li-like systems.
U. I. Safronova et al. (1993)	Screening theory for transition energies of highly charged ions.
Shabaev (1993a)	Finite-nuclear-size corrections to energy levels of multicharged ions. Expressed in terms of nuclear $\langle r^n \rangle$; n=2,4. Z=1-100.
Simionovici et al. (1993)	$n = 3$ transitions of Na-like Pb.
Trigueiros et al. (1993)	$4p^4$ configuration of Ge-like Kr V.
Verner et al. (1993)	Subshell ionization energies of He-Zn.
Vidolova-Angelova (1993)	Sr I.
Vosko and Chevary (1993)	Electron shell filling of Ln anions: Lu^- is $5d^1 6s^2 6p^1$.
P. Wang et al. (1993)	Kα satellites of Al. BP.
W. J. Wang (1993)	FS levels in N-like Co-Zn. MCDF.
Z. W. Wang et al. (1993)	$1s^2 np$ states; n=2-5 for Li-like systems.
L. Yang (1993)	K-shell excitation and ionization of Na.
Zilitis (1993)	Rydberg levels of Li-like ions, Z=3-16.
Anisimova et al. (1994)	Two-el. atoms with p and d electrons.
Biémont et al. (1994a)	Resonance transitions of Cl.
Bieroń et al. (1994)	Ground-state energy of He, He-like Ho, U, with and without negative-energy projection operators.
M.-K. Chen and Chung (1994)	Core-excited doublet states of Li.
K. T. Cheng et al. (1994)	Singlet states for He-like ions, $n = 2$, $4 \leq Z \leq 92$.
X.-L. Cheng et al. (1994)	Ni-like ions, $72 \leq Z \leq 79$.
Chevary and Vosko (1994)	Ground-state configuration for Tm^-.
Chou et al. (1994a)	Zn-like ions.
Chou et al. (1994b)	Be-like ions.
Dembczynski et al. (1994)	Odd-parity levels of Pb I.
Dzuba et al. (1994)	Dy.
Eliav et al. (1994a)	Au^+, Au. DFB+CCSD.
Eliav et al. (1994b)	Ground states of atoms with 2-5 electrons. DFB+CCSD.
Eliav et al. (1994c)	Li-Fr. DFB+CCSD.
Eliav et al. (1994d)	Xe ground-state energy and pair correlations.
Eliav et al. (1994e)	E111 (eka-Au). Ground state $d^9 s^2$, not $d^{10} s^1$.
Flambaum et al. (1994)	Chaotic spectrum of Ce.
Fritzsche and Grant (1994a)	P II $^5S_2 - ^3P_J$ intercombination transitions.
Fritzsche and Grant (1994b)	Be-like ions, 2s^2 $^1S_0$-$2s3p$ $^3P_1$ intercombination transitions.
Froese Fischer (1994)	C II and C III.
Gaigalas et al. (1994)	O isoelectronic sequence.
Gębarowski et al. (1994)	$3s^2 3p - 3s^2 3d$ transitions for Al-like sequence.
Gleichmann and Hess (1994a)	Hg at DK MRCI level.
Haftel and Mandelzweig (1994a)	Rel., finite-size and QED corrections for the 2 ^1S state of He.
Haftel and Mandelzweig (1994b)	He ground state.
Hsu et al. (1994)	^4P series of Li. Autoionization.
Itoh et al. (1994)	Relativistic effects on multiplet terms of lanthanide ions.
Lindroth (1994)	Doubly excited states of He with a finite, discrete spectrum.

Reference	Comments
Pershina et al. (1994b)	Energy levels of M^{q+}; M=V-Db(E105); q=0, 5.
Plante et al. (1994)	n=1,2 states of He-like ions, Z=3-100.
U. I. Safronova (1994)	$2p^{-1}nl$ and $2s^{-1}nl$ levels of Ne-like ions. Z=20-60, n=3-6, l=0-3.
U. I. Safronova and Bruch (1994)	Transition and Auger energies of Li-like ions. $1s2lnl'$ configurations.
U. I. Safronova et al. (1994a)	$1s2s^22p^n$ and $1s^22s^22p^n$ configurations as function of Z. Compares the SUPERSTRUCTURE and MZ codes.
U. I. Safronova and Nilsen (1994)	Autoionization states of Li-like ions with large n.
U. I. Safronova et al. (1994b)	Ne-like ions, $2p^{-1}nl$, $2s^{-1}2p^{-1}nl$, $n = 3-6$, $l = 0-3$. Z=20-60.
U. I. Safronova et al. (1994c)	Two-electron doubly-excited states.
Shabaev and Artemyev (1994)	Relativistic nuclear recoil corrections for multicharged ions.
Tolstikhina et al. (1994)	K x-ray satellite energies for Ar^{17+} upon metallic surface.
Vilkas et al. (1994)	O-like sequence.
Ynnerman et al. (1994)	$2p-2s$ energies in Li-like U.
Zilitis (1994)	Al-like ions.
Anisimova et al. (1995)	Ne $2p^5nd$ levels, n=3-8.
Avgoustoglou and Beck (1995)	Ne-like $2p^53s$ levels, Z = 10-92.
M. H. Chen et al. (1995)	$n = 2$ states of Li-like ions.
Connerade et al. (1995)	Level statistics in complex spectra.
Datta and Beck (1995)	FS of ^{139}La II, $(5d+6s)^2$.
Dzuba et al. (1995)	Fr.
Eliav et al. (1995a)	Rf (E104) ground state $7s^26d^2\ ^3F_2$. DFB+CCSD.
Eliav et al. (1995b)	The f^2 Pr^{3+} and U^{4+}. DFB+CCSD.
Eliav et al. (1995c)	Yb, Lu, Lr. Ground state of Lr(E103) $7s^27p_{1/2}\ ^2P_{1/2}$.
Eliav et al. (1995d)	Hg, E112 (eka-Hg). Ground states of the latter mono- and dication d^9s^2 and d^8s^2, respectively.
Fritzsche et al. (1995)	Cl-like ions, $3p^43d$ configurations.
Gribakina et al. (1995)	The Ce atom as a 'chaotic' system. Recall Flambaum et al. (1994).
W. Huang et al. (1995)	Ne I excited levels.
Johnson et al. (1995)	$2s_{1/2}-2p_{3/2}$ transitions in Li-like to Ne-like U. Includes one-loop QED corrections.
Komninos et al. (1995)	^2D level of Al.
Kornienko et al. (1995)	Effect of CI on SO parameter of Ln ions.
Kukla et al. (1995)	He-like Ar^{16+}, $1s2s\ ^3S$ - $1s2p\ ^3P$. Uncalculated QED terms estimated as $0.15(Z\alpha)^4$ a.u.
Liaw (1995)	Bound states of Ca^-. No EA in Brueckner approximation.
Y. Liu et al. (1995)	Photoexcitation of Ni $3p$ level. SO, exchange, hybridization.
Mårtensson-Pendrill et al. (1995)	Low-Z $1s2p$ states of He-like systems.
I. Martin et al. (1995)	Ag I isoelectronic sequence. QDT.
Neale and Wilson (1995)	Influence of core polarization on FS splittings of Kr VIII. MCDF.
Pal'chikov and von Oppen (1995)	Spin-spin mixing of $1s3s\ ^3S_1$ and $1s3d\ ^3D_1$ states in the He-like sequence. Z=2-100.

Reference	Comments
Polasik (1995)	$K\beta_{1,3}L^0M^r$ x-ray spectra of Mo, Pd, La. MCDF.
Wijesundera et al. (1995)	Ground state of Lr. MCDF.
Yan and Drake (1995)	$2s$, $2p$ and $3d$ levels of Li.
H. Y. Yang and Chung (1995)	Be-like systems, $Z = 3 - 10$, core-excited $1s2s2p^2(^2P)$ and $1s2p^3(^5S)$ states.
Ynnerman and Froese Fischer (1995)	Be-like systems, Z=6...42. The E1 $2s^2\ ^1S_0 - 2s2p\ ^1P_1$ and the forbidden $2s^2\ ^1S_0 - 2s2p\ ^3P_1$ energies and TP.
Zilitis (1995)	$3p$, $3d$ levels of Al-like systems. Several crossings, non-monotonical $3d$ FS splittings.
Aspromallis et al. (1996)	Be^- $1s^22s2p^2\ ^4P$ fine structure.
Avgoustoglou and Liu (1996)	Ne-like $[2p^53d]_{J=1}$ levels, Z = 10-92.
Aymar et al. (1996)	Review on multichannel Rydberg spectroscopy.
Charro et al. (1996)	Triplet-triplet transitions in Be-like ions.
K. T. Cheng and Chen (1996)	$2s - 2p_{3/2}$ transitions in Li-through-F-like U.
Chou et al. (1996)	S-like ions, $Z \geq 18$.
Dzuba et al. (1996bc)	Tl.
Eliav et al. (1996a)	Ba, Ra. NR ground states of both cations are d^1, not s^1. Relativistic M^+ are s^1.
Eliav et al. (1996c)	Tl, E113 (eka-Tl). Low-lying $6d$ hole states found. 'Is eka-Tl a transition element?'
Indelicato (1996)	Correlation and negative continuum effects for the M1 transition in two-electron ions. MCDF.
Ishikawa and Koc (1996)	Li-like systems, $Z \leq 20$.
Ivanov and Ivanova (1996)	Sturmian orbital method for radiation physics of atoms and ions. $2l3l$, $2l4l$ levels of Ne-like systems.
E. Johnson et al. (1996)	Am. MCDF.
W. R. Johnson et al. (1996)	E1 transitions for Li-like (Z=3-100), Na-like (Z=11-100) systems and the alkali atoms K-Fr.
W.-Y. Liu et al. (1996)	Rydberg levels of Cs in strong magnetic field.
W. C. Martin and Sugar (1996)	ds^2p levels of Zr-Rf. Includes CI with $(d+s)^3p$.
Pak et al. (1996)	D-line of Na-like ions. Includes a phenomenological Lamb-shift model.
Quinet (1996)	Pd II.
Rakowitz and Marian (1996)	FS splitting of Tl ground state. Different radial distributions of $np_{1/2}$ and $np_{3/2}$ important.
M. S. Safronova et al. (1996a)	$n = 2$ levels of Be-like systems, Z=4-100.
M. S. Safronova et al. (1996b)	$n = 2$ levels of B-like systems, Z=5-100.
M.-L. Tan et al. (1996a)	Cu-like Ag^{50}.
M.-L. Tan et al. (1996b)	Bi^{54+}.
Thøgersen et al. (1996)	FS of Ir^- and Pt^-. MCDF.
Umemoto and Saito (1996)	Electron configurations of superheavy elements, Z=121-131. g-electrons appear at Z=126.
Vijayakumar and Gopinathan (1996)	SO splittings of TM atoms and ions. Cr ... Au. DFT/DF.
Visscher et al. (1996a)	FS splitting of the Cl atom. Effect of triples important.
W. J. Wang et al. (1996)	O-like ions, Z=52-79. MCDF.
Wijesundera et al. (1996)	FS levels of ns^2np Ca^- - Ba^-.
Ali (1997)	FS splitting of Ga-like atoms
Beck (1997a)	$3d^4$ systems Xe^{32+}, ... Gd^{42+}. J = 2,3.

Reference	Comments
Beck (1997b); Beck and Datta (1993)	FS for $(d+s)^3$ states of La I, Zr II, Hf II.
Bhatia and Drachman (1997)	Rydberg states of Li.
Biémont et al. (1997a)	Lowest $5g-6h$ supermultiplet of Fe II.
Bonnelle et al. (1997)	nd x-ray emissions (n=3-5) in U^{4+} compounds. MCDF.
M. H. Chen and Cheng (1997a)	$2s^2$ and $2s2p$ ($J=1$) states of Be-like ions, $Z=10...92$.
M. H. Chen and Cheng (1997b)	$3s^2$ and $3s3p$ states of Mg-like ions, $Z=12...42$.
Chou and Johnson (1997)	Cu-, Ag- and Au-like ions.
Dzuba and Gribakin (1997)	M^-, M=Ca-Ra.
Elander and Yarevsky (1997)	Antiprotonic helium, $\bar{p}He^+$. Relativistic and QED effects.
Ishikawa and Koc (1997)	B-In, Cu-Au.
Johnson et al. (1997b)	Mg I, Al II, Al I, Tl II, Tl I, Pb I, Bi II, Bi I.
Johnson et al. (1997c)	n=3 states of B-like Na.
Jönsson and Froese Fischer (1997b)	Mg-like $3s^2$-$3s3p$ transitions. $Z \leq 18$.
Korobov and Bakalov (1997)	Metastable states of antiprotonic He.
Rosberg and Wyart (1997)	Au II.
M. S. Safronova et al. (1997ab)	$n=3$ levels of Be-like systems, Z=4-30,54.
Sekiya et al. (1997)	$6s$ and $4f$ ionized states of Ln; Ln=Ce-Lu.
Shukla et al. (1997)	FS splitting of Pb-like systems, Pb$\rightarrow Po^{2+}$.
Vilkas et al. (1997)	Be, Be-like Ne.
Wahlgren et al. (1997)	SO splitting of the Tl atom. BP/DK.
Yan and Drake (1997)	FS of Li $2p$ state. Includes QED.
Bhatia and Drachman (1998c)	Binding energy of Ps^-.
Biémont et al. (1998)	$s-p$ and $s-d$ transitions of Fr. Tests on Rb, Cs.
Bruch et al. (1998a)	$3lnl'$ states of Na-like Cu.
Bruch et al. (1998b)	$3lnl'$ states of Na-like Fe.
G.-X. Chen and Ong (1998a)	Fe XXIII.
Dzuba and Johnson (1998)	Ba. Test new MBPT+CI B-spline method.
Eliav et al. (1998b)	Bi, E115 (eka-Bi). DFB+CCSD.
Eliav et al. (1998c)	La, Ac, E121 (eka-Ac). Ac^{2+} and Ac^+ have $7s^1$ and $7s^2$ ground states, respectively. La^{2+} and La^+ have $5d$ and $5d^2$, respectively.
Feldmann et al. (1998)	S states of 3-electron atoms. BP Hamiltonian and explicitly correlated wave functions lead to individual divergencies which, however, cancel.
Fritzsche et al. (1998)	P-like iron-group ions, Z=22-32, $3s3p^4$ and $3s3p^23d$ configurations.
Froese Fischer et al. (1998a)	O isoelectronic series, Z=9-18. $2p^4-2p^33s-2s2p^5$ transitions.
Froese Fischer et al. (1998b)	Li isoelectronic series, Z=3-8.
Godefroid et al. (1998)	Review on atomic structure calculations.
Jönsson and Froese Fischer (1998)	$2s^2$-$2s2p\ ^3P_1$ intercombination transition in C III.
Jönsson et al. (1998)	Be isoelectronic sequence. Z=7-28,...42.
Kohstall et al. (1998)	Si-like ions, Z=16-36, $3s3p^3$, $3s^23p3d$.
Lindroth et al. (1998)	H^- resonances converging to $H(n=2)$ threshold.
W.-J. Liu and Dolg (1998)	$f^{n+1}d^0 \rightarrow f^nd^1$ transitions of La-Yb. Several density functionals compared.
W.-J. Liu et al. (1998c)	IPn; n=1-4, $5f-6d$ excitation energies of Ac-No. DFT and PP methods compared.

Reference	Comments
Pyykkö et al. (1998)	Estimated valence-electron Lamb shifts for Li-E119, Cu-E111. About -1% of kinetic Dirac shifts for large Z.
Qu et al. (1998a)	Li-like core-excited $1s2snp\ ^4P$ states, Z=4-8.
Qu et al. (1998b)	Li core-excited $1s2snl$ states.
Rakowitz et al. (1998)	Ir^+. A new SO operator tested.
M. S. Safronova et al. (1998)	Na-like systems, Z=11-16.
U. I. Safronova et al. (1998a)	$n = 3$ states of B-like systems, Z=6-30.
Sanoyama et al. (1998)	Low-lying levels of Ln^{3+}; Ln=Ce-Eu, Tb-Yb. Compare different PP. Semi-core correlation $(4d, 5s, 5p)$ important.
Santos et al. (1998a)	$2s_{1/2} - 2p_{3/2}$ transitions of Li-like to Ne-like Bi, Th and U. Includes QED via Welton potential.
Shabaev (1998a)	QED theory of nuclear recoil in atoms to all orders in αZ.
Taïeb et al. (1998)	Signature of relativistic effects in atom-laser interactions at ultrahigh intensities.
Vilkas et al. (1998a)	B-like ions, Z=10 ... 42. MCDF.
Vilkas et al. (1998b)	O-like Fe. Low-lying states.
Yan and Drake (1998)	Relativistic and QED energies in Li $2s$, $2p$ states and in Li^+.
Yan et al. (1998)	$2s$ and $2p$ states of Li-like systems. Z=3-20 + extrapolations.
Beck (1999)	Energies of the nearly Z-independent $J = 3 \rightarrow 2$ transition of the $3d^3_{3/2}3d_{5/2}$ systems W^{52+} and Bi^{61+}.
Bhatia and Drachman (1999)	C IV excitation energies.
Drake (1999)	High-precision theory of He.
Dzuba et al. (1999a)	Positron bound to Cu atom. Affinity 170 meV.
Dzuba et al. (1999bc)	Sensitivity of atomic data to space-time variation of natural constants.
Froese Fischer (1999)	Review on transitions in lighter atoms. BP/Dirac.
Gayasov and Joshi (1999)	$4d^9 - 4d^8 5p$ in La XIII and Ce XIV
Petit (1999)	U I. $f^3ds^2 + f^3d^2s$.
Quinet et al. (1999)	Ln II.
M. S. Safronova et al. (1999)	Alkali metals, Na-Fr.
U. I. Safronova et al. (1999c)	E1 transitions between $n = 2$ states in B-like ions. Z=6-100.
Tachiev and Froese Fischer (1999)	Be-like systems, Z=4-12. $2s2p$, $2p^2$, $2s3s$, $2s3p$, $2s3d$ excited levels. BP.
Vilkas et al. (1999)	O-like systems, Z=8-60.
Yan and Ho (1999)	Relativistic effects in positronium hydride (PsH, e^-e^+p).

Table 5.3: Auger and autoionization processes.

Reference	Comments
M. H. Chen (1993a)	Angular distributions in resonant Auger decay. Intermediate coupling.
Cornille et al. (1993)	Autoionization rates for Be-like systems.
d'État et al. (1993)	X-ray spectroscopy of highly charged ions, interacting with surfaces. A review.
Fritzsche (1993)	Xe resonant $4d^{-1}6p$ Auger spectra. Angular distribution.
Kotochigova and Lambropoulos (1993,1994)	Multiphoton ionization and autoionization of Si I.
Lohmann (1993)	Correlation effects on Hg Auger spectra. MCDF.
Mäntykenttä (1993)	Channel interaction and relaxation effects in Xe $N_{4,5}OO$ Auger transitions.
Mäntykenttä et al. (1993)	Electron correlation in $4d$ hole states of Ba studied by Auger and photoelectron spectroscopy.

Reference	Comments
Pindzola (1993)	Parity-violation effects on Auger-electron emission from highly charged atomic ions. $2s^2$ $J=0$ of U^{90+}.
Tulkki et al. (1993a)	Auger decay of $2p_{3/2}^{-1}ns$ resonances in Ar.
Tulkki et al. (1993b)	Effect of channel interaction, exchange and relaxation on the angular distribution and spin polarization of Auger electrons from noble-gas atoms.
Tulkki and Mäntykenttä (1993)	Subshell-dependent relaxation in the Auger effect.
M. H. Chen and Reed (1994)	Angular Distribution of Auger electrons after electron-impact ionization. Be-like ions, Z=12...42.
Flambaum et al. (1994)	Chaotic spectrum of Ce.
Hsu et al. (1994)	^4P series of Li. Autoionization.
Ivanov and Safronova (1994)	Correlation and relativistic effects for $3l3l'$ autoionization states.
Kivimäki et al. (1994)	Kr and Xe satellite Auger spectra.
Lohmann and Fritzsche (1994)	KLL Auger spectra of alkali atoms.
M. S. Safronova et al. (1994)	Autoionization states of Be-like ions. Z-dependence.
U. I. Safronova and Nilsen (1994)	Autoionization states of Li-like ions with large n.
Tulkki et al. (1994)	Effect of initial-state–final-state CI on anisotropy of resonant Auger decay of Kr $3d^{-1}5p$ and Xe $4d^{-1}6p$ states.
Aksela et al. (1995)	Resonant Auger decay of Xe $4d_{3/2,5/2}^{-1}6p$.
Cole et al. (1995)	Auger parameter shifts from free atoms to solids, in alkali and alkaline earth metals.
U. I. Safronova et al. (1995)	Radiative and autoionization TP for $1s2l"nl - 1s^2n'l'$; $n, n' = 2, 3$ transitions for He-like ions.
Aksela et al. (1996a)	Review on resonant and nonresonant Auger recombination.
Aksela et al. (1996b)	Resonantly excited $3d_{3/2,5/2}^{-1}5p$ states of Kr.
Aksela et al. (1996c)	Resonantly excited $4d_{3/2,5/2}^{-1}6p$ states of Xe.
Aksela and Mursu (1996)	Resonant $2p^{-1}4s$ and $3p^4 4s$ states of Ar.
Flambaum et al. (1996)	'Narrow chaotic compound autoionizing states in atomic spectra. Simultaneous excitation of several electrons $\Rightarrow$ dense spectrum $\Rightarrow$ chaos. Ce.
Gel'mukhanov et al. (1996)	Auger spectra for molecular-field-split core levels. H_2S.
M. H. Chen et al. (1997)	K-shell Auger and radiative transitions for C-like systems, $6 \leq Z \leq 54$.
U. I. Safronova et al. (1997)	Autoionizing states of Li-like ions with high n.
M. Cohen et al. (1998)	Excitation autoionization rate from ground and excited levels. Li-like to S-like Ar.
Indelicato et al. (1998)	K-, L-, and M-shell IP for elements with Z=10-100. 'Auger shifts' included.
Qu et al. (1998a)	Li-like core-excited $1s2snp$ ^4P states, Z=4-8.
Qu et al. (1998b)	Li core-excited $1s2snl$ states. Autoionization.
U. I. Safronova and Johnson (1998)	Autoionizing rates for doubly excited $2lnl'$ states of He-like ions.
U. I. Safronova et al. (1998b)	Autoionization rates for $1s^{-1}2p^2$, $1s^{-1}2s^{-1}2p^3$ and $1s2p^4$ states of B-like ions, $Z = 6 - 54$.
Cornille and Dubau (1999)	Autoionization of $1s2s^22p^2$, $1s2s2p^3$ and $1s2p^4$ states of B-like ions, Z=6-54.
Safronova and Shlyaptseva (1999)	Autoionization rates and energies for C-, N-, O-, and F-like ions with $Z = 6 - 54$.

Reference	Comments
Santos et al. (1999)	Radiative and non-radiative decay rates of $2s$ (L_1) hole states in Yb and Hg. MCDF and MBPT.
L.-R. Wang et al. (1999)	Doubly excited states in photoionization of Zn. Autoionization resonances.

Table 5.4: Ionization potentials and electron affinities.

Reference	Comments
Arnau et al. (1992)	EA of Group-13 atoms, Al-Tl.
Chakravorty et al. (1993)	IP for 3-18 electrons, Z=3-36, .. Includes Dirac-Breit +QED.
K. T. Chung et al. (1993)	IP of Be-like systems, Z=4-10,15,20.
Datta and Beck (1993)	Possible EA(Ln) with $4f$-electron attachment. Negative results for the nf^{14} Tm^- and Md^-.
Fricke et al. (1993)	IPn of Ta, E105 (Db). MCDF.
Hughes and Kaldor (1993)	EA of F-At. Fock-space CC method.
Ilyabaev and Kaldor (1993)	Relativistic coupled-cluster approach for open-shell atoms. IP of Li, C, O, F, Na.
Vosko and Chevary (1993)	Electron shell filling of Ln anions: Lu^- is $5d^1 6s^2 6p^1$.
Woon and Dunning (1993)	EA of Al-Cl. Includes SO.
Datta and Beck (1994)	Th^- can have both $6d^2 7s^2 7p$ and $6d^3 7s^2$ bound states. EA 0.365 and 0.189 eV, respectively.
Dinov et al. (1994)	EA of Ce. $6p$ or $5d$ attachment.
Dzuba and Gribakin (1994)	Correlation-potential method for negative ions.
Eliav et al. (1994a)	IP and EA of Au. DFB+CCSD.
Eliav et al. (1994c)	IP and EA of alkali metals, Li-Fr. DFB+CCSD. Average IP error 0.09%, EA error 4-9%.
Eliav et al. (1994e)	IP and EA of E111 (eka-Au).
Mitas (1994)	PP QMC calculation of IP and EA of Fe.
Sundholm and Olsen (1994)	EA of Ca. PT relativity.
Dinov and Beck (1995a)	EA of $6p$ electrons in Pr.
Dinov and Beck (1995b)	EA for U. $7p$ attachment.
Eliav et al. (1995a)	EA of Rf (E104) 6.01 eV. DFB+CCSD.
Eliav et al. (1995c)	IP and EA for Yb, Lu, Lr.
Eliav et al. (1995d)	IP of eka-mercury, E112.
García de la Vega (1995)	Relativistic corrections to EA of Li-I. Mostly BP without SO.
Liaw (1995)	Bound states of Ca^-. No EA in Brueckner approximation.
Sundholm (1995)	EA of Sr and Ba. PT relativity.
Tatewaki et al. (1995)	$6s$ and $4f$ IP Cs, Ba, La-Lu. DF.
Yan and Drake (1995)	IP of Li.
J.-M. Yuan (1995)	Without relativity, $d_{3/2}$ state of Ba^- would be bound.
Dinov and Beck (1996)	EA for Pa. $7p$ attachment.
Eliav et al. (1996a)	IP of Ba, Ra.
Eliav et al. (1996b)	EA of E118. The first rare gas with an EA. Both relativity and correlation essential.
Eliav et al. (1996c)	IPn; n=1-4 and EA of Tl, E113 (eka-Tl).
Glushkov et al. (1996)	EA of Ca and Sr.
Salomonson et al. (1996)	EA of Ca and Sr.
Thøgersen et al. (1996)	FS of Ir^- and Pt^-. MCDF.
Wijesundera et al. (1996)	FS levels of ns^2np Ca^- - Ba^-.
Andersen et al. (1997)	Negative alkaline-earth ions.

Reference	Comments
Avgoustoglou and Beck (1997)	EA of Ca-Ba, Yb.
Dzuba and Gribakin (1997)	FS of EA of Ca-Ra.
Eliav et al. (1997)	EA of B-Tl.
Forstreuter et al. (1997)	IPn of Ln, Ln=Ce→Lu. DFT tested.
Ishikawa and Koc (1997a)	IP of B-In, Cu-Au.
Ishikawa and Koc (1997b)	IP of Zn, Cd.
Jursic (1997)	EA for several metal atoms from DFT and HF-MP4, QCISD.
King (1997,1999)	High-precision calculations on the Li atom. Review.
Koga et al. (1997)	IP and EA for $Z \leq 54$. Relativistic and mass corrections included. PT.
Miadoková et al. (1997)	Standardized basis sets for highly correlated DK calculations of electric properties. IP of Li-Fr, Be-Ba.
Neogrády et al. (1997)	IP and EA of Cu-Au. R/NR. SR DK CCSD(T).
Sekiya et al. (1997)	$6s$ and $4f$ ionized states of Ln; Ln=Ce-Lu.
Z. W. Wang and Ge (1997)	IP of Li-like Sc-Zn.
Wijesundera (1997)	EA of B-Tl. MCDF.
Dzuba and Gribakin (1998)	Yb^- $6p_{1/2}$ is a resonance at 0.02 eV and not a bound state.
Eliav et al. (1998b)	IP and EA of Bi, E115 (eka-Bi).
Eliav et al. (1998c)	IP and EA of La, Ac, E121 (eka-Ac).
Glushkov et al. (1998)	EA of Si-Pb.
Indelicato et al. (1998)	K-, L-, and M-shell IP for elements with Z=10-100. The $1s$ Lamb shift of heavy elements.
W.-J. Liu and Dolg (1998)	IPn of La-Yb; n=1-4. Several density functionals compared.
W.-J. Liu et al. (1998c)	IPn; n=1-4, $5f-6d$ excitation energies of Ac-No. DFT and PP methods compared.
O'Malley and Beck (1998)	EA of Sn.
Wijesundera and Parpia (1998)	EA of C, N, P.
Pyykkö et al. (1998)	Lamb shifts of ns valence electrons of neutral Li-E119, Cu-E111 estimated.
Vilkas et al. (1998b)	IP of Zn.
Yan and Drake (1998)	Relativistic and QED contributions to IP of Li.
Biémont et al. (1999)	IP for Li–Sn.
Chaudhuri et al. (1999a)	IP of Li-Fr, B-Tl.
Chaudhuri et al. (1999b)	IP of Li, Na, Be, Mg.
G.-X. Chen and Ong (1999b)	EA of B-Tl, C-Pb.
de Oliveira et al. (1999)	EA of H, B-F, Al-Cl. PT relativity.
Dzuba et al. (1999)	Cu. EA 1.218 eV (exp. 1.236 eV), positron affinity 0.170 eV.
Godefroid and Froese Fischer (1999)	Isotope shift in oxygen EA. FS of the anion.
Iliaš and Neogrády (1999)	IP of Zn-Hg.
Johnson et al. (1999)	IPn; n=1-6 of Cr-Sg(E106).
Labzowsky et al. (1999b)	Lamb shifts of ns valence electrons of neutral Li-E119, Cu-E111, Hg^+, Tl^{2+} calculated.
Lim et al. (1999)	IP of the alkali atoms Li-E119. CCSD(T).
J. M. L. Martin and de Oliveira (1999)	EA of H-Cl (Z=1-17). Very precise. PT.
Norquist et al. (1999)	EA of Ru.
O'Malley and Beck (1999)	EA of La. $6p$ or $5d$ attachment.

Reference	Comments
M. S. Safronova et al. (1999)	IP of Na-Fr.
Sundholm et al. (1999)	EA of Ga and In, 0.297(13) and 0.374(15) eV, respectively.

Table 5.5: Supercritical ($Z > 137$) systems.

Reference	Comments
Momberger et al. (1991, 1993)	Electron-positron pair production in collision of two naked nuclei. U+U.
Rumrich et al. (1991,1993)	Pair creation in relativistic heavy-ion collisions.
Ionescu et al. (1993)	Collective excitations of the QED vacuum.
Maier et al. (1993)	Time-dependent 1D Dirac square well: Pair production.
Müller-Nehler and Soff (1994)	Electron excitations in superheavy quasimolecules.
J. Thiel et al. (1994,1995)	Electron-positron pair creation in relativistic heavy-ion collisions.
Eichler (1995)	Charge transfer from the negative-energy continuum. Alternative pair-production mechanism.
Eichler and Belkacem (1996)	Gauge transformations for coupled-channel calculations of pair production.
Ionescu and Eichler (1996)	Pair creation in heavy-ion collisions as a charge-transfer process.
Alscher et al. (1997)	Electron-positron pair production in heavy-ion collisions.
V. I. Matveev et al. (1997)	LCAO solution of two-centre Dirac equation. Supercritical limit obtained.
Olsen and Kunashenko (1997)	Channeling of relativistic electrons in crystals. Electron-positron production. Dirac equation in cylindrical coordinates.
Dietz and Pröbsting (1998)	Electron-positron production in intensive laser fields.
Segev and Wells (1998)	Pair production in ultrarelativistic heavy-ion collisions.
Busic et al. (1999)	Electron-positron production with electron capture. U^{92+}-U^{92+}
Ionescu and Belkacem (1999)	Relativistic collisions of highly-charged ions. Time evolution of QED vacuum solved numerically.
Matrasulov et al. (1999)	Dirac electron in the field of two opposite charges. $Z < 137$ and $Z > 137$ considered.

Table 5.6: Electromagnetic transition probabilities.

Reference	Comments
Liaw (1992b)	E1 oscillator strengths for alkali-like ions.
Y. Zhang et al. (1992b)	Na-like Nb.
Baluja et al. (1993)	Allowed transitions in Fe XXI.
Beideck et al. (1993)	Au II $5d^9 6p$ lifetimes.
Bugacov and Shakeshaft (1993)	Multiphoton transitions in strong fields. Inclusion of photon momentum $\Leftrightarrow$ no dipole approximation. H^- photodetachment.
S. Cheng et al. (1993)	Branching ratio for M1 decay of $2s$ H-like Kr.
X.-L. Cheng et al. (1993)	Ni-like ions, $47 \leq Z \leq 54$.
Chung (1993)	Theory of transition rates of few-electron ions.
Curtis (1993)	$nsnp$ states of Cd- and Hg-like systems.
Harra et al. (1993)	He-like Ni. $1s2l$, $1s3l$. MCDF. E1,2 and M1,2 TP.
Judd (1993a)	Spin-dependent E1 radiation.
Judd (1993b)	Two-photon absorption in Ln compounds.

Reference	Comments
Kwato Njock et al. (1993)	E1 radial integrals in semiclassical Coulomb approximation.
La John and Luke (1993)	Spin-forbidden transitions in P II.
Lavin and Martin (1993)	B isoelectronic sequence.
Lavin et al. (1993)	Singlet-singlet transitions for Zn- and Cd-like systems.
Liaw (1993a)	Cs.
Liaw (1993b)	Alkali metal atoms in the Brueckner approximation.
Marcinek and Migdalek (1993a)	Al I isoelectronic series.
Marcinek and Migdalek (1993b)	Ga I isoelectronic series.
Marketos and Zambetaki (1993)	Two-photon non-resonant transitions on O III.
Marketos et al. (1993)	O III TP.
Migdalek and Stanek (1993)	$5s^2 - 5s5p$ transitions in the Sr sequence.
Papp et al. (1993)	L-x-ray intensity ratios. DF/DS/experiment.
Radozycki and Faisal (1993)	Multiphoton ionization of strongly bound relativistic electrons in very intense laser fields.
Tiwary (1993); Tiwary and Kandpal (1994)	R/NR oscillator strengths in Na sequence.
Q.-S. Wang and Stedman (1993)	Derives from FW an E1 term of form $q\mathbf{s} \cdot \dot{\mathbf{A}} \times \mathbf{p}/2m^2c^2$. Applications to lanthanides.
Aashamar and Luke (1994a)	Sextet levels of Cr II.
Aashamar and Luke (1994b)	The $4p\ ^{6,4}P^o_J$ and $4p\ ^6D^o_J$ levels of Cr II.
Biémont et al. (1994)	TP for the Ga $4s^24p$ isoelectronic sequence.
Biémont et al. (1994c)	$4s - 4p$ transitions in neutral P.
S. Cheng et al. (1994)	M1 decay of $1s2s\ ^3S$ He-like Kr.
W.-Y. Cheng and Huang (1994b)	Z-dependence of oscillator strengths for Be-like ions.
X.-L. Cheng et al. (1994)	Ni-like ions, $72 \leq Z \leq 79$.
Cornille et al. (1994)	Radiative data for Ne-like ions.
Gębarowski et al. (1994)	$3s^23p - 3s^23d$ transitions for Al-like sequence.
Kwato Njock et al. (1994b)	Oscillator strengths of Li-like ions from exact QDT.
Land et al. (1994)	Selection rules for dipole radiation from a relativistic bound state.
Lavin et al. (1994)	FS transitions in Cu isoelectronic sequence, $Z \leq 92$.
Lavin and Martin (1994)	Triplet-triplet transitions in Cd-like ions.
Marinescu et al. (1994b)	Two-photon excitation of Rb, $5\ ^2D$.
Marketos (1994)	E2 transitions between even O III levels.
P. Martin et al. (1994)	Triplet-triplet transitions in Zn-like ions. QDT.
Padma and Deshmukh (1994)	Oscillator strengths for $2p$ ionization near threshold.
Parente et al. (1994)	M1 and E2 hyperfine quenching of Ti-like ions, $3d^4$, J=4, Z=55-92.
Schneider et al. (1994b)	M1 hfs-level lifetime in H-like $^{209}Bi^{82+}$.
Tong et al. (1994a)	Relativistic effect on atomic radiative processes. A 'zero' of the H-like E1 transition matrix element found.
Tong et al. (1994b)	TP for O III.
Vilkas et al. (1994)	E1 TP for O-like sequence.
H. L. Zhang and Sampson (1994a)	Oscillator strengths for B-like ions, Z=8-92. $\Delta n = 0, n = 2$.
H. L. Zhang and Sampson (1994b)	Oscillator strengths for B-like ions, Z=8-92. n = 2 - n = 3.
Aboussaïd et al. (1995)	Hyperfine-induced transitions in He-like ions. $1s2p\ ^3P_0$, Z = 9, 11, 13.
Beck and Datta (1995)	Nb II $4d^35p$ [$^5G_3$ and $^3D_3$] $\rightarrow (4d+5p)^4$ TP.
Dzuba et al. (1995)	Fr.

Reference	Comments
Fritzsche et al. (1995)	Cl-like ions, $3p^4 3d$ configurations.
Mendoza et al. (1995)	Al I sequence.
Nana Engo et al. (1995)	Relativistic semiclassical dipole matrix elements for $nlj \rightarrow n'l'j'$ transitions in non-hydrogenic ions.
Mårtensson-Pendrill (1995)	Tl I $7s \rightarrow 6p$ E1, $6p_{3/2} \rightarrow 6p_{1/2}$ E2 TP.
U. I. Safronova et al. (1995)	Radiative and autoionization TP for $1s2l''nl - 1s^2n'l'$; $n, n' = 2, 3$ transitions for He-like ions.
Uylings and Raassen (1995)	Accurate calculation of TP using orthogonal operators. Include valence correlation in transition matrix. $5d^9 \rightarrow 5d^8 6p$ in Hg IV ... Bi VII.
Ynnerman and Froese Fischer (1995)	Be-like systems, Z=6...42. The E1 $2s^2\ {}^1S_0 - 2s2p\ {}^1P_1$ and the forbidden $2s^2\ {}^1S_0 - 2s2p\ {}^3P_1$ energies and TP.
Bharadvaja and Baluja (1996)	E1 TP in Li, F, Na and Cu isoelectronic series. Parametric fit.
Brage et al. (1996); Brage et al. (1999)	Tl II, Tl III.
Fleming et al. (1996)	Resonance line of Be sequence.
Indelicato (1996)	Correlation and negative continuum effects for the M1 transition in two-electron ions. MCDF.
Johnson et al. (1996)	E1 TP for Li-like (Z=3-100), Na-like (Z=11-100) systems and the alkali atoms K-Fr.
Luke (1996)	Quartet multiplets in V I.
Quinet (1996)	Pd II.
W. J. Wang et al. (1996)	O-like ions, Z=52-79. MCDF.
Ali (1997)	M1 and E2 transitions of Ga-like systems.
Biémont (1997a,1997b)	E1, E2, E3; M1, M2, M3 TP for Ni-like and Pd-like systems.
Biémont et al. (1997b)	$4d^9 5p$ levels of Ag II.
M. H. Chen et al. (1997)	K-shell Auger and radiative transitions for C-like systems $6 \leq Z \leq 54$.
M.-K. Chen (1997)	E1 TP for He-like ions, Z=3...10.
Chou and Johnson (1997)	Cu-, Ag- and Au-like ions.
Curtis et al (1997)	Methods for intermediate-coupling amplitudes. $2s^2 - 2s2p$ transitions in Be-like, $6s^2 6p^2 - 6s^2 6p7s$ in Pb-like systems.
Henderson et al. (1997)	Au-like systems, Z=79-83.
Indelicato (1997)	E2 and two-electron one-photon E1 transitions of $2s3p$ ${}^3P_0$ states of Be-like ions. MCDF.
Johnson et al. (1997a)	M1 hfs of $2\ {}^3P$ levels of He-like ions, Z=2-51. Hfs-induced quenching of the levels.
Jönsson and Froese Fischer (1997b)	E1,E2 and M1,M2 TP for Mg-like $3s^2$-$3s3p$ transitions.
Kotochigova et al. (1997)	Giant resonance in the $4d - 4f$ spectrum of Gd^{3+}, $4d^9 4f^8$. MCDF.
Kovalik (1997)	Effects of relativity on the KL_1L_2 (${}^3P_0$) transition rate of ${}^{159}_{65}Tb$.
Lavin et al. (1997)	FS transition in Al isoelectronic sequence.
Marketos and Nandi (1997)	O II levels.
Nana Engo et al. (1997)	Supersymmetry-inspired R and QR quantum-defect theory. Applications on transition matrix elements.
Porsev (1997)	E1 lifetimes of low-lying odd-parity levels of Sm.
Quinet (1997)	Cr II.
Sampson and Zhang (1997)	Collision strengths for hfs transitions.

Reference	Comments
Skripnikova and Zapriagaev (1997)	Correlation and relativistic corrections to TP in He-like ions.
Szymanowski et al. (1997)	Two-photon bound-bound amplitudes in H-like atoms.
Velasco et al. (1997)	Neutral F.
H. L. Zhang and Sampson (1997)	Oscillator strengths for $(n=2)-(n=3)$ transitions in C-like ions, Z=9-54.
Avgoustoglou and Beck (1998)	$np^6 \rightarrow np^5(n+1)s$ E1 resonance lines of Ne–Xe.
Beck (1998)	Cs II $5p^5$ $6p$ level lifetimes.
Biémont et al. (1998)	$s-p$ and $s-d$ TP of Fr.
Derevianko et al. (1998)	Negative-energy contributions to En and Mn TP in He-like ions. Significant for M1 $3\,^3S_1 \rightarrow 2\,^3S_1$.
Fritzsche et al. (1998)	P-like iron-group ions, Z=22-32, $3s3p^4$ and $3s3p^23d$ configurations.
Froese Fischer and Rubin (1998)	E2 and M1 TP for Fe IV.
Froese Fischer et al. (1998a)	O isoelectronic series, Z=9-18. $2p^4-2p^33s-2s2p^5$ transitions.
Froese Fischer et al. (1998b)	Li isoelectronic series, Z=3-8.
He et al. (1999)	FS transition of Br. M1.
Jönsson and Froese Fischer (1998)	$2s^2$-$2s2p\ ^3P_1$ intercombination transition in C III.
Jönsson et al. (1998)	Be isoelectronic sequence. Z=7-28,...42.
Y.-K. Kim et al. (1998)	Failure of MCDF at NR limit. The wave function for specific J may not reduce to the LS limit. Be-like example. Effects on TP discussed.
Kohstall et al. (1998)	Si-like ions, Z=16-36, $3s3p^3$, $3s^23p3d$.
Lagmago Kamta et al. (1998)	Semiclassical KG dipole matrix elements.
Marinescu et al. (1998)	Radiative transitions of Fr.
Maul et al. (1998)	'Stark quenching' in laser fields for PNC observations. One- and two-photon transitions of Be-like $2s2p\ ^3P_0$ levels.
O'Malley and Beck (1998)	M1 decay rates of Sn$^-$, $5p^3$.
Pal'chikov (1998)	En and Mn TP of arbitrary n for H-like atoms. Z=1-92.
Prosser (1998)	Energy spectrum of H atom in photon field. Line broadening.
Romero and Aucar (1998, 1999)	Relativistic corrections to generalized oscillator strength sum rules.
M. S. Safronova et al. (1998)	Na-like systems, Z=11-16.
Schmitz et al. (1998)	3D relativistic calculations of strong-field photoionization. Phase-space averaging, classical method.
Shabaev (1998b)	TP between M1 hfs levels of H-like ions related to the bound-electron g-factor.
Shi et al. (1998ab)	$4p-5s$ excitations of Kr.
Siegel et al. (1998)	E1 TP for the Na sequence, Na I - Ca X.
Yan et al. (1998)	$2p-2s$ TP of Li-like systems. Z=3-20 + extrapolations.
Beck (1999)	$M1$ decay rate of the nearly Z-independent $J=3 \rightarrow 2$ transition of the $3d^3_{3/2}3d_{5/2}$ systems W^{52+} and Bi^{61+}.
J. Chen et al. (1999)	High-order harmonics for H-like atom in ultrastrong laser field.
Donnelly et al. (1999)	Oscillator strengths of Cu$^+$.
Froese Fischer (1999)	Review on transitions in lighter atoms. BP/Dirac.
Henderson et al. (1999)	$5d^96p$ levels in Hg III.
Horodecki et al. (1999)	Forbidden lines in Pb I, $6s^26p^2$. MCDF.
Kohstall et al. (1999)	Cl-like ions, $3s3p^6$, $^2S_{1/2}$.

Reference	Comments
Z.-S. Li et al. (1999)	21 levels of Ge I. $4p4d$, $4p5d$, $4p6s$.
Merkelis et al. (1999)	M1 and E2 transitions in in the N I sequence.
Nahar (1999)	E1 FS transitions of Si-like Fe XIII. 307 683 transitions.
Norquist et al. (1999)	M1 decay of Ru^-, $d^7s^2\ ^4F$.
O'Malley and Beck (1999)	E1 TP of La^-. $6p$ or $5d$ attachment.
Ozdemir and Karal (1999)	Oscillator-strengths from relativistically corrected HF.
Porsev et al. (1999b)	Low-lying levels of Yb.
Pradhan (1999)	TP for Fe V, XXIV, XXV. BP R-matrix method.
Quinet et al. (1999)	Ln II.
M. S. Safronova et al. (1999)	Alkali metals, Na-Fr.
U. I. Safronova et al. (1999a)	Be-like ions. $2l_12l_2 - 2l_33l_4$ TP. Z=6-100.
U. I. Safronova et al. (1999b)	M1 transitions in Be-like ions, Z=4-100. Importance of negative-energy states discussed.
U. I. Safronova et al. (1999c)	E1 transitions between $n = 2$ states in B-like ions. Z=6-100.
U. I. Safronova et al. (1999d)	$2l_12l_2 - 2l_32l_4$ lines in Be-like ions. $Z = 4 - 100$.
Santos et al. (1999)	Radiative and non-radiative decay rates of $2s$ (L_1) hole states in Yb and Hg. MCDF and MBPT.
Santos et al. (1998b)	Two-photon decay rates of $2s$-level in H-like ions. Z=1-100.
Tachiev and Froese Fischer (1999)	Be-like systems, Z=4-12. $2s2p$, $2p^2$, $2s3s$, $2s3p$, $2s3d$ excited level lifetimes. BP.
Yan and Ho (1999)	Relativistic effects in positronium hydride (PsH, e^-e^+p). Two-photon annihilation rate calculated.
H. L. Zhang and Sampson (1999)	Oscillator strengths for the 105 Δn=0 transitions with $n = 2$ in the N-like ions, Z=12-92.

Table 5.7: Polarizabilities, dispersion and screening constants.

Reference	Comments
Feiock and Johnson (1968)	Magnetic shielding factors. Use Dirac current!
Bondarev and Kuten (1993)	Polarizability tensors for H-like $ns_{1/2}$ and $np_{1/2}$ states. Due to M1 hfs.
Fuentealba and Reyes (1993)	Polarizabilities and hyperpolarizabilities of Li-Cs.
Kellö et al. (1993)	Polarizabilities of K-Fr. PT relativity.
Le et al. (1994)	DC Green function. Polarizability of H-like atoms.
Schwerdtfeger et al. (1994b)	Polarizability of Hg.
van Wijngaarden and Li (1994)	Polarizabilities of Cs s, p, d, f states.
Z.-W. Wang and Chung (1994)	Dipole polarizabilities of Li-like systems, Z=3-50.
Baluja (1995)	Dipole polarizability of H-like ions.
Kellö and Sadlej (1995a)	Polarizabilities of M^{q+}; M=Zn-Hg; q=0-2.
Nicklass et al. (1995)	Dipole and quadrupole polarizabilities of Ne-Xe.
Szmytkowski and Alhasan (1995)	Polarizabilities for alkaline-earth-like systems.
Hättig and Hess (1996)	Dynamic multipole polarizabilities, dispersion coefficients for Ar-Rn. SR DK.
Johnson and Cheng (1996)	Polarizabilities of He-like ions. MCDF. R/NR.
Kellö and Sadlej (1996a)	DK CCSD(T) polarizabilities of Cu-Au, Zn-Hg and their cations.
Kellö et al. (1996b)	Polarizabilities of M^-; M=Cu-Au. R/NR. DK/PT.
Neogrády et al. (1996,1997)	QR CCSD(T) polarizabilities for M,M^+; M=Cu-Au.

Reference	Comments
Dzuba et al. (1997a)	Dipole polarizability of Cs.
Henderson et al. (1997)	Exp. contributions to dipole polarizabilities of Au I - Bi V.
Miadoková et al. (1997)	Standardized basis sets for highly correlated DK calculations of electric properties. Polarizability of K-Fr, Li^+-Fr^+, Ca-Ba.
Szmytkowski (1997)	DC Sturmians and the DC Green function. Relativistic polarizability of the H-like atom, Z=1-137.
Visscher et al. (1997)	$\alpha(\omega)$ of Hg. RRPA.
Hättig and Hess (1998)	Dynamic multipole polarizabilities, dispersion coefficients for F^- to I^-. SR DK.
Coutinho et al. (1999)	Two definitions for electric polarizability, with or without the vacuum background.
Derevianko et al. (1999a)	Dipole polarizabilities and dispersion constants. Na-Fr.
Iliaš and Neogrády (1999)	Dipole polarizability of M^+; M=Zn-Hg.
Lim et al. (1999)	Polarizabilities of alkali metals Li-E119. For Fr and E119, substantial SO effects and small total values. CCSD(T).
Porsev et al. (1999b)	Electric polarizabilities for low-lying levels of Yb.
M. S. Safronova et al. (1999)	Electric polarizabilities of Na-Fr.

Table 5.8: Electric and magnetic hyperfine properties.

Reference	Comments
Acker et al. (1966)	M1 and E2 hfs for muonic atoms. Fermi nuclear potential.
Moss and Sadler (1986)	Electric quadrupole moments of one-electron atoms. In $j = \frac{1}{2}$ states due to hfs. Resolves earlier inconsistencies.
M.-K. Chen (1993)	M1 hfs of the muonic ^{3}He.
Dembczyński et al. (1993)	Co $(3d+4s)^9$ M1, E2 hfs. $Q(^{59}\text{Co})$ 410(10) mb.
Dougherty et al. (1993)	M1 hfs of Cu-Au. Core polarization analyzed.
Finkbeiner et al. (1993)	M1 hfs energy and lifetime for Bi^{82+}.
Jönsson (1993)	MCDF approach to atomic hfs. B-Tl. Li-like F.
Jönsson and Froese Fischer (1993)	M1 and E2 hfs of low-lying states in Be, B, C.
Kronfeldt et al. (1993a)	Isotope shifts of Re I.
Kronfeldt et al. (1993b)	Fine and hyperfine structure of Er I, $4f^{11}5d6s6p$.
Marques et al. (1993a)	Hyperfine quenching of Be-like $1s^22s2p\ ^3P_0$.
Marques et al. (1993b)	Hyperfine quenching of Mg-like $3s3p\ ^3P_0$.
Schneider et al. (1993b)	M1 hfs of $^{209}Bi^{82+}$.
Shabaeva (1993)	Ground-state M1 hfs of Li-like Bi.
Young et al. (1993)	M1 and E2 hfs in Zr II.
G. X. Chen (1994)	Sc II, $3d^2$. Core-polarization effects.
Dzuba et al. (1994)	M1 and E2 hfs in Dy.
Mårtensson-Pendrill et al. (1994)	Yb^+ hfs, isotope shifts.
Parente et al. (1994)	M1 and E2 hyperfine quenching of Ti-like ions, $3d^4$, J=4, Z=55-92.
Schneider et al. (1994b)	M1 hfs-level lifetime in H-like $^{209}Bi^{82+}$.
Scofield and Nilsen (1994)	M1 hfs for $2p^{-1}3p$ and $2p^{-1}3s$ levels of Ne-like systems, Z=17-59.
Shabaev (1994a)	M1 and E2 hfs of H-like ions. Finite-nucleus corrections.
Yakhontov and Amusia (1994)	M1 hfs in the ${1s_{1/2}}^{(e)}\ {2s_{1/2}}^{(\mu)}$ state of the exotic $(^4He^{2+}$-μ^--$e^-)^0$ and $(^3He^{2+}$-μ^--$e^-)^0$ atoms.
Aboussaïd et al. (1995)	M1 and E2 hyperfine-induced transitions in He-like ions.

Reference	Comments
	$1s2p\ ^3P_0$, $Z = 9, 11, 13$.
Datta and Beck (1995)	M1, E2 hfs of ^{139}La II, $(5d+6s)^2$, J=2.
Dinov and Beck (1995b)	M1, E2 hfs of U and U^-.
Fedorov et al. (1995)	M1 and E2 hfs for the (K and L) inner shells, Z=10-100. DF.
Jönsson (1995)	Large-scale MCDF calculations. Thesis.
Kroger and Kroger (1995)	Program for angular coefficients of relativistic one-electron hfs parameters.
Labzowsky et al. (1995)	Introduce the 'dynamic proton model' for hfs of H-like ^{209}Bi. Interaction between valence electron and valence proton at QED level.
Loginov (1995)	CI effects on atomic hfs at semiempirical level. Examples Sc II and Y II.
Mårtensson-Pendrill (1995)	Magnetic moment distributions in ^{203,205}Tl.
Pilkuhn (1995)	An M1 hyperfine operator between two Dirac particles, both having anomalous magnetic moments.
Shabaev et al. (1995, 1997a,1998d)	Hfs of H-like and Li-like ions.
Shabaeva and Shabaev (1995)	Interelectronic contribution to hfs of Li-like ions. $1/Z$ expansion, including QED effects. Z=5-100.
H. Y. Yang and Chung (1995)	Be-like systems, $Z = 3-10$, core-excited $1s2s2p^2(^2P)$ and $1s2p^3(^5S)$ states.
Young et al. (1995)	M1 and E2 hfs in Nb II.
X. Yuan et al. (1995a)	M1 hfs of $7p$ states of Ra^+.
X. Yuan et al. (1995b)	M1 hfs of Ca^+ and Sr^+. Results compared with trends in Groups 1, 2, 11.
Bieroń et al. (1996a)	M1, E2 hfs of $2s$ and $2p$ states of Li.
Bieroń et al. (1996b)	M1, E2 hfs of nd^2 levels of Sc^+, Y^+.
Bigeleisen (1996)	Nuclear size and shape effects in chemical reactions. Uranium redox reactions as example.
Brage et al. (1996); Brage et al. (1999)	M1 hfs in Tl II, Tl III. Large MCDF.
Crespo Lopez Urrutia et al. (1996)	Observe F=3-4 transition of H-like $^{165}Ho^{65+}$. Result deviates from tabulated magnetic moment.
Dinov and Beck (1996)	M1, E2 hfs for Pa^-.
Jönsson et al. (1996)	HFS92: A program for atomic hfs calculations. MCDF.
O'Malley and Beck (1996)	M1 and E2 hfs of Cs II and Ba III $(5d+6s+6p)$ levels.
Yan et al. (1996)	FC term of M1 hfs in Li $2s, 2p, 3s$, Be^+ $2s$. Includes QED.
Beck (1997b); Beck and Datta (1993)	M1 and E2 hfs for $(d+s)^3$ states of La I, Zr II, Hf II.
Bieroń et al (1997)	E2 hfs of $3d^2$ and $3d4p$ levels of Sc. $Q(^{45}Sc)$ -231(4) mb.
Childs (1997)	Matrix elements of hfs operators in SL and jj representations for s^N, p^N and d^N configurations.
Friar et al. (1997)	Nuclear sizes and isotope shifts. The nuclear physics perspective.
Friar and Payne (1997a)	Higher-order nuclear-polarizability corrections for hydrogen (H or D).
Friar and Payne (1997b)	Higher-order nuclear-size corrections for hydrogen.
Johnson et al. (1997a)	M1 hfs of $2\ ^3P$ levels of He-like ions, Z=2-51. Hfs-induced quenching of the levels.
Jönsson and Froese Fischer (1997a)	SMS92: a program for relativistic isotope shift calculations.

Reference	Comments
King (1997,1999)	High-precision calculations on the Li atom. Review.
Labzowsky et al. (1997ac)	M1,E2,M3 hfs of the $2p_{3/2}$ state of H-, Li-, B- and N-like $^{209}_{83}$Bi. Interaction between valence electron and valence proton treated at QED level.
Owusu et al. (1997a)	M1 hfs in excited S states of K, Fr.
Owusu et al. (1997b)	M1 hfs in K-like system (K-Sc^{2+}) ground states.
J. R. Persson (1997)	Hfs of Yb^+.
Sampson and Zhang (1997)	Collision strengths for hfs transitions.
Shabaev et al. (1997b)	Ground-state M1 hfs splitting of H-like ions. Z=49-83.
Tulub et al. (1997)	Simulate the QED effects in many-electron atoms by a renormalized nucleus (right radius, changed charge inside). Application on M1 hfs of H-like Bi.
Bakalov and Korobov (1998)	Hfs levels of antiprotonic helium, $^4He^+\bar{p}$.
Bieroń et al. (1998)	E2 hfs of $4d5s^2$ levels of Y. $Q(^{90}Y)$ -125(11) mb.
Bouazza et al. (1998)	Zr I $(4d+5s)^4$ hfs. $Q(^{91}Zr)$ -230(20) mb.
Crespo Lopez Urrutia et al. (1998)	Observe M1 hfs transition of H-like Re. QED corrections from Shabaev (1994) a few tenths of per cent. Bohr-Weisskopf effects extracted.
Esquivel et al. (1998)	M1 hfs of Li, $2s^1$.
Godefroid et al. (1998)	Review on atomic structure calculations.
Gustavsson and Mårtensson-Pendrill (1998a)	Review on hyperfine anomalies (nuclear volume effects on M1 hfs).
Gustavsson and Mårtensson-Pendrill (1998b)	Nuclear magnetic moments should be remeasured. Present diamagnetic screening corrections for bulk are too inaccurate. Better values will enable QED tests on highly-ionized atoms.
Karpeshin et al. (1998)	TP between hfs levels of the H-like $^{229}Th^{89+}$. ($I=5/2$ nuclear ground state and $I=3/2$ 3.5 eV isomer). The 28 MT magnetic field at nucleus mixes the *nuclear* states.
O'Malley and Beck (1998)	M1 hfs of Sn^-, $5p^3$.
M. S. Safronova et al. (1998)	M1 and E2 hfs of Na-like systems, Z=11-16.
Shabaev (1998b)	TP between M1 hfs levels of H-like ions related to the bound-electron g-factor.
Shabaev et al. (1998e)	Ground-state M1 hfs splitting and lifetime for Li-like ions, Z=49-83. Includes QED.
Sunnergren et al. (1998)	Radiative corrections to M1 hfs of H-like systems.
Tomaselli et al. (1998)	M1 $1s$ hfs in H-like atoms with one-hole nuclei. ^{165}Ho...^{207}Pb.
van Lenthe et al. (1998)	M1 hfs of Cu-Au at DFT level. R/ZORA/NR.
Bieroń et al. (1999a)	E2 hfs of $3d^24s$ and $3d^3$ Ti^+. $Q(^{49}Ti)$ 247(11) mb.
Bieroń et al. (1999b)	M1 and E2 hfs of $2s$ and $2p$ states of Be^+, F^{6+}.
Derevianko et al. (1999b)	Off-diagonal, $6s-7s$ M1 hfs in Cs.
Kozlov and Porsev (1999)	M1 and E2 hfs of low-lying levels of Ba.
Norquist et al. (1999)	M1 and E2 hfs of Ru^-. MCDF.
Porsev et al. (1999a)	M1 and E2 hfs for low-lying levels of Yb.
M. S. Safronova et al. (1999)	M1 hfs of alkali atoms, Na-Fr.
Shabaev (1999)	Hfs of highly charged (H- and Li-like) ions. By combining them, nuclear factors can be eliminated. TP. QED included.

Table 5.9: Average radii and magnetic g-factors.

Reference	Comments
Moss and Sadler (1986)	Electric quadrupole moments of one-electron atoms. In $j = \frac{1}{2}$ states due to hfs. Resolves earlier inconsistencies.
L. Liu and Li (1991b)	Amplitudes of wave functions at nuclei for ions.
Anthony and Sebastian (1993)	Relativistic corrections to g-factors of He-like systems.
Fricke et al. (1993)	Radii for atoms and ions of Ta, E105 (Db).
Lindroth and Ynnerman (1993)	g-factors for Li, Be^+, Ba^+.
Marketos (1993)	Alkali atom g-factors. Li-Cs.
J. Persson (1993)	g-factors of np^2 systems.
Dzuba et al. (1994)	g-factors in Dy.
L. Liu et al. (1994)	Amplitudes of wave functions at nuclei for ions. DS.
Yan (1994)	g-factor ratio for ^{3}He and ^{4}He in the 2 $^3S_1$ state.
Yan and Drake (1994)	g-factors for helium 2 $^3S_1$, ^3P_J, $^1P_1$ and 3 ^3P_J states.
W. Huang et al. (1995)	Ne I excited levels.
Land and Horwitz (1995)	Zeeman effect for a relativistic bound state.
Seth et al. (1995)	Trends of atomic orbital energies and radii for Groups 11 (Cu-E111) to 15 (As-E115). Relativistic and shell-structure effects.
W.-Y. Liu et al. (1996)	Rydberg levels of Cs in strong magnetic field.
Gonzalo and Santos (1997)	Radiative corrections to Zeeman effect of 2 ^3P states of He.
Persson et al. (1997a)	Radiative corrections to electron g-factor in H-like ions.
Beck (1998)	Cs II $5p^- \ 6p$ level g-factors.
Shabaev (1998b)	TP between M1 hfs levels of H-like ions related to the bound-electron g-factor.
van Lenthe et al. (1998)	g-factors of Cu-Au at DFT level. R/ZORA/NR.
E. Johnson et al. (1999)	Ionic radii from $\langle r^n \rangle$ for M^q; M=Cr-Sg(E106); q=4+ to 6+.
Labzowsky et al. (1999a)	Relativistic and QED corrections to ns electron g factors of neutral K-Fr, Ba^+.

Table 5.10: Compton profiles, momentum distributions and spin densities.

Reference	Comments
Bergstrom et al. (1993)	Compton scattering from bound electrons in a fully relativistic independent-particle approximation.
Timms and Cooper (1993)	Electron momentum distribution in Pb.
Scherer et al. (1994)	Low-energy Compton scattering by a Dirac proton with anomalous magnetic moment. FW used.
Rocchi and Sacchetti (1995)	Radiative corrections to Compton cross section.

Table 5.11: Photon scattering, photoionization, x-ray scattering factors. Radiative recombination (the inverse photoelectric effect).

Reference	Comments
Gorshkov et al. (1967)	Screening effects in K-shell photoionization.
Mikhailov and Polikanov (1968)	Central potential. PT by reducing the Dirac equation to a Riccati equation. Used for K-shell relativistic photoeffect.
Boyle et al. (1993)	Photoionization cross section of atomic W.
T.-N. Chang (1993)	MBPT of atomic structure and photoionization.
S. H. Kim (1993)	Thomson scattering by a relativistic electron.
Kotochigova and	Multiphoton ionization and autoionization of Si I.

Reference	Comments
Lambropoulos (1993,1994)	
Kutzner et al. (1993)	Inner-shell photoionization of alkaline earth atoms.
Z. W. Liu and Kelly (1993)	Multiphoton ionization in strong laser fields.
J. Luo et al. (1993)	Scattering operator for elastic and inelastic resonant x-ray scattering. Applied on Ln, An, TM.
Pratt and Kim (1993)	Non-dipolar effects in atomic photoionization.
Schaphorst et al. (1993)	Multielectron inner-shell photoexcitation of Kr.
Tulkki (1993)	Outer-shell photoionization of Ar, K^+, Ca^{2+}. Combined effect of relaxation and channel interaction.
Verner et al. (1993)	Subshell photoionization cross sections of He-Zn.
Basavaraju et al. (1994)	Elastic scattering of 81-keV γ rays.
Chi and Huang (1994)	Photoionization of Mg.
Crawford and Reiss (1994)	Stabilization of relativistic photoionization by circularly polarized light.
Kutzner and Radojevic (1994)	Relaxation and interchannel coupling in inner-shell photoionization of atomic Yb.
Kutzner et al. (1994)	Inner-shell photoionization of Group 12 atoms.
Kutzner and Vance (1994)	Photoionization of Pd.
L'vov and Milstein (1994)	Elastic scattering of photons in a Lorentz scalar potential. Virtual spin-0 particle-antiparticle pairs involved
Rao et al. (1994)	L x-ray fluorescence, Z=46-51.
Rez et al. (1994)	DF x-ray scattering factors. $Z = 2 - 92$.
Ron et al. (1994)	Relativistic, retardation and multipole effects in photoionization cross sections. Z, n, and l dependence.
Shabaev (1994b)	QED theory of electron recombination with highly charged ions.
Tseng (1994)	Pair production by photons near threshold. Positron energy spectrum for Z=1-82.
C.-M. Wu et al. (1994)	Photoionization of Sr above $5p_{3/2}$ threshold. MC RRPA.
Filipponi and Di Cicco (1995)	K-edge X-ray absorption of Mo, Rh→Sn. EXAFS.
R. N. Lee and Milstein (1995); R. N. Li and Mil'shtejn (1995)	Delbrück scattering in a screened Coulomb field.
Y. Liu et al. (1995)	Photoexcitation of Ni $3p$ level. SO, exchange, hybridization.
Miecznik et al. (1995)	SO effects in photoionization of Al I.
Pratt and LaJohn (1995)	Multipoles beyond dipole in photoionization.
Rao et al. (1995)	M x-ray fluorescence, Z=78-92.
Tseng (1995)	Pair production by photons near the $2mc^2$ threshold. Energy-angle distribution. Z=1 ... 82.
Yoo et al. (1995)	Photoionization of Bi.
Froelich and Weyrich (1996)	Relativistic corrections to inelastic scattering of photons by atomic electrons.
Ichihara et al. (1996)	Radiative electron capture and photoelectric effect at high energies. Bare projectile, low-Z atom.
Keitel (1996)	Ultra-energetic electron ejection in relativistic atom. Laser field interaction.
Meyerhofer et al. (1996)	Relativistic mass-shift effects during high-intensity laser–electron interactions. Multiphoton scattering.
J.-H. Wang et al. (1996)	X-ray scattering factors for He-Ar. DF.

Reference	Comments
Chi (1997)	Near-threshold photoionization of Mg.
Donnelly et al. (1997)	$3p$ photoabsorption of Cr^+.
Kylstra et al. (1997)	1D Dirac model atom in an intense laser field.
Rao et al. (1997)	Rayleigh and Compton scattering cross sections for low, medium and high Z and $23.18 \leq E \leq 30.85$ keV.
Z.-W. Su and Coppens (1997)	X-ray elastic scattering factors. Z=1-54.
Tseng (1997b)	Pair-production polarization correlations of intermediate-energy photons on atoms.
Bhatia and Drachman (1998b)	Optical properties of helium, including BP-level relativity.
J.-J. Chang (1998)	Similarities between Dirac-Coulomb and Schrödinger-Coulomb radial functions. Photoionization of H.
Donnelly et al. (1998)	$3p$ photoabsorption of Mg^+.
Eichler et al. (1998)	Alignment caused by photoionization and in radiative electron capture into excited states of high-Z H-like ions.
Gorczyca et al. (1998)	SO effects in the photoionization excitation of Ne.
Kahane (1998)	Incoherent photon scattering functions. Z=1-110. DF.
Lagutin et al. (1998)	Relativistic effects in inner-shell atomic photoabsorption.
J. C. Liu et al. (1998)	SO components of resonant satellite photoionization of Ca^+.
Matrasulov (1998,1999)	Chaotic ionization of an H-like atom in intensive monochromatic field.
Mohan et al. (1998)	Ground-state photoionization of Ne-like Fe XVII.
Mohr (1998)	QED and the fundamental constants.
Wills et al. (1998)	SO effects in parity-unfavoured photoionization of Ne. Doubly excited resonances.
H. L. Zhang (1998)	Relativistic calculations of photoionization cross sections. He-like, Li-like systems.
Brinzanescu et al. (1999)	Radiative recombination: exact relativistic theory and the NR dipole approximation. Results for naked Ne and U nuclei.
Chakraborty et al. (1999)	Near-threshold photoionization for Ne-like systems, $Z = 10 - 15$.
Derevianko et al. (1999c)	Photoelectron angular distribution, He-Xe. Non-dipole effects.
Haque et al. (1999)	Photoionization of Ne-like Fe.
Haque and Pradhan (1999)	Photoionization Mg-like Fe.
Ionescu et al. (1999)	Inner-shell photoionization at relativistic energies. Vacuum-assisted processes most probable at high E. 'Vacuum spark'.
V. K. Ivanov et al. (1999)	Many-body effects in negative ion photodetachment. Review.
Krainov (1999)	Tunneling ionization of atoms by intensive laser radiation. Electron energies and angular distributions.
Luc-Koenig et al. (1999)	Multiphoton ionization dynamics of Ba. Raman couplings, dynamical Stark shifts.
Panek et al. (1999)	Electron scattering in powerful laser field. Angular and polarization effects. Electron handled as laser-dressed KG particle.
Trail and Bird (1999b)	Accurate structure factors from PP methods. Si.
Vrejoiu et al. (1999)	Retardation corrections to angular distributions of free-bound photoionization transitions in H-like atoms.
L.-R. Wang et al. (1999)	Doubly excited states in photoionization of Zn. Autoionization resonances.

Table 5.12: Electron and positron scattering. Dielectronic recombination.

Reference	Comments
Fontes (1992); Fontes et al. (1993a)	Effect of the generalized Breit interaction on electron-impact excitation cross sections.
Bartschat (1993); Bartschat et al. (1994)	Low-energy electron scattering from Cs.
Bettega et al. (1993, 1995)	MH_4 molecules; M=Ge-Pb. Electron scattering, elastic and inelastic. Local-density norm-conserving PP used.
Braidwood et al. (1993)	Satellite structure of Xe valence shell by electron-momentum spectroscopy.
M. H. Chen (1993b)	Dielectronic recombination for Ni-like Ta.
M. H. Chen and Reed (1993a)	Electron-impact ionization of Kr^{24+}, Kr^{25+} and Xe^{43+}.
M. H. Chen and Reed (1993b)	Effect of relativity and M2 transitions on the resonance contribution to electron-impact ionization. Li-like ions, Z=18...54.
Fontes et al. (1993b)	Method for electron-impact ionization for ions of any complexity.
Furst et al. (1993)	SO effects in excitation of noble gases by spin-polarized electrons.
C. Hofmann et al. (1993)	Distorted-wave scattering solution of Dirac electron from a Coulomb potential.
Hou et al. (1993)	Elastic electron or positron scattering from Ar.
Ivanov et al. (1993ab)	QED-based theory of decaying atomic states. Application for electron-collision strengths. Ne-like ions.
Khare and Raj (1993)	Elastic electron scattering by Ar. Spin polarization.
Koike (1993)	Inelastic electron scattering by Ln and An atoms. Quadrupole interaction important. Analogy to molecular rotation.
Kuo et al. (1993)	Positron-impact ionization of H-like ions.
Popov and Kuz'mina (1993ab)	Eikonal approximation for relativistic (e,2e) experiments.
Qiu et al. (1993)	Dielectronic recombination for Ne-like Ti from low-lying F-like states.
Raeker et al. (1993)	Charge-cloud distribution of heavy atoms after excitation by polarized electrons. Hg, Tl.
Reed and Chen (1993)	Polarization of line emission from H-like and He-like ions, following electron impact.
Salvat and Mayol (1993)	Partial wave analysis for electron and positron elastic scattering. Program.
Srivastava et al. (1993)	Electron-impact excitation of Hg.
Szmytkowski (1993a)	Low-energy positron scattering on alkaline-earth atoms.
Szmytkowski (1993b)	Elastic positron scattering from Hg.
Szmytkowski (1993c)	Elastic positron scattering from Zn and Cd.
Thumm (1993)	Dirac R-matrix method for scattering of slow electrons from alkali-metal-like targets.
Thumm et al. (1993)	Relativistic effects in spin polarization for low-energy electron-Cs scattering.
Thumm and Norcross (1993)	Angle-differential and momentum-transfer cross sections for low-energy electron-Cs scattering.
J.-M. Yuan and Zhang (1993a)	Spin polarization of electrons, elastically scattered from Ar in the Ramsauer-Townsend region.
J.-M. Yuan and Zhang (1993b)	Spin polarization of electrons, elastically scattered from alkaline earth atoms in the Ramsauer-Townsend and low-lying shape resonance regions.

Reference	Comments
H. L. Zhang and Sampson (1993)	Improvements of distorted-wave approach to electron impact excitation of ions.
Ast et al. (1994)	Electron-impact K-shell ionization of Ag, Au.
Baluja and Gupta (1994)	Electron scattering from iron at 10-5000 eV.
Bartschat (1994)	Low-energy electron scattering from Tl.
Berrington (1994)	Electron-impact excitation of Be-like systems, Z=4-28.
M. H. Chen and Reed (1994)	Angular distribution of Auger electrons after electron-impact ionization. Be-like ions, Z=12...42.
M. H. Chen et al. (1994)	Effect of Coster-Kronig transitions on electron-impact excitation rates for F-like ions.
W.-Y. Cheng and Huang (1994a)	Polarization correlations of radiation from electron-impact excited atoms.
Dasgupta and Whitney (1994)	Z-scaled data for dielectronic recombination from O-like to F-like ions.
Dzuba and Gribakin (1994)	Correlation-potential method for electron scattering.
Faulkner (1994)	Scattering matrices for non-spherical SR potential.
Fontes et al. (1994)	Effect of generalized Breit interaction on electron-impact excitation cross sections.
Glass (1994a)	Relativistic continuum distorted wave theory for electron capture.
Glass (1994b)	Asymmetric theories of relativistic electron capture.
Johnson and Guet (1994)	Elastic electron scattering from Xe,Cs^{+},Ba^{2+}.
Keller and Whelan (1994)	Plane wave Born approximation for relativistic (e,2e) processes.
Keller et al. (1994)	Distorted-wave Born approximation for inner-shell (e,2e) processes.
Khare et al. (1994)	L3-shell ionization of Xe and Au by electron and positron impact.
Kumar et al. (1994a)	Electron scattering from Zn and Pb. Spin polarization and cross sections.
Kumar et al. (1994b)	Electron scattering from heavy alkaline earths.
Rez et al. (1994)	DF mean inner potential for electron scattering. $Z = 2 - 92$.
Rosenberg (1994a)	Infrared radiative corrections to potential scattering of a Dirac electron.
Rosenberg (1994b)	Minimum principle for potential scattering of Dirac electron.
Szmytkowski (1994)	Elastic positron scattering from Kr and Xe.
Szmytkowski and Sienkiewicz (1994a)	Spin polarization of slow electrons, elastically scattered from Zn-Hg.
Szmytkowski and Sienkiewicz (1994b)	Spin polarization of slow electrons, elastically scattered from Xe.
Szmytkowski and Sienkiewicz (1994c)	Spin polarization of slow electrons, elastically scattered from Sr, Ba.
Tang and Dorignac (1994)	Electron scattering factors for High-resolution electron microscopy (HREM) image simulation. DS potential.
H.-G. Teng et al. (1994a)	Dielectronic recombination of He-like systems.
H.-G. Teng et al. (1994b)	Dielectronic recombination of F-like systems.
S. Wang et al. (1994)	Cross sections for electron scattering by ground-state Ba. Elastic scattering and $6s6p\ ^1P_1$ excitation.
Young and Norrington (1994)	Solution of relativistic asymptotic equations in electron-ion scattering. Program.
Zeman et al. (1994a)	Relativistic distorted-wave calculation of electron impact excitation of Cs.
Zeman et al. (1994b)	Intermediate energy electron impact excitation of Cs.

Reference	Comments
H. L. Zhang and Pradhan (1994)	Electron-impact excitation. B-like ions. Z=8 ... 26.
H. L. Zhang and Sampson (1994a)	Electron-impact collision strengths for B-like ions, Z=8-92. $\Delta n = 0, n = 2$.
H. L. Zhang and Sampson (1994b)	Electron-impact collision strengths for B-like ions, Z=8-92. $n = 2$ - $n = 3$.
Baylis and Sienkiewicz (1995)	Represent electron-scattering polarization data by 'polarization trajectories' on a Poincaré sphere.
M. H. Chen and Scofield (1995); M. H. Chen and Scofield (1998)	Relativistic effects on angular distribution and polarization of dielectronic satellite lines of H-like ions, Z=9...92.
Fontes et al. (1995)	Electron-impact ionization cross sections for $U^{90+,91+}$.
Horbatsch and Shapoval (1995b)	Relativistic two-particle scattering resonances in the Tamm-Dancoff approximation.
Khare and Wadehra (1995)	K-shell ionization by electron impact.
Kisielius et al. (1995)	Electron-impact excitation of H-like ions.
Mohan et al. (1995)	Electron-impact FS transitions in Cu XX.
Moores and Reed (1995abc)	Electron-impact ionization of high-Z H-like to Be-like ions.
Pindzola et al. (1995)	Photorecombination of highly charged U ions. (Inverse photoionization: $e^- + U^{90+} \rightarrow U^{89+}(1s^2 2s) + h\nu$).
Prinz et al. (1995)	SO effects of the continuum electrons in (e,2e) experiments.
Rosenberg (1995)	Electron scattering by heavy hydrogenic ion.
Srivastava et al. (1995a)	Excitation of the $np^5(n+1)p\ ^3D_3$ states of Ar-Xe by spin-polarized electrons.
Srivastava et al. (1995b)	Electron excitation of Yb.
Srivastava et al. (1995c)	Electron excitation of Cu.
Szmytkowski (1995)	Relativistic multi-channel variable phase method for asymptotic equations of electron-atom and electron-ion scattering.
Whelan et al. (1995)	Triple differential (e,2e) cross sections for Au and U at relativistic impact energies.
J.-M. Yuan (1995)	Low-energy electron scattering from Ca-Ba, Yb. Intra-atomic relativistic effects.
Zeman et al. (1995a)	Electron impact excitation of Cs.
Zeman et al. (1995b)	Test of LS approximation for electron-impact excitation of Cs.
H. L. Zhang and Pradhan (1995a)	Relativistic and radiation damping effects in electron-impact excitation of highly charged ions.
H. L. Zhang and Pradhan (1995b)	Relativistic and electron-correlation effects in electron-impact excitation of Fe^{2+}.
H. L. Zhang and Sampson (1995)	Relativistic collision strengths for optically allowed $\Delta n = 0$ transitions between magnetic sublevels of highly charged ions. Extremely large angular momenta (of order of 190) needed for the free electron.
Ait-Tahar and Grant (1996)	B-like Fe. Dirac R-matrix approach.
Ast et al. (1996)	Binary peak in (e,2e) collisions.
G.-X. Chen (1996);G.-X. Chen and Y.-B. Qiu (1997b)	Fully relativistic distorted-wave Born procedure for electron-impact ionization. Ni-like Gd, U.
Keller et al. (1996a)	Relativistic 1st-order Born approximation for inner-shell (e,2e) processes.
Keller et al. (1996b)	(e,2e) processes with spin-polarized electrons.

Reference	Comments
McCann et al. (1996)	Energy dependence of relativistic nonradiative electron capture.
Srivastava et al. (1996a)	Excitation of the 6 $^{1,3}P_1$ states of Hg by polarized electrons.
Srivastava et al. (1996b)	Excitation of the 5 $^{1,3}D_2$ states of Cd by polarized and unpolarized electrons.
Zeman et al. (1996)	Relativistic effects of generalized Stokes parameters for electron impact excitation of one valence electron atoms.
Ait-Tahar et al. (1997)	Cs. Dirac R-matrix approach.
Cai et al. (1997)	Electron-impact excitation of Cu-like ions, $Z \leq 82$.
G.-X. Chen and Qiu (1997a)	Electron-impact ionization of F-like Se.
Ehlotzky et al. (1997)	Review on electron-atom collisions in laser fields.
El Messaoudi et al. (1997)	K-shell ionization by electron impact. Effect of Breit interaction.
Kaur et al. (1997)	Electron-impact excitation of Mg and Zn.
Keller and Dreizler (1997)	Triply differential cross sections for electron-atom bremsstrahlung.
Labzowsky et al. (1997b)	Non-resonant QED corrections to radiative electron capture of highly charged ions.
Pelc and Horwitz (1997)	Complete sets of states in relativistic scattering theory.
Pindzola and Griffin (1997)	Electron-impact ionization of W ions.
Schaffer and Pratt (1997)	Relativistic partial-wave calculations of triply differential electron-atom bremsstrahlung.
Sienkiewicz and Baylis (1997)	Spin polarization of slow electrons, elastically scattered by Kr.
Tseng (1997a)	Unpolarized triply differential cross section of electron bremsstrahlung. Z=13,79.
Whelan et al. (1997)	Relativistic, energy-sharing (e,2e) collisions in coplanar constant $\Theta_{1,2}$ geometry.
Zeman et al. (1997)	Relativistic calculation of superelastic electron-alkali atom scattering.
H. L. Zhang and Sampson (1997)	Collision strengths for $(n = 2) - (n = 3)$ transitions in C-like ions, Z=9-54.
Ancarani et al. (1998)	Effect of Coulomb boundary conditions for (e,2e) processes on the K-shell.
C. Y. Chen et al. (1998)	Electron-impact ionization of Ne-like ions, $22 \leq Z \leq 39$.
G.-X. Chen and Ong (1998b)	Electron-impact excitation of Fe XXIII.
G.-X. Chen and Ong (1998c, 1999a)	Electron-impact excitation of F-like Se.
J. N. Das and Dhar (1998)	Triple differential cross sections of K-shell ionization by electrons for medium-heavy atoms.
C.-Z. Dong et al. (1998)	Dielectronic recombination of Fe^{25+}.
Fontes (1998)	Electron-impact ionization of lowest $J = 0, 2$ levels of Kr, Xe. Role of the $5p^5 5d$ configuration and SO coupling.
Griffin et al. (1998)	R-matrix electron-impact excitation cross sections in intermediate coupling: MQDT approach.
Keller et al. (1998)	Three-body effects in relativistic (e,2e) processes.
Keller and Dreizler (1998)	Interpretation of (e,2e) experiment in coplanar asymmetric geometry.
Mazevet et al. (1998)	Semirelativistic DWBA for ionization of closed-shell atoms at intermediate energies.
Y.-Z. Qu et al. (1998)	Dielectronic recombination for H-like systems, Z=2-79.

Reference	Comments
Sauter et al. (1998)	Spin asymmetry in (e,2e) processes. Z-dependence. Cu, Ag, Au.
Zeman et al. (1998)	Relativistic distorted-wave calculation of inelastic electron-alkali atom scattering.
Badnell and Griffin (1999)	Correlation resonances in electron-impact excitation of Ni^{4+}.
Beier et al. (1999)	QED effects on radiative electron capture.
G.-X. Chen and Pradhan (1999)	Electron-impact excitation of Fe^{5+}.
Fontes et al. (1999a)	Fully relativistic $1s$ ionization cross sections.
Fontes et al. (1999b)	Fully relativistic $2s, 2p$ ionization cross sections. $Z - N \geq 3, Z \leq 92, N \leq 12$.
Fontes et al. (1999c)	Effect of generalized (energy-dependent) Breit interaction on electron-impact ionization to specific magnetic sublevels. Large for Xe.
J. F. Gao et al. (1999)	Direct and indirect relativistic effects on electron scattering from Cs and Au atoms. ('Direct' for incident electron, 'indirect' for atomic electrons.) Direct found more important for Cs, indirect for Au.
Griffin et al. (1999)	Electron-impact excitation of Mg-like ions.
X.-Z. Guo et al. (1999)	Low-energy electron-impact excitation of Kr $4p^5 5s$. Sensitive to relativistic effects in target.
Nefiodov et al. (1999b)	QED effects of radiative interference in recombination of electrons with heavy multicharged ions. Dielectronic recombination of He-like U.
H. L. Zhang and Sampson (1999)	Collision strengths for the 105 Δn=0 transitions with $n = 2$ in the N-like ions, Z=12-92.

Table 5.13: Particle-atom collisions.

Reference	Comments
Cusson et al. (1985)	Time-dependent 3D Dirac equation in heavy-ion scattering.
Hofstetter et al. (1992)	Ionization of hydrogen by relativistic heavy projectile.
Porter (1992)	High-velocity projectile-z^3 term in modified Bethe-Bloch stopping-power theory. (z projectile nuclear charge).
Avdonina and Pratt (1993)	Bremsstrahlung from ions in a model potential.
Baltz et al. (1993)	Pair production in heavy-ion collisions.
Dzuba et al. (1993)	Interaction between slow positrons and atoms. Ne-Rn.
Ichihara et al. (1993)	Electron capture in relativistic atomic collisions. Single-electron targets and bare projectiles.
Kürpick et al. (1993,1995abc)	Ion-atom collisions using molecular codes.
Musakhanov and Matveev (1993)	Inelastic collisions of multiply charged ions with atoms.
De Cesare et al. (1994)	^{4}He-ion induced L-shell ionization, Z=46-70.
Halabuka et al. (1994)	Inner (L,M) shell ionization by heavy projectiles. DF.
V. I. Matveev and Musakhanov (1994ab)	Inelastic collisions of multicharged ions with atoms.
Šmit and Orlić (1994)	Adiabatic L-shell ionization by protons.
Baltz (1997)	Exact Dirac calculation of ionization and pair production induced by ultrarelativistic heavy ions.
Ghilencea et al. (1994)	Coulomb Born approximation for double Bremsstrahlung.
Deco et al. (1995)	Scattering involving Coulomb potentials. Ion-atom collisions.

Reference	Comments
Halabuka et al. (1995)	Proton-induced alignment. DF.
Momberger et al. (1995)	Numerical, momentum-space solution of time-dependent Dirac equation for Au^{79+} or U^{92+} impinging on U^{91+}.
Dzuba et al. (1996a)	Positron scattering and annihilation. H, He-Xe.
Ichihara et al. (1996)	Radiative electron capture and photoelectric effect at high energies. Bare projectile, low-Z atom.
Kürpick (1996)	High-energy H^+-Ne collision handled using molecular DF-LCAO code.
Lindhard and Sørensen (1996)	Relativistic theory of stopping for heavy ions.
Vargas et al. (1996)	Electronic energy loss of low-energy protons, channeled in single-crystal Au $\langle 100 \rangle$ direction.
Moiseiwitsch (1997)	Virial theorem for electron capture from H-like atom.
X. Feng et al. (1998)	Positron scattering from Rb, Cs. Relativistic close-coupling calculations. DF bound states.
Fricke et al. (1998)	Review on ion-atom and ion-solid collisions.
Pivovarov (1998)	Coherent excitation of H-like heavy ions penetrating through a crystals.
V. I. Matveev and Matrasulov (1999)	Single and double K-vacancy production by collision with highly charged ions.

Table 5.14: Atom-atom collisions and interatomic potentials.

Reference	Comments
Ichihara et al. (1993)	Electron capture in relativistic atomic collisions. Single-electron targets and bare projectiles.
Marinescu et al. (1994a)	Interaction of two alkali atoms. Includes retardation.
J. Thiel et al. (1994,1995)	Electron-positron pair creation in relativistic heavy-ion collisions.
Eichler and Meyerhof (1995)	Relativistic atomic collisions. Book.
Thumm et al. (1995)	Highly-charged ion–C_{60} collisions. DS.
Kunz et al. (1996)	Hg-Hg interaction potential. Damped dispersion +repulsion. DK.
Alscher et al. (1997)	Electron-positron pair production in heavy-ion collisions.
Hickman et al. (1997)	FS effects in O^+-O collisions.
Khabibullaev et al. (1998)	Inelastic collisions of relativistic highly charged ions with heavy atoms. K-vacancy production.
Marinescu et al. (1998)	Francium-francium potentials, C_n.
Segev and Wells (1998)	Pair production in ultrarelativistic heavy-ion collisions.
Busic et al. (1999)	Electron-positron production with electron capture. U^{92+}-U^{92+}.
Ionescu and Belkacem (1999)	Relativistic collisions of highly-charged ions. Time evolution of QED vacuum solved numerically.
Schulze et al. (1999)	MO x-ray spectra in U^{92}–Pb collisions.

Table 5.15: Nuclear and mesonic processes involving electronic wave functions.

Reference	Comments
Band et al. (1992a)	Internal conversion in ^{229}Pa.
Band and Trzhaskovskaya (1993a)	Internal conversion at $E_\gamma = 72$ keV in Re-187.
Band and Trzhaskovskaya (1993b)	Internal conversion coefficients for 35 low-energy nuclear transitions with $E_\gamma \leq 3$ keV.
Watanabe et al. (1993)	Electron energy spectrum and asymmetry in decay of bound 1s muons. ^{16}O ... ^{209}Bi.

Reference	Comments
Greub et al. (1995)	Decays of muonic and pionic atoms into electronic atoms are enhanced by high Z. For Z=80, the relativistic increase is a factor of 50.
C. R. Hofmann and Soff (1996)	Internal-pair creation in extended nuclei.
Yamanaka and Ichimura (1999)	Nuclear polarizability contributions to $1s$ energies of H-like ^{208}Pb and ^{238}U. Transverse polarization is even more important than longitudinal one.

Table 5.16: Parity-violation effects in atoms and molecules.

Reference	Comments
Lewis and Blinder (1975)	Stark-induced anapole magnetic fields in atoms. (Anapole = PNC-induced nuclear moment of wrong parity.)
Flambaum (1993)	Dynamic enhancement of PNC in systems with a dense spectrum of excited states.
Flambaum and Hanhart (1993)	Magnetic interaction between R atomic electrons and PNC nuclear moments.
Flambaum and Sushkov (1993); Sushkov (1993)	Proposed observation of nuclear anapole moments as NMR shifts in a laser beam.
Johnson et al. (1993)	PNC in atomic Cs.
Khriplovich (1993)	Fundamental symmetries in atomic physics.
Mårtensson-Pendrill (1993b)	PNC effects in atomic systems.
Pindzola (1993)	Parity-violation effects on Auger-electron emission from highly charged atomic ions. $2s^2$ $J=0$ of U^{90+}.
Sandars (1993)	P and/or T violation. Review.
Dzuba et al. (1994)	P- and PT-odd weak interactions in Dy.
Kozlov and Ezhov (1994); Kozlov (1997)	YbF. Dipole moment of the odd electron.
Shukla et al. (1994a)	Atomic Rb.
Shukla et al. (1994b)	CC calculation of PT violation in atoms and molecules.
Bednyakov et al. (1995)	Parity-conserving weak interactions for highly charged ions. For Z=83 $1s$, 10^{-5} and 10^{-7} eV for Lamb shift, M1 hfs, respectively.
Dzuba et al. (1995)	PNC in Fr.
Kozlov and Labzowsky (1995)	PNC in diatomics. MF; M=Ba,Yb,Hg,Tl,Pb.
Kulkarni et al. (1995)	Relativistic mean-field approach to anapole moment. Atomic, parity-violating hyperfine transitions.
Malhotra et al. (1995)	PNC in Ba^+.
Schäfer and Reinhardt (1995)	VP as test of C and CPT invariance. Muonic $4f_{7/2}$ and $5g_{9/2}$ levels in ^{209}Pb.
Dunford (1996)	PNC in high-Z He-like ions.
Funakubo et al. (1996)	Numerical approach to CP-violating Dirac equation.
Sapirstein (1996b)	PNC effects in atoms.
Kozlov et al. (1996)	Nuclear anapole moment and the M1 transitions in Bi.
Kozlov and Yashchuk (1996)	P- and PT-odd effects in diatomic van der Waals molecules.
Maul et al. (1996)	Prospects for PNC experiments with highly charged ions.
Titov et al. (1996)	YbF. P,T-odd spin-rotational Hamiltonian.
Bouchiat and Bouchiat (1997)	Review on parity violation in atoms.
B. P. Das (1997)	PNC E1 transition of Yb.
Dzuba et al. (1997a)	PNC in Cs.

Reference	Comments
Flambaum and Murray (1997)	A radial (monopole) magnetic field can induce a T-invariance-violating electric dipole moment. Xe, Cs, Hg, Tl. TlF, HgF, YbF.
Khriplovich and Lamoreaux (1997)	CP violation without strangeness.
Kozlov et al. (1997)	Enhancement of the electron dipole moment in BaF.
Laerdahl et al. (1997b); Quiney et al. (1998a)	TlF. PT-odd interactions. DF.
Lazzeretti and Zanasi (1997)	PNC in HEEH; E=O,S.
Parpia (1997)	TlF. DF.
Zolotorev and Budker (1997,1999)	PNC in relativistic hydrogenic ions. Circular dichroism on the $1s \to 2s$ transition due to interference between the M1 and PNC-induced E1 amplitudes.
Bakasov et al. (1998)	Parity-violating potentials as function of structure for H_2O_2, H_2S_2, ... alanine. 'Electroweak quantum chemistry'. PT.
Bruss et al. (1998,1999)	PNC-induced energy shifts in inhomogeneous electric fields.
Ceulemans et al. (1998)	Molecular anapole moments
Geetha et al. (1998)	Nuclear-spin-dependent PNC transitions in Ba^+, Ra^+.
Kiyonaga et al. (1998)	1- and 2-electron SO contributions to PNC energy shifts of amino acids and helical alkanes.
Kulkarni et al. (1998)	Relativistic mean-field calculation of parity-violating observables in Fr.
Maul et al. (1998)	'Stark quenching' in laser fields for PNC observations. One- and two-photon transitions of Be-like $2s2p\ ^3P_0$ levels.
Mosyagin et al. (1998)	YbF molecule. Electron dipole moment experiment.
Parpia (1998)	YbF. DF.
Quiney et al. (1998b)	PT-odd effects in YbF $^2\Sigma$.
Zanasi and Lazzeretti (1998)	Stabilization of natural L-enantiomers of α-amino acids via parity-violating effects.
Bednyakov et al. (1999)	Electroweak radiative corrections.
Byrnes et al. (1999)	Enhancement of electron dipole moment for Au, Fr.
Flambaum (1999)	Large P and T invariance violation in Ra atom due to high Z, and close-lying even/odd states.
James and Sandars (1999)	Parametrized nuclear size and shape for atomic PNC.
Laerdahl and Schwerdtfeger (1999)	PNC for chiral molecules HEEH; E=O-Po. Z^5 behaviour. For a single heavy centre, as in HTeOH, the effect is small. DF.
Lazzeretti et al. (1999)	Stabilization of d-camphor via weak neutral currents.

Chapter 6

Symmetry

Table 6.1: Group theory and symmetry aspects.

Reference	Comments
Titov (1992)	Matrix elements of the $U(2n)$ generators in SO basis.
Fowler and Ceulemans (1993)	SO coupling coefficients for icosahedral molecules.
Hecht and Barron (1993)	Time-reversal and Hermiticity characteristics of polarizability and optical-activity operators.
Altmann and Herzig (1994)	Comprehensive point-group theory tables up to D_{10d}, I_h.
J. Meyer (1994)	Erratum to J. Meyer et al. (1989)[5277].
Aucar et al. (1995a)	Operator representations in Kramers bases.
Koizumi and Sugano (1995)	Geometric phase in two Kramers doublet molecular system.
Aucar (1996)	Time reversal asymmetry and ground-state average values.
Balasubramanian (1996)	The icosahedral double group, I_h.
Cao et al. (1996)	Use of symmetry in the DF method.
Jensen et al. (1996)	MCDF for molecules. Formalism. Incorporate time-reversal and D_{2h} symmetries.
J. Meyer et al. (1996)	New version of program TSYM by J. Meyer et al. (1989)[5277].
Visscher (1996)	Construction of double-group symmetry functions. D_2^*, D_{4h}^*.
Fleig et al. (1997)	Use of time-reversal symmetry between Kramers pairs in MC DK calculations, especially CASSCF.
J. Meyer (1997)	Addendum to J. Meyer et al. (1989)[5277].
Sjøvoll et al. (1997b)	Determinantal approach to SO CI. Time-reversal and double-group symmetry.
Kettle (1998)	Double, triple and quadruple groups. (The latter two correspond to 'spins' of 1/3 and 1/4, respectively).
Boya and Byrd (1999)	Clifford periodicity from finite groups.
La Cour Jansen et al. (1999)	Evaluation of SO matrix elements in a spin-adapted basis.
Saue and Jensen (1999)	Quaternionic symmetry in DF calculations. D_{2h} and subgroups.
Yabushita et al. (1999)	SO CI using graphical unitary group approach.

Chapter 7

Molecular Calculations

In this chapter we first list the technical developments by method. Finally, in Table 7.10, we review the molecular production calculations by the heavy element, or group of elements in the compound.

Table 7.1: One-electron systems.

Reference	Comments
Moss (1993a)	Rovibrational studies of HD^+. Includes R and QED[3701] corrections.
Moss (1993b)	$2p\sigma_u$-$1s\sigma_g$ spectrum of D_2^+.
Moss (1993c)	Dito for H_2^+.
Rutkowski et al. (1993)	Relativistic virial theorem for diatomic molecules. H_2^+. Large-R and small-R limits.
Düsterhöft et al. (1994)	$NiPb^{109+}$.
Franke (1994)	Relativistic effects from Direct Perturbation Theory. Th_2^{179+}.
C. Müller (1994)	Finite-element solution of time-dependent Dirac equations.
Sundholm (1994)	Fully numerical (2D) solutions of 2nd-order Dirac equation for H_2^+, ... Th_2^{179+}, Fm_2^{199}. Accurate for small Z.
Momberger et al. (1995)	Numerical, momentum-space solution of time-dependent Dirac equation for Au^{79+} or U^{92+} impinging on U^{91+}.
Parpia and Mohanty (1995a)	H_2^+, H_3^{2+}, H_4^{3+}. Th_2^{179+}.
Pöschl and Dietrich (1995)	One-electron two-center Dirac problem using a diatomic basis of H-like spinors.
Cencek and Kutzelnigg (1996)	Accurate H_2^+.
Chauhan and Raina (1996)	High-energy electron scattering from two-centre potentials.
Gorvat et al. (1996)	One-electron two-centre Dirac problem.
Lazur et al. (1996abc)	The one-electron two-centre problem.
Molzberger and Schwarz (1996)	Effects of different order in α^2 on energies of H-like atoms. Compare Dirac with 1st- and 2nd-order DK.
W. H. E. Schwarz et al. (1996)	H_2^+. Interpretation of relativistic effects on chemical bonding.
Franke (1997)	One-electron systems using DPT.
V. I. Matveev et al. (1997)	LCAO solution of two-centre Dirac equation. Supercritical limit obtained.
LaJohn and Talman (1998)	Th_2^{179+}. A minimax STO solution.
Rutkowski et al. (1998)	DPT on H_2^+-like systems.
Alexander and Coldwell (1999)	Th_2^{179+}.
Balinsky and Evans (1999ab)	Stability of one-electron molecules. 'No-pair' Hamiltonian.
Matrasulov et al. (1999)	Dirac electron in the field of two opposite charges. $Z < 137$ and $Z > 137$ considered.
Rutkowski (1999)	Th_2^{179+} at $R = 2/90$ au. Several states. DPT.
Yan and Ho (1999)	Relativistic effects in positronium hydride (PsH, e^-e^+p).

The following Table 7.2 contains both DF-LCAO and fully numerical calculations. For further production results, see Table 7.10.

Table 7.2: Four-component, full Dirac calculations on molecules.

Reference	Comments
S. N. Datta (1993)	DF/HF comparisons for Li compounds.
Dyall (1993a)	DF/HF for MO; M=Ge-Pb. Comparisons with PP, PT.
Dyall (1993b)	PtH, PtH^+, PtH_2. DF/HF. Comparisons with PP, DK.
Ishikawa (1993)	Comment on 'segmented contractions' of relativistic Gaussian basis sets (Matsuoka (1992a), paper 5228).
Malli et al. (1993ab, 1994)	Universal Gaussian basis for DF calculations.

Reference	Comments
Mohr and Soff (1993)	Nuclear-size correction to electron SE. $1s, 2s, 2p_{1/2}$, $Z \leq 100$.
Pisani et al. (1993)	Discuss relativistic effects in atoms and molecules using LCAO calculations on H-like systems as example.
Visscher (1993,1994)	Correlated, fully relativistic methods.
L. Yang et al. (1993)	Fully numerical (FEM), fully relativistic H_2 to 10-figure accuracy at DF level.
Baeck and Lee (1994)	Effect of the magnetic part of the Breit term on the $^2\Pi$ states of hydrides (HX or HX^+; X=Be-F, Mg-S).
de Jong et al. (1994)	The MOLFDIR package.
Düsterhöft et al. (1994)	Numerical solution of Dirac equation for $NiPb^{109+}$.
Dyall (1994a)	Separation of spin effects for DCB Hamiltonian. Rg (Ne-Rn).
Dyall (1994b)	MP2 theory based on DF. Up to two open shells. O_2.
Dyall (1994c)	Review on DF/GTO for polyatomic molecules.
Kaldor and Hess (1994)	DK CC on Au, AuH.
Malli and Styszyński (1994)	ThF_4. DF.
Merenga and Andriessen (1994)	Improvement of MOLFDIR.
Nieuwpoort et al. (1994)	The MOLFDIR molecular code.
Pisani and Clementi (1994a)	DF calculations on closed-shell molecules. Hydrides.
Pisani and Clementi (1994b)	H_2E; E=O-Po. DF/HF.
Visscher et al. (1994)	The MOLFDIR package.
Dijkstra et al. (1995)	DF + Madelung potential model for $[EuO_6]^{9-}$ in crystals.
Kaldor (1995,1997)	Reviews on the Fock-space CC approach.
Kotochigova and Tupitsyn (1995)	Generalized Valence Bond approach, based on numerical DF basis orbitals. Large-R quasimolecules Na_2, In_2.
Minami and Matsuoka (1995)	GTO basis for $_{86}$Rn-$_{94}$Pu. DF.
Parpia and Mohanty (1995b)	H_2,HF,HCl,H_2O,NH_3,CH_4. DF/HF.
Pisani and Clementi (1995a)	HE; E=O-Po. DF/HF.
Pisani and Clementi (1995b)	DF calculations on closed- and open-shell molecules.
Saue (1995)	Four-component molecular methods. Thesis.
Visscher et al. (1995)	The MOLFDIR package.
Cao et al. (1996)	Use of symmetry in the DF method.
Dyall and Faegri (1996)	GTO basis for DF. Rn.
Jensen et al. (1996)	MCDF for molecules. Formalism. Incorporate time-reversal and D_{2h} symmetries.
Jorge et al. (1996,1997ab); Jorge and da Silva (1998)	GTO-basis sets.
Malli and Styszyński (1996)	UF_6. DF.
Mohanty and Parpia (1996)	AgH. DF/HF.
Saue et al. (1996)	DF on HX; X=I,At,E117.
Visscher (1996)	Construction of double-group symmetry functions.
Visscher et al. (1996a)	Relativistic, unrestricted CC method, including noniterative connected triples.
Dyall (1997); Dyall and Enevoldsen (1999)	'Normalized elimination of small components'. Can make light atoms non-relativistic.
Flocke et al. (1997abc); Flocke (1997)	Symmetric-group approach to relativistic CI.

Reference	Comments
Ishikawa et al. (1997)	Finite-nucleus models for GTO spinors.
Ladik (1997)	DF equations for solids.
Laerdahl et al. (1997a)	DF-based MP2. A 'direct' (no disk), Kramers-restricted code. CuF-AuF.
Merenga (1997)	Applications of MOLFDIR on cluster models of Ce-containing crystals.
Parpia (1997)	DF on TlF.
Saue et al. (1997)	Principles of direct (no disk) DF. Application on CsAu.
Sjøvoll et al. (1997a)	PdH. SO CI at DK level or DF level.
Tatewaki and Matsuoka (1997)	GdF. DF. Gd $5d \rightarrow$F $2p$ donation, F $2s$ - Gd $5p$ hybridization.
Visscher (1997)	Eliminate small-component integrals $(SS\|SS)$ by a simple point-charge correction.
Visscher and Dyall (1997)	DF atomic calculations using different nuclear charge distributions.
Visscher et al. (1997)	Relativistic RPA for frequency-dependent polarizabilities, $\alpha(\omega)$. H_2O, SnH_4, Hg.
de Jong (1998)	DF applications on I_2, UF_6, UO_2^{2+}.
Düsterhöft et al. (1998)	Numerical solution of Dirac-Slater equations for C_2.
Dyall (1998b)	Optimized DZ basis for $4p, 5p, 6p$ elements. Finite nucleus. DF/HF.
Hu et al. (1998)	Band-structure version of the DF code MOLFDIR. Application to a 1D chain of Se.
Ishikawa et al. (1998)	DF finite-field theory of NMR shielding.
Laerdahl et al. (1998)	DF (+MP2) for MF, MH, MH_3; M=La,Lu,Ac,Lr. Lanthanide and actinide contractions.
Malli and Styszyński (1998)	$RfCl_4$. DF.
Parpia (1998)	YbF. Unrestricted DF.
Quiney et al. (1998a)	TlF. DF. Small-r solutions for PNC purposes.
Quiney et al. (1998c)	H_2O. PT or self-consistent Breit term.
Skaane (1998)	Development and applications of the BERTHA code.
Watanabe and Matsuoka (1998)	DF with a frozen-core approximation. HI, ThO.
Dyall (1999)	ThO_2, PaO_2^+, UO_2^{2+}. Bonding and bending.
Hada et al. (1999)	HX;X=H,F-I. Proton NMR shifts.
Hu et al. (1999ab)	Fits of GTO basis functions. Hg, HgO, Se chain.
Kaldor and Eliav (1999)	High-accuracy calculations for heavy and superheavy elements.
Kullie et al. (1999)	DF-FEM on diatomics. H_2,LiH,BH,HF,N_2,CO,Ar,HCl.
Quiney et al. (1999)	The BERTHA molecular DF code. Review.
Saue and Jensen (1999)	Quaternionic symmetry in DF calculations. D_{2h} and subgroups.
Suzuki and Nakao (1999)	Fully relativistic LCAO approach for crystals. Au, InSb.

Table 7.3: Molecular all-electron calculations using 'no-pair' or other transformed Dirac Hamiltonians. Perturbation theories included.

Reference	Comments
Fraga and Karwowski (1974)	Two nuclear-mass-dependent terms in the BP equation. Of same order as the hyperfine interaction.
Klotz (1985)	Ab initio calculation of spin-forbidden transitions.
Bearpark et al. (1993)	SO interactions from SCF wave functions.
Dyall and Faegri (1993)	Finite-nucleus effects at PT level discussed.
Gould and Battle (1993ab)	Spin-dependent unitary-group approach to the BP Hamiltonian.
Pizlo et al. (1993)	The DK approach tested on Au and AuH.
Rohse et al. (1993)	H_2, H_3^+. DPT.
van Lenthe et al. (1993)	Relativistic regular two-component Hamiltonians. Tests on the U atom.
Dyall (1994a)	Separation of spin terms from the Dirac Hamiltonian.
Gleichmann and Hess (1994a)	DK tested on Hg atom at MRCI level.
Gleichmann and Hess (1994b)	LiHg. DK including SO. MRCI.
Kutzelnigg et al. (1994,1995)	Relativistic HF based on DPT.
Park and Almlöf (1994)	Pt_2, M_2, MH; M=Ag,Au. Test two-electron terms in DK (HF, MP2).
Sadlej and Snijders (1994)	Spin separation in the regular Hamiltonian approach. Cp. Dyall (1994a).
Shukla and Banerjee (1994)	A four-component relativistic DF method for valence electrons only. An effective core-valence potential constructed. Tested on atoms, Li-S.
van Leeuwen et al. (1994)	ZORA solutions for H-like atoms are scaled Dirac ones.
van Lenthe et al. (1994a)	Relativistic total energy using regular approximations. For Au_2, AuH, ZORA is close to frozen-core.
van Lenthe et al. (1994b,1996c)	Reviews on regular Hamiltonians.
Faas et al. (1995)	The ZORA formalism applied to the DF equation. Xe, Rn. The deep core densities can be improved by scaling.
Ottschofski and Kutzelnigg (1995)	DPT tested on atomic ground states of He-Rn.
Sadlej et al. (1995)	Four-component regular Hamiltonians.
van Lenthe et al. (1995)	Solves the 2nd-order Dirac equation for the large component in a Slater basis. Tests on U atom.
van Wüllen (1995)	DFT by DPT. Tests on $M(CO)_n$, MH; M=Ag,Au; Au_2.
Wolff et al. (1995)	Self-consistent SO and Zeeman terms in otherwise NR ab initio calculations. Magnetic structure factors for $CsCoCl_5$.
Hada et al. (1996)	^{183}W NMR shielding in WX_6; X=F,Cl, WO_4^{2-}. DK.
Hess et al. (1996)	A mean-field SO method ('AMFI') for correlated wavefunctions.
Kellö and Sadlej (1996a)	Basis sets for highly correlated DK calculations for Groups 11 and 12 (Au, Hg).
Kutzelnigg (1996)	Theory of stationary DPT (Direct Perturbation Theory) of relativistic corrections.
Molzberger and Schwarz (1996)	Effects of different order in α^2 on energies of H-like atoms. Compare Dirac with 1st- and 2nd-order DK.
Nasluzov and Rösch (1996)	DFT code within SR DK approximation with analytical energy gradients. Tested on Au_2,AuH,AuCl, $M(CO)_6$; M=Mo,W; $[M(PH_3)_2O_2]$; M=Pd,Pt.

Reference	Comments
Neogrády et al. (1996)	Polarized basis sets for high-level correlated calculations of electric properties. Polarizabilities for M,M^+; M=Cu-Au.
Rösch et al. (1996)	DK DFT molecular calculations. Review.
Rutkowski (1996)	Basic aspects of regular PT.
Rutkowski and Schwarz (1996)	Effective DPT Hamiltonian for near-degenerate states. Formalism.
Snijders and Sadlej (1996)	Perturbation versus variation treatment of regular relativistic Hamiltonians. Latter preferred.
van Lenthe (1996)	The ZORA equation. Thesis.
van Lenthe et al. (1996a)	Solve the Dirac equation using the FW transformation and large components only. Tests on U atom.
van Lenthe et al. (1996b)	SO effects in closed-shell molecules via ZORA. Tested on several diatomics containing I to Bi.
van Wüllen (1996b)	DFT with DPT. Tests on TM carbonyls.
Barysz (1997)	Compares Pauli PT, DPT and Dirac levels for H-like systems.
Barysz and Sadlej (1997)	Expectation values $\langle O \rangle$ in approximate two-component theories. Example $O = 1/r$.
Barysz et al. (1997)	Two- or one-component Hamiltonians to arbitrary order in α^2.
Boettger (1997)	SR LCGTO DK code tested on atoms. Ce, Au, Pb, Pu.
Dyall (1997); Dyall and Enevoldsen (1999)	'Normalized elimination of small components'. Can make light atoms non-relativistic (local α=0).
Fleig et al. (1997)	Use of time-reversal symmetry between Kramers pairs in MC DK calculations, especially CASSCF.
Franke (1997)	One-electron systems using DPT.
Geipel and Hess (1997)	DK approach for solids (in crystal HF). Ag, AgX; X=F-Br. Bond contraction. Metal stabilized, salts destabilized by relativity.
Havlas and Michl (1997)	SO effects in triplet tetramethyleneethane biradical.
Klopper (1997)	1st-order DPT relativistic corrections to correlated calculations.
Kutzelnigg (1997)	Relativistic 1-electron Hamiltonians 'for electrons only' and the variational treatment of the Dirac equation. Relations to 'regular approximation', DK and minimax principle.
Miadoková et al. (1997)	Standardized basis sets for highly correlated DK calculations of electric properties. IP of Li-Fr, Be-Ba. Polarizability of K-Fr, Li^+-Fr^+, Ca-Ba.
Ottschofski and Kutzelnigg (1997)	DPT tested on He-like systems. r_{12} wave functions. Z=1-120.
Philipsen et al. (1997)	CO adsorption on (111) surfaces of Ni,Pd,Pt. ZORA.
Rutkowski and Kozlowski (1997)	Relativistic H atom in static, uniform **B**. DPT.
Sjøvoll et al. (1997a)	PdH. SO CI at DK level or DF level.
Sjøvoll et al. (1997b)	Determinantal approach to SO CI.
Sundholm and Ottschowski (1997)	DPT incorporated into the atomic MCSCF code LUCAS. Tested on Zn-Hg, Ne-Rn.
van Lenthe et al. (1998)	ZORA calculations of molecular g-tensors.
Wahlgren et al. (1997)	SO splitting of the Tl atom. BP/DK.
Császár et al. (1998)	Relativistic corrections to potential energy surface and rovibrational levels of water.
Dyall (1998a)	Attempted search for a variational α^2 method.

Reference	Comments
Franke and van Wüllen (1998)	Comparisons of PT and DPT at MP2 level. He-Ar, MH; M=Cu-Au.
Gagliardi et al. (1998a)	A two-centre implementation of DK. The MAGIC code.
Havlas et al. (1998); Havlas and Michl (1998)	SO coupling in organic biradicals. Correlated ab initio calculations including the BP Hamiltonian. Carbenes and silylenes.
Kellö and Sadlej (1998a)	Picture-change effect on expectation values. DK and spin-free Pauli. EFG in HX; X=Cl-At as example.
H. Müller et al. (1998)	HF. Spectroscopic accuracy with CCSDT1-r_{12}. PT relativity.
Rutkowski et al. (1998)	DPT on H_2^+-like systems.
Sjøvoll et al. (1998a)	PtH. Include SO during CI or after CI stage.
Sjøvoll et al. (1998b)	New way to treat the **p**-dependent terms in DK calculations. Tests on Br_2.
van Lenthe et al. (1998)	Calculation of g-tensors and M1 hyperfine interactions at ZORA level. Exact relations between Dirac and ZORA for H-like atoms. Tests on TiF_3, 5- or 7-atom coinage-metal clusters.
van Wüllen (1998)	DFT with ZORA. Introduces a model potential for use in the kinetic-energy operator. Tests on MM', MX; M=Cu-Au, X=H,F,Cl.
Wahlgren et al. (1998)	Keep one-centre DK terms, only. Tests on Au_2, UO_2^{2+}.
Balinsky and Evans (1999ab)	Stability of one-electron molecules. 'No-pair' Hamiltonian.
Dyall and van Lenthe (1999)	An infinite-order relativistic approximation (cp. ZORA).
Griesemer and Tix (1999)	Claims that the N-electron no-pair model is magnetically instable for all values of α.
Havlas and Michl (1999)	Zero-field and SO splittings in triplets of m-xylylene.
Kutzelnigg (1999a)	Effective Hamiltonians for degenerate and quasidegenerate DPT. Relationship to FW transformation.
Kutzelnigg (1999b)	Relativistic corrections to magnetic properties. DPT.
La Cour Jansen et al. (1999)	Evaluation of SO matrix elements in a spin-adapted basis.
J. M. L. Martin and de Oliveira (1999)	Atomization energies of many 1st-row -to- 3rd-row (H-Cl) molecules. PT relativity.
J. M. L. Martin and Taylor (1999)	Atomization energy of SiF_4. PT relativity.
Miadoková et al. (1999)	MF; M=K-Fr. Electric properties. PT/DK.
Nakajima and Hirao (1999)	'RESC' scheme: eliminate small components, use $T = \sqrt{p^2c^2 + m^2c^4} - m$ kinetic energy. Tests on Ag, Au, AgH, AuH.
Nakajima et al. (1999)	Combine DFT and RESC. Tests on MH,MX; M=Ag,Au.
Rutkowski (1999)	Iterative solution of one-electron DPT equations.
Schwerdtfeger et al. (1999)	EFG in HCl and CuCl. Accuracy of the DFT approach. DK.
Tatchen and Marian (1999)	Test approximate SO Hamiltonians on NC_nH^+; N=4,5, NC_4N^+. AMFI approximation excellent.
van Lenthe et al. (1999)	Analytical energy gradients for geometry optimization using ZORA. Tests on W, Os and Pt carbonyls.
Visscher and van Lenthe (1999)	Separation between SR and SO effects done in three different ways with three different results. U^{91+}.
Wolff et al. (1999)	NMR shifts using ZORA and DFT. ^{199}Hg in LHgL'.

Table 7.4: Methodological density-functional theory papers.

Reference	Comments
W. F. Schneider et al. (1991)	DV DFT studies of 3-coordinate An complexes. Review.
L. Yang et al. (1993)	Fully numerical (FEM), fully relativistic DS for light diatomics, LiH, Li_2, BH, C_2.
Bastug (1994)	Accurate DS total energies for molecules.
Onoe et al. (1994ab); Onoe (1997)	Analyses relativistic effects on DVM-DFT MO:s for Cu_2, AuH, Pb_2, XF_6;X=S-Po, Mo-W,Re→Pt, U,Np,Pu.
Schreckenbach et al. (1995)	Implementation of analytical energy gradients in SR DFT methods. Tests on $M(CO)_5$; M=Fe-Os, $M(CO)_6$; M=Cr-Mo.
Hirata (1996)	Applications of DS-DVM in actinide chemistry.
Malkin et al. (1996)	NMR shieldings including SO.
Mayer et al. (1996)	M_2; M=Cu-Au, AuH, AuCl. Various functionals, finite nucleus effects studied.
Nasluzov and Rösch (1996)	DFT code within SR DK approximation with analytical energy gradients. Tested on Au_2,AuH,AuCl, $M(CO)_6$; M=Mo,W; $[M(PH_3)_2O_2]$; M=Pd,Pt.
van Wüllen (1996a)	Can the common PP be used in DFT? Compared with AE. A: Yes.
Baerends and Gritsenko (1997)	DFT on molecules reviewed.
Fricke et al. (1997)	Review on the DFT DVM applications on superheavy elements.
Menchi and Bosin (1997)	DFT PP in QMC.
Malkina et al. (1998)	NMR shieldings including SO using AMFI.
Schreckenbach and Ziegler (1998)	DFT calculation of NMR shifts and ESR g-tensors. Review.
Ellis and Guenzburger (1999)	Review on the DFT DVM.
Schreckenbach et al. (1999)	DFT calculations on actinide compounds.
van Wüllen (1999)	Status of all-electron DFT on molecules.
Varga et al. (1999)	Latest description of the DV DFT methods. Tests on M_2; M=Cu-Au.

Table 7.5: Relativistic or quasirelativistic scattered-wave (multiple-scattering) calculations.

Reference	Comments
Parsons and Till (1993)	Introduction of non-muffin-tin corrections permits bond-length optimizations. Octahedral MF_6; M=Mo,W,U.
Vijayakumar and Gopinathan (1993)	SW method with the authors' Ξ functional. Tests on PES of SF_6, MnO_4^-.
Subramanian and Ramasami (1995)	MX_4; M=Zr,Hf; X=Br,I. PES. Adjustment of spheres discussed.

The next Table 7.6 contains reviews and methodological papers on relativistic or quasirelativistic pseudopotential (effective core potential) calculations. See also Table 4.8.

Table 7.6: Pseudopotential (effective core potential) methods.

Reference	Comments
Vijayakumar et al. (1993)	An efficient, relativistic MRCI method. Tests on Pb_2, Bi_2.
Schwerdtfeger et al. (1995a)	Accuracy of the PP approach tested on InCl, $InCl_3$. Small core better.
Sekkat et al. (1995)	Effective potentials for *molecular* EH_3 groups, E=N-As. Tested on AH_3-EH_3, A=B-Tl.

Reference	Comments
Seijo and Barandiarán (1996)	Use of embedded cluster models, based on PP in study of laser materials. M:M'O; M=Sc→Zn; M=Mg-Sr. Tl^+:$KMgF_3$.
Titov (1996)	A two-step PP method proposed. Restores the core-region orbitals for hfs properties or PNC applications.
Menchi and Bosin (1997)	DFT PP in QMC.
Schimmelpfennig et al. (1998ab)	The 'atomic-mean-field integral (AMFI)' approach to SO CI.
Skylaris et al. (1998)	Efficient method for PP integrals. Uses projection operators. Part of the MAGIC code.
Balasubramanian (1998)	ECP techniques for heavy-atom molecules.
Takahashi et al. (1999)	SCF method with SO. D_{2h} or subgroups.
Yabushita et al. (1999)	SO CI using graphical unitary group approach. Implementation with PP SO operators. COLUMBUS package.

Table 7.7: Semiempirical methods.

Reference	Comments
Christoffersen and Hall (1966)	Orbit-orbit coupling in aromatic molecules. Effect on magnetic susceptibility of benzene.
Minaev (1983)	SO effects in molecular spectroscopy and chemical kinetics.
Bengtsson and Hoffmann (1993)	Pb_2X^{3+} structural units; X=F-I. Pb...Pb discussed.
Breza (1993ab,1994)	Cluster models for Y-Ba-Cu-O superconductors. CNDO/1.
A. Datta et al. (1993)	Cluster models for $YBa_2Cu_3O_7$. REX.
Garcia et al. (1993)	Model electron-hopping hamiltonian for IP of Rg_n, Hg_n.
Milletti (1993)	$M_2X_4(PH_3)_4$; M=Mo,W; X=Cl,Br. Fenske-Hall. Influence of halide ligands on M-M quadruple bond.
Reindl and Pastor (1993)	Pb_n^{2+}.
von Grünberg et al. (1993)	Using DIM for semiempirical band structures. Ar-Xe crystals.
Boča (1994)	$[Au(Pt\text{-}Bu_3)_4]^{2+}$ bonding. CNDO/1. Various phosphines compared.
Boča et al. (1994)	Cu(II) dimethylpyrazole complexes. INDO/1.
Burdett et al. (1994)	Au...Au interactions in $X_nA(AuPR_3)_m$ molecules.
Calhorda and Veiros (1994)	Au(I)...Au(I) interactions. Chains, rings. EHT.
Cory et al. (1994)	INDO approach for actinide compounds. AnX_n; n=4,6; An=Th-Pu. $An(COT)_2$, amide/imide complexes.
Dionne and Allen (1994)	Bi-O-Fe hybridization for magneto-optical states in $Y_{3-x}Bi_xFe_5O_{12}$.
Lawrence and Apkarian (1994)	SO states of I atoms in crystalline Kr and Xe.
Turi Nagy et al. (1994)	A SR INDO approach. Tested on tetrahedral XY_4 molecules.
Akamova et al. (1995)	MNDO on linear clusters of Hg_n, n=2-4.
Boudreaux and Baxter (1995); Boudreaux and Baxter (1997)	$Sm(Cp^*)_n^{q+}$ systems.
Braun et al. (1995); Braun et al. (1996)	$[Th_{12}N_6Cl_{41}]$ clusters. EHT.
Z.-C. Dong and Corbett (1995)	Tl_n; n=12,13 models for Zintl compounds. REX/EHT.
Endo et al. (1995)	Si and C NMR shifts in SiX_4, $CH_{4-n}X_n$; X=Cl-I; n=1-4. PM3+SO.

Reference	Comments
Estiú et al. (1995)	Ln bis(octaethylporphyrinate); Ln=Ce,Eu. INDO-S/CI.
Watkins and Williams (1995)	Semiempirical model for substitutional impurities Ni^-, Pd^-, Pt^- and Au^0 in Si. g-tensor.
Boča (1996)	$[Pt(CO)(AuPPh_3)_8]^{2+}$. CNDO/1.
Böckmann et al. (1996)	SO coupling in organic molecules. MNDOC-CI. Triplet-state reactivity.
Braun and Simon (1996)	Octahedral $Th_6H_xBr_{15}$ clusters, x=5,7. EHT band structure with parameters of Pyykkö et al. (1989).
Breza (1996)	Cluster models for La-Ba-Cu-O superconductors.
Hammer et al. (1996)	CO on M(111); M=Ni,Cu,Pd,Ag,Pt,Au; Ru(0001),Cu_3Pt. Simple model for d-band hybridization proposed.
Kane et al. (1996)	PbS nanoclusters. Up to 912 atoms. SO included.
Michl (1996)	SO coupling in biradicals. 2-electron-2-orbital model.
Thiel and Voityuk (1996)	Extension of MNDO to d orbitals. Cl-I, Zn-Hg. Tested on heats of formation of a 575 molecule set.
Belkhiri et al. (1997)	UF_5, AnO_2^{n+}, An=Th-Am. Rel. effects on bond angles.
Calhorda et al. (1997)	Au(I)...Au(III) interactions. EHT.
J. A. González et al. (1997)	NMR ^1J coupling constants in Me_3XY; X=C-Pb; Y=F,Cl. MNDO-level correlated approach.
Muinasmaa et al. (1997)	Carboxylic acid complexes of M^{2+}; M=Mg,Zn,Cd,Pb. PM3.
Sasaki et al. (1997)	SO-induced electron-spin polarization in donor-acceptor complexes containing heavy halogen atoms. ESR.
D. P. Huang and Corbett (1998)	$TaAs_4^{7-}$ and $(TaAs_4Tl_2)^{5-}$ clusters. EHT.
Julg (1998)	Absorption spectra of Pb^+ and Pb^{3+} in amazonite, KSi_3AlO_8. $6s \rightarrow 6p$ handled with simple MO model.
Lobayan and Aucar (1998ab)	NMR spin-spin coupling constants within PM3 parametrization, CLOPPA scheme. As-Bi, Ga-Tl, Si-Pb.
Minaev and Ågren (1998)	SO coupling in oxygen-containing diradicals. Reaction paths. MNDO CI.
Naumkin (1998)	DIM model for Rg·X_2 systems. Ar·I_2 potential.
Neese and Solomon (1998)	INDO/S-CI for ZFS and g-factors in TM complexes. $FeCl_4^-$ as example. Results compared with BP PT ab initio ones.
Omary and Patterson (1998)	$[Ag(CN)_2]_3$ chain model for solid Tl$[Ag(CN)_2]$. EHT.
W.-J. Li et al. (1999b)	Introduce CI to A. J. Stone's (1963)[2777] theory of the g-matrix. Applications on $[MOX_n]^{m-}$; M=V,Cr,Mo; X=F-Br. INDO.
Olenev et al. (1999)	$(P_2Hg_6)^{8+}$ model for solid Hg_2PCl_2. EHT.
Seo and Hoffmann (1999)	Structures of solid elements P-Bi. EHT.
Seong and Anderson (1996)	Water dissociation on Pt anodes. Semiempirical model.

Table 7.8: Relativistic crystal-field theory.

Reference	Comments
Richter et al. (1992,1995)	Crystal-field parameters of $SmCo_5$ obtained from a solid-state ab initio calculation.
Burdick et al. (1993)	Correlation effects on two-photon absorption intensities for Eu^{2+}.
Burdick and Reid (1993)	Dito for Gd^{3+} in LaF_3.
Chatterjee and Buckmaster (1993ab)	Relativistic crystal field theory.
Hubert et al. (1993)	Eu^{3+} and Am^{3+} in cubic-symmetry site of ThO_2.
Lulek (1993)	SO coupling for f electrons in a crystal field.
Parrot and Boulanger (1993, 1997)	SO for d^5 ions in crystals.

Reference	Comments
Bonča and Gubernatis (1994)	QMC for magnetic impurity with SO and crystal-field splitting. Single-impurity Anderson model.
Steinbeck et al. (1994)	Obtain crystal-field parameters for the $4f$ states of Er and Dy in Ag or Au from solid-state DFT calculations.
Kornienko et al. (1995)	Effect of CI on SO parameter of Ln ions.
Schamps et al. (1995)	LaO. Ligand field model tested against PP MRCI.
Dushin et al. (1996)	Crystal-field theory for trivalent Er, U, Md.
Steinbeck et al. (1996)	Obtain crystal-field parameters for the $4f$ states of Sm in Fe and Co intermetallics from solid-state DFT calculations.
Bencheikh and Schamps (1997)	Relativistic ligand field theory. SO in diatomics.
Brooks et al. (1997)	Crystal field excitations of Ln are combined from $4f$ excitations and a cloud of shielding conduction electrons. Spin Hamiltonians of TmSb and PrSb reproduced *ab initio*.
D.-D. Dai et al. (1998)	Crystal-field parameters from DFT for diatomic CeO, CeF.
S. Edvardsson and Klintenberg (1998)	Shielding parameters for crystal-field theory from DF atomic calculations.
Rahman (1998)	Crystal-field excitations in UO_2.
Ren et al. (1998)	LnF; Ln=Pr$\rightarrow$Yb. Low-lying levels.
Smentek (1998)	Spectroscopy of Ln ions in crystals.
Smentek and Hess (1998)	Odd-rank crystal-field parameters. Determine intensities. Are related to each other.
Zhorin and Liu (1998)	f-elements in $LaCl_3$.
Smentek (1999)	Relativistic contributions to E1 transitions of Ln ions in crystals. SO part.

Table 7.9: Relativistic theories of molecular properties.

Reference	Comments
Ågren and Vahtras (1993)	SO coupling in molecular Auger spectra. Water.
Aucar and Oddershede (1993)	Nuclear spin-spin coupling in Polarized Propagator approach.
Pegarkov (1993)	Adiabatic states of diatomic molecules with large SO coupling.
Zhu et al. (1993)	SO and rotational autoionization in HCl and DCl.
Dubernet and Hutson (1994)	Atom-molecule van der Waals complexes for open-shell atoms. Different angular-momentum coupling cases (cp. Hund's cases).
Serebrennikov and Steiner (1994); Steiner and Serebrennikov (1994)	Theory of spin-rotational relaxation for electrons with $S^{\text{eff}} = \frac{1}{2}$. 'Adiabatic rotation of effective spin'.
Ågren et al. (1996)	SO effects in molecules.
Ballard et al. (1996)	Proton NMR shifts in HX; X=F-I. 'No-pair' + SO.
Fukui et al. (1996)	Relativistic effects in NMR shielding. Lowest order.
Fukui and Baba (1998)	NMR shielding at no-pair level.
Kaupp et al. (1998b)	A simple interpretation for SO-induced NMR shifts, analogous to FC spin-spin coupling. Applied on iodo-organic compounds. ^{1}H and ^{13}C shifts.
Aucar et al. (1999)	'Diamagnetic' (A^2) term for magnetic properties in Dirac theory.
Baba and Fukui (1999)	Gauge-origin-independent relativistic effects in NMR shifts.

The following Table 7.10 contains production results for molecules, classified by the heaviest element, or group of elements in them. The methods of calculation are explained in Table 10.1 in the Appendix. The last column gives the way of including relativistic effects. For series of chemical elements, a hyphen, '-', goes down a column and an arrow, →, from left to right in the Periodic Table.

Table 7.10: Production results for molecules.

Element		Compounds	Ref.	Method
1	H	H_2, H_3^+.	9423	DPT
		Accurate relativistic corrections for H_2^+, H_2.	7140	PT
		Accurate relativistic corrections for H_3^+. Rovibrational analysis given.	7141,8166	PT
		H_2 ground state. $R \leq 12$ au.	10223	PT
		H_2 excited states.	10224,10225	PT
		H_2 ground state. 2D FEM. DF/HF.	10268	DF
2	He	Spin-orbit and spin-spin coupling in He_2, He_2^-.	6960	PT
3	Li	DF/HF comparisons for Li compounds.	7398	
		LiH, Li_2. DS/HFS.	10268	DFT
4	Be	Be_2. Explicitly correlated. Rel. D_e decrease -0.56%	7816	PT
		Be_2. Rel. D_e decrease -0.42%	8831	PT
5	B	BH. DS/HFS.	10268	DFT
		BF. BF_3, whose atomization energy reduced by 0.7 kcal/mol.	6865	DK
6	C	C_2. DS/HFS.	10268	DFT
		C_{60}^{x+}, x=0-7	6851	DFT
7	N	N_2. Phosphorescent A $^3\Sigma_u^+ \rightarrow$ X $^1\Sigma_g^+$ decay, Vegard-Kaplan band.	9092	PT
		NH^+ FS.	9148	PT.
8	O	O^+-O collisions.	8019	PT
		O_2^{2+} predissociation.	7567	PT
		O_2^+, O_2^-. FS splitting. Up to CCSD(T) level.	10126	DF+corr.
		CO $V(R)$	6815	PT
		OH. SO splitting. NO^+ radiative lifetime. H_2O^+ Renner-Teller effect.	6873	PT
		H_2O. Rovibrational levels. CCSD(T), QZ.	7307	PT
		H_2O.	9333	DFB
		HCO SO coupling.	9188	PT
9	F	HF. R/NR.	8833	PT
		HF. PT relativity, CCSDT1-r_{12}.	8994	PT
		BF. BF_3, whose atomization energy reduced by 0.7 kcal/mol.	6865	DK
		FO. FS splitting. Up to CCSD(T) level.	10126	
		C_2F_4	6867	DK
		CF_3, CF_4, C_2F_4, :$CFCF_3$, Heat of formation.	7468	PT
		SO corrections to F+D_2→DF+D.	8048	PT
10	Ne	H^++Ne high-energy collision.	8500	DFT
		BNe. Lifetime of C $^2\Delta$. MRCI.	9793	PT
13	Al	AlF,AlCl. EFG, Q(Al) derived.	8334	DK
		AlH_2, AlAr.	9170	DF+CCSD(T)
		AlNe. SO. CCSD(T).	10271	PT
		AlX_4^-; X=H,F-I. ^{27}Al NMR shifts.	9022	PT

Element		Compounds	Ref.	Method
14	Si	SiH_4. Heats of formation in Si chemistry.	7258	DK
		SiH_4.	8401	DPT
		SiF, SiF^+. Accurate D_0.	9393	PT
		SiF_4 atomization energy.	8834	PT
15	P	PH_n, PH_n^+; n=1-3	6788	PP
		PO_n, PO_nH; n=1-3	6862	DK
16	S	S_n^q; n=2,3; q=0,-1.	7988	NR+SO
		SH (A $^2\Sigma^+$). Fragmentation. SO.	9387	PT
		SO, SO_2. Atomization energies.	6866	PT
		Adsorbed sulfur oxide anions.	9090	
17	Cl	HCl 2p PES interpreted.	7615, 7695 7696	DF+CI
		ClO. FS splitting. Up to CCSD(T) level.	10126	
		H+Cl_2 →HCl+Cl. Cl+HCl→ClH+Cl. Rel. effects on reaction energy.	10121	MP2: R/DK/NR
18	Ar	Ar_2^+, Ar_3^+.	7797	PP
		BAr. External heavy-atom effect by Ar on the SO. MRCI.	9792	PT
		AlAr.	9170	DF+CCSD(T)
		GaAr, $GaAr^+$.	9160	PP,PT
		MAr^+; M=Fe,Co,Ni,Cu,Zn; Ar = argon	6853	DFT
		Ar^{8+}–C_{60} collision system.	9968	DFT
		Ar·I_2 ground-state potential. Linear and T.	8496	DK
		DF-level thermochemistry for Pople's G2/97 test set of 148 species, up to Ar (Z=18. Atomization energies.	8311	DPT
19	K	KF, KCl. EFG, Q(K) derived.	8328	DK
		NaK. Photodissociation.	9388	PT
20	Ca	$CaCl^{2+}$.	10227	PP
		CaX^{2+}, X=Br, I.	6645	PP
21	Sc	ScN.	7377	PP
22	Ti	TiF^{n+}; n=1-3.	9618	PP
		TiCl, TiH, TiH^+.	6861	DK
		TiF_3.	6896	DFT
		$H_2Ti(\mu\text{-}H)_2TiH_2$. SO.	10190	PT
24	Cr	Cr_2.	6673, 6675	PT
			9430	PT
		Cr_2.	9838	PP
		$Cr_2(O_2CH)_4$.	6674	PT
26	Fe	FeO^+ + H_2.	7376	PT
		FeL^+; L=Xe,CO_2,N_2,CH_4.	7457	PP
		Fe^+ + small alkanes.	8006	DFT
		FeCO, FeL^+; L=CO,H_2O,CH_4.	9395	PT
		$Mn^{2+/3+}$ and $Fe^{2+/3+}$ (aq) redox potentials and pK_a.	8614	DFT
		$[FeHgCp_2]^{2+}$, $[FeCp(C_5H_4)Hg]^+$ ferrocene derivatives.	8870	DFT
27	Co	CoH, CoH^-	7748	DK
		$HCo(CO)_4$. SO-induced radiationless transitions.	7997, 9390	PT
28	Ni	Ni^{2+}:MgO. Electronic spectrum.	8703	PP
		Ni_2^+: symmetry breaking	8890	PT
		NiH_2	6835	PT
		MH^+, MCH_3^+, MCH_2^+, M = Sc→Ni.	6972	PT

Element		Compounds	Ref.	Method
		NiC. Low-lying states. X $^1\Sigma^+$.	9729	PT
		Ni(C_2H_4 + ferrocene.	9245	PT
		$M(CO)_x$; M=Cr,Fe,Ni.	9208	PT
		Octahedral Ni(II).	9389	PT
		NiF_6^{4-}. Zero-field splitting.	9392	PT
29	Cu	Cu_2, Cu_2^+, CuH, CuF, CuCl.	8761	PP
		Cu_3, X $^2E'$, SO included. Geometric phase.	8421	PT
		M_2, MH; M=Ni,Cu.	9288	PT
		CuH.	8401	DPT
		CuCl. Low-lying excited states.	9804	PT
		CuX; X=Cl-I. Low-lying excited states.	9805	PT, DF+CI
		CuX; X=F-I. Lifetimes of excited $^1\Sigma$ states.	9357	PP
		CuF. EFG. CCSD(T).	9191	DF, DK
		CuX, Cu_2X; X=O-Po.	8762	PP
		AlCu.	6863	PT
		CuM, Cu_2M, CuM_2; M=Si-Sn.	9263	PP
		$[MO_2]^{3-}$; M=Fe→Cu in A_3MO_2 crystal field.	10167	DFT
		$(Cu_2S)_n(PR_3)_m$.	7411	PP
		$(Cu_2Se)_n$, $(Cu_2Se)_n(PR_3)_m$.	7412, 9581–9583	PP
		Cluster models for $Cu_2O(100)$. H_2O adsorption.	9079	PT
		Cu + CH_2N_2.	7311	PP
		CO_2 + $CuH(PH_3)_2$.	9536	PP
		$L_n(Cu_2O_2)L_n$ oxyhemocyanin models, L=NH_3,PH_3.	8697	DFT+PP
		Models of blue copper proteins. CASPT2. Test case $Cu(NH_3)_2(SH)(SH_2)^+$. Level shifts $\leq 1\%$.	9243, 9244	PT
30	Zn	Zn_2.	7339	PP
		ZnH.	8164	PP
		ZnH^+, ZnH_2, Zn_2H, HZnZnH, H_2Zn_2, H_4Zn_2.	7878	PP
		MX_2; X=F,Cl; M=Ca→Zn.	10168	DFT
		MAr^+; M=Fe,Co,Ni,Cu,Zn; Ar = argon	6853	DFT
31	Ga	GaAr, $GaAr^+$.	9160	PP,PT
		Ga_2P_2	7675	PP
		Ga_2P, GaP_2, their ions	7677	PP
		Ga_3P_2, Ga_2P_3	7678	PP
		Ga_3P, GaP_3	7681	PP
		GaF. Q(Ga) obtained.	9189	DK, DF+CCSD
		GaX_n; X=F,Cl; n=1-3. SR corr.	6858	PT, DK
		R_2Ga-GaR_2.	7286	PP
32	Ge	Ge_5.	7357	PP
		GeH_n; n=1-4, Ge_2H_n; n=1-6. Heat of formation.	9394	PT
		$GeCl^+$.	7368	PP
		GeCl.	8631	PP
		GeF.	8632	PP
		GeF^+.	10241	PP
		GeO, GeS. EFG, Q(Ge) derived.	8329	DK
		$GeCH^+$.	8149	PP+DFT
33	As	As_2, As_2^+.	9530	PP
		AsH.	6643	PP

Element		Compounds	Ref.	Method
		AsF.	8655	PP
		AsX_2, AsX_2^+, X = Cl, Br.	8574	PP
		As speciation in minerals and aqueous solution.	10002	PP
		EH_3 effective potential, E=N-As.	9659	PP
		$GaAs_2$, Ga_2As.	6724	MS
		Ga_2As_3.	6718	MS
		Ga_3As_2, Ga_2As_3.	8646	PP
34	Se	$Se_n^{0/-1}$; n=2,3.	7989	PP
		Me_2SeX_2; X=F-At.	8252	DFT
		$ClSeNSeCl^+$, a new cation.	7063	PP
		$(Cu_2Se)_n$, $(Cu_2Se)_n(PR_3)_m$.	7412, 9581–9583	PP
		Selenoiminoquinones: Se ... O interaction.	6820	PP
35	Br	Br_2. Lowest levels, tests new DK method.	9769	DK
		HBr.	9681	PP
		$CaBr^{2+}$.	6645	PP
		AsH^+, SeH, HBr^+, BrO, BrF^+, $NaBr^+$, Br_2^+. SO splittings.	6963	PP
		BrCl. EFG.	7741	DK
		BrO_3.	9140	PP
		PH_2Br.	7052	PP
		2nd-order SO for GeH_4, ... Br_2.	6963	PP
		Ni-Br collision. Very short R.	8503	DFT
		$C_2H_4 \cdots X^-$, halopyridines; X=F-Br. Heavy-atom effects on phosphorescence using AMFI SO.	9496	PT
36	Kr	KrF, KrF^+, KrF_2.	8036	PP
37	Rb	RbF, RbCl. EFG, Q(Rb) derived.	8330	DK
		MM'; M,M'=Li-Rb. Electric properties.	10053	PT
38	Sr	Sr_2.	7032	PP
		$SrAr^+$.	10238	PP
		SrF. RI-MP2.	10192	PP
		MO, OMO, MO_2; M=Be-Sr.	6677	PP
		L→Sr^{2+}; L = [R_3P=O], amides, pyridine.	6921	PP
		OPR_3 complexes of Sr^{2+}, $Sr(NO_3)_2$. R=H,Me,Ph.	8085	PP
39	Y	Y_n, n=1-4.	7354	PP
		YC_n; n=2-6.	9450, 9451	PP
		$MY_6I_{12}^{2-}$; M=Ru,Os.	6715, 6716	MS
40	Zr	Zr_3.	7353	PP
		Zr_5.	8767	PP
		ZrF^{n+}; n=1-3.	9618	PP
		MCl; M=Ti,Zr. Ground states $^4\Phi$, $^2\Delta$, respectively.	9532	PP
		Zr_2 + ethene, butadiene. Diels-Alder.	9448	PP
		ZrI_4, $Zr(TeSiH_3)_4$.	8248	DV DFT
		ZrX_6^{2-}; X=F-I.	7922, 9343	PP
		MO; M=Ti,Zr. IP.	8714	DFT
		Bridge bonding of N_2 to M_2 systems, M=Ti,Y,Zr.	6970	PT
		ZrN_2-L complexes.	6847	PP
		Cp_2MR^+; M=Zr. Ziegler-Natta catalysis.	8797	DFT
		$Cp_2ZrCH_3^+$ as model polymerization catalyst.	7777	PP
		Ethylene polymerization catalyzed by M-chelating	9002	PP

Element		Compounds	Ref.	Method
		alkoxide; M=Ti, Zr.		
		(p2n2)Zr($\mu-\eta^2$-N_2)Zr(p2n2) + H_2. (p2n2)=$(PH_3)_2(NH_2)_2$	6846	PP
		$(NH_2)_2M$=NH; M=Ti,Zr CH_4 adducts	7312	PP
		Cl_2Zr=E; E=MH_2 (M=C-Sn), MH (M=N-Sb), M (M=O-Te)	6910	PP
		$Zr(OH)_2(H_2O)_2Cl_2$. Distortion from C_{2v} intrinsic.	7656	PP+DFT
		L_nZr=NX + CH_4	7318	PP
		Zr@C_{28}	7077	PP
41	Nb	Nb_n; n=8-10	7898	PP
		Nb_3O, Nb_3O^+.	10265	DFT
		NbN	7719	PP
		MO, MO_2, MO_2^+; M=Ti,V,Zr,Nb. Reactions with C_2H_4, C_2H_6	7967	PP
		NbO. IP.	8714	DFT
		NbCO	9932	PP
		$NbCl_5$	7868	DK
		Nb_4C_4, Nb_5C_6, Nb_6C_7	7103	PP
		[Cp_2NbH_3] + Lewis acid A;	7111,7112	PP
		A = $HBO_2C_2H_2$, BF_3, BH_3. H_2 elimination		
		Nb_n^+ + C_6H_6; n=1,2.	9460	PP
42	Mo	$MoCl_5$	7644	PP
		MoC. Low-lying states.	9728	PT
		MoC_4	8611	PP
		MCO, MCS; M=Y-Mo	8191	PP
		MoO. IP.	8714	DFT
		MoX_n^+; X=O,S; n=1-3.	8477	PT
		Mo_2 + ethene, butadiene. Diels-Alder.	9446	PP
		M_n + N_2; M=Nb, Mo; n=1-4.	6916	DFT
		Solid MoO_3. Adsorption of CO, H_2O on (100) surface.	9153	PP
		M_8C_{12}; M=Y→Mo.	8662	PP
		$Mo_2Cl_4(PH_3)_4$.	8702	DFT+PP
		$MMo(O_2CH)_4$;M=Cr,Mo. $Mo_2X_4(PH_3)_4$; X=Cl-I.	9757	PP
		MoO_4^{2-}, MoS_4^{2-}, $MoOCl_4$.	9758,9759	PP
		$M(CO)_6$; M=Cr-W. $M(CO)_5L$; L=N_2,CS,NO^+. DFT by	10083,10084	DFT
		DPT. For comments on the method, see.[10076]		
		$M(CO)_6$; M=Cr,Mo.	10084	PP
		MoO_2X_2; X=F-I.	9998	PP
		H/D exchange in $CpMoH_3(PMe_3)_2$.	6583	PP
		$Mo_2Cl_4(PH_3)_4$ models for diphosphine complexes.	7428,7429	PP+DFT
		[$Mo_2Cl_4(PH_3)_4$].	8174	PP
		$L_4Mo\underline{\underline{4}}MoL_4$; L=$PH_3$,Cl.Mo-Mo quadruple bond.	6964	PP
		MCSCF required. UV, PES interpreted.		
		Mo(VI)H_4O +PH_3, Mo(VI)H_2O_2 +PH_3.	9246	PP
		$(NH_3)_2(SH)_2$Mo(VI)O_2 +H_2O.	9247	PP
		^{95}Mo chemical shifts.	7078,7079	DFT
		CpMXY, CpMXYZ; M=Co,Ru,Rh,Mo; X,Y,Z=Cl,Me,	8873	DFT
		NO,CO. Functionals compared.		
		Bispentalene dimolybdenum, $Mo_2(C_8H_6)_2$. Mo-Mo 234 pm.	7248	DFT
		MO_2X_2 + CH_3OH→$MX_2(OH)_2$+CH_2O. M=Cr,Mo;X=Cl.	7435	DFT+PT
43	Tc	$TcX(NR)_3$	6908	PP
		$[TcYX_4]^q$; Y=N,O; X=Cl,Br. Opt. transitions	6795	DFT

Element	Compounds	Ref.	Method
	$M_2(O_2CH)_4$, $M_2Cl_4(PH_3)_4$; M=Nb→Tc	7287	PP
44 Ru	RuN	9351	PP
	$[RuO_4]^{n-}$, n=0-2. Ground state for $n = 2$?	7408	DFT
	RuCO	9935	PP
	$M(CO)_5$; M=Fe,Ru	10084	PP
	$Y_2(PX_3)_2Ru{=}CZ_2$ alkylidenes; X,Y,Z=H,Me,F-Cl.	6909	PP
	MAr_n^+; M=Nb,Rh; n=4,6. Ar = argon	6929	PP
	$[Ru(NH_3)_6]^{2+/3+}$ redox pair	7060	PP
	$RuHX(CO)(PR_3)_2$; X=F-I, OH, OPh, SPh,...	9289	PP
	$Ru(H)_2(CO)_4$. Classical dihydride.	8577	DFT+PP
	Ru silylenes $(Cp)L_2Ru{=}SiX_2^+$	6708	DFT,PP
	$[Ru(P\text{-}P)_2\text{"}H_3\text{"}]^+$, P-P = diphosphine ligand	8842	PP
	$[Ru(bpy)_3]^{2+}$, bpy=bipyridine.	7402	DFT
	Cp_2M; M=Mn→Ni, Ru. Hyperpolarizabilities.	8856	DFT
	Cp_2M; M=V...Ru.	8869	DFT
	$Cp_2Ru_2(\mu\text{-}H)_4$. Anal. 2nd derivatives.	7311	PP
	$[Ru(H\cdots H)Cp(H_2PCH_2PH_2)]^+$. Nuclear dynamics.	7823	PP
	$LRh\text{"}H_4\text{"}$; L=Cp,Tp.	7824	PP
	Fluxional $TpRu(PPh_3)\text{"}H_2SiR_3\text{"}$ complexes. Tp=Hydridotris(pyrazolyl)borate.	9055	PP
	$Ru(CO)(PH_3)_2$ + benzaldehyde.	8852	PP
	$MO_2X_2 + CH_3OH \rightarrow MX_2(OH)_2 + CH_2O$. M=Ru;X=O.	7435	DFT+PT
45 Rh	Rh_4.	7358	PP
	Rh_5.	8765	PP
	Rh_n^+, n=3-5.	8769	PP
	RhC.	9934	PP
	RhN. Low-lying states.	9730	PT
	$RhCO^q$; q=-1,0,+1.	10349	DFT+PP
	$RhCl_6^{3-}$	9391	PP
	$Rh_2 + H_2$.	7134	PP
	Rh_3 + CO.	8766	PP
	$Rh^{0/+/-}$ + CO, $+N_2$.	7655, 8874	PP
	$Rh^+ + CH_4$.	10202	PP
	Butadiene + Rh_n; n=1,2.	7365	PP
	$CH_4 + Rh_n$.	6745	DFT
	Benzene + Rh^+, Rh_2^+.	8771	PP
	Rh alkene complex NMR shifts, UV spectra. CASSCF.	6614	PT
	$CpRhCO + CH_4$.	9802	PP
	$Cp^*Rh(PMe_3)$ + thiophene.	9569	DFT+PP
	$CO_2 + [RhH_2(PH_3)_3]^+$ or $RhH(PH_3)_3$	9536	PP
	$RhXL + CH_4$	7627	PP,DFT/MR-CI
	$[HRh(CO)_4]$ +CO $+H_2$.	9240	PP
	$[RhCl(CO)_3]_2$. $d^8 - d^8$.	9073	PP
	$[CpRh(C_2H_4)(\eta^2\text{-}C_2H_5)]^+$. H exchange.	8663	PP
	$[[Rh_2Cp_2(\mu\text{-}CH_2)_2]_2(\mu\text{-}S_4)]^{2+}$.	8878	PP
	$M_2(HNCHNH)_4$, $M_2(HNNNH)_4$; M=Nb →Rh.	7287	PP
	$RhCl(PH_3)_2$ as activator of SiH, SiSi and CH bonds.	8415	PP
	$[Rh(PH_3)_2]^+$ complexes of N-alkenylamides.	8394	PP
	$Rh(CO)_2^-$, $Rh(CO)_2F_2^{3-}$.	8746	PP

Element	Compounds	Ref.	Method
	$Rh_6(CO)_{16}$.	8284	PP
46 Pd	Pd_2.	9019	DFT+PP
	Pd_n; n=2,4. R/NR	10239	DFT
	Pd_5.	7362	PP
	Pd_n; n=1-6.	10062	PP
	Pd_n; n=1-6 + S, Cl.	10063	PP
	Pd_n; n=1-6. Ethylene adsorption.	7646	DFT
	Pd_n; $n \leq 147$.	8482	DFT
	PdH.	7714	DK
	PdH.	9768	DK/DF+CI
	MO, MCH_2; M=Y→Pd.	9742	PT
	MO_n; n=1-4; M=Y→Pd.	9743	PT
	MH, MO; M=Ni,Pd.	10154	PP
	PdC.	9931	PP
	MHX; M=Y→Pd; X=F,Cl.	9746	PT
	$[PdCl_4]^{2-}$. Cl exchange.	7409	DFT
	$Pd_3 + H_2$.	7363	PP
	$Pd_n + H_2$, n=2,3.	7136	PP
	PdMgO, PdOMg.	8715	PP
	M_4 on MgO(001) surface; M=Ni,Pd.	8860	DFT
	Pd, Pd_4 on MgO(001) surface.	10302	DFT
	PdCO, $PdCH_3$.	7743	PP
	$Pd(CO)_2$, $Pd(CO)_2F_2^{2-}$.	8746	PP
	$M(CO)_4$; M=Ni,Pd.	10084	PP
	CO insertion into M-H and M-CH_3 bonds, M = Y→Pd.	6969	PT
	$Pd(NO)(CH_3)(PH_3)$ + CO substitution reaction.	7744	PP
	Reductive elimination of $Pd(XH_3)(\eta^3\text{-}C_3H_5)(PH_3)$; X=C-Sn.	6959	PP
	Small hydrocarbons +M, M=Y→Pd.	7128	PT
	M + C_2H_2; M=Y→Pd.	9744	PP
	MX_2; M=Y→Pd; X=H,F,Cl.	9745	PP
	MH_2, MHX; M=Y→Pd; X=F,Cl.	9747	PP
	M-CH_3, M-C_2H_3, M-C_2H; M=Y→Pd.	9748	PT
	$MH_n + H_2$, CH_4; M=Y→Pd.	9750	PP
	$PdCl_2(H_2O)_n$, $PdHCl(H_2O)_n$.	9751	PP
	$CH_4 + M^+$; M=Y→Pd.	6971	PP
	Pd_2 + CO.	7366	PP
	Pd_2, PdCu or Cu_2 + CO or NO.	9415	PP
	$Pd_2 + C_6H_6$. $Pd_2 + C_6H_6$/Ag.	8020	PP
	$PdCl_4^{2-}$.	9839	PP
	$Pd(H)_2(Cl)(NH_3)^-$, $Pd(H)_2(NH_3)_2$.	8911	PP
	$Pd_2(\mu\text{-Br})(\mu\text{-}C_3H_5)(PH_3)_2$, $PdCl(\eta^3\text{-}C_3H_5)(PH_3)$.	9540	PP
	M=CH_2^+; M=Y→Pd.	7627	PP
	$PdCH_2I^+$. Strong Pd-C bond.	9627, 9628	DFT,PP
	$[L_2MCH_3]^+$ + ethylene reaction.	7776	PP
	Nucleophilic attack on cationic η^3-allylpalladium complexes.	7940	PP
	Pd acetate catalyzed Wacker reaction.	8464	PP
	Pd acetate catalyzed acetoxylation of ethene.	8465	PP

Element	Compounds	Ref.	Method
	Solid $(Me_4N)[M(dmit)_2]_2$; M=Ni,Pd, 'dmit' = 2-thioxo-1,3-dithion-4,5-dithiolate.	8938	PP
47 Ag	Ag_2.	9502,10329	PP
	Ag_n, Ag_n^+.	7006–7009	PP
	Ag_3 hfs.	7050	DSW
	Ag_6.	7810,7811	PP
	Ag_7.	6722	MS
	Ag_7.	7024	DFT,PP
	Ag_n; $n \leq 9$.	9287	PP
	M_n; M=Ru,Pd,Ag; n=55,135,140.	8183	PP
	M_4 on MgO(001) surface; M=Cu,Ag.	8860	DFT
	AgH.	8949	DF
	AlAg.	6863	PP
	AgF.	9358	PP
	AgF. RI-MP2.	10192	PP
	Solid AgX; X=Cl-I.	9076,9077 10135	PP
	$(Ag_nBr_p)^{\pm}$; $n, p \leq 2$.	9335	PP
	AgO^+, $[AgO\{N(=CH_2)(CH_3)\}_2]^+$.	7321	
	AgHe.	8158	PP
	HeAgHe.	9220	PP
	AgCO.	8798	DK
	$AgNH_3^{0/+1}$.	9405	PP
	$M_n(NH_3)$; M=Cu,Ag; n=1-4.	7148	PP,DFT
	M-NO_2; M=Cu,Ag.	9421	PP
	$Ag(CN)_2^{q-}$; q=1,2.	9382	PP
	M_2X_2; M=Cu,Ag; X=Cl-I. $[Ag_2Br_2](PH_3)_3$.	7596	PP
	Ag_2-L; L=NH_3, CO, C_2H_4, H_2O.	7030	PP
	M^+L; M=Cu,Ag; L=H_2O, OH^-, CO.	7884	PP
	Sulfur oxide anions adsorbed on Ag cluster models	9090	PP
	Ag^+-L for 18 ligands.	8753	PP
	$Ag(C_6H_6)_n^+$; n=1,2.	8754	PP
	Ag^+ + benzene, other hydrocarbons.	8031	PP
	M^+-C_5H_5N N-M complexes; M=Cu,Ag. HF/MP2/B3LYP.	10272	PP
	$(Ag^+)_n$ + butadiene complex, n=1,2.	9108	PP
	Ag_2 + olefin + O, O_2. Oxidation, epoxidation model.	9023	PP
	Ag_4 + C_6H_6.	8020	PP
	$[Ag_4(\mu_4\text{-E})]^{2+}$, its diphosphine complex. E=S-Te.	10157	PP
	Ag_4Ph_4	6897	DFT
	$[Ag(NHCHNH)]_2$, $[Ag(dmtp)(NO_3)]_2$.	7597	PP
	Bimetallic cluster $[Ag_{13}\{Fe(CO)_4\}_8]^{n-}$, n=0-5.	6627	DFT
	M-$(Mg_{13}O_{12})$ surface model; M=Rb,Pd,Ag.	7690	PP
	M_n-MgO(001); n=2,4; M=Pd,Ag.	7692	DFT
	Ag_2/'AgBr'.	9715	PP
	HCHO + Ag_2.	9718	PP
	MgO(100)/Ag(100) and MgO(110)/Ag(110) interfaces.	9300	PP
	Solid DCNQI-M; M=Cu,Ag; DCNQI=dicyanoquinonediimine.	8939	PP

Element		Compounds	Ref.	Method
		Pd_2 $+C_6H_6$. Pd_2 $+C_6H_6/Ag$.	8020	PP
48	Cd	Cd_2.	7337	PP
		CdH^q; q=-1,0,+1.	7600	DFB+CCSD
		CdRg, Rg=He-Xe.	7342	PP
		MO, MO_2; M=Zn,Cd.	7208	PP
		CdE; E=S-Te. Nanocrystalline clusters. SO.	9988, 9990	PP
		MX_2, $(MX_2)_2$; X=H,F-I.	8309	PP
		CdX_2; X=F,Cl. $CdF_2Cl_2^{2-}$. EFG. CCSD(T).	8002	PT
		Cd-(en), Cd-NH_3 exciplexes. 'En'=ethylenediamine.	9925	PP
		Cd + CH_4.	9359	PP
		Cd + H_2.	9360	PP
		Cd^+ + C_6H_6.	8031	PP
		$M^{2+}L$; M=Zn,Cd; L=H_2O, OH^-, CO.	7884	PP
		M^{2+}:M'O; M=Cu,Ag; M'=Mg-Sr. Impurities.	9171	PP
49	In	InR, $(InR)_4$; R=H,Me.	10042	PP
		MeM; M=Ga,In.	8024	DFT+PP
		In_2P_2.	7676	PP
		In_3P_2, In_2P_3.	7679	PP
		In_2P, InP_2.	7680	PP
		In_3Sb_2, In_2Sb_3, their cations.	7682	PP
		$In_2As_2^n$, n = -1, 0, +1.	7683	PP
		InH, InF, InCl.	8600	PP
		InCl, $InCl_3$.	6859	DK, PP
			9639	PP
		InX_n, X=H,Cl,-CH_3; n=1-3. SR.	6860	DK
		MX_4^-; M=Ga-In; X=Cl-I. Ga, In NMR shifts.	9926	PT
		$InMH_6$, MBH_6; M=B-In.	6907	PP
		X_3M–D; M=Al-In; X=F-I; D=YH_3,YX_3,X^-; Y=N-As. Donor-acceptor complexes.	9972	PP
		Univalent ligands MeM, (Cp)M, (Cp*)M, $(H_3Si)_2NM$; M=B-In. Their $Fe(CO)_4$ complexes.	8755	DFT+PP
		In^+ complexes with H_2O, CH_3OH, other organics	6731	PP
		Hydration of Group 2 - Group 13 ions. $M(H_2O)_6^{n+}$. M=Ca→Ga, Sr→In.	6616, 6617	PP,PT
50	Sn	Solid C - Sn. Cohesive energy.	9175	PP
		SnH.	6639	PP
		MH; M=C-Sn, Te. Test SO PP program.	9924	PP
		SnF^+.	6793	PP
		SnF.	7352	PP
		Sn_5.	7359	PP
		M_5^+, M=Ge,Sn.	7361	PP
		MH_2; M=C-Sn. SO splitting.	8434	PP
		SnH_4.	7599	DF+CCSD
		SnH_4. Freq. dep. polarizability.	10129	RRPA
		MH_3 + MH_4; M=Si-Sn.	7369	PP
		MH_4; M=C-Sn. Infrared intensities.	7437	PP
		M_2, CuM, Cu_2M, CuM_2; M=Si-Sn.	9263	PP
		MX_2; M=Si-Sn; X=F,Cl.	9741	DFT
		ML_2; M=Ge,Sn; L=H, $P(SiH_3)_2$.	7508	PP

Element	Compounds	Ref.	Method
	MOH^+, HMO^+; M=C-Sn.	9422	PP
	$Sn[E(SiH_3)_2]_2$; E=N, P.	7508	PP
	HZnMH; M=C,Sn.	9556	PP
	Isomers of vinyl stannane.	8818	PP
	$(CO)_5Cr{=}MH_2$; M=C-Sn.	8153	DFT
	Reductive elimination of $Pd(XH_3)(\eta^3\text{-}C_3H_5)(PH_3)$; X=C-Sn. Hypervalent behaviour of X.	6959	PP
	M_8C_{12}; M=Si-Sn.	6988	PP
	σ-bond activation of $(HO)_2B\text{-}XH_3$ by $M(PH_3)_2$; X=C-Sn; M=Pd,Pt.	9533	PP
51 Sb	SbO.	6637	PP
	SbH.	6644	PP
	SbI.	7387	PP
	Sb_2.	7390	PP
	Ozone-like, Sb_3^{3-} compounds. M_3Sb_3; M=Na-Rb. $Na_4Sb_3^+$.	7938	PP
	Sb_4 and isoelectronic species.	7939	PP
	In_3Sb_2, In_2Sb_3, their cations.	7682	PP
	Solid EM semiconductors; E=B-In; M=N-Sb. Cohesive energy.	9176	PP
	Sb speciation in sulfidic solutions.	10000	PP
	SbF.	7031, 8572	PP
	SbF_2, SbF_2^+.	8573	PP
	SbX_2, SbX_2^+; X=Cl,Br.	6792	PP
	$(Sb_nO_m)^q$, q=0,1.	8239	PP
	CH_3EH_2, $CH_2EH_2^+$; E=N-Sb.	8265	PP
52 Te	H_2Te.	8455	PP
	H_2X; X=O-Te. H and X NMR shifts. 2nd and 3rd order SO effects. SOO.	10059	PT
	H_2X; X=O-Te. NMR spin-spin coupling. 2nd and 3rd order SO effects. SOO.	10060	PT
	Tetra-atomic $(X_2)(Y_2)$ and (XY)(XY) clusters; X,Y=O-Te.	9107	PP
	MF_6, MF_6^{2-}; M=Se,Te.	8399	PP
	MCl_6, MCl_6^{2-}; M=Se,Te.	8400	PP
	$W(PH_3)_4E_2$; E=O,S,Te. W(IV), d^2.	8248	DV DFT
	$TeCl_4$.	8448, 8449	PP
	CH_2EH^+; E=S-Te.	8265	PP
	$CH_2(EH)_2$; E=O-Te.	9547	PP
	CdE; E=S-Te. Nanocrystalline clusters. SO.	9988, 9990	PP
	$Te(SR)_2$; R=H,Me. $Te(SH)_4$ thermodynamically unstable.	7716	PP
	$HX\text{-}CH_2\text{-}CHO$; X=S-Te. No significant X...O interactions.	8813	PP
	$[C(ER)_3]^+$; E=O-Te.	9084	PP
	X_2Y_2; X=N-Sb; Y=O-Te. Planar or butterfly?	8889	PP
	CH_3 + CH_3EH; E=S-Te.	9776	PP
	$C_6H_4S_2Te$.	9622	PP
	$HESi(SiH_3)_3$; E=O-Te. Photoelectron spectrum.	9970	PP
	Vibr. freq. of TM chalcogenides; M=Ti...Os; Ch=O-Te.	7325	PP
	^{125}Te NMR shifts.	9483	SR DFT
53 I	I_2, I_2^-, I_3^-, I_5^-.	7374, 9713	PP
	I_2 excited states.	9952	PP
	I_2. HF to CCSD(T) level.	7418	DF

Element	Compounds	Ref.	Method
	I_n^-; n=3,5.	9897	PP
	X_2; X=F-I. 7-VE PP.	7472	PP
	X_2, HX; X=F-I	7473	PP
	I_2, HI. Spin-orbit effects via ZORA.	10077	DFT
	Interhalogens up to IBr. Up to CCSD.	7416	DF
	Interhalogens up to IBr. Up to CCSD(T).	9186	PT
	HX, XY, X_2 complexes with NH_3. X,Y=F-I	10338	PP
	IF	7106	PP
	IF and ICl. SR transition moments for B $0^+(^3\Pi) \rightarrow$ X 0^+ $(^1\Sigma^+)$.	10303	PP
	I_2^-, ICl^-. Ground and excited states.	8846	PT
	X_3^+, XY_2^+; X,Y=F-I	8621	PP
	Ar·I_2 ground-state potential. Linear and T.	8496	DK
	I·CO_2, I·OCS van der Waals complexes.	9557	PP
	Solid MX; M=Li ... Rb; X=F ... I.	7486	PP
	$Na_nX_m^q$; X=Cl-I; q=0,+1. n=1-3, m=1-2.	10203	PP
	CaI^{2+}.	6645	PP
	CuX^-; X=F-I. Cu_5-I^-.	8097	PP
	GaX_2, GaX_3, X=Cl-I.	7349	PP
	HX, CH_nX_{4-n}; X=F-I. SO-induced ^{1}H, ^{13}C NMR shifts.	8774	DFT
	HX, CH_3X; X=F-I. SO-induced ^{1}H, ^{13}C NMR shifts. Importance of two-electron SO terms.	8775	DFT
	HX, CH_3X. ^{13}C NMR shifts. SO.	10221	DFT
	HX; X=Cl-I. EFG.	8326	PT,DK
	HX; X=Cl-At. Picture-change effect on EFG.	8327	
	HX, CH_3X; X=F-I. ^{1}H and ^{13}C NMR shifts.	9026	PT
	HX, CH_3X; X=F-I. H, C and X NMR shifts. 2nd and 3rd order SO effects.	10059, 10061	PT
	HX; X=F-I. NMR spin-spin coupling. 2nd and 3rd order SO effects. SOO.	10060	PT
	HX; X=F-I. NMR shielding and spin-spin coupling.	10124	RRPA
	EFG in HX; X=Cl-I. MP2 to CCSD(T). R/NR.	10125	DF+corr.
	HI, CH_3I.	8371	PP
	HI. Frozen-core DF test.	10182	DF
	Pseudopot. vs. all-electron on Br, I compounds. G2 set.	7839	PP
	CH_3I SO splitting.	9920	PP
	CX_3^+; X=F-I. ^{13}C NMR shifts.	8298	DFT+PP
	CX_4, CX_4^-; X=Cl-I. PES.	9458, 9461	PP
	CF_3I Rydberg levels. SO.	9918	DFT
	CD_3X; X=F-I. Deuterium EFG.	10058	PP
	EtX; X=F-I.	8098	PP
	IF^+, ICF_n^+; n=1-3.	7674	DFT+PP
	CF_3X^-; X=Cl-I.	9459	PP
	AlX_4^-; X=H,F-I. ^{27}Al NMR shifts.	9022	PT
	PX_4^+; X=F-I ^{31}P NMR shifts. SO.	8292	DFT+PP
	CF_3IF_n; n=0,2,4,6. ^{13}C and ^{19}F NMR shifts.	8300	DFT+PP
	Iodo-organic compounds. ^{1}H and ^{13}C NMR shifts. SO analyzed.	8301	DFT
	I_2HSi-SiH_2I.	8193	PP

Element	Compounds	Ref.	Method
	$I(N_3)_2^+$.	9996	PP
	X-N_3; X=H,F-I.	9624	PP
	ICN, INC.	9549	PP
	HOX; X=F-I.	7840	PP
	XO_n^-; X=Cl-I; n=1-4.	9080	PP
	H_2CI_2 isomers.	7838	PP
	PI_2, PI_2^+.	8575	PP
	$P_2I_5^+$.	6746	PP
	BX_2, BX_2^+, X=Br,I.	8593	PP
	$I(N_3)$.	7098	PP
	$PdCH_2I^+$.	9628	PP
	IF_4^-, IF_5^{2-}.	7231	PP
	AlX_4^-; X=F-I. Al NMR shift.	9022	PP
	SiX_4, $SiXI_3$. Si NMR shift.	9025	PP
	TiX_4; X=Br-I. PES.	9020	PP
	TiX_4; X=F-I. Ti NMR shift.	9024	PP
	NbX_6^-, $NbCl_5X^-$; X=F-I. Nb NMR shift.	9024	PP at X
	IOF_5.	7226	PP
	IOF_6^-, IF_7, XeF_5^-.	7228	PP
	TeF_7^-, IF_7, XeF_7^+ are D_{5h}.	7230	PP
	$XONO_2$; X=F-I.	7570	PP
	CH_2X^+; X=F-I.	8265	PP
	S_N2 reaction X^- + CH_3X, X=F-I.	7434	DFT
	$I^- \cdot CH_3I$.	8066	PP
	Reactions of iodo-lithio-ethenes.	7116, 7117	PP
	$C_6H_4X^-$ (benzyne + halide); X=F-I.	8973	PP
	$C_6H_6{\cdot}I_2$ complex.	8877	PP
	Vibr. freq. of 50 inorg. molecules as test of PP.	7102	PP
	$I(^2P_{3/2})$+O_2(a $^1\Delta_g$). $\leftrightarrow$ $I(^2P_{1/2})$+O_2(X $^3\Sigma_g^-$).	8244	PT
	Hydrated halogenide complexes. AE/PP.	8519	PP
54 Xe	RgH^+; Rg=Ar-Xe.	8735	PP
	RgH; Rg=Ne-Xe.	8359	PP
	XeH. Low-lying states.	10023	PP
	XeX^q; X=F-I; q=-1,0,+1.	9617	PP
	XeO, XeS. Low-lying states.	10256	PP
	$RbXe^+$.	6791	PP
	BRg^+; Rg=Ar-Kr.	8436	PP
	$(Rg)_2$; Rg=Ne-Xe	7088, 9437	PP
	$(Rg)_2$; Rg=Kr,Xe	8940	PP
	CdRg, Rg=He-Xe.	7342	PP
	$RgNO^+$; Rg=He-Xe	8592	PP
	$RgAu^+$, $RgAuRg^+$; Rg=He-X. First Au-Rg bonds predicted.	9303	PP
	RgBeO; Rg=Ar-Xe	10096	PP
	HXeH	9231, 9488	PP
	RgF_2; Rg=Kr,Xe	7093	PP
	XeF_2.	8800	PP
	XeF_4	7094	PP
	XeF_n; n=2,4,6.	9848	DFB

Element	Compounds	Ref.	Method
	$[XeOF_5]^-$	7229	PP
	$XeHXe^+$	9058	PP
	$RgHRg^+$; Rg=Ar-Xe.	8738	PP
	$Xe_2H_3^+$ ($HXe{\cdot\cdot}H{\cdot\cdot}XeH^+$).	8734, 8737	DFT+PP
	HXeCl.	8196, 9232	PP
	HXeOH.	9227	PP
	HXeSH.	9229	PP
	HKrCN, HXeCN, HXeNC.	9228	PP
	FXeX; X=N_3,NCO,OCN.	9623	PP
	F_2C=C-Xe.	9552	PP
	HXeX; X=Cl-I. HKrCl.	9230	PP
	$RgHX^+$; Rg=Ar-Xe; X=Cl-I.	8739	PP
	$C_6F_5Rg^+$; Rg=He-Xe.	7786	PP
	XeF^+, $XeCF^+$, $XeCF_2^+$. Cleavage of C-F bonds.	7674	DFT+PP
	XeH_2 - H_2O dihydrogen-bonded (O-H··H-Xe) complex.	8736	PP
	$M(CO)_5$-Rg; M=Cr-W; Rg=Ar-Xe.	7576	PP, DFT
55 Cs	MH; M=Li-Cs.	7471	PP
	MF, M=K-Cs.	7533, 8599	DF/PT/HF;PP
	CsF. Q(Cs) obtained.	9190	DK CCSD(T)
	CsAu.	9573	DF/HF
	ML; M=Li-Cs; L=H,Me,NH_2,OH,F,Cp.	8561	PP
	M_nX_m; M=Li-Cs; X=Se,Te, $n, m \leq 2$.	8095	PP
	M_nX_m; M=Li-Cs; X=As,Sb, $n \leq 2$, $m \leq 4$.	8096	PP
	CsRg; Rg=Ne-Xe.	7569	PP
	M^+-CO; M=Li-Cs.	7691	PP
	M^+-L; M=Li-Cs; L=adenine, guanine.	7087	PP
56 Ba	Ba_2.	6657	PP
	Ba_n, n=2-14.	7033	PP
	Solid Ba.	10318	PP
	BaLi.	6656	PP
	BaRg; Rg=He-Xe.	7336, 7338 7341	PP
	BaH^+.	6658	PP
	BaI^{2+}.	6646	PP
	MX_2; X=H,F; M=Ca-Ba.	7101	PP
	MX_2; M=Sr,Ba; X=F-I. All floppy; SrF_2 and BaF_2 to $BaBr_2$ are bent.	8937	PP
	MCN, MNC; M=Be-Ba.	8570	PP
	$M(CN)_2$, M=Be-Ba.	8266	PP
	MH, M_2H_4; M=Mg-Ba.	8304	PP
	M^{2+}-L; M=Mg-Ba; L=adenine, guanine.	7087	PP
	$MCl_2 \cdot nH_2O$; M=Sr,Ba; n=1,2. Madelung potential.	8961	PP
	Hydrated M^{2+} + adenine, thymine base pairs; M=Mg-Ba.	9809, 9810	PP
	$Ba_6Li_3O_2$ cluster with ligands.	6986	DFT
57 La	La_n; n=2-13. Used to derive potential for MD simulations.	7628	DFT-DVM
	LaO, LaO^+.	8819, 9585	PP
	MO^q, MO_2^q, MO_3^q species; M=Y,La.	6680	DFT,PP
	LaX_3, X=F,Cl.	7454	PP
	LaX_3; X=Cl-I.	8451	PP

Element		Compounds	Ref.	Method
		MF_6^{3-}; M=Sc,Y,La.	7922	PP
		Cp_2LaL; L=$CH(SiH_3)_2$.	7453	DFT,PP
		LaC_2.	9453	PP
		LaC_n^+; n=12,13.	9454	PP
		LaC_n; n=2-6.	9455	PP
		LaC_n; n=2-8.	10234	PP
		LaC_2, LaC_2^+.	10236	PP
		LaC_3^{q+}; q=1-3.	10235	PP
		MC_n; n=3-6; M=Y,La.	9846	PP
		La@C_{82}.	9266	PP
58	Ce	CeO.	7450, 9531	PP
		CeO, CeF.	7346	DFT
		CeO^+, $CeO_2^{0/+1}$.	7986	PP
		MO_8^{12-}; M=Zr,Ce. Models for solid MO_2.	9017	DV DFT
		Ce^{n+} in alkaline-earth fluorides.	10133	DF
		CeX_n, n=3,4; X=F,Cl.	8569	PP
		$NaCeCl_4$.	8261	PP
		$Ce(Cp)_3$, $4f^1$.	8255	DV DFT
		Ce@C_{28}. f-Orbital covalency.	9435	DFT
62	Sm	Sm(III)-catalyzed olefin hydroboration.	8490	PP
		Ethylene insertion into Sm-C bond of $H_2SiCp_2SmCH_3$.	8413	PP
		SmF_6^{4-} cluster models.	7510	
63	Eu	Diatomic EuE; E=O-Te. $5d$ important.	7345	DFT
		$[EuO_6]^{9-}$ +Madelung cluster model for Eu^{3+} in Ba_2GdNbO_6. The f^6 spectrum.	7460	DF
64	Gd	GdF. DF. Gd $5d \to$F $2p$ donation, F $2s$ - Gd $5p$ hybridization.	9948	DF
		GdF_2.	9949	DF
		$[Gd(H_2O)_9]^{3+}$.	7281	PP
		$M(C_6H_6)_2$; M=Y,Gd.	7455	PP
		Gd(III) polyamino carboxylates.	7282	PP
67	Ho	Ho_2.	9048	PP
		C-F bond activation by Ln^+; Ln=Ce,Ho.	8011	PP
69	Tm	LnF_n; Ln=Er,Tm; n=1-3.	8603	PP
70	Yb	Yb_2.	10172	PP
		Yb_n; n=3-7.	10173	PP
		LnO; Ln=La ... Yb. YbF. R/NR.	10164	DFT
		LnO, LnF; Ln=Ce ... Yb.	7483	PP
		LnO, LnS; Ln=Eu,Yb.	8691	DFT
		YbF. PNC experiment.	8982, 9974	PP
		YbF. PNC.	9162, 9332	DF
		YbH,YbF,YbO.	8689	PP
		L$\to Ln^{3+}$, Ln=La,Eu,Yb; L = [R_3P=O], amides, pyridine.	6921	PP
		Dithiophosphinate Ln^{3+} complexes. Ln=La ...Yb.	6992	PP
		$Ln(C_8H_8)_2$; Ln=Nd,Tb,Yb. Excited states. Lanthanocenes are $4f^n\pi^3$, actinocenes $5f^{n-1}\pi^4$.	8687	PP
		$Ln(C_8H_8)_2^-$; Ln=Ce,Nd,Tb,Yb. Excited states.	8688	PP
71	Lu	LnH, LnF, LnO; Ln=La,Gd,Yb,Lu. Lanthanide contraction.	10165	DFT
		LnH, LnF, LnO; Ln=La,Lu. Lanthanide contraction.	8485	PP

Element	Compounds	Ref.	Method
	LnO, $LnO^{\pm}$, LnO_2, LnO_2^-, LnO_3^-, $(LnO)_2$. Ln=Ce-Lu.	10218	DFT
	MF, MH, MH_3; M=La,Lu,Ac,Lr. Lanthanide and actinide contractions.	8550	DF+MP2
	LnX_3; Ln=La-Gd,Lu; X=F,Cl.	8568	PP
	LnX_3; Ln=La,Gd,Lu; X=F-I. DFT vs. MCSCF or CISD+Q.	6590,6591	PP,DFT
	LnX_3; Ln=La-Lu; X=F,Cl.	8229,8230 10021	PP
	LnX_3; Ln=Ce-Lu; X=F-I.	7327	PP
	$Ln(C_6H_6)_2$, Ln=La ... Lu.	8047	PP
	$Ln(C_6H_6)_2$, Ln=La ... Lu. R/NR. Dominated by Ln $5d \rightarrow \pi^*$ back donation.	8721	DFT
	Ln-CO, Ln-OC; Ln=La,Gd,Lu.	8046	DFT
	$M_2@C_{80}$; M=Sc,Y,La,...Lu.	8406,8408	PP
	$M@C_{82}$; M=Sc,Y,La,...Lu.	8407	PP
72 Hf	R/NR orbital energies for M_6, M=Ti-Hf.	6589	DFT
	Hf_3.	7367	PP
	MH_n; M=Zr,Hf; n=1-4.	7206	PP
	MCl_4; M=Ti-Hf. Vibr. freq.	9489	PP
	MF_4; M=Ti-Hf. PES.	7535	DFT
	$HfCl_4$. PES.	8251	DFT
	MX_4; M=Ti-Hf; X=Cl,Br.	8694	DFT
	MX_4; M=Zr,Hf; X=Br,I. PES.	9871	SR SW
	HfX_6^{2-}; X=F-I. Salts	7923	PP
	MN, NMN, $M(N_2)$, $M(\mu\text{-}N)_2M$; M=Ti-Hf.	8507	PP
	MMe_2Cl_2; M=Ti-Hf. Bent's rule for bond angles	8208	PP
	$[MMe_6]^{2-}$; M=Ti-Hf. Non-octahedral	8289	PP
	Cp_2M=E; M=Ti-Hf; E=O-Te.	6911	PP
	$H_2(X)M\text{-}NH_2 \rightarrow H_2M\text{=}NH + HX$; M=Ti-Hf; X=H,Cl, Me,$NH_2$,$SiH_3$.	7320	PP
	$H_3M\text{-}EH_3$, $H_2M\text{=}EH_2$; M=Ti-Hf; E=C-Sn.	7324	PP
	$[MCp(CO)_4]^-$; M=Ti-Hf. ^{13}C NMR shifts.	8283	PP
	$M(C_6H_6)_2$; M=Ti-Hf.	8375	PP
	$Hf@C_{28}$.	10024,10025	PP
	$Hf@C_{28}H_4$.	10026	PP
73 Ta	MO; V-Ta.	9349	PP
	TaL_5; L=H,Cl,Me. Which structure?	8260	PP
	TaF_6^-.	7922	PP
	Heptoxides $M_2O_7^{4-}$, M=V-Ta.	6667	PP
	Ta + CO.	8764,9938	PP
	CMO^-, OMCCO, $(C_2)MO_2$, $M(CO)_n$; n=1-6; M=Nb,Ta.	10348	DFT+PP
	$Ta^+ + CH_4$.	9553	PP
	TaC, TaC^+.	8770	PP
	TaC_2^+.	8768,9457	PP
	TaC_n^+; n=7-13.	9456	PP
	MN_n; M=Nb,Ta; n=1-3.	10345	DFT+PP
	$[MMe_6]^-$; M=V-Ta. Non-octahedral	8289	PP
	$[Ta(OH)_2L'(H_2)L]^+$; L=PH_3; L'=F^-,Br^-,I^-,CO,CN^-, ...	6871	PP
	$M(NH_2)(\text{=}NH)_2$; M=V-Ta imido adducts of CH_4.	7312,7314	PP

Element	Compounds	Ref.	Method
	$Cl_3Ta{=}E$; E=MH_2 (M=C-Sn), MH (M=N-Sb), M (M=O-Te).	6910, 6911	PP
	$[Cp_2TaH_3]$.	7111	PP
	$[MCp(CO)_4]$; M=V-Ta ^{13}C NMR shifts.	8283	PP
74 W	W_3.	6790	PP
	Octahedral M_6, M=Mo,W.	6714	MS
	$W_6X_8X_6^{2-}$; X=Cl-I.	6717	MS
	WF_4. Lowest triplet state.	7921	PP+DFT.
	WL_6; L=H,F,Me. Which structure?	8260	PP
	WX_6; X=H,F.	9944	PP
	MF_6; M=Mo,W.	9169	SR SW
	WH_6. C_{3v}, not O_h!	9719	PT
	WH_6.	8013	PP
	WMe_6. Distorted, C_3 structure.	8286	PP
	$[MMe_6]$; M=Cr-W. Non-octahedral	8289	PP
	WMe_nCl_{6-n}; n=1-3.	8290	DFT+PP
	WF_4 triplet state	7921	PP
	$[MS_6]^{8-}$ cluster models for MS_2 solids; M=Ta,W	7384	MS
	MO_2Cl_2; M=Cr-W. Vibr. freq.	9489	PP
	$[M_2Cl_9]^{3-}$; M=Mo,W.	9841	DFT
	$WO_nS_{4-n}^{2-}$; n=0-4, WX_6; X=F,Cl,CO. ^{183}W NMR shifts.	9420	DFT
	MS_4^{2-}; M=Mo,W. Excited states.	9872	SR SW
	$MS_4^{2-/3-}$; M=Mo,W. Redox energy in vacuum.	10306	DFT,PP
	$M_3X_4^{4+}$; M=Mo,W; X=O,S.	8622, 8623	PP
	$M_2O_7^{2-}$; M=Cr-W.	6667, 8895	PP
	MO_2^-; M=Cr-W.	10346	DFT+PP
	Solid WO_3.	7271	PP
	M + CO; M=Mo,W.	9937	PP
	$M(CO)_6$; M=Cr-W.	7378, 7573, 7574	PP
	$M(CO)_6$; M=Mo,W.	9037	SR DFT
	$M(CO)_6$; M=Cr-W.	8616, 8617 9613	DFT
	$M(CO)_6$; M=Cr-W. Electronic spectra.	9432	DFT
	$M(CO)_6$; M=Cr-W	8209	PP
	MO_4^{2-}, $M(CO)_6$; M=Cr-W. O NMR shifts.	9611	SR DFT
	$M(CO)_6$; M=Cr-W. C and O NMR shifts.	8296	DFT+PP
	$M(CO)_6$; M=Cr-W. C and O NMR shifts. SR/NR.	9482	DFT
	$M(CO)_6$; M=Cr-W. HOMO electron density.	9424	PP
	$W(CO)_6$ as test of ZORA geometry optimization.	10076	DFT
	$M(CO)_6$; M=Cr-W.	10088	DFT
	$(CO)_5M{=}CH_2$; M=Mn^+,Mo,W.	8153	DFT
	$M(CO)_5$-Rg; M=Cr-W; Rg=Ar-Xe.	7576	PP, DFT
	$W(CO)_n^+$; n=1-6.	7080	PP
	WCl_4L, WCl_5L^-, $W(CO)_5L$; L=acetylene, ethene.	9239	PP
	$[Cl_3M{\equiv}N]_n$; M=Mo,W; n=1-6 rings.	9604	PP
	$M(CO)_5CX$; M=Cr-W; X=O-Se.	6915	DFT
	$M(CO)_5L$; M=Cr-W; L=CO, SiO, CS, N_2, NO^+, CN^-, NC^-, HCCH,CCH_2, CH_2, CF_2, H_2.	7572	PP
	$[(HO)_3W{\equiv}P]$, $[(HO)_3W{\equiv}P{\rightarrow}W(CO)_5]$, $[(thf)(HO)_3W{\equiv}P{\rightarrow}W(CO)_5]$.	8468	PP

Element	Compounds	Ref.	Method
	$W(CO)_5NH_3$.	10309	PP
	MN, NMO, $MNO^{\pm}$; M=Mo,W.	6679	PP
	MO_2^{2+}, MN_2, MP_2; M=Mo,W.	9315	PP
	Supported nitrido dimers and trimers of W.	7214	PP
	$WC_2H_2^+$.	7819	PP
	Mixed-ring sandwich $[M(\eta^7-C_7H_7)(\eta^5-C_5H_5)]$; M=Ti,V-Ta. Cp bonding more ionic, cycloheptatrienyl more covalent.	8249	DFT
	Cp_2M, Cp_2MO, $Cp_2MCl(OSiH_3)$; M=Mo,W.	8743	PP
	$[MCpMe(CO)_3]$; M=Cr-W ^{13}C NMR shifts.	8283	PP
	Metallacyclophosphazenes $MCl_3[N_3(PH_2)_2]$; M=Cr-W. Metallacyclothiazenes $MCl_2[N_3S_2]$.	9881	DFT+PP
	$[M(CO)_5L]$; M=Cr-W; L=PH_3, PX_3 ^{31}P NMR shifts	8285	PP
	$[M(CO)_5L]$; M=Cr-W; L=OH_2,NH_3,PH_3,PMe_3,N_2,CO;OC,CS, CH_2,CF_2,CCl_2,NO^+. Discusses 'trans effect'.	10087	DFT
	$[Cp(CO)_2M]_2(\mu-E)$; M=Cr,W; E=S-Te.	8179	PP
	$[WL_2(\mu-CR)]_2$; L=H,Me,F,OH; R=H,F,Me.	8177	PP
	$(CO)_5M{=}CH_2$; M=Cr-W Fischer carbenes.	8151	DFT
	Low-valent (Fischer) carbenes $[(CO)_5W{=}CH_2]$, ... high-valent (Schrock) carbenes $[X_4W{=}CH_2]$,...; X=F-I.	10148	PP
	Low-valent (Fischer) carbynes $[F(CO)_4W{\equiv}CH]$, ... high-valent (Schrock) carbenes $[X_3W{\equiv}CH]$,...; X=F-I.	10149	PP
	$M(PMe_3)_4X_2 + H_2$; M=Mo,W; X=F-I.	7968	DFT+PP
	$M(=NH)_3$; M=Mo,W imido adducts of CH_4.	7312	PP
	$W(OH)_2(=NH) + CH_4$.	7313	PP
	$Cl_4W{=}E$; E=MH_2 (M=C-Sn), MH (M=N-Sb), M (M=O-Te).	6910,6911	PP
	d^1 systems MEX_4^{z-}; M=V,Cr-W,Tc-Re; E=O,N; X=F-Br. g-tensors.	9173	DFT
	$[(N_3,N)WL]$, L=Cl, P-BH_3, $\equiv$E; E=N-Sb. $(N_3,N) = N(CH_2CH_2NSiMe_3)_3$.	9588	PP
	MO_2X_2; M=Cr-W; X=F-Br.	6911	PP
	$W(CO)_3(PH_3)_2(H_2)$. Classical or non-classical?	9985	PP
	$M(CO)_n(PH_3)_{5-n} + H_2$; M=Cr-W; n=0,3,5.	9986	PP
	W-Al bonds and other TM–main-group bonds.	7750	PP
	Phosphinidenes $M(CO)_5$-PR; M=Cr-W; R=H,Ph,OH,NH_2.	7577	DFT
	$[Cp_2MH_3]^+$; M=Mo,W.	7111	PP
	$X_4M(HCCH)$; M=Mo,W; X=F,Cl vinylidene rearrangement.	9826,9827	PP
	$M(C_6H_6)_2$; M=Cr-W.	8375	PP
	$[W(OH)_2(C_6H_6)(Ph_3)] + N_2H_2$ $4e$ reduction.	8845	PP
	$W(PH_3)_4E_2$; E=O,S,Te. W(IV), d^2.	8248	DV DFT
	Trans-$M(PH_3)_4E_2$; M=Mo,W; E=O,S,Te. W(IV), d^2.	8366	DFT
	$M_6S_8L_6$ clusters, M=Mo,W.	6713	MS
	$W_2(OR)_6$ + alkynes;R=H. Relativistic effects crucial.	7657	DFT
75 Re	$HM(CO)_5$, M=Mn,Re.	8210	PP
	ReH.	7348	PP
	ReN, ReN_2.	10345	PP
	Heptoxides M_2O_7, M=Mn-Re.	6667	PP
	MS_4^{n-}; M=Mo,W,Re. ESCA spectra.	8991	SR SW
	$MS_4^{1-/2-}$; M=Tc-Re. ReS_4^{3-}. Redox energy in vacuum.	10306	DFT,PP
	$ReNi_{12}H_2 + H_2$. Re-doped heterogeneous catalyst model.	9899	DFT

Element	Compounds	Ref.	Method
	$Re_6S_8X_6^{4-}$; X=Cl-I.	6719	MS
	$Re_6Se_8X_6^{4-}$; X=Cl,I.	6721	MS
	d^1 systems MEX_4^{z-}; M=V,Cr-W,Tc-Re; E=O,N; X=F-Br. g-tensors.	9173	DFT
	d^2 complexes with $[M(NAr)_2]$ cores; M=W,Re. Isolobality of MCp_2 and $M(NR)_2$.	10215	SW DFT
	M + CO; M=Tc-Re, Ta.	9938	PP
	Addition of $LReO_3$, L=O^-,Cl,Cp, to C_2H_4.	7445	
	Epoxidation of ethene by $MeRe(O)_2(O_2)$, $MeRe(O)(O_2)_2$ +water.	10233	PP
	MO_3, MO_3^+, $CpMO_3$, $M(CO)_3$; M=Mn-Re.	9917	PP
	MO_3X; M=Mn-Re; X=F-Br,Me.	6911	PP
	RMO_3; R=Me,Ph,Cp; M=Tc,Re. PES.	8437–8439	DFT
	$LReO_3$; L=Me,Ph,Cl,F,OH,NH_2. Oxygen transfer to PPh_3. $(CH_3ReO)_2(\mu\text{-}O)_2$.	7827, 7829	DFT+PP
	$CpReO_3$, MCl_4E; M=Cr-W,Re; E=O-Te: Metal-chalcogen multiple bonds.	7851	DFT+PT
	Metallacyclophosphazenes $MCl_3[N_3(PH_2)_2]$; M=Tc-Re.	9881	DFT+PP
	$[Cp(CO)_2M]_2(\mu\text{-}E)$; M=Mn,Re; E=S-Te	8179	PP
	$L_4Re\overset{4}{\equiv}ReL_4$	6962, 6965	PP
	$HRe(CO)_3(DAB)$; DAB=1,4-diaza-1,3-butadiene. Optical spectrum. MCSCF required for the quadruple bond.	7912	PP
	Allylic rearrangement of allyloxo metal oxo complexes. M=Ta,Re.	6899	PP
	Re(O) acetylides, $ReOCl_4^-$, $ReOF_5$.	7317	PP
	$[(C_6H_6)M(CO)_3]^+$: M=Mn-Re.	7262	PP
	$[MMe_6]^+$; M=Tc,Re. Non-octahedral.	8289	PP
	Lewis basicity of $Cl_2(PH_3)_3ReN$-L complexes.	10150	PP
	$M(H)(CO)_3(H\text{-}DAB)$; M=Mn,Re; H-DAB = 1,4-Diaza-1,3-butadiene. SO effects on the M-H bond homolysis.	7370	PP
	MH_n^q, MMe_n^q; M=Zr,TcTa,W,Re. Molecular shapes.	8566	PP
76 Os	R/NR orbital energies for M_6, M=Fe-Os.	6589	DFT
	OsN.	9353	PP
	OsO_4. 5p SO influence on valence PES.	9306	PP
	OsO_4. Vibrations.	9489, 9997	PP
	OsO_4. RI-MP2	10192	PP
	MO_4, M=Ru, Os. PES.	6712	MS
	MO_4, M=Ru, Os. PES.	7876	PP
	OsO_4, OsO_3F_2, ..., OsF_8.	10094	PP
	OsO_4. SO effects on PES.	7096	DFT
	MO_4; M=Ru-Os. O NMR shifts.	9611	SR DFT
	$MS_4^{0/1-}$; M=Ru-Os. OsS_4^{2-}. Redox energy in vacuum.	10306	DFT,PP
	^{17}O NMR shifts in MO_4; M=Fe-Os, MnO_4^-; M=Mn-Re,	8299	DFT+PP
	MO_4^{q-}; M=V...Os. Excited states. MO_4^{2-}; M=Cr-W.	9847	DFT
	Nonlinear optical properties of MO_4^q; M=Ti...Os.	7323	PP
	Vibr. freq. of TM chalcogenides; M=Ti...Os; Ch=O-Te.	7325	PP
	$OsO_4(NR_3)$.	10043	PP
	OsCO.	9936	PP

Element	Compounds	Ref.	Method
	MCO^+, $M(CO)_n^q$; M=Ru,Os; n=1-4; q=-1,0.	10347	DFT+PP
	$Os(CO)_4H_2$, metallacyclophanes.	10179	PP
	$M(CO)_5$; M=Fe-Os.	7575	PP
	$M(CO)_5$; M=Fe-Os.	8209	PP
	$M(CO)_5$; M=Fe-Os.	9613	DFT
	$M(CO)_5$; M=Fe-Os. DFT by DPT.	10085	DFT
	$M(CO)_5$, $M(CO)_4(C_2H_2)$; M=Fe-Os.	7405	PP
	$Os(CO)_5$ as test of ZORA geometry optimization.	10076	DFT
	$[M_nC(CO)_m]^q$; M=Fe,Rh,Os interstitial carbides. ^{13}C NMR shifts.	8288	PP
	$M_2(CO)_9$, $M_3(CO)_{12}$; M=Fe-Os. HF/MP2/DFT compared.	8083	PP,DFT
	$H_2M(CO)_4$, M=Fe-Os.	8210	PP
	$OsH_3X(PH_3)_2$; X=Cl,I.	7250	PP
	$[MH_3(PMe_3)_4]^+$; M=Fe-Os	8150	DFT
	$MH_4(PH_3)_3$; M=Fe-Os. Non-classical hydrides of Os due to R.	8613	DFT
	$OsH(Ph)(CO)(PH_3)_2$ + CO.	9384	PP
	MH^+, M=Fe-Os. Compare several PP.	8601	PP
	$M(CO)_5$, M=Fe-Os.	8617	DFT
	$M(CO)_4(C_2H_4)$, M=Fe-Os.	8618	DFT
	Cis-OsO_2F_4.	7227	PP
	MEX_n^q; M=Mo,W,Re,Os; E=N,O; X=F,Cl; n=3-5.	9053	PP
	Os_2Cl_8, $Os_2Cl_8^{2-}$. How short can a M-M bond be?	8984	PP
	Metallacyclophosphazenes $MCl_3[N_3(PH_2)_2]$; M=Ru-Os.	9881	DFT+PP
	OsO_4-catalyzed dihydroxylation of olefins.	7379	PP
	Cp_2M; M=Fe-Os.	7266	PP
	$CpM(CO)^-$ + CH_4; M=Ru,Os.	9860	PP
	Pd and Pt bis(carbene) complexes. PES.	7877	DFT
	Bis(dihydroquinidine)-3,6-pyridazine·OsO_4-catalyzed dihydroxylation of styrene.	10044	PP
	$[M(SiR_3)H_3(PH_3)_3]$; M=Ru,Os.	8075	PP
	$Os(SiR_3)Cl(CO)(PPh_3)_2$; R=F,Cl,OH,Me.	8076	PP
	$[Os(NH_3)_4(OAc)(H_2)]^+$.	7296	PP
	$[Os(NH_3)_4L(\eta^2\text{-}H_2)]^{q+}$: H-D spin-spin coupling.	6771, 7100	PP
	$M_3(CO)_9(\mu\text{-}H)_3(\mu_3\text{-}CH)$; M=Ru,Os.	9402	PP
	$[P(CH_2CH_2PH_2)_3M(D)(D_2)]^+$; M=Ru, Os. DQCC.	6782	PP
	$OsCl_2H_2(PPr_3^i)_2$.	8840	PP
	Trans-$[LM(H_2PCH_2CH_2PH_2)_2(\eta^2\text{-}H_2)]^{n+}$; M=Ru,Os; n=1,2; L=H^-, CH_3^-,F^-,CF_3^-, CN^-,Cl^-,Br^-,CO,NCH, NH_3,PH_3. Dihydrogen acidity.	10246	DFT+PP
	$ML_n(L')_{3-n}$ + CH_4; M=Ru,Os; L,L'=PH_3,CO. Competing σ and π approaches found.	9861	DFT+PP
	$OsH_2X_2L_2$; X=Cl-I; L=PH_3.	7917	PP
	$OsH_3(BH_4)(PR_3)_2$.	7427	PP
	$[Os(PR_3)_3$"H_5"$]^+$.	8841	PP
	$[Os(PR_3)_3H_4]$ and other $[ML_n(H\cdots H)]$.	8843, 8844	PP
	$MY_6I_{12}^{2-}$; M=Ru,Os.	6715	MS
	$(CH_3)_2M(\equiv CH)(X)$, $(CH_3)M(=CH_2)_2(X)$, M=Mo,W;Ru,Os; X=Cl,CH_3,CF_3,SiH_3,SiF_3.	7217	PP+DFT
77 Ir	H_2 addition to $Ir(PR_3)_2(CO)X$.	6584–6586	PP

Element	Compounds	Ref.	Method
	MOC, MCO, M=Rh, Ir.	7351	PP
	Ir_3.	7364	PP
	IrC.	9933	PP
	IrN.	9352	PP
	IrO, OIrO, $Ir(O_2)$, $(O_2)IrO_2$, Ir_2O, Ir_2O_2.	7241	PP
	$Ir_2 + H_2$.	7133	PP
	$Ir^+ + CH_4$.	9192	PP
	YIrC, $YIrC_2$.	9452	PP
	MF_6^{2-}; M=Co-Ir. $d-d$ spectrum.	10127	DF/HF
	$M(CO)_6^n$, M=Cr→Fe, Tc→Ru, Hf→Ir. Dissociation energy.	7578, 9901	PP
	$CpM(PH_3)(CH_3)^+$ + C-H bonds. M=Rh,Ir.	9855	PP
	$CpM(PH_3)$ + C-H bonds. M=Rh,Ir.	9863	DFT+PP
	CpML + C–H bonds; M=Rh,Ir; L=CH_2,CO,SH_2,PH_3. Singlet-triplet gap.	9865, 9866	DFT+PP
	$[Ir(CN)_5]^{3-}$. R/NR.	9067	DFT-DVM
	$CpM(CO) + CH_4$; M=Rh,Ir.	9860	PP
	Trans-$Ir(Cl)(PH_3)_2 + CX_4$; X=F-I.	9864	DFT+PP
	Ir(*trans*-$(PH_3)_2$)(CO)(*i*-Pr_2SiO).	10057	PP
	Cis-$[M(CO)_2I_2]^- + CH_3I$.	7890	PP
	$IrXH_2(PR_3)_2$; X=Cl-I; R=H,Me.	7249, 7564	PP
	$Ir(PH_3)_2(X) + CH_4$; X=H,Cl.	7315	PP
	$(Cp)ML + CH_4$; M=Rh,Ir; L=CO, SH_2, PH_3.	9857	
	Dito. M = Ru^-,Os^-,Rh,Ir,Pd^+,Pt^+.	9858	PP
	$M(X)(PH_3)_2$ + C-F bonds; M=Rh,Ir; X=CH_3,H,Cl.	9856	DFT+PP
	$IrClH_2(PH_3)_3$.	7980	PP
	$[Ir(H)_2(PR_3)_2]^+$.	7268	PP
	$Ir(H)_2(CO)L$; L=$C_6H_3(CH_2PH_2)_2$. *Cis/trans* isomerization.	8626	DFT+PP
	$[IrH_3(PH_3)Cp]^+$.	8167	PP
	Octahedral L_5M-(CH)=CHR alkenyl complexes. M=Ru,Re,Ir.	7216	PP
78 Pt	PtH, PtH^+, PtH_2.	7522	DF/HF, DK.
	PtH.	10128	DF+CI
	PtH.	7712, 7713	DK,PP
	PtH. Mean-field spin-orbit.	8801	PP
	PtH.	9767	DK+SO
	Pt_2.	9158	DK
	PtH, PtH^+, Pt_2, Pt_2H.	10369	PP
	MH; M=La, Hf→Pt.	10219	PP
	PtH, $PtO^{0/+1}$, $PtCH_2^+$. $Z_{\rm eff}$ SO operator.	7990	PP
	PtH_2^+. SO effects.	7645	DK-RASCI
	MH_x^{2-}, MCl_y^{2-}; M=Ni-Pt; x=2,4,6; y=4,6 in crystal field.	8640	DFT
	Solid A_2MH_2; A=Li,Na; M=Pd,Pt.	8645	DFT
	MXe, M=Ni-Pt.	7086	PP
	Pt_n; n=2-12.	10243	PP
	Pt_3, Pt_4.	9478	PP
	M_4, M=Pd,Pt.	7355	PP
	Pt_n^q; n=1-4; q=0,±1.	7739	DFT
	Pt_n; n=2-6,13.	10270	PP
	M_{13}; M=Pd,Pt. SO.	10184	PP

Element	Compounds	Ref.	Method
	Pt_n; n=9-13. $RePt_n$; n=8-12. Models for Pt(100). H_2 dissociation.	9898	DFT
	Pt_nH_m, $Pt_nH_m^-$. Models for on-top H adsorption at platinum electrodes.	10244	DFT+PP
	$Pt_{13}H_n$; $n \leq 20$.	10183	PP
	$ZrPt_3$ cluster.	10158	PP
	MO, MCO; M=Ni-Pt.	7237, 7238	DFT
	M_2CO; M=Pd,Pt.	9085	
	CO on Pt(110).	7817	PP
	MO; M=Ni-Pt.	9349	PP
	PtN.	7356, 8235	PP
	MXe; M=Ni-Pt.	7086	PP
	$[MX_6]^{2-}$; X=F-Br; M=Ti...Pt. M-X bond energies. Different DFT compared.	7410	DFT
	$[MX_6]^{q-}$; M=Ta→Pt; X=F-Cl; q=0-2. Trends in redox potentials studied.	8757	DFT
	Pt_3Au, PtAu.	7350	PP
	$[Pt_3Fe_3(CO)_{15}]^{n-}$; n=0-2.	9833	DFT
	$M_3 + H_2$; M=Pd,Pt.	7363	PP
	Pt_3/O, Pt_4/O adsorption models.	8422	PP
	PtSn + H_2.	7135	PP
	Cu/Pt_9, Ag/Pt_9. Surface models.	9417	PT
	PtAl, Pt/Al_9, Pt_4Al_4.	9418	PP
	Mixed-metal clusters with $[Pt_2Au]^{3+}$ and $[Pt_2Hg]^{4+}$ cores.	9995	SR SW
	PtC_n^+; n=1-16. Structures.	8913	DFT/PP
	S/Pt_{12}.	9419	PP
	PtL, ZrL, PtZrL, ZrPtL; L=H,CH_3.	9385	PP
	$PtCN^-$, $PtNC^-$.	7711	PP
	Pt + CO.	9444, 9445	PP
	CO adsorption on M; M=Ni-Pt. Relativistic trends.	9121	DFT
	Pt-CO interactions in mordenite zeolites.	10252	DFT
	TM core level shifts upon adsorption on Al_{37} model of Al(100). 'TM'=Ne,K,Cu-Au,Ni-Pt. Covalent interaction.	9122	DFT
	$Pt(CO)_3$, $Pt_3(CO)_6$.	7897	PP
	LM_n chemisorption models. L=HS-, MeS-; n=10,11; M=Pd,Pt.	9660	PP
	$M(CO)_3L$; M=Ni-Pt; L=Co, SiO, ...	7572	PP
	$M_2(\mu\text{-}PR_2)$ core; M=Pd,Pt; PH_3, organic ligands.	8876	DFT+PP
	$M(CO)_4$, M=Ni-Pt.	7575, 8209, 8617	PP
	$Pt(CO)_4$ as test of ZORA geometry optimization.	10076	DFT
	$M(CO)_4$; M=Ni-Pt. DFT by DPT.	10085	DFT
	Pt + H_2.	7459	PP
	H_2 + M, M=Re→Pt.	7899	PP
	$PtCH_2^+$	7993	PP
	$[Pt(CH_3CN)Cl_3]^-$ complex. with alkane and bis(diphenylphosphino)alkane ligands.	9152	PP
	$[Pt_2(\mu\text{-}O_2CCH_3)_4(H_2O)_2]^{2+}$.	9842	SR SW DFT

Element	Compounds	Ref.	Method
	$[(\eta^3\text{-}CH_2CCPh)Pt(PPh_3)_2]^+$.	7870	DFT
	M + C_2H_4; M=Pd,Pt. Ethylene activation.	8924	PP
	$Pt^{0,+1}$ + benzene.	9447	PP
	M + CH_4; M=Re→Pt.	9900	PP
	ML_2 + CH_4; M=Pd,Pt; L=CO,PH_3; L_2=$PH_2CH_2CH_2PH_2$.	9862	PP
	M, M_2 reacting with H_2, CH_4; M=Pd,Pt.	7310	PP
	CH_4 + Pt^+.	7995, 9179	PP
	CH_4 + (eda)$PtCH_3^+$; eda = $H_2N\text{-}CH_2\text{-}CH_2\text{-}NH_2$.	7983	DFT
	CH_4 + Pt photodissociation. Embedded cluster model for surface.	10207, 10208	PP
	Ethyl to ethylene conversion on Pt.	10208	PP
	$PtCH_2^+$.	7993	PP
	CH_4 + $Pt^{0/+/-}$.	7935	PP
	CH_4 + Pt_n; n=1, ...,10.	6618	PP
	Pt_2 + CO.	9449	PP
	$PtCH_2^+$ + NH_3.	6736	PP
	M^+-L; M=Ni,Pt; L=glycine, formate, formamide.	6843	PP
	MO^+ + C-H or C-C bonds. M=Os,Pt.	9620	PP
	CO and O on a Pt(111) slab surface.	6967	PP
	CO adsorption on M(111); M=Ni-Pt.	9234	ZORA
	$[PtCl_2(CO)_2]_2$ dimer: $d^8 - d^8$ interaction.	6754, 9073	PP
	PtI_2.	8058	PP
	$Pt_2(PH_3)_4(\mu\text{-}S)_2$. Complexes with Ga(III), In(III), Tl(I), Pb(II).	9930	PP
	$Tl_2Pt(CN)_4$.	7481	PP
	$[Pt(NH_3)_2]^{2+}$ interaction potential with H_2O.	8988	PP
	$\eta^2\text{-}C_{60}\text{-}Pt(PH_3)_2$.	8414	PP
	^{13}C NMR shifts in M-olefin complexes, M=Cu,Rh,Ag,Pt.	7977	PP
	$C_{59}M$; M=Ir,Pt.	9265	DFT+PP
	$[M(PH_3)_2O_2]$; M=Pd,Pt.	9037	SR DFT
	$[(PH_3)_2PtSe]_2^{n+}$; n=0,2.	6903	PP
	$[Pt(NH_3)_3(\text{adenine})]^{2-}$. Force field determination.	8454	PP
	$Pt(PX_3)_2$; X=H,F. The nature of the Pt-P bond.	7658	PP
	$[M(PH_3)_2(\eta^2\text{-}C_2X_4)]$; M=Ni-Pt; X=H,F,CN. Geometry, bonding.	9078	DFT
	$M(PH_3)_2$ + $BX_2\text{-}BX_2$; M=Pd, Pt; X=H, OH.	9534	PP
	$Pt(PH_3)_2$-catalyzed hydrosilylation of ethene.	9535	PP
	$Pt(PR_3)_2$-catalyzed alkene, alkyne diboration. Pd/Pt?	7308, 7309	PP
	$M(PH_3)_2$ + Si-X bonds; M=Pd,Pt; X=H,Si.	9537	PP
	$PtCl_2(PH_3)_2$ + $SnCl_2$.	9413	PP
	σ-bond activation of $(HO)_2B\text{-}XH_3$ by $M(PH_3)_2$; X=C-Sn; M=Pd,Pt.	9533	PP
	$Pt(H)(PH_3)_2(SnCl_3)(C_2H_4)$. Olefin insertion.	9414	PP
	$MH(\eta^{1,3}\text{-}C_3H_5)(PH_3)$; M=Pd,Pt.	9539	PP
	Diimine-M(II) complexes (M=Ni-Pt), and their zirconocene complexes in olefin polymerization.	8999	PP
	$Pt(PR_3)_2$ + H_2.	8853	PP
	PtL_2 + H–OCH_3. L_2=$(CO)_2$, $(PH_3)_2$, diphosphine.	9859	PP
	$PtH(SiH_3)(PH_3)$ + C_2H_2.	9878	PP

Element	Compounds	Ref.	Method
	$PtH_3(PH_3)_2^+$. Dihydrogen complex.	7918	PP
	$CpM(CO)^+ + CH_4$; M=Pd,Pt.	9860	PP
	$[M(PH_3)_2]_nC_{60}$; M=Pd,Pt; n=1,2,6.	6982	PP
	$M(PH_3)_2X_2$; M=Ni-Pt; $X_2=O_2,C_2H_2,C_2H_4$.	8618	DFT
	Pt(II)(OOH) complexes.	7659	PP
	MM'$(PH_3)_4$; M,M'=Pd,Pt. $d^{10}-d^{10}$.	9538	PP
	A Pt_2(II,III) bridged model system $[Pt_2(\mu\text{-}C_6F_5)_2(CF_3)_4]^-$.	10055	SR DFT SW
	$Pt_2(dta)_4X_2$; X=Br,I. 'dta'=$CH_3CS_2^-$. Peierls distortions in the chain.	9409	PP
	Dimers of $M(dmit)_2$; M=Ni-Pt, '$dmit^{2-}$' = 2-thioxo-1,3-dithiole-4,5-dithiolato.	9433	DFT
	$Pt_2[NHCHN(C(CH_2)(CH_3))]_4$. d^8-d^8.	9039	PP,DFT
	$[Pt_2Cl_2(CO)_4]$, $[Pt_2Cl_4(CO)_2]^{2-}$. d^9-d^9 bond.	9877	PP
	MM2 force field derived for Pt(II) square complexes.	7319	PP
	Ethylene polymerization catalyzed by diimine-M(II); M=Ni-Pt.	9002	PP
	$Pt(PH_3)_2(R)$, R=olefin (C_2H_4, C_8H_{10}, ..., $C_{11}H_{16}$).	10041	PP
	$(X_3P)_2Pt$=PR phosphinidenes.	6909	PP
	$Pt(P(t\text{-}Bu)_3)_2 + H_2$, Diels-Alder reaction of acrolein + isoprene, etc.	9896	PP
	$(C_2H_4)MCl_x(NH_3)_{3-x}$; M=Ni-Pt; x=1-3.	9844	PP
	$M[(CHNH)_2]R^+$ + ethene; M=Ni-Pt; R=Me,Et.	9845	PP
	$[Pt(NH_3)_2L_2]^{2+}$; L=Cl^-, H_2O.	7126	PP
	'Carboplatin' versus 'cisplatin'.	9994	PP
	Trans-$[(CH_3NH_2)_2PtCl_2]$ + adenine, thymine.	7125	PP
	$M_3(CO)_6^{2-}$ Cini clusters (M=Ni,Pt).	6905	PP
	HCN synthesis from CH_4+NH_3 on Pt^+.	7456	PP
79 Au	Au_2	6850	DFT
	M_2; M=Cu-Au.	10090	DV DFT
	M_2; M=Cu-Au, AuH, AuCl.	8868	DFT
	M_2, MH, MCl; M=Cu-Au. R/NR. Various functionals, finite nuclei compared.	9892	SR DFT
	CuAg, CuAu, AgAu. Electric properties.	8324	PT
	Au_3	6723	MS
	Au_n; n=2-6	7048,7049	PP
	M_n; n=5,7; M,M'=Cu-Au. g-factors, hfs at ZORA level.	10078	DFT
	Au_{13}.	8560	DFT
	Au_n; $n \leq 75$.	7812	PP
	Au_n; n=2,6,55. $AuPR_3$, $MeAuPR_3$. Naked and clad $[Au_6]^{m+}$.	7927	DFT
	Au_n; $n \leq 147$.	7928,8482	DFT
	Octahedral M_6, M=Ag,Au.	6714	MS
	Au_n^q; n=1-4; q=-1,0,+1.	9669	PP
	R/NR orbital energies for M_6, M=Cu-Au.	6589	DFT
	$(Au)_\infty$: infinite 1D chain.	9087	DFT+PP
	Au_8 on MgO surface. Catalyst for CO oxidation.	9550	PP
	$(Au_n^+)CH_3OH$; $n \leq 15$. Car-Parrinello study of adsorption.	9463	DFT+PP
	$Au_3(PH_3)_4L_n$; n=0-3; L=-$InCl_2H_2O$.	7796	DFT

Element	Compounds	Ref.	Method
	Also Au_3^q, $Au_3(PH_3)_4^q$; q=0,+1. $Au_3(PH_3)_3^+$.		
	MM'; M=Rh, ...Au	7962,7963	PP
	Au_6^q; q=0-4. Au_6X^q; X=B,C,N. $[(LAu)_nX]^q$; n=4-6.	7930	DFT
	$Au_{12}Pd$	6720	MS
	M_nAu; M=Na,Cs; $n \leq 9$.	7998	DFT
	AuH.	9261	DK
	AuH.	8243	DK CC
	MH; M=Cu-Au. Dipole moments.	8323	PT
	MH; M=Cu-Au. Dipole moments.	8325,8333	DK
	MH; M=Cu-Au. Comparison between methods	7257	PP
	MH; M=Cu-Au. Comparisons at MP2 level.	7747	DPT,PT
	MH; M=Ag,Au. Test new method.	9015	RESC
	MH, MCl; M=Ag,Au. Test new method.	9016	DFT+RESC
	M_2, MH; M=Ag,Au. Test two-electron terms.	9158	DK
	Au_2, AuH as tests of the ZORA approximation.	10073	DFT
	MM', MX; M=Cu-Au, X=H,F,Cl. ZORA/PT with DFT.	10087	DFT
	MH; M=Ag,Au, Au_2 as tests of DFT by DPT.	10083	DFT
	Au_2, AuH. Spin-orbit effects.	8595	PP
	Au_2, AuH. Spin-orbit effects via ZORA.	10077	DFT
	Au_2,AuH,AuCl.	9037	SR DFT
	AuF, AuF^+, AuF_2, Au_2F_2.	9643,9644	PP
	MF; M=Cu-Au.	8551	DF+MP2
	MF_n^-; M=Cu-E111; n=2,4,6.	9677	PP
	MF; M=Cu-Au.	8099	DK
	AuCl.	6781,9633	PP
	AuX; X=F-I.	7530	DF
	Au_2Cl_6. PES.	8125	DSW
	MI, M_2I^+, M_3I^{2+}, M_4I^{3+}; $[I(MCl)_m]^-$; m=1-4; M=Cu-Au.	6647	PP
	AuI_n^-; n=2,4	9897	PP
	$[(AuCl_3)_2(\mu\text{-}Cl)]^-$. Occurs in room-temperature molten salt.	9614	PP
	Diatomic CsAu.	9573	DF/HF
	SiM; M=Cu-Au. All ground states $^2\Sigma$. $SiCu^\pm$.	10037,10038	DK
	$[XMX]^{3-}$; M=Cu-Au; X=O-Se in crystal field.	8636,8637	DFT
	BM; M=Cu-Au.	6839	DK
	AlM; M=Cu-Au. Electric properties.	10054	DK
	AuX^+; X = Be,Mg, C, Si. $Au{\equiv}C^+$ triple bond found.	6836	DK
	$RgAu^+$, $RgAuRg^+$; Rg=He-Xe predicted.	9303	PP
	$ArAu^+$, $XeAu^+$ handled with polarization and dispersion.	9373	
	Au_2O, MO_n; n=1-3; M=Ag,Au. AuO_2 observed.	7242	PP
	$Au{\equiv}C^+$. Do other Au≡L exist?	9316	PP
	SAu_3^+, $S(AuPH_3)^+$. $d^{10}-d^{10}$.	9305	PP
	$(XAuPH_3)_2$; X=H,F-I,CH_3,-CCH,CN,-SCH_3. $d^{10}-d^{10}$.	9308,9311 9309	PP
	$MeAuPR_3$; R=H,Me,Ph.	7929	DFT
	$ClAuPH_3$, $ClAuPH_2Ph$, $ClAuPPh_3$. Eigenvalue spectrum.	8571	SR SW
	$(ClAuPH_3)_n$ isomers, n=2,4.	9312	PP
	$H_2C[P(Ph)_2AuX]_2$, $HC[P(Ph)_2AuX]_3$; X=Cl,I. $d^{10}-d^{10}$.	9869	PP

Element	Compounds	Ref.	Method
	$H_3PS[Au(PH_3)_n{}^{n+}$; n=1-3. $H_2P(S)S[Au(PH_3)]_2{}^+$. $d^{10}-d^{10}$.	9295	PP
	$[(PR_3)_2M]^+$, $[PR_3MCl]$; M=Ag,Au; R=H,Me.	7034	DFT
	Au(I)...Au(III) interactions modelled by $[S(AuPH_3)_n(AuH_3)_{3-n}]$.	7108	DFT
	M_2Se, M_2I^+, M=Ag,Au.	6648	PP
	$H_3PAuC{\equiv}CAuPH_3$ complex with $CHCl_3$	7650	PP
	MCO; M=Cu-Au.	9635	PP
	$M(CO)^+$; M=Ag,Au.	7378	PP
	MCO^+; M=Cu-Au.	8745	PP
	$AuCO^+$. Accurate.	7382	PP
	H_2 on M(111); M=Ni,Cu,Pt,Au. 'Why gold is the noblest of all metals'. Relativity not mentioned.	7953	PP+DFT
	Adsorption of M on cluster models of NaCl(100); M=Cu-Au.	8883	PP
	CO on M(111); M=Ni,Cu,Pd,Ag,Pt,Au; Ru(0001),Cu_3Pt. Simple model for d-band hybridization proposed.	7952	PP+DFT
	NO_2 on Au atom, Au clusters.	8722	DFT+PP
	ClM(CO), M=Cu-Au.	6687	PP
	$M(CO)_2^+$, $[M(CO)_2F_2]^-$; M=Cu-Au. 'Nonclassical' metal carbonyls.	8746	PP
	$[M(CO)_n]^{x+}$; M^{x+}=Cu^+-Au^+, Zn^{2+}-Hg^{2+}; n=1-6.	8748	PP
	$M(CO)_n^+$; MCN, $M(CN)_2^-$; M=Ag,Au; n=1-3.	10095	PP
	$[Au(CN)_2]^-$ to $[Tl(CN)_2]^+$, $[Au(CO)_2]^+$ to $[Tl(CO)_2]^{3+}$. DFT, MP2, CCSD(T).	8211	PP
	Bimetallic clusters: $[M_4\{Fe(CO)_4\}_4]^{4-}$; M=Cu-Au.	6627	DFT
	MR; M=Cu-Au; R=Me,Ph.	6688	PP
	MBH_4; M=Cu-Au.	9000	PP
	Au speciation in aqueous solution.	10001	PP
	Decomposition of Au(I) compounds.	10003	PP
	AuSH, $Au(SH)_2^-$ + H_2S, H_2O	9049	PP
	Cyclic $[Se_nAu_2]^{2-}$; n=5,6: $d^{10}-d^{10}$.	6769	PP
	$(M_2S)_n$; n=1,2; M=Ag,Au: $d^{10}-d^{10}$.	6776	PP
	$[XAuPH_3]_2$; X=H,Cl. $d^{10}-d^{10}$.	9487	PP
	Pt_3Au, PtAu.	7350	PP
	Solid MCl; M=Ag,Au. $d^{10}-d^{10}$.	7485	PP
	$M{=}CH_2^+$, $M\text{-}CH_3^+$; M=Cu-Au.	8059	PP
	MCH_2^+; M=La,...Au.	8124	PP
	MCH_2^+; M=Fe,...Au.	7456	PP
	MCH_2^+; M=Ni,Pd,Ir→Au. HCN synthesis from methane and ammonia.	7987	DFT
	Au+ethylene interaction.	8887	PP
	ClM=M'R_2; M=Cu-Au; M'=C-Ge. R/NR for Au.	6991	PP
	MCH_3^+; M=Sc-Cu,La,Hf-Au.	8044	PP
	C_6H_6 + M^+; M=Cu-Au.	7383	PP
	C_2H_4 + M^+; M=Cu-Au.	8012	PP
	H_2O + Au^+.	8061	PP
	$M^+(H_2O)_n$; n=1-4; M=Cu-Au.	7673	PP
	Au_n^+; n=1-15 + CH_3OH.	9463	PP
	$[Me_3P\text{-}Au]^+$ + alkyne + alcohol.	9953	PP
	$[E(AuPH_3)_4]^+$; E=N-As. $d^{10}-d^{10}$.	8615	PP

Element	Compounds	Ref.	Method
	$X(AuPH_3)_n^{m+}$; X=C,N,O,P,S; n=1-6. $[HC(AuPH_3)_4]^+$. $d^{10}-d^{10}$.	9317	PP
	$[OAu_3]^+$, $[O(AuPH_3)_3]^+$, dimers.	7239	DFT
	$[CH_2(PH_2AuPH_2)_2CH_2]^{2+}$.	7686	PP
	$[M_2(H_2PCH_2SH)_2]^{2+}$; M=Cu-Au.	7687	PP
	Rings $[M_2(PH_2CH_2PH_2)_2]^{2+}$, $[M_2(NHCHNH)_2]$, $[M_2(SCHS)_2]$, $[M_2X_4]^{2-}$; M=Cu-Au; X=Cl-I.		
	$[Au_2(PH_2CH_2PH_2)_2]Cl_2$. $[Au_2Te_4]^{2-}$. $d^{10}-d^{10}$.	9310	PP
	$Au_2X_2C_2H_2(PH_2)_2$; X=Cl-I. *Cis* has $d^{10}-d^{10}$. Photochemistry of cis-to-trans conversion.	9638	PP
	Cl_nAuSMe_2; n=1,3. Inversion at sulfur.	10009	PP
	$[Cp_2Ti(C{\equiv}CH)_2]Au$-R.	8417	DFT+PP
	$[Cl_2Ti(C{\equiv}CH)_2]MCH_3$ 'tweezers'; M=Cu-Au.	8450	PP
	$(Et_3P)Au$(2-Thiouracil).	9835	PP
	$AuPt^-$, Au_2, $AuHg^+$.	10199	PP
	$AuPt_n + H_2$; n=1,2.	7306	PP
	$AuPt_2$, Au_2Pt + ethene.	7305	PP
	$[(AuPH_3)_6Pt(PH_3)]^{2+} + H_2$. Model for dihydrogen activation.	10245	DFT+PP
	$Au(C_2H_4)$.	8887	PP
	$[Au(III)(C_6H_4S_2)_2]^-$. Aromatic Au(III) dithiolene.	9592	PP
	Au^+-L; L=H_2O, CO, NH_3,C_2H_4,...	8010,8060,8061	PP
		9619	PP
		9621	PP
	$ClMPH_3$, MPH_3^+; M=Cu-Au.	8357	DFT+PP
	M^+-L; M=Cu-Au; L=adenine, guanine.	7087	PP
	M^+-L; M=Cu,Au; L=glycine, formate, formamide.	6843	PP
	M^+-L; M=Cu-Au; L=C_2H_4, N_2, HCN, C_2H_2, N_2O, FCN, HCNO, HN_3, HCCF, CH_2N_2, CH_3CN, C_2F_2, FCNO, CH_3CNO	6844	PP
	$[M_2(\mu\text{-}Y)(\mu\text{-}XR)L_4]$, $[M_2(\mu\text{-}XR)_2L_4]$; M=Rh→Pd, Ir→Au; X=O-Te; Y=Cl,S: d^8-d^8.	6755,6756	PP
	$[Au(\overline{CNHCH=CHN}\,H)_2Cl_2]^+$, $[Au(\overline{CNHCH=CHN}\,H)_2]^+$. carbenes.	9367	PP
	M_5Ph_5; M=Cu,Au.	6897	DFT
	Au_9-SCH_3.	6872	PP
	Au_{12}/H_2S, $HSCH_3$.	9662	PP
	M_{16}/SH, SCH_3; M=Ag,Au.	9665	PP
	$Au_{38}(SCH_3)_{24}$. Capacitance.	7944	PP
80 Hg	Hg_2.	7476,7700,9642	PP
		7340,9879,9880	PP
	Hg_2 from interaction potentials.	8497	DK
	Hg_2, Hg_2^+, Hg_3.	6854	DFT
	Hg_n.	7477	PP
	Hg_n.	9765	SR DFT
	M_2; M=Zn-Hg.	10296	PP
	M_n, $n \leq 6$, M=Zn-Hg.	7703,9587	QMC+PP
	M_2^{2+}, M=Zn-Hg. Their compounds.	8308	PP
	Hg_2^{2+}. Interpretation of relativistic effects.	9630	DFT

Element	Compounds	Ref.	Method
	HgX_2; X=Cl-I. ^{199}Hg NMR shifts.	9021	PT
	MX, MX_2, M_2X_2; M=Zn-Hg; X=F-I. Molecules and solids.	8644	DFT
	Hg_2^{2+}, Hg_2L_2; L=F-I. Molecules and solids.	8633, 8638, 8643	DFT
	Hg_n^{2+}; n=2,3. Complexes with Cl, C_6H_6.	10047	PP
	$Hg_n + H_2/O_2/F_2$. Dissociation models for surface of liquid mercury. Potential functions.	9663, 9664	PP
	$AuHg^+$.	10199	PP
	HgCd.	6947	PP
	HgH.	6638	PP
	HgH, HgH^+.	7972, 8983	PP
	MH; M=La→Hg.	7131	PP
	MH^+, MH_2; M=Zn-Hg. Vibrational spectra.	7878	PP
	LiHg.	7835, 7906, 7907	DK
	NaHg.	7905	DK
	CsHg. Green bands.	9272, 9273	DK
	HgO.	8064, 8065	DF
	MS; M=Zn-Hg. Electric properties.	9364	DK
	HgX_2; X=Cl,CN.	6615	PP
	HgX, HgX_2, Hg_2X_2; X=H,F,Cl,CH_3,CF_3. Hg_2^q; q=0,+1,+2.	9636	PP
	MR_2; M=Zn-Hg; R=Me,Ph.	6688	PP
	LHgL'; L=Cl-I,Me,CN. ^{199}Hg NMR shifts using ZORA.	10222	DFT
	$Hg(CF{=}CF_2)_2$.	6810	PP
	Photodecomposition of Hg methyl complexes.	10004	PP
	HgI_2.	8058	PP
	HgF_4.	8305, 8306	PP
	HgX_n; n=2,4; X=F,Cl.	8690	PP
	MO_4^q; M=Ir-Hg. Extreme oxidation states ($d^{0/2/4}$). HF only.	7433	PP
	HgX_2, $(HgX_2)_2$; X=H,F-I. R/NR.	8307, 8309	PP
	Solid MF_2; M=Cd,Hg.	8309	PP
	HgL_2; L=F,Cl,Oh,Sh,CN. $HgCl_n^{2-n}$; n=1-4. $Hg_3S_2(SH)_2$ model for cinnabar. Hydration.	10005	PP
	$MeHg^+$, Me_2Hg, MeHgX (X=Cl-I), $MeHg(PH_3)^+$, $[MeHg(PH_3)_3]^+$.	6816–6818	PP
	RHg, RHgH; R=Me, CH_2CH-, MeO(CO)(MeC)CH-.	9018	PP
	$CH_n(HgX)_{4-n}$; X=Cl,CN; n=0-4. RHgH; R=Me, Et,... . ^{13}C and ^{1}H NMR shifts.	8297	DFT+PP
	TM carbonyl ^{13}C NMR shifts; TM=Hf-Hg.	10222	
	Mercury sulfides.	9843	
	Solid $2HgS{\cdot}SnBr_2$	9481	PP
	$(HgPMe)_n$, n=4-6,8,12	6600	PP
	$M(CO)_2^{2+}$, $[M(CO)_2F_2]$; M=Zn-Hg	8746	PP
	$[FeHgCp_2]^{2+}$, $[FeCp(C_5H_4)Hg]^+$ ferrocene derivatives.	8870	DFT
	$H_2 \rightarrow$ 2H on Hg surface.	9661	PP
	M^{2+}-L; M=Cd-Hg; L=adenine, guanine	7087	PP
	M^+-L; M=Cd,Hg; L=glycine, formate, formamide	6843	PP
	$XH_4 + M(^3P) \rightarrow HMXH_3$ (X=C,Si; M=Zn-Hg)	6654	PP
	Hydrated M^{2+} + adenine, thymine base pairs; M=Zn-Hg	9809, 9810	PP
	Hg* ($^3P_1$) + H_2, CH_4, C_3H_8, SiH_4	9752	PP
	FHgX + CH_4; X = NH_2, NO_2, N_3,...	7326, 7329	PP

Element	Compounds	Ref.	Method
81 Tl	Tl_2.	8363	PP
	Tl_2, TlH. Spin-orbit effects	8595	PP
	TlH.	8364,9348 9344,9590	PP
	TlH, TlH_3.	7956	PP
	MH, MH_3, M_2H_6; M=In,Tl	8084	PP
	M_2H_2; M=B-Tl	10012	PP
	TlCl	8629	PP
	TlF. PT-odd interactions.	8552,9161 9330	DF
	TlX; X=F-I	7530	DF/HF
	TlH, TlX; X=F,I. Spin-orbit effects via ZORA.	10077	DFT
	TlX, TlX_3; X=F-I.	9641	PP
	MX_3, MH_2X; X=F-I; M=B-Tl.	7751	PP
	TlX_2^+, $TlX_2^+ \cdot 4H_2O$; X=Cl,CN.	6615	PP
	$[Au(CN)_2]^-$ to $[Tl(CN)_2]^+$, $[Au(CO)_2]^+$ to $[Tl(CO)_2]^{3+}$. DFT, MP2, CCSD(T).	8211	PP
	TlAr.	8598	PP
	MO_2^-; M=In,Tl.	6678	PP
	M_2O_2; M=Al-Tl.	6703	PP
	M_2O_3; M=Ga-Tl.	6704	PP
	$Tl_2E_2^{2-}$; E=Se,Te.	7017	PP
	$TlMTe_3^{3-}$.	7018	PP
	TlCp, $TlCp_2^-$.	6706	PP
	MCp, $(MCp)_2$, MCp_3, $MCp^{\pm}$, MH, $(MH)_2$. Evidence for a Tl(I)-Tl(I) attraction.	9314	PP
	$Tl_2Pt(CN)_4$.	7481	PP
	$TlN(SiMe_3)_2$.	7926	PP
	$H_3M\text{-}EH_3$; M=B-Tl. Uses a molecular EH_3 effective potential. E=N-As.	9659	PP
	Tl^+ in alkali halides. Stokes shifts. Lifetimes.	6924,9407	PP
82 Pb	Pb_2.	7389,10103	PP
	Pb_5.	7360	PP
	M_5^{2-}; M=Sn,Pb.	10048	PP
	Pb_n; n=3-14.	8960	PP
	$Pb_9^{3-/4-}$.	7114	PP
	Na_6M, M=Mg,Pb. Pb inside, Na outside.	6626	DFT
	MH_2, MH_4; M=Si-Pb, MO; M=Ge-Pb.	6813	PP
	MH_2; M=C-Pb. $^1A_1$ and $^3B_1$. SO.	8854	PP
	MH_3^-,MH_4,MH_5^-; M=Si-Pb.	8941	PP
	MH_4; M=C-Pb.	9831	PP
	MH_4; M=Ge-Pb. Electron scattering, elastic and inelastic.	6927,6928	PP
	MH_4; M=C-Pb. Rotational excitation by electron impact.	10089	AE,PP
	MO, MH_4, MCl_4; M=C-Pb. R/NR.	10166	DFT
	$Pb_nH_m^{0/+1}$.	7957	PP
	M_4H_4; M=C-Pb.	9814	PP
	PbH_2, PbH_4.	7956	PP
	Aromatic or polyhedral compounds with Ge, Sn, Pb skeletons.	9008,9009 9010	PP

Element	Compounds	Ref.	Method
	EM_6; E=C-Pb; M=Li-Cs. Hypermetallation is ubiquitous.	8821, 10138	PP
	M_nPb_m; M=Li,Na; m=1,4.	8958, 8959	PP
	MH_4; M=Sn,Pb. Low-energy electron scattering.	6928	PP
	MH_4, MH_3Me; M=Ge-Pb.	7984	PP
	MH_5^-; M=Si-Pb.	8941	PP
	MMe_2Cl_2; M=C-Pb. Bent's rule for bond angles.	8208	PP
	$M_3H_3^+$; M=C-Pb.	8178	PP
	PbBr.	6900	PP
	PbI.	6901	PP
	MF_2; M=Ge-Pb.	7347	PP
	PbI_2.	8058	PP
	Solid $Pb(N_3)_2$.	10294	PP
	MX_3^+, MH_2X^+; M=C-Pb; X=F-I.	7751	PP
	MO; M=Ge-Pb.	7480	DF,PP,PT
	MO, MS; M=Si-Pb. Electric properties.	8321	PT
	MO, MS; M=Si-Pb. Electric properties.	8332	DK
	MX_2, MX_4; M=C-Pb; X=F-I. LSDA, BLYP, B3LYP functionals compared.	7633	PP+DFT/HF
	PbX; X=O-Te. R/NR.	7530	DF
	PbO. R/NR.	8855	DF
	PbO. Dipole moment.	7480	PP
	PbO. Test of new PP.	10213	PP
	PbO, PbX_2.	7530	DF,ESC
	PbX; X=O,Te. Spin-orbit effects via ZORA.	10077	DFT
	Solid PbS.	8902	PP
	MX_2, MX_2^+; M=Ge-Pb.	6666	PP
	Me_3PbX; X=Cl-I, -OMe, -SMe, -SeMe, $-NEt_2$. PbX_4; X=Cl-I. ^{207}Pb NMR shifts.	9420	DFT
	Pb_nO_m; n=1-4; m=1-4.	7207	PP
	PbR_nX_{4-n}, PbR_nX_{2-n}. Why org. Pb(IV), inorg. Pb(II)?	8303	PP
	MH_3-Cl; M=C-Pb. 'Charge-shift bonding'.	9740	PP
	M_2H_6; M=Si-Pb. Core polarization important.	9056	PP
	$M_2E_3^{2-}$; M=Sn,Pb; E=S-Te.	7018	PP
	M_mPb_n; M=Li-K; m=2-7; n=1,4.	6660, 7154	PP
	H_2MO, Me_2MO; M=C-Pb.	8262	PP
	$H_2C{=}MH_2$; M=C-Pb.	8152	DFT
	$H_2M{=}MH_2$; M=C-Pb.	8397	PP
	$H_2M{=}MH_2$, H_2MO; M=Si-Pb.	8263	PP
	Coordination of stannocene, plumbocene.	6705	PP
	Group 14 metalloles: $C_4H_4MH_2$, $C_4H_4MH^{\pm}$, $C_4H_4M^{2-}$, ...	7847	PP
	M-C bonds; M=Ge-Pb. H_3M-Y; Y=H, A, ... $ABCDH_3$.	6845	PP
	p-Toluene-XY_3^+ complexes; X=C-Pb; Y=H,Cl,Me.	6841	PP
	H_3M'-MH_3^+, H_3M-Y bond energies, M,M'=C-Pb. 68 ligands, Y.	6842	PP
	$(Me_2NCH_2)_2$ adducts of MCp_2, M=Sn,Pb	6705	PP
	Mono- and bis-amidinate ($[HNCHNH]^-$) complexes of Si-Pb. Valence isoelectronic ligands studied.	9605	PP
	Diphosphanylmethanide ($[R_2PCHPR_2]^-$) M(II) complexes; M=Si-Pb.	9606	PP

Element	Compounds	Ref.	Method
	HMOOH, MeMOOH isomers; M=C-Pb. Shift from M(IV) to M(II).	9396,9397	PP
83 Bi	Bi_2.	7388,10103	PP
	Bi_2. Spin-orbit effects via ZORA.	10077	DFT
	M_2; M=Tl,Pb,Bi.	6852	DFT
	M_4, M_6; M=P-Bi. Benzene analogues, prismanes etc.	8405	PP
	M_n^-; M=Sb,Bi.	7814,8281	DFT
	Bi_5^{3+}.	8092	PP
	M_5^{3+}; M=Sb,Bi.	10048	PP
	Bond lengths for many small inorganic molecules, up to BiH. A test on the authors' PP.	7966	PP+DFT
	BiH.	6634,7461	PP
	MH; M=Tl→Bi.	10213	PP
	MH_3; M=As-Bi.	7051	PP
	MH_4^-, MF_4^-; M=P-Bi.	8942	PP
	MH_5; M=P-Bi.	8943	PP
	MF_5, MH_6^-, MF_6^-, MH_7^-, MF_7^-; M=P-Bi.	8944	PP
	BiF.	6636	PP
	BiI.	6635	PP
	BiN.	6640,6642 6633	PP
	BiX; X=N-Sb.	7530	DF/HF
	BiO.	6641	PP
	BiS.	8682	PP
	R-Bi=Bi-R.	9983	PP
	R-M=M-R; M=As-Bi.	7286	PP
	MF_3; M=N-Bi.	9637	PP
	MX_3; M=N-Bi; X=F-I.	9640	PP
	MF_7^{2-}; M=As-Bi.	7498	PP
	$(RhBi_7)Br$. $5c-4e$ bonding.	10248	PP
	MMe_5; M=Sb,Bi.	7924	PP
	BiOH, HBiO.	8341	PP
	PMH_2; M=As-Bi.	8760	PP
	$Bi_3O_4^+$, $Bi_5O_7^+$.	8376	PP
	$[H_2E\text{-}EH_2]_2$; E=As-Bi. Chain fragments.	8398	PP
	Bi_2H_4. $(C_4H_4Bi)_2$. Chain fragments. HF-level.	8711	PP
	$[Bi_4Co_9(CO)_8(\mu\text{-}CO)_8]$.	10366	DFT
	$H_3M{=}CH_2$; M=N-Bi. Their reactions.	9014	PP
	Test of Troullier-Martins (1991) PP on R_e, ω, and D_e of a large number of small molecules, $Z \leq 83$.	7175	PP+DFT
84 Po	M_3; M=Se-Po	6789	PP
	HE; E=O-Po.	9257	DF/HF
	H_2E; E=O-Po.	9256	DF/HF
	M_2H_2; M=Tl→Po.	7957	PP
	$[HE\text{-}EH]_2$; E=Se-Po. Chain fragments.	8398	PP
	PNC for chiral molecules HEEH; E=O-Po. HTeOH.	8553	DF.
	EF_6^{2-}; E=Se-Po. Structure.	8302	PP,DPT
	CuX, Cu_2X; X=O-Po.	8762	PP
	PoX; X=C-Sn.	7530	DF/HF

Element		Compounds	Ref.	Method
85	At	X_2; X=F-At. DF to DF+CCSD(T).	10122	
		HX; X=F-At. Up to CCSD(T). R/NR.	10130	DF+corr
		MH; M=Tl→At.	7462, 9651	PP
		X_2, XH; X=F-At.	7473	PP
		HX, HX^+; X=F-At.	9651	PP
		AtX; X=B-In.	7530	DF/HF
		$(HX)_2$; X=F-At.	7085	PP
		AtN_3.	8390	PP
		CH_3X, CX_4; X=Br-At.	7957	PP
		XF_3; X=Cl-At.	9634	PP
		XF_6^-; X=Cl-At. Structure.	8302	PP,DPT
		Me_2SeX_2; X=F-At.	8252	DFT
86	Rn	Rn_2	7954	PP
		Xe_2, XeRn, Rn_2.	9486	PP
		$FeRg^+$; Rg=Ar-Rn.	7994	PP
		$CoRg^+$; Rg=Ar-Rn.	7992	PP
		RgX_n; Rg=Kr,Xe,Rn; X=F,Cl. Molecules and solids.	8641	DFT
		RgF_6; Rg=Ar-Rn. Structure.	8302	PP,DPT
		$MgRn^+$.	7232	PP
		$Rn{\cdot}H_2O$ complex.	8594	PP
87	Fr	MF; M=K-Fr. Electric properties.	8901	PT/DK
88	Ra	$M^{2+}(H_2O)_n$; M=Mg-Ra; n=1-6	7836	PP
90	Th	Th_2^{179+} model.	7746	DPT
		ThH_n; n=1-4.	9806	DFT
		ThO.	10181, 10182	DF
		ThO	7804, 8486	PP
		ThO^+	7275	PP
		NThO.	10350	DFT+PP
		CThO. Bent triplet.	10352	DFT
		ThN, ThN_2, ThN_4, Th_2N_2.	8505	DFT
		ThF_4. Assumed R 215 pm.	8786	DF
		$M(C_8H_8)_2$; M=Ce,Th	7478, 7479	PP
		$Th(Cp)_3$. R $6d^1$, NR $5f^1$ ground state.	8255	DV DFT
		H+MCl_3; M=Hf,Th. Bond energy. SR/NR.	10204	DFT
		Cp_3ThOCH_3. PES.	7913	DFT
		$Cl_2\overline{Th(CH_2PMeCH_2)}$	7451	PP
		$Th(S_2PMe_2)_4$.	8148	DFT
91	Pa	PaX_6^{2-};X=F-I, $5f^1$. Optical transitions.	8253	DFT
		MX_6^-; M=Nb,Ta,Pa,Db; X=F,Br.	9196	DV DFT
		$Pa(COT)_2$ $5f^1$.	8255	DV DFT
92	U	UH_n; n=1-4, U_2H_2, U_2H_4.	9807	DFT
		UF^{n+}; n=1-3.	9615	PP
		UO_2^+, UO_2^{2+}.	7273, 7274	PP
		UO_2^{2+} excited states.	10340, 10354	PP
		UO_2^{2+}, $HOUO^{2+}$, $U(OH)_2^{2+}$.	10066	PP
		$UO_2^{2+}{\cdot}L_n$; L=F^--Cl^-,OH^-,H_2O. Equatorial coordination.	10022	PP
		UO_2^{2+} with coordination of H_2O, OH^-. Simulates strongly alkaline solutions. Evidence for $[UO_2(OH)_4]^{2-}$.	10152	PP
		ThO_2, PaO_2^+, UO_2^{2+}. Bonding and bending.	7529	DF

Element	Compounds	Ref.	Method
	UN, UN_2, U_2N_3, U_2N_2.	8505	DFT
	UN, UO, NUO, NUO^+, NUO_2, $OU(\mu-N)_2UO$.	8506	DFT
	CUO, CUO^-, OUCCO, $(\eta^2\text{-}C_2)UO_2$, $U(CO)_n$; n=1-6.	10351	DFT
	NUO^+ isoelectronics, UO_3, UO_6^{6-}, UF_6,		
	$(OUO)^{2+}L_n$; L=CO_3^{2-}, NO_3^-,n=3; L=F^-, n=2-6.	9307	PP
	NUO, NUO^+.	10350	DFT+PP
	Uranyl nitrate complexes.	9082	DVM DFT
	NUO^+, NUS^+.	7991	PP
	MN_2; M=Mo,U.	7065	DFT
	Cluster models for solid UC.	8499	DV DFT
	$[UO_2X_4]^{2-}$; X=OH,F,Cl.	9608, 9609	DFT+PP
	Both *trans*- and *cis*-uranyls found.		
	UF_4. PES.	8990	DFT
	UCl_4	7925	PP
	UF_5.	9100	DVM DFT
	UF_6.	9104	DVM DFT
	UF_6.	9169	SR SW
	$An(C_6H_6)_2$; An=Th,U.	8047	PP
	UF_6. Assumed R 199.9 pm.	8787	DF
	UF_6^q, q=0,-1. R/NR.	7415	DF
	$M(CO)_6$; M=Cr,W,U,Sg.	9030	PP
	$U@C_{28}$.	7151	PP
	$An@C_{28}$; An=Pa,U.	10341	PP
	Cp_3AnL; An=Th,U; L=Me, BH_4.	7452	PP
	OPR_3 complexes of UO_2^{2+}, $UO_2(NO_3)_2$. R=H,Me,Ph.	8085	PP
	$U(C_8H_8)_2$. Excited states.	8687	PP
	Lanthanocenes are $4f^n\pi^3$, actinocenes $5f^{n-1}\pi^4$.		
	Bispentalene complexes $[An(C_8H_6)_2]$; An=Th,U.	7247	DFT
	$[(NH_2)_3(NH_3)U]_2(N_2)$. Actinide dinitrogen complex.	8256	DFT
	Mainly $U \rightarrow N_2$ π^* backbonding.		
93 Np	$An(NH_2)_3$; An=U,Np.	7978	PP
	PaX_6^{2-}, UX_6^-; X=F-I, NpF_6. $5f^1$ optical spectra.	8254	DV DFT
94 Pu	Pu_2.	6702	PP,DFT
	AnH_3; An=Th,U,Pu. Orbital energies.	9603	DV DFT
	AnF_6; An=U→Pu.	7979, 9609	DFT+PP
	$(AnF_6)_n$; n=1,2; An=U,Pu.	7799	PP
	PuN, PuN_2.	8505	DFT
	AnO_2^{2+}; An=U,Pu. Their nitrates, sulfates.	7297	PP
	AnO_2^{2+}; An=U,Pu.	8136	PP
	$AnO_2^{2+}\cdot nH_2O$; An=U,Pu; n=0,4-6.	9808	PP
	n=5 most stable.		
	PuO_2^{2+} excited states.	8815	PP
	$Pu(H_2O)_n^{3+}$ clusters, n=6...12.	6966	PP
	n = 8 or 9 preferred.		
	$An(C_8H_8)_2$; An=U-Pu.	7151	PP
	$An(C_6H_3R_3)_2$; An=Th,U,Pu; R=Me,^{t}Bu.	8620	DFT
	$(AnO_2)(NO_3)_2\cdot 2(TEP)$; An=U,Np,Pu; TEP=triethyl	8026	DFT-DVM
	phosphate.		
95 Am	AnO_2^{2+}, $HOAnO^{2+}$, $An(OH)_2^{2+}$; An=U-Am.	10065	PP

Element		Compounds	Ref.	Method
		$An(Ch)_2$; An=Th-Am; Ch=η^7-C_7H_7. Also UCh_2^-.	8619	DFT
		$An(Bz)_2$; An=Th-Am; Bz=C_6H_6.	8620	DFT
103	Lr	AnH, AnF, AnO; An=Ac,Lr. Actinide contraction.	8485	PP
		MF, MH, MH_3; M=Ac,Lr. Lanthanide and actinide contractions.	8550	DF+MP2
104	Rf	$RfCl_4$. R/NR.	8788	DFB
105	Db	MO; M=Nb,Ta,E105	7484	PP
		MCl_5; M=Nb,Ta,Db. Bonding trends.	9198	DV DVM
		MCl_4; M=Zr-Rf, MCl_5; M=Nb-Db, MCl_6; M=Mo-Sg.	9199	DV DFT
		MOX_3; M=V-Ta,Pa,Db; X=Cl,Br.	8114,8122	DFT
		MX_6^-; M=Nb,Ta,Pa,Db; X=F,Br.	9196	DV DFT
		$M(OH_2)_6^{5+}$; M=Nb,Ta,Pa,Db.	9194	DV DFT
		$M(OH)_nCl_{6-n}^-$, $MOCl_4^-$, $MOCl_5^{2-}$; M=Nb,Ta,Pa,Db.	9195	DV DFT
106	Sg	Halides of Groups 4-6.	8118,8122	DFT
		$M(CO)_6$; M=Cr-W,U,Sg.	9030,9035	PP
		$[MO_4]^{2-}$; M=Cr-Sg.	9200	DV DFT
		$MOCl_4$; M=Mo-Sg.	9201	DV DFT
		MO_2Cl_2; M=Cr-Sg.	9202	DV DFT
111		MH; M=Cu-E111.	9684	PP
		MF_n^-; M=Cu-E111; n=2,4,6.	9677	PP
		MX; X=H,F-Br,O,Au. M_2; M=Au,E111. SO effects.	8695	DFT
112		$(E112)H^+$, $(E112)F_2$, $(E112)F_4$.	9683	PP
113		(113)H, (113)F.	7955	PP
		$(113)X_n$; n=1,3,5; X=H,F-I.	9685	PP
114		MH; M=Sn-E114.	9031	PP
		MX_2, MX_4, M=C-E114, X=H,F,Cl.	9679	PP
117		(117)H.	7955	PP
		XH; X=I,At,E117.	9572	DF
		XH; X=Br-E117. $7p8s$ hybridization!	9034	PP
118		RgF_n; n=2,4; Rg=Xe,Rn,E118.	7958	PP
		RgF_4; Rg=Xe-E118.	9031,9032	PP

Chapter 8

Solid-State Theory

Table 8.1: Band-structure calculations.

Reference	Comments
Bucher (1990)	Cohesive properties of AgCl. Van der Waals important.
Z. W. Lu et al. (1991b)	Long-range order in PtX; X=Ni,Cu,Rh,Pd. In PtNi, Pt $6s$ stabilization promotes ordering, not phase separation.
Troullier and Martins (1991)	PP for plane-wave calculations. R/NR Cu.
Cooper et al. (1992)	Surface electronic structure and chemisorption. U, Pu.
Richter et al. (1992,1995)	Crystal-field parameters of $SmCo_5$ obtained from a solid-state ab initio calculation.
Saalfrank (1992)	Quantum size effects in thin lead films. Pb(111).
Shik et al. (1992)	Methods for electronic structure and magnetic properties of substitutional impurities in regular crystals.
Solov'ev (1992)	AnC, AnN; An=U→Pu. Electron structure and magnetic properties.
Ahuja et al. (1993)	High-pressure phases of Ti, Zr, Hf.
Aldén et al. (1993)	Core-level shifts at surface for $4d$ and $5d$ metals.
Antonov et al. (1993a)	Optical properties of cubic $5d$ metals.
Antonov et al. (1993b)	Optical properties of hcp $5d$ metals.
Antonov et al. (1993a)	Optical properties of actinides.
Bauer et al. (1993)	$GdAl_2$. ^{27}Al Knight shift, ^{157}Gd E2 hfs.
Bose et al. (1993)	Semiconducting $CsSnBr_3$.
Braun and Borstel (1993)	Photoemission from GaAs(110).
de Mello et al. (1993)	Zr, Hf. Electric field gradient.
Eibler et al. (1993)	Unreconstructed Au(001) surface.
B.-S. Fang et al. (1993)	SO effects on electronic structures of Nb(001).
Fernández Guillermet et al. (1993)	Cohesive properties of $5d$-metal carbides and nitrides with NaCl structure.
Fiorentini et al. (1993)	Reconstruction mechanism of fcc TM (001) surfaces. Close-packed quasihexagonal reconstruction occurs for $5d$ metals Ir,Pt,Au, but does not occur for the $4d$ metals Rh,Pd,Ag. Driven by relativistically induced $5d$ charge depletion on surface.
Hao et al. (1993)	Surface electronic structure of γ-U.
Hemstreet et al. (1993)	SO PP for solids. III-V semiconductors.
Hjelm and Calais (1993)	SCF calculations of Zeeman splittings in metals.
Hjelm et al. (1993)	Induced magnetism in U metal. L and S parallel, opposite to Hund's third rule.
Jäger et al. (1993)	Ca_3AuN is an $Au^{-}N^{3-}\cdot 2e^{-}$ auride subnitride. Compared with Al_2Au, CsAu.
Jepson and Anderson (1993)	Electronic structure of hcp Yb.
Johansson and Brooks (1993)	Theory of cohesion in lanthanides and actinides.
Koshibae et al. (1993)	Cuprates with SO interaction.
Lovatt et al. (1993)	Relativistic spin-polarized scattering theory for space-filling potentials.
Massidda et al. (1993)	HF-LAPW approach.
Mendez et al. (1993)	Transfer matrix method for 1D band structures.
Moruzzi and Marcus (1993)	Trends in bulk moduli of $3d$ and $4d$ metals. Relativistic corrections counterproductive.
Ozoliņš and Körling (1993)	Structural properties of $3d, 4d, 5d$ TM. GGA.
Papanikolau et al. (1993)	Also $5d$ and some sp impurities (B,C,N) in alkali

Reference	Comments
	metals can be magnetic.
Pick and Mikušik (1993)	Pd and Pt overlayers on W(011).
Postnikov et al. (1993)	Ferroelectric structure of $KNbO_3$ and $KTaO_3$.
Romanov (1993)	Origin of linear terms in quasi-2D dispersion law. SO.
Safonov (1993)	Crystal-field anisotropies handled via the metric tensor of the space-time, for Dirac electrons.
Shick and Gubanov (1993)	An impurities in Th; An=U,Np,Pu.
Smelyansky et al. (1993)	Conduction-electron g-factors in noble metals.
Stahler et al. (1993)	Magnetic K-edge absorption in $3d$ elements. Relation to magnetic structure.
Thole et al. (1993)	Multiplet FS in photoemission of Gd and Tb $5p$ levels.
Timms and Cooper (1993)	Electron momentum distribution in Pb.
Weinberger et al. (1993ab)	Effective pair interactions in the Au-Pd system.
J. H. Xu et al. (1993)	Tight-binding theory of rhombohedral As,Sb,Bi including SO.
L. H. Yang et al. (1993)	Effect of $(n-1)p$ semicore banding on lattice constants of K-Cs. Increase by 2-3 %. Comes from hybridization with valence orbitals, not from direct overlap.
Agassi and Restorff (1994)	Inclusion of SO in band-structure calculations along a symmetry axis. PP. Cylindrical coordinate multipoles. PbSe[111].
Ahuja et al. (1994a)	Magnetism of Gd and Tb.
Ahuja et al. (1994b)	Optical properties of PdO and PtO.
Ahuja et al. (1994c)	Fermi surface of noble metals (Cu,Ag,Au).
Ahuja et al. (1994d)	Influence of semicore states on high-pressure phases of transition metals (Y, ... Pt).
Basu et al. (1994)	Localization of rel. electrons in a 1D disordered system.
Drchal et al. (1994)	Random alloys and their surfaces. Tested on fcc $Cu_{75}Au_{25}$.
Faulkner (1994)	Scattering matrices for non-spherical SR potential.
Jeon et al. (1994)	X-ray absorption studies of d occupancies of $4d/5d$ transition metals, compounded with Group III/IV ligands.
Kalpana et al. (1994)	Phase stability of BaE; E=S-Te.
Kaupp and von Schnering (1994c)	Crystal HF for MF_2; M=Cd, Hg. R/NR for Hg.
Kirchhoff et al. (1994a)	Solid Te. Polymorphs under pressure.
Kirchhoff et al. (1994b)	Solid AgCl.
Kirchner (1994)	DFT studies of adsorption to metal surfaces. Thesis. Ar/Ag(111), Ag/Si(111), O/Ge(001), Sb/Ge.
Markendorf et al. (1994)	Hcp TM ($3d-5d$, Sc,...,Os). NMR spin-lattice relaxation.
Munzar and Christensen (1994)	Sn/Ge superlattices.
K.-T. Park et al. (1994)	MO; M=Pd,Pt,Ag.
Reinisch and Bross (1994)	Total energy and Fermi surface of gold.
Shick and Gubanov (1994)	An impurities in bcc Fe; An=U,Np,Pu.
Singh (1994ab)	Relativistic effects in Zn, Cd and Hg. Mercury crystal structure due to relativity.
Söderlind et al. (1994,1995)	Ce, Th→Pu. Electronic properties using GGA.
Steinbeck et al. (1994)	Obtain crystal-field parameters for the $4f$ states of Er and Dy in Ag or Au from solid-state DFT calculations.
Svane (1994,1996)	Ce. α phase has delocalized $4f$ electrons.
Weinberger et al. (1994)	Cu-Au alloys.

Reference	Comments
Würde et al. (1994)	Surface electronic structure of Pb(001), (110), (111). $6s, 6p$ bands decoupled.
Antonov et al. (1995)	Magneto-optical properties of ferromagnetic metals.
Boisvert et al. (1995)	Diffusion on (100), (111) surfaces of Ag, Au, Ir. Ag, Ir (111) barriers equal melting point, Au one is twice it.
Brooks et al. (1995)	Review on trends from TM to Ln and An.
Burdett and Sevov (1995)	Stability of oxidation states of Cu. Oxides. SR LMTO. Also Ag, Au.
Bzowski et al. (1995a)	Ag or Au overlayers on Ru(001).
Bzowski et al. (1995b)	Au-M intermetallics; M=Al-In, Sn→Te.
Costa Cabral and Silva Fernandes (1995)	Dynamics and structure of molten CsAu. Au^- dimers at $R \sim 3$ Å found.
Dronskowski (1995)	In_2ThBr_6. $6d$ bonding for Th.
Dufek et al. (1995)	Iron compounds. Yield a new $^{57}Fe^*$ nuclear quadrupole moment.
Fast et al. (1995)	Elastic constants of hexagonal transition metals.
Grin et al. (1995)	Hyperbolic lone pair structure in $RhBi_4$.
Hammer and Nørskov (1995)	H_2 on M(111); M=Ni,Cu,Pt,Au. 'Why gold is the noblest of all metals.' Relativity not mentioned.
M. Heinemann et al. (1995)	β-PbO_2. Semimetal.
Hota et al. (1995)	PbTe. ^{207}Pb NMR shifts including SO effects.
Johansson et al. (1995)	Fcc Th. Stabilized by $5f$.
Krasovska et al. (1995)	RuO_2. Optical and photoelectron spectra.
Kwon et al. (1995)	Solid Ar and Kr under high pressure. $V/V_0 = 1$ to 0.1.
Y. Liu and Allen (1995)	Sb, Bi. TB+SO.
Medvedeva et al. (1995,1996)	δ-Bi_2O_3 with or without oxygen defects.
Z. W. Lu et al. (1995)	M_3X; M=Pd,Pt; X=Sc-Cu. Spin-polarization-induced structural selectivity.
Miyazaki et al. (1995)	DCNQI-(Cu,Ag) systems.
Paulus et al. (1995)	Group-4 semiconductors (C-Sn, diamond structure). HF + correlation from clusters. PP.
Sandratskii and Kübler (1995,1999)	Magnetic structures of uranium compounds. Effects of relativity (SO) and symmetry.
Terpstra et al. (1995)	PbO. red and yellow. Pb $6s^2$ hybridizes with O $2p$ and does not form an 'inert pair'.
Willatzen et al. (1995)	II-VI zinc blende materials. SO parameters and electron g-factor.
C. J. Wu et al. (1995)	Ta (100) and (110) surfaces.
R.-Q. Wu (1995)	Bimetallic interface Pd/Ta(110).
Ahuja et al. (1996)	Electronic and optical properties of HgI_2.
Becker et al. (1996)	Pu_3X; X=Al-Tl. Covalency between Pu df and X p.
Bei der Kellen and Freeman (1996)	Self-consistent relativistic full-potential Korringa-Kohn-Rostoker total-energy method. Pd, Ir, Pt, Au. InSb etc.
Charpentier et al. (1996)	Ba, α-Ce, Th.
X.-J. Chen et al. (1996)	II-VI semiconductors, Zn-Hg, S-Po. Cation-d–anion-p hybridization.
Cortona and Villafiorita Monteleone (1996)	Cohesion and structure of MO; M=Mg-Ba.
Delin et al. (1996)	Optical properties of MC, MN, MO; M=Ti-Hf.
de Wijs et al. (1996)	First-principles MD of liquid Mg_3Bi_2.

Reference	Comments
C. M. Fang et al. (1996)	$Pb_{1/3}TaS_2$.
Gasche et al. (1996)	Optical properties of Th, Pa, U.
Grosch and Range (1996)	MAu; M=Li-Cs. LiAu metallic, others ionic. The NaCl, CsCl and NaTl structures compared.
Grosch and Range (1996b)	MAu_5; M=Na-Cs.
Hjelm et al. (1996)	WO_3, MWO_3; M=Li,Na,H, H_2WO_3. Optical properties. H forms a hydroxide, alkalis give electrons to $5d$ band.
Karzel et al. (1996)	ZnO, ZnSe. Mössbauer isomer shift, (wurzite) EFG.
Mehl (1996)	Structure of Hg.
Mejías (1996)	Adsorption of Cu-Au on NaCl(001) surface. Cluster models.
Mian et al. (1996)	Solid galena, PbS. HF. Different PP, correlation corrections compared.
F. Nogueira et al. (1996)	Transferability of *local* PP, based on solid-state electron density. Good for Na, useful for K-Cs, Mg, Al-Tl, Sn-Pb, poor for Li, Be, Ca-Ba.
Paulus et al. (1996)	Cohesive energies for III-V semiconductors (BN to InSb, with ZnS structure). Localized-orbital CCSD. PP.
Pick (1997)	CO chemisorption on Pt(111).
Richter et al. (1996)	U_2Pd_2M; M=In,Sn. Giant magnetoresistance.
Shick et al. (1996)	Relativistic Green's function method for electric and magnetic properties of random alloys. Ferromagnetic $3d$ metals. Fcc alloys $Co_{50}M_{50}$; M=Ni,Pt.
Springborg and Albers (1996)	Electronic structure of Pt in polyyne, $[Pt(PH_3)_2C_6]_x$.
Steinbeck et al. (1996)	Obtain crystal-field parameters for the $4f$ states of Sm in Fe and Co intermetallics from solid-state DFT calculations.
Weinberger et al. (1996)	'Band structure' and electrical conductivity of disordered layered systems.
Akella et al. (1997)	High-pressure phases of U.
Amirthakumari et al. (1997)	High-pressure phases of alkali halides.
Asokamani et al. (1997)	$AgGaX_2$, X=S-Te.
Brooks et al. (1997)	Crystal field excitations of Ln are combined from $4f$ excitations and a cloud of shielding conduction electrons. Spin Hamiltonians of TmSb and PrSb as examples.
Celestini et al. (1997)	'Hexatic' order on surface of liquid Au? Here MD simulation. Relativistic $5d-6s$ shifts shorten surface bonds of Ir, Pt, Au. See Fiorentini et al. (1993).
Ebert et al. (1997c)	Relativistic band structure of disordered magnetic alloys.
C. M. Fang et al. (1997)	ReS_2, $ReSe_2$, TcS_2.
Fehrenbach and Schmidt (1997)	Test the relativistic kinetic energy operator in spline-augmented plane-wave calculations.
Fleszar et al. (1997)	One-electron excitations and the plasmon in Cs metal. $5p$ semicore correlation shifts plasmon energy. 'Why is Cs yellow?'
Geipel and Hess (1997)	DK approach for solids (in crystal HF). Ag, AgX; X=F-Br. Bond contraction. Metal stabilized, salts destabilized by relativity.
Hwang et al. (1997)	Pr.
Karki et al. (1997)	$AgGaSe_2$. Elasticity and lattice dynamics.
Knöpfle et al. (1997)	U_3Sb_4. Symmetry properties of intra-atomic spin

Reference	Comments
	and angular-momentum densities.
Kollár et al. (1997)	α-Pu. GGA works.
Ladik (1997)	DF equations for solids.
M.-S. Liao et al. (1997)	CH_4+M(111); M=Ni-Pt,Cu.
M.-S. Liao and Schwarz (1997b;1999)	Hg_2L_2; L=F-I. Molecules and solids.
Lorenzana et al. (1997)	PbF_2: high-pressure phases.
Mishra et al. (1997)	Bi_2Se_3, Bi_2Te_3 including SO. Electronic structure and thermoelectric properties.
Pénicaud (1997)	δ-Pu, Es. Electron localization.
C. Persson and Lindefelt (1997)	SiC polytypes. SO important on hole masses.
Philipsen et al. (1997)	CO adsorption on (111) surfaces of Ni,Pd,Pt.
Ravindran et al. (1997)	Optical properties of monoclinic SnI_2.
Raybaud et al. (1997ab)	TM sulfides. Crystal and electronic structure. Catalytic activity related to TM-S bond energy.
Schmidt and Springborg (1997)	Linear or zig-zag chains of Tl,Pb,Bi.
Sekine et al. (1997)	TiC and UC: a comparison.
Severin et al. (1997)	Core M1 hfs of ferromagnetic Fe, Co, Ni.
Söderlind and Eriksson (1997)	Pa. Pressure-induced phase transitions.
Söderlind et al. (1997)	Structures of Pu.
Solanki et al. (1997)	Optical properties of HgI_2
Stachiotti et al. (1997)	ReO_3 and related oxides.
Svane et al. (1997)	Hfs parameters in tin compounds.
Terpstra et al. (1997)	Pb_3O_4, a mixed-valence compound.
Victora (1997)	AgX; X=Cl-I. Band structure.
Vitos et al. (1997)	Atomic volume of α-phase Fr$\rightarrow$Pu. Itinerant $5f$ found in Th to Pu.
Wei and Zunger (1997)	PbE; E=S-Te. The Pb $6s$ band below Fermi level leads to band structure anomalies.
Younk and Kunz (1997)	Solid $Pb(N_3)_2$. Band structure supports Gilman detonation model.
Zeng et al. (1997)	Ba under high pressure. Hexagonal close-packed phase. Transfer from $6s$ to $5d$ crucial in the decrease of c/a ratio.
Ahuja et al. (1998a)	High-pressure phases of CsH.
Ahuja et al. (1998b)	High-pressure phases of Sr.
Boettger (1998)	SR LCGTO. Applied on Au.
Boettger (1998b, 1999)	Phases of Mo. R/NR.
Boucher and Rousseau (1998)	Cs_3Te_{22}. Te-Te contacts ca. 340 pm. Two $CsTe_6$ slabs + four Cs Te_8 layers in unit cell.
Boykin (1998)	SO effects in tight-binding models. GaSb.
Deb and Chatterjee (1998)	Electronic structure and bonding of Ag_2O.
Delin et al. (1998ab)	Cohesive properties of lanthanides. Effect of generalized gradient corrections.
Delin et al. (1998c)	PbCh; Ch=S-Te.
S.-Q. Deng et al. (1998)	EHT band structure of Hg. Superconductivity discussed.
Doublet et al. (1998)	Electrical properties of TiX_2; X=S-Te. QR/NR.
Dudešek et al. (1998)	GaN. Ga-$3d$-to-N-$2s$ bonding.
Hu et al. (1998)	Band-structure version of the DF code MOLFDIR. Application to a $1D$ chain of Se.

Reference	Comments
Huhne et al. (1998)	Relativistic KKR method. Bcc Fe, fcc Co, fcc Ni.
Kokko and Das (1998)	Ground-state properties of 3*d* and 4*d* metals. GGA.
Krasovska et al. (1998)	The colour of sulphur. Electronic.
M.-S. Liao et al. (1998)	CN^- adsorption on M(100) electrodes; M=Cu-Au.
M.-S. Liao and Zhang (1998b)	CH_4 dissociation on Ru,Rh-Ir,Ni-Pt,Cu-Au.
Löken et al. (1998)	$Tl_2Au_4S_3$. Au-Au and Au-S interactions.
Lundin et al. (1998)	Hard MO_2; M=Ru,Os. Latter harder.
Maehira et al. (1998)	UC Fermi surface.
Miyagi et al. (1998)	Br_2, I_2 under high pressure.
Nemoshkalenko and Antonov (1998)	Computational methods in solid-state physics. Book.
Nunes et al. (1998ab)	Solid AgX; X=Cl-I. Pressure-induced phase transitions.
Ozoliņš et al. (1998)	Cu-Ag, Ag-Au, Cu-Ag and Ni-Au intermetallics. Phase diagrams and structures.
Papakondylis and Sautet (1996)	α-MoO_3. Crystal HF. H_2O, CO adsorption.
Patnaik et al. (1998)	PbTe. Nuclear spin-spin coupling.
Pavone et al. (1998)	Sn. $\alpha \leftrightarrow \beta$ phase transition. Entropy driven, due to different vibrational spectra.
Ravindran et al. (1998)	High-pressure phases of Ce.
Sághi-Szabó et al. (1998)	Piezoelectricity in $PbTiO_3$.
Seifert et al. (1998)	DFT with Car-Parrinello for Zintl systems. Liquid NaSn.
Seshadri et al. (1998)	Metal-metal bonding and metallic behaviour of some ABO_2 Delafossites. $AgMO_2$; M=Fe,Co,Ni. $MCoO_2$; M=Pd,Pt.
Sigalas et al. (1998)	Average interstitial electron densities for alkali, noble and transition metals. Related to cohesive energy.
Söderlind (1998)	Theory of the crystal structures of Ce, early An. Review.
Söderlind and Moriarty (1998)	Ta up to 10 Mbar.
Strange (1998)	Relativistic QM with condensed-matter and atomic physics applications. Book.
Vogel et al. (1998)	Solid AgX; X=Cl-I.
Yamagami (1998)	All-electron spin-polarized relativistic LAPW method. Bcc Fe, hcp Gd, U monochacogenides.
Yamaguchi and Miyagi (1998)	Molecular solid I_2 under pressure. SR. Transition from molecular phase to monatomic body-centred orthorombic (bco) phase.
Yoo et al. (1998)	Phase diagram of U at high p and T.
Y. Zhang et al. (1998)	Scheelite materials AMO_4; A=Ca,Pb; M=Mo,W.
Alatalo et al. (1999)	Truncated pseudopotentials for alloy calculations. Cut-offs in momentum space introduced. Pd, Pd-Al alloys as example.
Asato et al. (1999)	Full-potential KKR calculations for metals and semi-conductors. Lattice parameters, bulk moduli agree with other FP methods. Fe,Ni,Cu; Rh,Pd,Ag.
Boettger et al. (1999)	Phases of U.
Bose (1999)	Electronic structure of liquid mercury.
Carlesi et al. (1999)	Cs under pressure.
de Wijs and de Groot (1999)	Amorphous WO_3. W-W bonds of <3 Å occur.
Fehrenbach (1999); Fehrenbach and Bross (1999)	SR spline augmented plane-wave method. DK. Ag, Au. Pd. GaN.

Reference	Comments
Filonov et al. (1999)	Isostructural Ru and Os silicides and germanides. Os has larger gaps than Ru.
Ge and King (1999)	CO chemisorption on Pt(110). Slab model.
Hemmingsen et al. (1999)	Structure and EFG:s in β-$Cd(OH)_2$.
Korzhavyi et al. (1999)	Vacancy formation energy in transition and noble metals.
Krasovskii (1999)	Fe-Ni compounds at the Earth's core conditions.
Lalić et al. (1999)	Hf_2Fe. EFG.
Landrum et al. (1999)	Ce(III) versus Ce(IV) nitrides and other solids. Odd electron in CeN more d than f.
Liubich et al. (1999)	Interstitial B in W.
Lowther et al. (1999)	Relative stability of ZrO_2 and HfO_2 structural phases.
Miyazaki and Ohno (1999)	Solid $(Me_4N)[M(dmit)_2]_2$; M=Ni,Pd.
Okamoto and Takayanagi (1999)	Infinite 1D chain of gold atoms. Structure and conductance.
Pacchioni et al. (1999)	TM core level shifts upon adsorption on Al_{37} model of Al(100). 'TM'=Ne,K,Cu-Au,Ni-Pt. Covalent interaction.
Rościszewski et al. (1999)	Cohesion of rare-gas crystals, Ne-Xe. Three-body contributions nearly 7% for Xe.
Rotenberg et al. (1999)	SO-induced surface band splitting in Li/M(110); M=Mo,W. A probe on surface potential gradient. Experimental.
Schmid et al. (1999)	Relativistic GGA tested on 5d TM. Band structure and cohesive properties. Pt, Au.
Senda et al. (1999)	Liquid K-Pb alloys. Pb_4^{4-} Zintl ions found.
Shick et al. (1999)	Electronic structure and phase transitions of Bi.
Shishidou et al. (1999)	5f orbital magnetic moment of US.
Sökeland et al. (1999)	Separability of relativistic electron propagators. Scattering matrices from Dirac-Green functions. Photoemission, PES.
Soliman and Abelraheem (1999)	Modification of band structure in intense laser fields. 1D model. New gaps introduced.
Springborg (1999)	PtS_2 chains. 3D K_2PtS_2.
Suzuki and Nakao (1999)	Fully relativistic LCAO approach for crystals. Au, InSb. R/NR.
Temmerman et al. (1999)	Electronic configuration of Yb compounds. Yb(II) versus Yb(III).
Tse et al. (1999)	K-Ag intermetallics at high pressure. Claims Ag-Ag bonding interactions at 554 pm, finds short K-K distances of 339 pm. Ag^+ form hexagonal layers, interlayer Ag-Ag 360 pm.
Voss et al. (1999)	WSe_2.
Watson et al. (1999)	Ab initio calculation reproduces the distorted structure of solid α-PbO. The 'sterically active lone pair' has, however, Pb 6s - O 2p hybridization. Cp. Terpstra et al. (1995).
Xie et al. (1999)	Ag. Thermal properties in quasiharmonic approximation.

Table 8.2: Relativistic theories of solid- and liquid-state phenomena.

Reference	Comments
Abrikosov and Gor'kov (1962)	SO interactions and the Knight shift in superconductors.
Appel (1965)	SO effect on Knight shift of main-group superconductors.
Ham (1965)	Quenching of SO by dynamical Jahn-Teller effects.
Akai et al. (1990)	Theory of hyperfine interactions in metals.
Alouani and Albers (1993)	Magnetism in linear MX chains, M=Ni,Pt, X=Cl-I.
Balcar and Lovesey (1993)	Magnetic neutron spectroscopic intensities of Sm.

Reference	Comments
Bonesteel (1993)	SO scattering and pair breaking in disordered Cu-O layer.
Chalker et al. (1993)	Correlations at mobility edge in a 2D system with SO scattering. Anderson metal-insulator transition.
Coehoorn and Buschow (1993)	Hfs field at Gd nuclei in intermetallic compounds.
Eberhart et al. (1993)	Mechanisms of fracture. Why is CuAu ductile but TiAl brittle. Transition state for decohesion sought.
Ebert and Akai (1993)	Relativistic effects on hfs of TM.
Ebert et al. (1993)	Magnetic x-ray dichroism in dilute and concentrated TM alloys.
Godfree and Staunton (1993)	Relativistic approach to magnetic anisotropy.
Halilov et al. (1993)	Photoemission from magnetic compounds.
J. Luo et al. (1993)	Scattering operator for elastic and inelastic resonant x-ray scattering. Applied on Ln, An, TM.
Ramazashvili (1993)	Electron states in planar 2D ferromagnet. SO effects.
Reinisch and Bross (1993)	Relativistic effects on the Compton profile of polycrystalline gold.
Roy and Basu (1993ab)	Relativistic electrical conduction in disordered systems.
Schmiedeskamp et al. (1993)	Spin polarization in $5d$ photoemission from a Tl film.
Taylor and Gyorffy (1993)	Ferromagnetic monolayer with model SO and dipole-dipole interactions.
X. D. Wang et al. (1993)	Circular magnetic x-ray dichroism in heavy rare-earth metals.
Ebert and Guo (1994)	MXD spectra using spin-polarized relativistic LMTO.
Gotsis and Strange (1994)	Relativistic theory of x-ray Faraday effects.
Grass et al. (1994ab)	Theory of photoemission from solids.
Jenkins and Strange (1994)	Relativistic, spin-polarized single-site scattering theory.
Kulakowski et al. (1994)	Magnetic anisotropy and magnetostriction of atom pairs. SO.
Macêdo (1994)	Influence of SO scattering on electron-wave propagation in a disordered, quasi-1D medium.
Meir and Wingreen (1994)	SO scattering and the Kondo effect.
Meservey and Tedrow (1994)	Spin-polarized electron tunneling.
Ostanin and Shirokovskii (1994)	New method for magnetocrystalline anisotropy energy in relativistic ferromagnets.
Rajagopal (1994)	DFT including EM fields in condensed matter.
Tamura et al. (1994)	Linear and circular dichroism in angle-resolved Fe $3p$ photoemission.
Tyson (1994)	Relativistic effects in x-ray-absorption fine structure.
Yildirim et al. (1994)	Symmetry, SO and spin anisotropies: A spin anisotropy can only arise for tetragonal symmetry, if both SO and Coulomb exchange is included. Magnetic insulators.
Campbell and Segre (1995)	SO-induced lattice instabilities for half-filled hopping Hamiltonian.
Capelle et al. (1995)	Relativistic theory of superconductivity.
Cole et al. (1995)	Auger parameter shifts from free atoms to solids, in alkali and alkaline earth metals.
Evangelou (1995)	Anderson transition in presence of SO coupling. 2D disordered systems considered.
Fluchtmann et al. (1995)	R full-potential photoemission theory for ferromagnetic materials.
Fuks et al. (1995)	Solid and liquid W using effective potential.

Reference	Comments
Jenkins and Strange (1995)	X-ray magnetic dichroism in random substitutional alloys of f-electron elements.
Kraft et al. (1995)	Magneto-optical Kerr spectra on UX surfaces, X=S-Te.
Tamura et al. (1995)	X-ray absorption near edge structure in metals. Relativistic effects and core-hole screening.
Újfalussy et al. (1995)	Magnetism of 4d and 5d adlayers on Ag(001) and Au(001). R/NR. Relativity kills magnetism for Ir mono- and Pt double layers.
Wolff et al. (1995)	Self-consistent SO and Zeeman terms in otherwise NR ab initio calculations. Magnetic structure factors for $CsCoCl_5$.
Zakharov and Cohen (1995)	Structural, electronic, vibrational and super-conducting properties of sulfur under high pressure.
Adrian (1996)	Conduction-electron spin relaxation in M_3C_{60}, M=K,Rb,
Continenza et al. (1996)	Optical conductivity of $YbCu_4AM$; M=Ag,Au.
Ebert (1996)	Review on magneto-optical effects in TM systems.
Ebert et al. (1996); Ebert et al. (1997b)	SO effects for Dirac equation and spin-dependent potentials divided into two parts, diagonal and off-diagonal. Applications to magneto-optical Kerr effect, SO-induced orbital moments.
Ebert and Schütz (1996)	SO-influenced spectroscopies of magnetic solids. Book.
Kuch et al. (1996)	Magnetic dichroism of perpendicularly magnetized Ni/Cu(001)
Rajagopal and Buot (1996)	Fundamentals of time-dependent DFT, including EM fields.
Roy (1996)	'Fibonacci lattice' of rectangular potential wells. Landauer resistance.
Vargas et al. (1996)	Electronic energy loss of low-energy protons, channeled in single-crystal Au $\langle 100 \rangle$ direction.
Ankudinov and Rehr (1997)	Relativistic theory of XAS and XMCD.
Arola et al. (1997)	Magnetic x-ray scattering from ferromagnetic Fe.
Capelle et al. (1997,1998)	Theory of dichroism for superconductors.
Ebert et al. (1997a)	Current density functional theory and spontaneous magnetization of solids. Large SO effects in Fe, Co. (Vignale-Rasolt theory.)
Grimaldi and Fulde (1997)	Screening of phonon-modulated SO interaction in metals. Comes from the spin-other-orbit term.
Kresse and Hafner (1997)	Metal/nonmetal transition in expanded fluid Hg. Occurs at 8.8 g cm^{-1}.
Moreau et al. (1997)	Relativistic effects in electron-energy-loss-spectroscopy on the Si/SiO_2 interface plasmon peak.
Oja and Lounasmaa (1997)	Nuclear magnetic ordering of simple metals.
Pollack et al. (1997)	Test of density-based local pseudopotential on 16 simple metals. Phonon spectra, bulk and shear moduli, resistivities.
Qian et al. (1997)	Persistent currents from competition between Zeeman coupling and SO interaction.
Arola and Strange (1998)	Magnetic x-ray scattering from single Cu crystals.
Delin et al. (1998ab)	Effect of GGA on lanthanide cohesive properties.
Chamarro et al. (1998)	SO effects in semiconductor nanocrystals. (CdS, ca. 2 nm).
Doll and Stoll (1998)	Ground-state properties of heavy alkali halides.
Doll et al. (1998)	Closed-shell interaction in solid AgCl and AuCl.

Reference	Comments
Jenkins and Strange (1998)	Magnetic x-ray circular dichroism in TM alloys.
E. Kim and Cox (1998)	Knight-shift anomalies in heavy-electron materials.
Kiselev et al. (1998)	Electron g factor in 1D and 0D semiconductor nanostructures (quantum wires and quantum dots).
Lyanda-Geller (1998)	Quantum interference and electron-electron interactions at strong SO coupling in disordered systems.
Petrilli et al. (1998)	EFG calculations using the projector augmented wave method.
Rajagopal and Mochena (1998)	SO interaction in many-body theory of magnetic electron systems. Includes two-electron SO.
Újsághy et al. (1998)	SO-induced magnetic anisotropy for impurities in metals. Finite-size effects in Kondo resistivity.
Bulgakov et al. (1999)	SO-induced Hall-like effect.
Capelle and Gross (1999a)	Dirac equation for superconductors.
Capelle and Gross (1999b)	Pauli equation for superconductors.
G.-H. Chen and Raikh (1999)	Exchange-induced SO effects in 2D systems. Effects on g-factor.
Ebert et al. (1999a)	Fully relativistic description of static M1 hfs in magnetic and nonmagnetic solids.
Ebert et al. (1999b)	Fully relativistic theory of magnetic EXAFS. Nonspherical potentials allowed. Ni, Pt, Fe_3Pt.
Ebert et al. (1999d)	Influence of SO and a current-dependent potential on residual resistivity of disordered magnetic alloys.
Huhne and Ebert (1999)	Fully relativistic description of magneto-optical properties of arbitrary layered systems.
Moroz and Barnes (1999)	SO effect on band structure and conductance of quasi-one-dimensional systems.

Chapter 9

Relativistic Effects and Heavy-Element Chemistry

Table 9.1: "Relativity and the periodic system". Periodic trends, reviews and pedagogical papers.

Reference	Comments
Petrucci (1989); Petrucci and Harwood (1993)	General chemistry textbook. Quotes relativistic effects.
Z. W. Lu et al. (1991b)	Long-range order in PtX; X=Ni,Cu,Rh,Pd. In PtNi, Pt 6s stabilization promotes ordering, not phase separation.
Haire and Gibson (1992)	Position of Pu among actinide metals and compounds.
D. C. Hoffman (1992)	Nuclear and chemical properties of E103-E105.
Sharpe (1992)	Inorganic chemistry textbook. Quotes relativity.
Yarkony (1992)	Spin-forbidden chemistry within BP approximation. Review.
Huheey et al. (1993)	Inorganic chemistry textbook. Quotes relativity.
Johansson and Brooks (1993)	Theory of cohesion in lanthanides and actinides.
Kaupp and Schleyer (1993a)	Why inorganic Pb(II) but organic Pb(IV)? Bent's rule for bond angles holds.
Kaupp and von Schnering (1993)	Proposal for the Hg(IV) compound HgF_4.
Kudo et al. (1993)	Review on strained, polycyclic compounds with M_n skeletons (M=Si-Pb).
Lambert et al. (1993)	'Inverted electronegativity' of Li and Na.
L. Li et al. (1993)	Experimental reactions of M_2; M=Cu-Au with O_2, N_2O, N_2, H_2, CH_4, CO, CO_2, C_2H_4 in the gas phase. Au_2 anomalous. Reasons discussed.
Norman and Koelling (1993)	Electronic structure of f electron metals.
Pisani et al. (1993)	Discuss relativistic effects in atoms and molecules using LCAO calculations on H-like systems as example.
Schwerdtfeger et al. (1993)	The Hg-Hg bond in inorganic and organometallic chemistry.
Umland (1993); Umland and Bellama (1996)	General chemistry textbook. Quotes the Dirac equation.
Eliav et al. (1994e)	Ground state of E111 (eka-Au) d^9s^2, not $d^{10}s^1$.
D. C. Hoffman (1994,1996); D. C. Hoffman and Lee (1999)	The heaviest elements.
Kaupp and von Schnering (1994a)	Dimerization of HgX_2; X=F-I. R: Weak C_{2h} dimers. NR: Covalent D_{2h} structures.
Kaupp and von Schnering (1994b)	Stability of Hg_2^{2+} attributed to differential solvation/aggregation effects in the condensed phase. Gas-phase $Hg_2X_2 \leftrightarrow HgX_2$+Hg shifted to right!
Klapötke and Tornieporth-Oetting (1994)	Textbook on main-group chemistry. Includes relativity.
M.-S. Liao and Schwarz (1994); M.-S. Liao and Schwarz (1997a); M.-S. Liao (1993)	Effective radii of monovalent coinage metals. For CN=2, Au(I) ca. 10 pm smaller than Ag(I).
W.-J. Liu and Li (1994)	Review on relativistic quantum chemistry.
Malli (1994ab)	Relativistic and electron correlation effects in molecules and solids.
Marian (1994)	Relativistic calculations on TM compounds.
Norman (1994,1997)	Periodicity and the p-block elements. School textbook. Mentions relativity.
Onoe et al. (1994ab); Onoe (1997)	Analyses relativistic effects on DVM-DFT MO:s for Cu_2, AuH, Pb_2, XF_6;X=S-Po, Mo-W,Re→Pt, U,Np,Pu.
Pershina and Fricke (1994a)	Highest chlorides of Rf→Sg.

Reference	Comments
Pershina et al. (1994a)	Physicochemical properties of Rf→Sg (E104-E106).
Pershina et al. (1994b)	Thermodynamic functions of Db (E105), deduced from atomic MCDF calculations.
Pershina et al. (1994c)	Complexation of Nb-Db in aqueous HCl solutions.
Ron et al. (1994)	Relativistic, retardation and multipole effects in atomic photoionization cross sections. Z, n, and l dependence.
Schwerdtfeger et al. (1994c)	Predict the existence of AuF.
Watson and Weinert (1994)	Charge transfer in gold–alkali-metal systems. 50/50 systems. Short Au-Au leads to metallic character for AuLi-AuK. AuRb, AuCs are semiconductors.
Wezenbeek et al. (1995)	H-MCl_3 bonds; M=Hf,Th. Relativity explains nearly half of the Th-H bond strength of 3.60 eV.
Hammer and Nørskov (1995)	H_2 on M(111); M=Ni,Cu,Pt,Au. 'Why gold is the noblest of all metals.' Relativity not mentioned.
Heinemann (1995)	Gas-phase ion-molecule chemistry. Experiment versus theory. Thesis.
Hess et al. (1995)	Review on SO effects and methods.
Holleman et al. (1995)	Inorganic chemistry textbook.
Kratz (1995)	Chemistry of the heaviest elements.
J. Li et al. (1995a)	$MH_4(PH_3)_3$; M=Fe-Os. Non-classical hydrides of Os due to relativity (cp. Pyykkö (1988a)[5680] p. 585).
Moc and Morokuma (1995, 1997)	Periodic trends in Group 15 (P-Bi). MX_5, MX_6^-, ...
Münzenberg (1995,1999)	Discovery of the heaviest elements.
Schädel (1995)	Chemistry of transactinides.
Seth et al. (1995)	Trends of atomic orbital energies and radii for Groups 11 (Cu-E111) to 15 (As-E115). Relativistic and shell-structure effects.
S. G. Wang and Schwarz (1995a)	LnH, LnF, LnO; Ln=La,Gd,Yb,Lu. Lanthanide contractions (La-Lu) 19, 10 and 5 pm, for H, F and O, respectively.
Yatsimirskii (1995)	Review on relativistic effects in chemistry.
Zumdahl (1995,1998)	General chemistry textbook. Quotes relativity.
Almlöf and Gropen (1996)	Review on relativistic effects in chemistry.
Bigeleisen (1996)	Nuclear size and shape effects in chemical reactions. Uranium redox reactions as example.
Cundari et al. (1996)	Review on pseudopotential methods.
Dolg and Stoll (1996)	Review on lanthanide chemistry.
Eliav et al. (1996b)	E118 (eka-Rn) first rare gas with electron affinity. Qualitative explanation the $8s$ LUMO stabilization.
Eliav et al. (1996c)	Is E113 (eka-Tl) a $6d$ transition metal?
Ionova et al. (1996a)	MOX_3; M=V-Ta,Pa,Db(E105); X=Cl,Br. Bonding trends.
Ionova et al. (1996b)	Atomic properties of Rf-Sg (E104-E106).
Ionova et al. (1996c)	Halides of Groups 4-6.
Ionova et al. (1996d)	Oxohalides of Group 6.
Ionova et al. (1996e)	Fugacity of halides and oxohalides of transition metals and transactinides. *Special comment:* According to V. Pershina, the preceeding papers Ionova (1996a-e) are unauthorized duplicates from her earlier papers.

Reference	Comments
Ionova et al. (1996f)	Sublimation enthalpy of 5d metals.
Kaupp et al. (1996b)	The 6th-row systems PoF_6^{2-}, AtF_6^-, RnF_6 octahedral, due to relativity. Their 5th-row analogues fluctuate.
Klapötke and Schulz (1996)	Quantum chemistry in main-group chemistry. Includes relativity.
Mackay et al. (1996)	Inorganic chemistry textbook. Includes relativity.
Musaev and Morokuma (1996)	Potential-energy surfaces of TM-catalyzed chemical reactions.
Neogrády et al. (1996)	QR CCSD(T) polarizabilities for M,M^+; M=Cu-Au. Ag > Au but $Ag^+ < Au^+$.
Pershina (1996)	Electronic structure and properties of transactinides and their compounds.
Roos et al. (1996)	Review multiconfiguration PT for molecules. Relativistic effects on optical properties of MCl_2; M=Sc,Cr,Ni.
Rösch et al. (1996)	DK DFT molecular calculations.
W. H. E. Schwarz et al. (1996)	Interpretation of relativistic effects on chemical bonding.
Seaborg (1996)	Evolution of the modern periodic table. Future sketched to E168.
Siegbahn (1996)	Electronic structure calculations for molecules containing transition metals.
Türler (1996)	Gas-phase chemistry of transactinides. Rf(E104), Db(E105).
Umemoto and Saito (1996)	Electron configurations of superheavy elements, Z=121-131. g-electrons appear at Z=126.
van der Lugt (1996)	Polyanions, especially Zintl anions, in ionic alloys A review.
Balasubramanian (1997ab)	Relativistic effects in chemistry.
Greenwood and Earnshaw (1997)	Chemistry of the elements. Textbook.
S. A. Cotton (1997)	Chemistry of precious metals. Treatise.
Fricke et al. (1997)	Review on superheavy elements. DVM.
Hess (1997)	Review on relativistic effects in heavy-element chemistry.
Janiak (1997)	Review on organothallium (I,II) chemistry.
Kaltsoyannis (1997a)	Review on relativistic effects in inorganic and organometallic chemistry.
Kaltsoyannis and Bursten (1997)	$Th(Cp)_3$. R $6d^1$, NR $5f^1$ ground state.
Küchle et al. (1997)	Lanthanide and actinide contractions. Depend on the ligand.
M.-S. Liao and Zhang (1997)	MH_x^{2-}, MCl_y^{2-}; M=Ni-Pt; x=2,4,6; y=4,6 in crystal field. Relativistic effects favour higher oxidation state.
Pacchioni et al. (1997)	CO adsorption on M; M=Ni-Pt. Relativistic trends.
Philipsen et al. (1997)	CO adsorption on (111) surfaces of Ni,Pd,Pt. Relativistic correction to CO/Pt adsorption energy 70% at SR level, 55% with SO included.
Pyykkö (1997)	Strong closed-shell interactions in inorganic chemistry. Review.
Schädel et al. (1997ab)	Experimental chemistry of Sg (E106).
Bartlett (1998)	Relativistic effects and the chemistry of gold.
Delley (1998)	Scattering-theory approach to scalar relativistic corrections to bonding. Local pseudopotential can include them.

Reference	Comments
Dolg (1998)	Lanthanides and actinides.
Günther et al. (1998)	Chromatographic extraction trend Zr>Rf>Hf in HCl/tributylphosphate.
Hess (1998)	Review on relativistic theory and applications in chemistry.
S. Hofmann (1998)	New elements – approaching Z=114.
Laerdahl et al. (1998)	MF, MH, MH_3; M=La,Lu,Ac,Lr. Lanthanide and actinide contractions. Relativity 10-30% of former, 40-50% of latter.
Mingos (1998)	'Essential trends in inorganic chemistry'. Mentions relativity.
Pershina (1998ab)	Hydrolysis of Group 5 cations in HCl: Nb,Ta,Pa,Db(E105).
Pershina et al. (1998)	Electronic structure of the transactinides.
Pyykkö et al. (1998); Labzowsky et al.(1999b)	Estimated valence-electron Lamb shifts for Li-E119, Cu-E111. About -1% of kinetic Dirac shifts for large Z. QED in chemistry about -1% of relativity in chemistry, for large Z.
Schwerdtfeger and Seth (1998)	Relativistic effects in superheavy elements.
Seth (1998)	Quantum chemistry of superheavy elements. Thesis.
Seth et al. (1998a)	Coinage metal fluoride anions for Cu-E111. E111 has the most stable +5 oxidation state.
Seth et al. (1998b)	Stability of oxidation state +4 in Group 14 (C-E114).
Söderlind (1998)	Theory of the crystal structures of Ce, early An. Review.
Autschbach (1999)	Trends of relativistic effects.
Cotton et al. (1999)	Inorganic chemistry textbook.
Fröhlich and Frenking (1999)	Bonding in transition-metal complexes. Carbonyls, carbenes, π-bonded systems, dihydrogen complexes.
Greiner and Gupta (1999)	Edited book on superheavy elements.
Han and Lee (1999)	SO effects and heavy RgF_n; n=2,4. Stabilization. $(E118)F_4$ obtains a T_d minimum due to SO.
Kaldor and Eliav (1999)	High-accuracy calculations for heavy and superheavy elements.
Kaltsoyannis and Scott (1999)	Textbook on f elements. Mentions relativity.
Kaupp (1999b)	Relationships between π bonding, electronegativity and bond angles in high-valent transition metal complexes. Bent's rule less valid for TM than for main-group elements.
Kedziora et al. (1999)	Relativistic effects on atomization energies. 148 species in Pople's G2/97 test set.
Kratz (1999)	Chemistry of the transactinides.
Leidler and Meiser (1999)	Physical chemistry textbook. Mentions relativistic effects.
W.-J. Liu and van Wüllen (1999)	MX; X=H,F-Br,O,Au. M_2; M=Au,E111. SO effects. Large atomic SO contributions due to E111 d^9s^2 ground state.
Mikheev and Rumer (1999)	Stabilization of Ln(II), An(II) in solutions, melts and clusters. Review.
Nash and Bursten (1999abc)	SO effects make $(118)F_4$ T_d, unlike D_{4h} XeF_4, RnF_4.
Nash and Bursten (1999d)	XH; X=Br-E117. $7p8s$ hybridization! Involves $7p_{3/2}$.
W. Paulus et al. (1999); Pershina and Bastug (1999)	Db(E105) chemistry in aqueous solution. Results lie between Nb and Ta.
Pershina and Fricke (1999)	Electronic structure and chemistry of the heaviest elements.
Pershina et al. (1999)	Redox potentials in acid solutions estimated for Cr-Sg.
Pettersson et al. (1999b)	New rare-gas-containing neutral molecules. Review.
Quiney et al. (1999)	Ab initio relativistic quantum chemistry.

Reference	Comments
Richardson et al. (1999ab)	HMOOH, MeMOOH isomers; M=C-Pb. Shift from M(IV) to M(II) in Group 14.
Schreckenbach et al. (1999)	DFT calculations on actinide compounds.
Seo and Hoffmann (1999)	Structures of solid elements P-Bi. Role of $s-p$ mixing.
Seth et al. (1999b)	Bonding trends in Group 13 (B-E113). MX_n; n=1,3,5. The $(E113)X_3$ are T-shaped, not D_{3h}. The $6d$ is involved.
Volkova and Magarill (1999)	Formation of polyatomic cations of mercury.
Watson et al. (1999)	The stereochemically active Pb(II) lone pair in solid α-PbO has $6s-2p$ hybridization.

Many of the papers in the following Table 9.2 also give data on relativistic changes of the force constant k_2 or the dissociation energy, D_e. C is the relativistic bond-length contraction, $C = R_{\mathrm{NR}} - R_{\mathrm{R}}$.

Table 9.2: Bond lengths, bond angles and potential-energy surfaces.

Reference	Comments
Barnes et al. (1993)	CO. C 0.013 pm. $\Delta\omega_e$ -1.3 cm^{-1}.
Bastug et al. (1993)	Au_2. DS (α=0.7) C 20.6 pm.
Dyall (1993a)	MO; M=Ge-Pb
Dyall (1993b)	PtH, PtH^+, PtH_2. PtH^+ C -2.5 pm. DF/HF.
Matsuoka et al. (1993)	PbO. C.
Schwerdtfeger and Ischtwan (1993)	TlX, TlX_3; X=F-I.
Bauschlicher et al. (1994)	AlCu C 4.2 pm. D_0 increase 0.225 eV.
Kaupp and von Schnering (1994a)	HgX_2; X=F-I, H. C.
Kaupp and von Schnering (1994c)	Solid HgF_2.
J. M. L. Martin and Taylor (1994)	HF. C -0.005 pm.
Park and Almlöf (1994)	Pt_2, MH,M_2; M=Ag,Au. Effect of two-electron DK terms.
Pisani and Clementi (1994b)	H_2E; E=O-Po. Bond-angle decrease for Po -1.66°, C 1.8 pm. DF/HF.
Pou-Amérigo et al. (1994)	CuH, NiH. PT relativity.
Schwerdtfeger and Bowmaker (1994)	MCO; M=Cu-Au.
Schwerdtfeger et al. (1994a)	MX_3; M=N-Bi; X=F-I. Inversion barrier.
Schwerdtfeger and Hunt (1999)	T-shaped transition state found for BiF_3.
Collins et al. (1995)	MH, M=Cu-Au. C at DF and DK level compared.
Schwerdtfeger (1995)	AuCl. R/NR for HF - QCISD(T).
Schwerdtfeger et al. (1995b)	AuF, AuF^+, AuF_2, Au_2F_2.
van Wüllen (1995,1996b)	C for carbonyls of Cr-W, Fe-Os, Ni-Pt.
S. G. Wang et al. (1995)	LnO; Ln=La ... Yb. YbF.
S. G. Wang and Schwarz (1995a)	LnH, LnF, LnO; Ln=La,Gd,Yb,Lu. C.
S. G. Wang and Schwarz (1995b)	MO, MH_4, MCl_4; M=C-Pb.
Saue et al. (1996)	XH; X=I,At,E117. DF/HF. C 0.6, -0.3, -12.9 pm, respectively.
Schwerdtfeger (1996)	Geometries of XF_3; X=Cl-At. Large bond angle effect.
Seth et al. (1996a)	HBr.
Seth et al. (1996b)	MH; M=Cu-E111. C.
Urban and Sadlej (1996)	C for AlM; M=Cu-Au. DK.
van Lenthe et al. (1996b)	SO effects on closed-shell diatomics. HI, ... Bi_2.
Visscher and Dyall (1996)	X_2; X=F-At. DF to DF+CCSD(T). C.

Reference	Comments
Visscher et al. (1996b)	HX; X=F-At. Up to CCSD(T). R/NR. C.
Barysz and Urban (1997)	BCu, BAg, BAu. C = 4.5, 16, 32.5 pm, respectively, at CASPT2 level.
D.-D. Dai and Li (1997)	Diatomic EuCh; Ch=O-Te. Small C (1-2 pm).
de Jong et al. (1997)	I_2.
Geipel and Hess (1997)	DK approach for solids (in crystal HF). Ag, AgX; X=F-Br. Bond contraction. Metal stabilized, salts destabilized by relativity.
Küchle et al. (1997)	Lanthanide and actinide contractions. Depend on the ligand (H,F,O). 6-11 pm for Ln, 11-17 pm for An. At NR level, expansion for the MO monoxides.
Laerdahl et al. (1997a)	MF; M=Cu-Au. C.
M.-S. Liao and Schwarz (1997a)	$[XMX]^{3-}$; M=Cu-Au; X=O-Se in crystal field. C.
W.-J. Liu et al. (1997b)	LnO, LnS; Ln=Eu,Yb. C.
Saue et al. (1997)	CsAu. DF/HF. C=41 pm.
Watanabe and Matsuoka (1997)	ThO. C -7.1 pm. DF/HF.
Barysz and Papadopoulos (1998)	NiH_2 X $^1A_1$ C 0.79 pm.
Barysz and Pyykkö (1998)	AuX^+; X=Be,Mg,C,Si. C 38.5, 31.2, 51.0 and 34.2 pm, respectively, at CASPT2 level.
Császár et al. (1998); Tarczay et al. (1999)	Potential-energy surface for water. The barrier to linearity. Relativity beats non-Born-Oppenheimer.
de Jong et al. (1998)	C for interhalogens up to IBr. HF to CCSD(T). All slightly negative.
Iliaš et al. (1998)	MF; M=Cu-Au. C, up to DK-CCSD(T) level.
H.-S. Lee et al. (1998)	SO effects on bonding in AuH,Au_2,TlH,Tl_2.
M.-S. Liao and Zhang (1998a)	RgX_n; Rg=Kr,Xe,Rn; X=F,Cl. Small C.
Malli and Styszyński (1998)	$RfCl_4$. C 6.7 pm. DF.
H. Müller et al. (1998)	HF at CCSDT1-r_{12} level. Effect on ω_e 23 cm^{-1}.
Quiney et al. (1998a)	TlF. DF/HF.
Seth et al. (1998a)	MF_n^-; n=2,4,6; M=Cu-E111. C at MP2 level.
Wahlgren et al. (1998)	Au_2 C as test of one-centre DK approximation.
Nash and Bursten (1999e)	$M(CO)_6$; M=Cr-W,U,Sg. Covalent radius for Sg 4 pm larger than for Mo or W.
Suzumura et al. (1999)	M_2, MH, MCl; M=Cu-Au. R/NR.
Turski (1999)	SiM; M=Cu-Au. C.

Table 9.3: Magnetic resonance parameters.

Reference	Comments
Ham (1965)	Dynamical Jahn-Teller effect in ESR. Partial quenching of SO.
Smelyansky et al. (1993)	Conduction-electron g-factors in noble metals.
Van de Walle and Blöchl (1993)	Hyperfine parameters in solids. Includes SR, uses δ-function FC term. PP.
Du and Li (1994)	Contribution of ligand SO to g-tensors in VX_2; X=Cl,Br.
Markendorf et al. (1994)	Hcp TM ($3d-5d$, Sc,...,Os). NMR spin-lattice relaxation.
Belanzoni et al. (1995)	g-Tensor, hfs of TiF_3.
Bündgen et al. (1995)	g-tensors in diatomic molecules. NO, O_2, SO.
Endo et al. (1995)	Si and C NMR shifts in SiX_4, $CH_{4-n}X_n$; X=Cl-I; n=1-4. PM3+SO.
Hota et al. (1995)	Solid PbTe. ^{207}Pb NMR shifts including SO effects. A 2% downfield effect via the 6s orbital.

Reference	Comments
Kaupp et al. (1995a)	Ligand NMR shifts in NMR complexes.
Kaupp et al. (1995b)	^{17}O NMR shifts in MO_4^q TM complexes.
Lushington et al. (1995); Lushington and Grein (1996)	Ab initio studies of g tensors on 2nd-row . molecules.
Nakatsuji et al. (1995a)	SO effects on NMR shifts. Si shifts in SiX_4; X=F-I, $SiXI_3$; X=Cl,Br. SO PP on X.
Nakatsuji et al. (1995b)	UHF PT method for SO effects on NMR shifts. 1-electron SO Hamiltonian only. HX, CH_3X; X=F-I. ^1H and ^{13}C NMR shifts.
Takashima et al. (1995)	MX_4^-; M=Ga-In; X=Cl-I. Ga, In NMR shifts. PT.
Willatzen et al. (1995)	II-VI zinc blende materials. SO parameters and electron g-factor.
Ballard et al. (1996)	Proton NMR shifts in HX, X=F-I.
Bastug et al. (1996)	Diamagnetic shielding for ^{209}Bi in $Bi(NO_3)_3$, Bi^{3+}. Almost equal. Lamb wrong. Recall Feiock and Johnson (1968).
Battocletti et al. (1996)	Influence of gradient corrections on M1 hfs in ferromagnetic Fe, Co, Ni.
Fowler et al. (1996)	EFG in BrCl. DK.
Fukui et al. (1996)	Relativistic effects in NMR shielding. Lowest order.
Hada et al. (1996)	^{183}W NMR shielding in WX_6; X=F,Cl, WO_4^{2-}. DK.
Hemmingsen and Ryde (1996)	EFG in Cd complexes.
Kaneko et al. (1996)	^{119}Sn NMR shifts in SnH_4, tin tetrahalides. PT SO. FC term dominates the shift.
Kaupp (1996a)	$[MCp(CO)_4]^-$; M=Ti-Hf. ^{13}C NMR shifts.
Kaupp (1996b)	^{13}C and ^{17}O NMR shifts in $Fe_2(CO)_9$, $Rh_6(CO)_{16}$.
Kaupp (1996c)	$[M(CO)_5L]$; M=Cr-W; L=PH_3, PX_3 ^{31}P NMR shifts.
Kaupp (1996f)	$[M_nC(CO)_m]^q$; M=Fe,Rh,Os interstitial carbides. ^{13}C NMR shifts.
Kaupp et al. (1996a)	$M(CO)_6$; M=Cr-W. C and O NMR shifts.
Kellö and Sadlej (1996b)	HX; X=Cl-I. Halogen Q.
Kutzelnigg et al. (1996)	IGLO method for NMR shift tensors. Review.
Malkin et al. (1996)	HX, CH_nX_{4-n}; X=F-I. SO-induced ^1H, ^{13}C NMR shifts.
Nakatsuji et al. (1996a)	HgX_2; X=Cl-I. ^{199}Hg NMR shifts. 1-electron SO only, FC operator on Hg.
Nakatsuji et al. (1996b)	AlX_4^-; X=H,F-I. ^{27}Al NMR shifts.
Ruiz-Morales et al. (1996)	$W(CO)_6$. C and O NMR shifts. SR/NR DFT.
Schreckenbach and Ziegler (1996)	DFT calculations of NMR shifts. (NR.)
K. Schwarz et al. (1996)	EFG calculations on solid borides. MB_2; M=Ti...Ta. MB_6; M=Ca-Ba.
Ehlers et al. (1997)	^{13}C shifts in $M(CO)_6^n$, M=Cr→Fe, Tc→Ru, Hf→Ir.
J. A. González et al. (1997)	NMR ^1J coupling constants in Me_3XY; X=C-Pb; Y=F,Cl. MNDO-level correlated approach.
Havlas and Michl (1997,1998, 1999); Havlas et al. (1998)	Zero-field splittings in organic biradicals.
Havlin et al. (1997)	^{13}C shielding tensors in metal-olefin complexes.
Kaupp et al. (1997a)	CX_3^+; X=F-I. ^{13}C NMR shifts.
Kaupp et al. (1997b)	^{17}O NMR shifts in MO_4; M=Fe-Os, MnO_4^-; M=Mn-Re, MO_4^{2-}; M=Cr-W.
Kirpekar et al. (1997)	Spin-spin coupling in XH_4; X=C-Sn. SO corrections only, no SR corrections. Found small.

Reference	Comments
Nakatsuji et al. (1997b)	TiX_4; X=F-I. Ti NMR shift. NbX_6^-, $NbCl_5X^-$; X=F-I. Nb NMR shift.
Oja and Lounasmaa (1997)	Nuclear magnetic ordering of simple metals. Covers theory of nuclear spin-spin coupling.
Pyykkö and Seth (1997)	H-like and DF-level relativistic correction factors for nuclear quadrupole coupling. Discusses 'SO tilting'.
Quiney et al. (1997)	Relativistic calculation of EM properties of molecules.
Ruiz-Morales et al. (1997)	^{125}Te NMR shifts. SR DFT.
Sasaki et al. (1997)	SO-induced electron-spin polarization in donor-acceptor complexes containing heavy halogen atoms. ESR.
Schreckenbach and Ziegler (1997)	SR DFT calculations of NMR shifts. Method described. ^{17}O shifts in MO_4^{n-}; M=Cr-W,...Os. $M(CO)_6$; M=Cr-W.
Svane et al. (1997)	Hfs parameters in tin compounds. $Q(^{119}\mathrm{Sn}(24\ \mathrm{keV};3/2+))$ obtained.
Vaara and Hiltunen (1997)	CD_3X; X=F-I. Deuterium EFG. PP.
van Lenthe et al. (1997)	ZORA calculations of molecular g-tensors.
Bruna and Grein (1998)	g-tensors of O_3^-, O_3Na, O_3Na.
Fukui and Baba (1998)	NMR shielding at no-pair level.
Ishikawa et al. (1998)	DF finite-field theory of NMR shielding. Applications on He-Xe, H_2, HF, HCl.
Kaupp and Malkina (1998)	$CH_n(HgX)_{4-n}$; X=Cl,CN; n=0-4. RHgH; R=Me, Et,.. ^{13}C and ^{1}H NMR shifts.
Kaupp et al. (1998a)	NMR of transition-metal compounds.
Kaupp et al. (1998b)	A simple interpretation for SO-induced NMR shifts, analogous to FC spin-spin coupling. Applied on iodo-organic compounds. ^{1}H and ^{13}C shifts.
Kellö and Sadlej (1998a)	Picture-change effects for EFG. HI, HAt as examples.
Kellö and Sadlej (1998b)	Q(K) from KF, KCl.
Lobayan and Aucar (1998ab)	NMR spin-spin coupling constants within PM3 parametrization, CLOPPA scheme. As-Bi, Ga-Tl, Si-Pb.
Malkina et al. (1998)	HX, CH_3X; X=F-I. SO-induced ^{1}H, ^{13}C NMR shifts. Importance of two-electron SO terms.
Minaev et al. (1998)	R-dependence of the SO-induced ^{1}H NMR shift in HX; X=Cl-I.
Neese and Solomon (1998)	INDO/S-CI for ZFS and g-factors in TM complexes. $FeCl_4^-$ as example. Results compared with BP PT ab initio ones.
Patnaik et al. (1998)	PbTe. Nuclear spin-spin coupling. δ-function FC term used?
Pernpointner and Schwerdtfeger (1998)	GaF. Q(Ga) obtained.
Pernpointner et al. (1998a)	CsF. Q(Cs) obtained. DK CCSD(T).
Pernpointner et al. (1998b)	CuF. PCNQM (point-charge nuclear quadrupole moment model) developed. DF or DK + CCSD(T).
Quiney et al. (1998b)	Hfs in YbF.
Quiney et al. (1998c)	NMR shielding in H_2O. DF.
Schreckenbach and Ziegler (1998)	DFT calculation of NMR shifts and ESR g-tensors. Review.
Soldner et al. (1998a)	EFG calculations for isolated molecules using WIEN95. Examples: $CdCl_2$, HgX_2; X=F,Cl.
Soldner et al. (1998b)	Hg EFG for mercaptides $Hg(SR)_2$; R=Me,Et,Pr.
Vaara et al. (1998)	Quadratic response calculation of SO contributions

Reference	Comments
	to NMR shift tensors. HX, CH_3X; X=F-I.
van Lenthe et al. (1998)	ZORA DFT calculations of g-tensors, hfs, in M_5, M_7, M,M'=Cu-Au.
Visscher et al. (1998)	EFG in HX; X=Cl-I. MP2 to CCSD(T). R/NR.
Wolff and Ziegler (1998)	DFT-GIAO NMR shifts including SO. HX, CH_3X, TM carbonyl ^{13}C shifts; TM=Hf-Hg.
Yamaguchi and Miyagi (1998)	Molecular solid I_2 under pressure. EFG and Mössbauer isomer shift.
Baba and Fukui (1999)	Gauge-origin-independent relativistic effects on NMR shifts.
Bruna and Grein (1999)	g-tensors for X_2^-, MX_2, $M_2X_2^+$; M=Li,Na; X=F,Cl.
Bühl (1999)	^{95}Mo NMR chemical shifts.
Bühl et al. (1999)	Review on DFT calculations on NMR shifts.
Cromp et al. (1999)	NMR chemical shifts in DFT. SO, rovibration effects. HBr.
Ebert et al. (1999a)	Fully relativistic description of static M1 hfs in magnetic and nonmagnetic solids. Knight shifts of $4d$ elements.
Hada et al. (1999)	HX;X=H,F-I. Proton NMR shifts.
Helgaker et al. (1999)	Review on ab initio calculations of NMR shielding and spin-spin coupling tensors.
Hemmingsen et al. (1999)	EFG in β-$Cd(OH)_2$. SO effect small.
Kaupp et al. (1999a)	PI_4^+ ^{31}P NMR shifts due to SO.
Kaupp et al. (1999b)	CF_3IF_n; n=0,2,4,6. ^{13}C and ^{19}F NMR shifts.
Kellö and Sadlej (1999a)	Q(Ge) from GeO, GeS.
Kellö and Sadlej (1999b)	Q(Rb) from RbF, RbCl. Summary of alkali Q.
Kellö et al. (1999)	Q(Al) from AlF, AlCl.
Kutzelnigg (1999b)	Relativistic corrections to magnetic properties. DPT.
W.-J. Li et al. (1999ab)	Introduce CI to A. J. Stone's (1963)[2777] theory of the g-matrix. Applications on $[MOX_n]^{m-}$; M=V,Cr,Mo; X=F-Br.
Moore (1999)	Magnetic shielding (NMR shift) for H-like Dirac atom. Closed-form expressions.
Patchkovskii and Ziegler (1999)	d^1 systems MEX_4^{z-}; M=V,Cr-W,Tc-Re; E=O,N; X=F-Br. g-tensors. DFT.
Pyper (1999ab)	Nuclear shielding in one-electron atoms.
Rodriguez-Fortea et al. (1999)	$WO_nS_{4-n}^{2-}$; n=0-4, WX_6; X=F,Cl,CO. ^{183}W NMR shifts. Me_3PbX; X=Cl-I, -OMe, -SMe, -SeMe, $-NEt_2$. PbX_4; X=Cl-I. ^{207}Pb NMR shifts. ZORA relativity. For W, paramagnetic contributions dominate. For Pb, SO-induced FC contributions dominate.
Schwerdtfeger et al. (1999)	EFG in HCl and CuCl. Accuracy of the DFT approach. DK.
Seth et al. (1999a)	EFG in HCl, MCl; M=Li-K. DK.
Vaara et al. (1999a)	HX, CH_3X; X=F-I. H_2X; X=O-Te. H, C and X NMR shift tensors. 2nd- and 3rd-order SO effects. SOO included.
Vaara et al. (1999b)	HX; X=F-I. H_2X; X=O-Te. NMR spin-spin coupling. 2nd- and 3rd-order SO effects. SOO.
Visscher et al. (1999)	HX; X=F-I. NMR shielding and spin-spin coupling.
Wolff et al. (1999)	NMR shifts using ZORA and DFT. ^{199}Hg in LHgL'. Solvent shifts using simple models.

Table 9.4: Electric dipole moments and molecular charge distributions. For electric field gradients, see previous table.

Reference	Comments
Dolg et al. (1993a)	Dipole moment of PbO. Both SR and SO effects included. The SO must be treated at correlated level. AE vs. PP tested.
Dyall (1993a)	MO; M=Ge-Pb. Dipole moment at DF/HF level.
Kellö and Sadlej (1993)	MO, MS; M=Si-Pb. Dipole moments. PT relativity.
Perera and Bartlett (1993)	Dipole moments of interhalogens, up to IBr. PT relativity, CCSD correlation.
Reinisch and Bross (1993)	Relativistic effects on the Compton profile of polycrystalline gold.
Pou-Amérigo et al. (1994)	Dipole moments of MH; M=Ni, Cu. PT relativity.
Schwerdtfeger and Bowmaker (1994)	MCO; M=Cu-Au. Dipole polarizabilities.
Kellö and Sadlej (1995b)	Dipole moments of MH; M=Cu-Au. PT relativity.
Kellö and Sadlej (1995c)	Electric properties of CuAg, CuAu, AgAu. PT relativity. Sign of CuAu dipole moment changed by R.
Schwerdtfeger (1995)	AuCl. R/NR dipole moment and polarizability for HF - QCISD(T).
Urban and Sadlej (1995)	MM'; M,M'=Li-Rb. Electric properties. PT.
Kellö and Sadlej (1996a); Kellö et al. (1996)	Dipole moments of MH; M=Cu-Au. SR DK CCSD(T). R/NR.
Urban and Sadlej (1996)	Electric properties for AlM; M=Cu-Au. DK.
D.-D. Dai and Li (1997)	Diatomic EuCh; Ch=O-Te. Dipole moment decrease.
de Jong et al. (1998)	Interhalogens up to IBr. HF to CCSD(T). Both μ and Θ.
Iliaš et al. (1998)	MF; M=Cu-Au. Dip. mom., up to DK-CCSD(T) level.
Kellö et al. (1998)	MO, MS; M=Si-Pb. Dipole moments. DK relativity. For PbO a large SO effect found.
Miadoková et al. (1999)	Electric properties of MF; M=K-Fr. NR/PT/DK.
Raptis et al. (1999)	Electric properties of MS; M=Zn-Hg. DK/NR.
Suzumura et al. (1999)	MH, MCl; M=Cu-Au. Dipole moments. SR/NR DFT.
Turski (1999)	SiM; M=Cu-Au. Dipole moment R/NR.

Table 9.5: Molecular energy levels and energy transfer.

Reference	Comments
Minaev et al. (1993)	Effect of intermolecular interaction on forbidden transitions of O_2. SO.
Minaev et al. (1995)	Singlet-triplet transitions of N_2.
Ågren et al. (1996)	Review on SO effects, handled with response theory. Predissociation, triplet bands, phosphorescence, external heavy-atom effects.
Minaev et al. (1996)	Magnetic phosphorescence of O_2 b $^1\Sigma_g^+$ - X $^3\Sigma_g^-$ red atmospheric emission band.

The following Table 9.6 gives available results on relativistic – especially spin-orbit – effects on chemical reactions. For further examples, see Table 7.10 and look for the symbol '+'.

Table 9.6: Relativistic effects on chemical reactions.

Reference	Comments
Minaev (1983)	SO effects in molecular spectroscopy and chemical kinetics. Thesis.
Yarkony (1992)	Spin-forbidden chemistry within BP approximation. Review.
Alexander et al. (1993)	SO branching in the photofragmentation of HCl.
Chang and Yarkony (1993)	Spin-forbidden processes in $N_2O \rightarrow N_2$+O(^3P).
L. Li et al. (1993)	Experimental reactions of M_2; M=Cu-Au with O_2, N_2O, N_2, H_2, CH_4, CO, CO_2, C_2H_4 in the gas phase. Au_2 anomalous. Reasons discussed.
Minaev and Lunell (1993)	Classification of SO effects in organic reactions.
Daniel et al. (1995)	SO-induced radiationless transitions in $HCo(CO)_4$.
Knuts et al. (1995)	SO effects in oxirane ring opening.
Minaev and Ågren (1995, 1996)	Reviews on SO-induced chemical reactivity and 'spin-catalysis'.
Riad Manaa (1995)	SH (A $^2\Sigma^+$). Fragmentation. SO.
Su (1995a)	Photochemical rearrangement of 3-substituted cyclopropenes to cyclopentadienes. SO.
Su (1995b)	Photorearrangements of cyclohexadienes.
Su (1995c)	Photochemical rearrangement of vinylcyclopropanes to cyclopentenes. SO. Triplet excited state, singlet ground state.
Visscher and Dyall (1995)	H+Cl_2 →HCl+Cl. Cl+HCl→ClH+Cl. Rel. effects on reaction energy. DF+MP2/DK/NR.
Ågren et al. (1996)	Review on spin catalysis.
Böckmann et al. (1996)	SO coupling in organic molecules. MNDOC-CI. Triplet-state reactivity.
Brouder et al. (1996)	X-ray magnetic circular dichroism. Fe K-edge.
Musaev and Morokuma (1996)	Potential-energy surfaces of TM-catalyzed chemical reactions.
Su (1996a)	SO coupling and triplet carbenic addition chemistry.
Su (1996b)	SO coupling in photochemical reaarangements of α,β-unsaturated cyclic ketones.
Su (1996c)	SO coupling in oxadi-π-methane rearrangements and related photochemical reactions.
Yarkony (1996)	CH (a $^4\Sigma^-$) + CO (X $^1\Sigma^+$). Surface of intersection, SO interaction and Kramers degeneracy.
Danovich and Shaik (1997)	SO coupling in oxidative addition of FeO^++H_2.
Alexander et al. (1998)	SO effects on F(^2P) + H_2.
Aquilanti et al. (1998)	SO effects on F + H_2.
Danovich et al. (1998)	SO effects in twist and pyramidalization of ethene. BP + MRCI study.
Kaledin et al. (1998)	I($^2P_{3/2}$)+O_2(a $^1\Delta_g$) ↔ I($^2P_{1/2}$)+O_2(X $^3\Sigma_g^-$). SO.
Minaev and Ågren (1998)	SO coupling in oxygen-containing diradicals. Reaction paths. MNDO CI.
Aoiz et al. (1999)	SO effects for F + H_2 → HF + H.
Daniel et al. (1999)	M(H)$(CO)_3$(H-DAB); M=Mn,Re; H-DAB = 1,4-Diaza-1,3-butadiene. SO effects on the M-H bond homolysis.
Honvault and Launay (1999)	SO corrections to F+D_2→DF+D.

Reference	Comments
S.-H. Lee and Liu (1999)	SO reactivity of Cl atom: In $Cl(^2P)+H_2 \rightarrow HCl+H$ the excited $Cl^*(^2P_{1/2})$ state is more reactive. Exp.
Riad Manaa (1999)	Photodissociation of NaK.

Chapter 10

Appendix

Table 10.1: List of acronyms and symbols.

Symbol	Meaning
A	Magnetic vector potential.
AE	All-electron (as contrasted to PP).
AMFI	Atomic mean field spin-orbit operator.
An	Actinide.
B	Magnetic field.
BP	Breit-Pauli.
BW	Bohr-Weisskopf effect.
CC	Coupled-cluster (method).
CCSD	Coupled clusters with single and double excitations.
COT	Cyclo-octatetraene, C_8H_8.
Cp	Cyclopentadiene, C_5H_5.
Cp*	Pentamethylcyclopentadiene, $C_5(CH_3)_5$.
nD	n-dimensional.
DC	Dirac-Coulomb.
DCB	Dirac-Coulomb-Breit (Hamiltonian).
DF	Dirac-Fock (='Dirac-Hartree-Fock').
DFB	Dirac-Fock-Breit.
DFT	Density functional theory.
DIM	Diatomics-in-molecule (a semiempirical model).
DPT	Direct perturbation theory.
DQCC	Deuteron quadrupole coupling constant.
DSW	Dirac Scattered-Wave.
DVM	Discrete variational method.
EA	Electron affinity.
ECP	Effective core potential (=PP).
EFG	Electric field gradient.
EM	Electromagnetic.
ESC	Elimination of small component.
Et	Ethyl, $-CH_2CH_3$.
EXAFS	Extended x-ray absorption fine structure.
FC	Fermi contact term.
FEM	Finite-element method.
FS	Fine structure.
FW	Foldy-Wouthuysen[1019] (-Tani[6210,6211]) transformation.
GGA	Generalized gradient approximation.
GTO	Gaussian type orbital.
GUGA	Graphical unitary group approach.
h_D	One-electron Dirac Hamiltonian.
Hfs	Hyperfine structure.
INDO	Intermediate neglect of differential overlap. A semiempirical MO method.
IP	Ionization potential.
j	Current density.
KG	Klein-Gordon (equation).
LAPW	Linearized augmented plane wave method.
LCAO	Linear Combination of Atomic Orbitals.
LDA	Local density approximation (in DFT).
LMTO	Linear muffin-tin orbital.

Symbol	Meaning
Ln	Lanthanide.
μ	Molecular dipole moment.
mb	Millibarn (10^{-31} m^2).
MBPT	Many-body perturbation theory.
MC	Monte-Carlo.
MCDF	Multiconfiguration Dirac-Fock.
Me	Methyl, -CH_3.
MP	Model potential (with nodes).
MPn	Møller-Plesset PT of order n.
MQDT	Multichannel quantum defect theory.
MRCI	Multireference CI.
MXD	Magnetic x-ray dichroism.
NR	Non-relativistic.
PES	Photoelectron spectrum.
Ph	Phenyl, -C_6H_5.
PP	Pseudopotential (=ECP).
PT	Perturbation theory.
π	$\mathbf{p} - e\mathbf{A}$.
Q	Nuclear quadrupole moment.
QDT	Quantum Defect Theory.
QED	Quantum Electrodynamics.
QM	Quantum Mechanics.
QMC	Quantum Monte Carlo.
R	Interatomic distance.
R	Relativistic.
RA	Regular Approximation.
RASCI	Restricted Active Space Configuration Interaction.
RESC	Relativistic scheme by Eliminating Small Components.
REX	Relativistic Extended Hückel.
RI	Resolution-of-the-identity, $\sum_n \|n><n\| = 1$.
ρ	Charge density.
RRPA	Relativistic random phase approximation.
SE	Self energy.
SO	Spin-orbit.
SOO	Spin-other-orbit.
SR	Scalar relativistic (spin-orbit averaged, 'quasirelativistic').
SW	Scattered wave (=multiple-scattering method, MS Xα).
Θ	Molecular quadrupole moment.
TB	Tight binding (= EHT for crystals).
TF	Thomas-Fermi.
TM	Transition metal.
TP	Transition probability.
UHF	Unrestricted Hartree-Fock (different orbitals for different spins).
VP	Vacuum polarization.
WB	Wood-Boring Hamiltonian (Wood and Boring 1978)[3037].
WKB	Wentzel-Kramers-Brillouin approximation, semiclassical ($\hbar \to 0$) approximation.
XAS	X-ray absorption spectra.
XMCD	X-ray magnetic circular dichroism.

Symbol	Meaning
Z	Nuclear charge.
ZFS	Zero-field splitting.
ZORA	Zeroth-order regular approximation.

Bibliography

[6578] Aashamar K. and Luke T.M. (1994a). Relativistic transition rates for sextet levels in Cr II. *Phys. Scr.*, **49**, 280–285.

[6579] Aashamar K. and Luke T.M. (1994b). Decay rates and lifetimes of the $4p\ ^{6,4}P_J^o$ and $4p\ ^6D_J^o$ levels of Cr II. *J. Phys. B*, **27**, 1091–1103.

[6580] Åberg T. (1993). Quantum electrodynamics of multiphoton ionization. *Phys. Scr.*, **T46**, 173–181.

[6581] Aboussaid A., Godefroid M.R., Jönsson P. and Froese Fischer C. (1995). Multiconfigurational Hartree-Fock calculations of hyperfine-induced transitions in heliumlike ions. *Phys. Rev. A*, **51**, 2031–2039.

[6582] Abrikosov A.A. and Gor'kov L.P. (1962). Spin-orbit interaction and the Knight shift in superconductors. *Sov. Phys. JETP*, **15**, 752–757.

[6583] Abugideiri F., Fettinger J.C., Pleune B., Poli R., Bayse C.A. and Hall M.B. (1997). Synthesis, structure and hydride-deuteride exchange studies of $CpMoH_3(PMe_2Ph)_2$ and theoretical studies of the $CpMoH_3(PMe_3)_2$ model systems. *Organometallics*, **16**, 1179–1185.

[6584] Abu-Hasanayn F., Goldman A.S. and Krogh-Jespersen K. (1993a). Computational study of the transition state for H_2 addition to Vaska-type complexes (trans-Ir$(L)_2$(CO)X): substituent effects on the energy barrier and the origin of the small H_2/D_2 kinetic isotope effect. *J. Phys. Chem.*, **97**, 5890–5896.

[6585] Abu-Hasanayn F., Krogh-Jespersen K. and Goldman A.S. (1993b). Factors influencing the thermodynamics of H_2 oxidative addition to Vaska-type complexes (trans-Ir$(PR_3)_2$(CO)X): Predictions from ab initio calculations and experimental verification. *Inorg. Chem.*, **32**, 495–496.

[6586] Abu-Hasanayn F., Krogh-Jespersen K. and Goldman A.S. (1993c). Theoretical study of primary and secondary deuterium equilibrium isotope effects for H_2 and CH_4 addition to *trans*–Ir$(PR_3)_2$(CO)X. *J. Am. Chem. Soc.*, **115**, 8019–8023.

[6587] Acikgoz I., Barut A.O., Kraus J. and Uenal N. (1995). Self-field quantum electrodynamics without infinities. A new calculation of vacuum polarization. *Phys. Lett. A*, **198**, 126–130.

[6588] Acker H.L., Backenstoss G., Daum C., Sens J.C. and De Wit S.A. (1966). Measurement and analysis of muonic X-ray spectra in spherical nuclei. *Nucl. Phys.*, **87**, 1–50.

[6589] Adachi H. (1997). Electronic state calculation of transition metal cluster. *Adv. Quantum Chem.*, **29**, 49–81.

[6590] Adamo C. and Maldivi P. (1997). Ionic versus covalent character in lanthanide complexes. A hybrid density functional study. *Chem. Phys. Lett.*, **268**, 61–68.

[6591] Adamo C. and Maldivi P. (1998). A theoretical study of bonding in lanthanide trihalides by density functional methods. *J. Phys. Chem. A*, **102**, 6812–6820.

[6592] Adkins G.S., Fell R.N. and Mitrikov P.M. (1997). Calculation of the positronium hyperfine interval. *Phys. Rev. Lett.*, **79**, 3383–3386.

[6593] Adkins G.S. and Sapirstein J. (1998). Order $m\alpha^6$ contributions to ground-state hyperfine splitting in positronium. *Phys. Rev. A*, **58**, 3552–3560.

[6594] Adkins G.S. and Shiferaw Y. (1995). Two-loop corrections to the orthopositronium and parapositronium decay rates due to the vacuum polarization. *Phys. Rev. A*, **52**, 2442–2445.

[6595] Adler S.L. (1994). *Quaternionic Quantum Mechanics and Quantum Fields.*. Oxford U. P., Oxford, 416 p.

[6596] Adrian F.J. (1996). Alkali spin-orbit interactions and the width of the conduction-electron spin resonance in superconducting alkali-metal fullerides. *Phys. Rev. B*, **53**, 2206–2209.

[6597] Agassi D. and Restorff J.B. (1994). Pseudopotential band calculations along a symmetry axis: spin-orbit interaction. *J. Phys: CM*, **6**, 4673–4684.

[6598] Ågren H. and Vahtras O. (1993). Spin-orbit coupling in molecular Auger spectra: propensity rules tested for water. *J. Phys. B*, **26**, 913–920.

[6599] Ågren H., Vahtras O. and Minaev B. (1996). Response theory and calculations of spin-orbit phenomena in molecules. *Adv. Quantum Chem.*, **27**, 71–162.

[6600] Ahlrichs R., von Arnim M., Eisenmann J. and Fenske D. (1997). $[(HgPt\mathrm{Bu})_4]_3$ – Synthesis, structure, and bonding. *Angew. Chem. Int. Ed. Engl.*, **36**, 233–235.

[6601] Ahuja R., Auluck S., Johansson B. and Brooks M.S.S. (1994a). Electronic structure magnetism and Fermi surfaces of Gd and Tb. *Phys. Rev. B*, **50**, 5147–5154.

[6602] Ahuja R., Auluck S., Johansson B. and Khan M.A. (1994b). Optical properties of PdO and PtO. *Phys. Rev. B*, **50**, 2128–2132.

[6603] Ahuja R., Auluck S., Söderlind P., Eriksson O., Wills J.M. and Johansson B. (1994c). Fermi surface of noble metals: Full-potential generalized-gradient-approximation calculations. *Phys. Rev. B*, **50**, 11183–11186.

[6604] Ahuja R., Eriksson O., Johansson B., Auluck S. and Wills J.M. (1996). Electronic and optical properties of red HgI_2. *Phys. Rev. B*, **54**, 10419–10424.

[6605] Ahuja R., Eriksson O., Wills J.M. and Johansson B. (1998a). Theoretical high-pressure studies of caesium hydride. *J. Phys. CM*, **10**, L153–L158.

[6606] Ahuja R., Johansson B. and Eriksson O. (1998b). Theoretical confirmation of the high-pressure orthorombic phase in strontium. *Phys. Rev. B*, **58**, 8152–8154.

[6607] Ahuja R., Söderlind P., Trygg J., Melsen J., Wills J.M., Johansson B. and Eriksson O. (1994d). Influence of pseudocore valence-band hybridization on the crystal-structure phase stabilities of transition metals under extreme compression. *Phys. Rev. B*, **50**, 14690–14693.

[6608] Ahuja R., Wills J.M., Johansson B. and Eriksson O. (1993). Crystal structures of Ti, Zr, and Hf under compression: Theory. *Phys. Rev. B*, **48**, 16269–16279.

[6609] Ait-Tahar S., Grant I.P. and Norrington P.H. (1996). Electron scattering by Fe XXII within the Dirac R-matrix approach. *Phys. Rev. A*, **54**, 3984–3989.

[6610] Ait-Tahar S., Grant I.P. and Norrington P.H. (1997). Dirac R-matrix modeling of spin-induced asymmetry in the scattering of polarized electrons from polarized cesium atoms. *Phys. Rev. Lett.*, **79**, 2955–2958.

[6611] Akai H., Akai M., Blügel S., Drittler B., Ebert H., Terakura H., Zeller R. and Dederichs P.H. (1990). Theory of hyperfine interactions in metals. *Progr. Theor. Phys. Suppl.*, **101**, 11–77.

[6612] Akamova L.V., Pinyaskin V.V., Tomilin O.B., Stankevich I.V. and Chistyakov A.L. (1995). End-group influence on the relative stability of linear clusters of mercury. *Zh. Strukt. Khim.*, **36**, 623–629.

[6613] Akella J., Weir S., Wills J.M. and Söderlind P. (1997). Structural stability in uranium. *J. Phys. CM*, **9**, L549–L555.

[6614] Åkermark B., Blomberg M.R.A., Glaser J., Öhrström L., Wahlberg S., Wärnmark K. and Zetterberg K. (1994). The metal-alkene bond in Rh(I)(β-diketonato)(alkene)$_2$complexes. Correlation of ^{103}Rh-NMR shifts with stability constants, alkene excitation energies and d-d absorption bands. An experimental and theoretical study. *J. Am. Chem. Soc.*, **116**, 3405–3413.

[6615] Åkesson R., Persson I., Sandström M. and Wahlgren U. (1994a). Structure and bonding of solvated mercury(II) and thallium(III) dihalide and dicyanide complexes by XAFS spectroscopic measurements and theoretical calculations. *Inorg. Chem.*, **33**, 3715–3723.

[6616] Åkesson R. and Petterson L.G.M. (1994b). Theoretical study of the mono- and di-hydrated divalent ions of the first-row transition metals. *Chem. Phys.*, **184**, 85–95.

[6617] Åkesson R., Pettersson L.G.M., Sandström M. and Wahlgren U. (1994c). Ligand field effects in the hydrated divalent and trivalent metal ions of the first and second transition periods. *J. Am. Chem. Soc.*, **116**, 8691–8704.

[6618] Akinaga Y., Taketsugu T. and Hirao K. (1997). Theoretical study of CH_4 photodissociation on the Pt(111) surface. *J. Chem. Phys.*, **107**, 415–424.

[6619] Aksela H., Aksela S. and Kabachnik N. (1996a). Resonant and nonresonant Auger recombination. In *VUV and Soft X-Ray Photoionization*, (Edited by Becker U. and Shirley D.A.), pp. 401–440, Plenum Press, New York.

[6620] Aksela H., Aksela S., Sairanen O.P., Kivimäki A., Naves de Brito A., Nõmmiste E., Tulkki J., Svensson S., Ausmees A. and Osborne S.J. (1994). Electron correlation in Xe $4d^{-1}_{5/2}6p$ to $5p^{-2}6p$ resonant Auger transitions studied by utilizing the Auger resonant Raman effect. *Phys. Rev. A*, **49**, R4269–R4272.

[6621] Aksela H., Jauhiainen J., Kukk E., Nõmmiste E. and Aksela S. (1996b). Electron correlation in the decay of resonantly excited $3d^{-1}_{3/2,5/2}5p$ states of krypton. *Phys. Rev. A*, **53**, 290–296.

[6622] Aksela H., Jauhiainen J., Nõmmiste E., Sairanen O.P., Karvonen J., Kukk E. and Aksela S. (1996c). Angular distribution of Auger electrons in the decay of resonantly excited $4d^{-1}_{3/2,5/2}6p$ states in Xe. *Phys. Rev. A*, **54**, 2874–2881.

[6623] Aksela H. and Mursu J. (1996). Electron correlation in the $3p^44s$ state of Ar studied via the $2p^54s$ resonant Auger spectra. *Phys. Rev. A*, **54**, 2882–2887.

[6624] Aksela H., Sairanen O.P., Aksela S., Kivimäki A., Naves de Brito A., Nõmmiste E., Tulkki J., Ausmees A., Osborne S.J. and Svensson S. (1995). Correlation effects in the resonant Auger decay of the Xe $4d^{-1}_{3/2,5/2}6p$ states studied by high resolution experiment and multiconfiguration Dirac-Fock theory. *Phys. Rev. A*, **51**, 1291–1303.

[6625] Alatalo M., Weinert M. and Watson R.E. (1999). Truncated pseudopotentials for alloy calculations. *Phys. Rev. B*, **60**, 7680–7683.

[6626] Albert K., Neyman K.M., Nasluzov V.A., Ruzankin S.P., Yeretzian C. and Rösch N. (1995). On the electronic and geometric structure of bimetallic clusters. A comparison of the novel cluster Na_6Pb to Na_6Mg. *Chem. Phys. Lett.*, **245**, 671–678.

[6627] Albert K., Neyman K.M., Pacchioni G. and Rösch N. (1996). Electronic and geometric structure of bimetallic clusters: Density functional calculations on $[M_4\{Fe(CO)_4\}_4]^{4-}$ (M = Cu, Ag, Au) and $[Ag_{13}\{Fe(CO)_4\}_8]^{n-}$ (n= 0-5). *Inorg. Chem.*, **35**, 7370–7376.

[6628] Alberto P., Fiolhais C. and Gil V.M.S. (1996). Relativistic particle in a box. *Eur. J. Phys.*, **17**, 19–24.

[6629] Alberto P., Fiolhais C. and Oliveira M. (1998). On the relativistic $L-S$ coupling. *Eur. J. Phys.*, **19**, 553–562.

[6630] Aldaya V., Bisquert J., Guerrero J. and Navarro-Salas J. (1993). Higher-order polarization on the Poincaré group and the position operator. *J. Phys. A*, **26**, 5375–5390.

[6631] Aldaya V. and Guerrero J. (1995). Finite-difference equations in relativistic quantum mechanics. *J. Phys. A*, **28**, L137–L145.

[6632] Aldén M., Skriver H.L. and Johansson B. (1993). Ab initio surface core-level shifts and surface segregation energies. *Phys. Rev. Lett.*, **71**, 2449–2452.

[6633] Alekseyev A.B. (1998). Relativistic configuration interaction calculations of the potential curves and radiative lifetimes of the low-lying states of bismuth nitride. *Chem. Phys.*, **225**, 247–258.

[6634] Alekseyev A.B., Buenker R.J., Liebermann H.P. and Hirsch G. (1994). Spin-orbit configuration interaction study of the potential energy curves and radiative lifetimes of the low-lying states of bismuth hydride. *J. Chem. Phys.*, **100**, 2989–3001.

[6635] Alekseyev A.B., Das K.K., Liebermann H.P., Buenker R.J. and Hirsch G. (1995a). Ab initio CI study of the electronic spectrum of bismuth iodide employing relativistic effective core potentials. *Chem. Phys.*, **198**, 333–344.

[6636] Alekseyev A.B., Liebermann H.P., Boustani I., Hirsch G. and Buenker R.J. (1993). Theoretical study of the energies and lifetimes of the low-lying states of bismuth fluoride. *Chem. Phys.*, **173**, 333–344.

[6637] Alekseyev A.B., Liebermann H.P., Buenker R.J. and Hirsch G. (1995b). Theoretical study of the electronic spectrum antimony oxide employing relativistic effective core potentials. *J. Chem. Phys.*, **102**, 2539–2550.

[6638] Alekseyev A.B., Liebermann H.P., Buenker R.J. and Hirsch G. (1996a). Spin-orbit configuration interaction study of potential energy curves and transition probilities of the mercury hydride molecule and tests of relativistic effective core potentials for Hg, Hg^+, and Hg^{2+}. *J. Chem. Phys.*, **104**, 4672–4684.

[6639] Alekseyev A.B., Liebermann H.P., Buenker R.J. and Hirsch G. (1996b). Ab initio study of the low-lying states of SnH. *Mol. Phys.*, **88**, 591–603.

[6640] Alekseyev A.B., Liebermann H.P., Buenker R.J. and Hirsch G. (1996c). Theoretical study of the low-energy BiN spectrum. *Chem. Phys. Lett.*, **257**, 75–81.

[6641] Alekseyev A.B., Liebermann H.P., Buenker R.J., Hirsch G. and Li Y. (1994). Ab initio relativistic configuration interaction calculations of the spectrum of bismuth oxide: Potential curves and transition probabilities. *J. Chem. Phys.*, **100**, 8956–8968.

[6642] Alekseyev A.B., Liebermann H.P., Hirsch G. and Buenker R.J. (1997a). Relativistic configuration interaction calculations of the potential curves and radiative lifetimes of the low-lying states of bismuth nitride. *Chem. Phys.*, **225**, 247–258.

[6643] Alekseyev A.B., Liebermann H.P., Hirsch G. and Buenker R.J. (1998). The spectrum of arsenic hydride: An *ab initio* configuration interaction study employing a relativistic effective core potential. *J. Chem. Phys.*, **108**, 2028–2040.

[6644] Alekseyev A.B., Liebermann H.P., Lingott R., Bludský O. and Buenker R.J. (1998). The spectrum of antimony hydride: An *ab initio* configuration interaction study employing a relativistic effective core potential. *J. Chem. Phys.*, **108**, 7695–7706.

[6645] Alekseyev A.B., Liebermann H.P., Lingott R., Buenker R.J. and Wright J.S. (1997b). Long-lived diatomic dications: potential curves and radiative livetimes for $CaBr^{2+}$ and CaI^{2+} including relativistic effects. *Mol. Phys.*, **91**, 777–787.

[6646] Alekseyev A.B., Liebermann H.P., Lingott R.M., Buenker R.J. and Wright J.S. (1999). Low-energy spectrum of the thermodynamically stable BaI^{2+} dication. *Spectrochim. Acta A*, **55**, 467–475.

[6647] Alemany P., Bengtsson-Kloo L. and Holmberg B. (1998). Angular bond flexibility and closed-shell metal-metal interaction in polymetal copper(I), silver(I) and gold(I) iodide complexes. A quantum chemical study. *Acta Chem. Scand.*, **52**, 718–727.

[6648] Alemany P., Novoa J.J. and Bengtsson L. (1994). A comparative study on the structure of M_2Se and M_2I^+(M=Ag, Au) using pseudopotentials and full ab initio methods. *Int. J. Quantum Chem.*, **52**, 1–8.

[6649] Alexander M.H., Pouilly B. and Duhoo T. (1993). Spin-orbit branching in the photo fragmentation of HCl. *J. Chem. Phys.*, **99**, 1752–1764.

[6650] Alexander M.H., Werner H.L. and Manolopoulos D.E. (1998). Spin-orbit effects in the reaction of $F(^2P)$ with H_2. *J. Chem. Phys.*, **109**, 5710–5713.

[6651] Alexander S.A. and Coldwell R.L. (1999). Relativistic calculations using Monte Carlo methods: One-electron systems. *Phys. Rev. E*, **60**, 3374–3379.

[6652] Ali M.A. (1997). Fine structure splitting and magnetic dipole and electric quadrupole transition probabilities between the ground levels of Ga-like ions. *Phys. Scr.*, **55**, 159–166.

[6653] Aliev T.M., Fainberg V.Y. and Pak N.K. (1994). New approach to the path integral representation for the Dirac particle propagator. *Phys. Rev. D*, **50**, 6594–6598.

[6654] Alikhani M.E. (1999). Theoretical study of the insertion reaction of zinc, cadmium, and mercury atoms with methane and silane. *Chem. Phys. Lett.*, **313**, 608–616.

[6655] AlJaber A.M. (1998). The fine structure of the N-dimensional hydrogen atom. *Nuovo Cim. B*, **113**, 651–657.

[6656] Allouche A.R. and Aubert-Frécon M. (1994). Electronic structure of BaLi. I. Theoretical study. *J. Chem. Phys.*, **100**, 938–944.

[6657] Allouche A.R., Aubert-Frécon M., Nicolas G. and Spiegelmann F. (1995). Theoretical study of the electronic structure of the Ba_2 molecule. *Chem. Phys.*, **200**, 63–77.

[6658] Allouche A.R., Spiegelmann F. and Aubert-Frécon M. (1993). Theoretical study of the low-lying electronic states of the BaH^+ molecular ion. *Chem. Phys. Lett.*, **204**, 343–349.

[6659] Almlöf A. and Gropen O. (1996). Relativistic effects in chemistry. In *Rev. Comp. Chem.*, (Edited by Lipkowitz K.B. and Boyd D.B.), vol. 8, pp. 203–244, VCH, New York.

[6660] Alonso J.A., Molina L.M., López M.J., Rubio A. and Stott M.J. (1998). Ab initio calculations for mixed clusters of lead and alkali elements, and implications for the structure of their solid and liquid alloys. *Chem. Phys. Lett.*, **289**, 451–456.

[6661] Alonso V. and De Vincenzo S. (1997). General boundary conditions for a Dirac particle in a box and their non-relativistic limits. *J. Phys. B*, **30**, 8573–8585.

[6662] Alonso V., De Vincenzo S. and Mondino L. (1997). On the boundary conditions for the Dirac equation. *Eur. J. Phys.*, **18**, 315–320.

[6663] Alouani M. and Albers R.C. (1993). Magnetic properties of the neutral MX chain systems. *Synthetic Metals.*, **55-57**, 3352–3357.

[6664] Alscher A., Hencjen K., Trautmann D. and Baur G. (1997). Multiple electromagnetic electron-positron pair production in relativistic heavy-ion collisions. *Phys. Rev. A*, **55**, 396–401.

[6665] Altmann S.L. and Herzig P. (1994). *Point-Group Theory Tables.*. Clarendon Press, Oxford, 704 p.

[6666] Al-Zahrani M.M. (1993). Spectroscopic constants, potential energy curves and energy separations of group-IV halides. *Diss. Abstr.*, **54**, 5649–B.

[6667] Amado A.M. and Ribeiro-Claro P.J.A. (1999). Ab initio calculations on some transition metal heptoxides by using effective core potentials. *J. Mol. Struct. (Theochem)*, **469**, 191–200.

[6668] Amirthakumari M., Pari G., Rita R. and Asokamani R. (1997). Phase transformation and metallisation studies of some of the alkali iodides through high pressure electronic structure calculations. *Phys. Stat. Sol. B*, **199**, 157–164.

[6669] Amore P., Barbaro M.B. and De Pace A. (1996). Relativistic Hamiltonians in many-body theories. *Phys. Rev. C*, **53**, 2801–2808.

[6670] Anandan J. and Mazur P.O. (1993). Geometric phase for the relativistic Klein-Gordon equation. *Phys. Lett. A*, **173**, 116–120.

[6671] Ancarani L.U., Keller S., Ast H., Whelan C.T., Walters H.R.J. and Dreizler R.M. (1998). Influence of Coulomb boundary conditions for (e,2e) processes on the K-shell of high-Z atoms. *J. Phys. B*, **31**, 609–623.

[6672] Andersen T., Andersen H.H., Balling P., Kristensen P. and Petrunin V.V. (1997). Structure and dynamics of the negative alkaline-earth ions. *J. Phys. B*, **30**, 3317–3332.

[6673] Andersson K. (1995). The electronic spectrum of Cr_2. *Chem. Phys. Lett.*, **237**, 212–221.

[6674] Andersson K., Bauschlicher Jr C.W., Persson B.J. and Roos B.O. (1996). The structure of dichromium tetraformate. *Chem. Phys. Lett.*, **257**, 238–248.

[6675] Andersson K., Roos B.O., Malmqvist P.Å. and Widmark P.O. (1994). The Cr_2 potential energy curve studied with multiconfigurational second-order perturbation theory. *Chem. Phys. Lett.*, **230**, 391–397.

[6676] Andrae D. (1997). Recursive evaluation of expectation values $\langle r^k \rangle$ for arbitrary states of the relativistic one-electron atom. *J. Phys. B*, **30**, 4435–4451.

[6677] Andrews L., Chertihin G.V., Thompson C.A., Dillon J., Byrne S. and Bauschlicher Jr C.W. (1996). Infrared spectra and quantum chemical calculations of group 2 MO_2, O_2MO_2, and related molecules. *J. Phys. Chem.*, **100**, 10088–10099.

[6678] Andrews L., Kushto G.P., Yustein J.T., Archibong E., Sullivan R. and Leszczynski J. (1997). Reactions of pulsed-laser-evaporated thallium atoms with O_2. Matrix infrared spectra of new TlO_2 species. Trends in group 13 dioxides and dioxide anions. *J. Phys. Chem. A*, **101**, 9077–9084.

[6679] Andrews L. and Zhou M.F. (1999). Reactions of laser-ablated molybdenum and tungsten atoms with nitric oxide. Infrared spectra of the MN, NMO, and M-η^1-$(NO)_x$ ($x = 1, 2, 3, 4$) molecules and $(NO)_2^+$ and $(NO)_2^-$ ions in solid argon. *J. Phys. Chem. A*, **103**, 4167–4173.

[6680] Andrews L., Zhou M.F., Chertihin G.V. and Bauschlicher Jr C.W. (1999). Reactions of laser-ablated Y and La atoms, cations and electrons with O_2. Infrared spectra and density functional calculations of the MO, MO^+, MO_2, MO_2^+, and MO_2^- species in solid argon. *J. Phys. Chem. A*, **103**, 6525–6532.

[6681] Anisimova G.P., Semenov R.I. and Tuchkin V.I. (1994). Inclusion of the spin-spin interaction in the energy operator of a two-electron atom with p and d electrons (in Russian). *Opt. Spektr.*, **77**, 695–699.

[6682] Anisimova G.P., Semenov R.I. and Tuchkin V.I. (1994). Inclusion of the spin-spin interaction in the energy operator of a two-electron atom with p and d electrons. *Opt. Spectr.*, **77**, 617–621.

[6683] Anisimova G.P., Semenov R.I. and Tuchkin V.I. (1995). Semiempirical fine-structure calculation for the Ne I atom. $2p^5nd$ (n=3-8) configurations. *Opt. Spektr.*, **79**, 443–452.

[6684] Anisimova G.P., Semenov R.I. and Tuchkin V.I. (1995). Semiempirical fine-structure calculation for the Ne I atom. $2p^5nd$ (n=3-8) configurations. *Opt. Spectr.*, **79**, 409–418.

[6685] Ankudinov A. and Rehr J.J. (1997). Relativistic calculations of spin-dependent x-ray-absorption spectra. *Phys. Rev. B*, **56**, 1712–1715.

[6686] Ankudinov A.L., Zabinsky S.I. and Rehr J.J. (1996). Single configuration Dirac-Fock atom code. *Comp. Phys. Comm.*, **98**, 359–364.

[6687] Antes I., Dapprich S., Frenking G. and Schwerdtfeger P. (1996). Stability of group 11 carbonyl complexes Cl-M-CO (M = Cu, Ag, Au). *Inorg. Chem.*, **35**, 2089–2096.

[6688] Antes I. and Frenking G. (1995). Structure and bonding of the transition metal methyl and phenyl compounds MCH_3 and MC_6H_5 (M = Cu, Ag, Au) and $M(CH_3)_2$ and $M(C_6H_5)_2$ (M = Zn, Cd, Hg). *Organometallics*, **14**, 4263–4268.

[6689] Anthony J.M. and Sebastian K.J. (1993). Relativistic corrections to the Zeeman effect in heliumlike atoms. *Phys. Rev. A*, **48**, 3792–3810.

[6690] Anthony J.M. and Sebastian K.J. (1994). Relativistic corrections to the Zeeman effect in hydrogenlike atoms and positronium. *Phys. Rev. A*, **49**, 192–206.

[6691] Antonov V.N., Bagljuk A.I., Perlov A.Y., Nemoshkalenko V.V., Antonov V., Andersen O.K. and Jepsen O. (1993a). Calculated optical properties of heavy metals. I. Relativistic formalism and cubic 5d metals. *Fiz. Niz. Temp.*, **19**, 689–703.

[6692] Antonov V.N., Bagljuk A.I., Perlov A.Y., Nemoshkalenko V.V., Antonov V., Andersen O.K. and Jepsen O. (1993b). Calculated optical properties of heavy metals. II. HCP 5d metals. *Fiz. Niz. Temp.*, **19**, 786–791.

[6693] Antonov V.N., Bagljuk A.I., Perlov A.Y., Nemoshkalenko V.V., Antonov V., Andersen O.K. and Jepsen O. (1993c). Calculated optical properties of heavy metals. III. Actinides. *Fiz. Niz. Temp.*, **19**, 792–804.

[6694] Antonov V.N., Perlov A.Y., Shpak A.P. and Yaresko A.N. (1995). Calculation of the magneto-optical properties of ferromagnetic metals using the spin-polarized relativistic LMTO method. *J. Magn. Magn. Mat.*, **146**, 205–207.

[6695] Aoiz F.J., Bañares L. and Castillo J.F. (1999). Spin-orbit effects in quantum mechanical rate constant calculations for the $F+H_2 \rightarrow HF+H$ reaction. *J. Chem. Phys.*, **111**, 4013–4024.

[6696] Aparicio J.P., Gaioli F.H. and Garcia Alvarez E.T. (1995). Interpretation of the evolution parameter of the Feynman parametrization of the Dirac equation. *Phys. Lett. A*, **200**, 233–238.

[6697] Appel J. (1965). Spin-orbit coupling and the Knight shift in nontransition-metal superconductors. *Phys. Rev.*, **139**, A1536–A1551.

[6698] Applebaum D. (1995). Fermion stochastic calculus in Dirac-Fock space . *J. Phys. A*, **28**, 257–270.

[6699] Aquilanti V., Cavalli S., De Fazio D., Volpi A., Aguilar A., Giménez X. and Lucas J.M. (1998). Hyperquantization algorithm. II. Implementation for the F+H_2 reaction dynamics including open-shell and spin-orbit interactions. *J. Chem. Phys.*, **109**, 3805–3818.

[6700] Arai A. (1993). Properties of the Dirac-Weyl operator with a strongly singular gauge potential. *J. Math. Phys.*, **34**, 915–935.

[6701] Arbatsky D.A. and Braun P.A. (1998). Symmetries of highly excited atomic hydrogen: Quadratic Zeeman splitting distorted by fine-structure effects. *Phys. Rev. A*, **58**, 1898–.

[6702] Archibong E.F. and Ray A.K. (1999). Possible existence of the plutonium dimer. *Phys. Rev. B*, **60**, 5105–5107.

[6703] Archibong E.F. and Sullivan R. (1995). An ab-initio study of the structures and harmonic vibrational frequencies of M_2O_2 (M = Al, Ga, In, Tl). *J. Phys. Chem.*, **99**, 15830–15836.

[6704] Archibong E.F. and Sullivan R. (1996). Molecular structures and harmonic vibrational frequencies of M_2O_3 (M = Ga, In, Tl). *J. Phys. Chem.*, **100**, 18078–18082.

[6705] Armstrong D.R., Beswick M.A., Cromhout N.L., Harmer C.N., Moncrieff D., Russell C.A., Raithby P.R., Steiner A., Wheatley A.E.H. and Wright D.S. (1998). Weakly bonded Lewis base adducts of plumbocene and stannocene: A synthetic and calculational study. *Organometallics*, **17**, 3176–3181.

[6706] Armstrong D.R., Herbst-Irmer R., Kuhn A., Moncrieff D., Paver M.A., Russell C.A., Stalke D., Steiner A. and Wright D.S. (1993). Bis(cyclopentadienyl)thallat(I), ein mit Stannocen isoelektronisches Anion. *Angew. Chem.*, **105**, 1807–1809.

[6707] Arnau F., Mota F. and Novoa J.J. (1992). Accurate calculation of the electron affinities of the group-13 atoms. *Chem. Phys.*, **166**, 77–84.

[6708] Arnold, Jr. F.P. (1999). Theoretical studies of the bonding in cationic ruthenium silylenes. *Organometallics*, **18**, 4800–4809.

[6709] Arola E. and Strange P. (1998). Application of relativistic scattering theory of x rays to diffraction anomalous fine structure in Cu. *Phys. Rev. B*, **58**, 7663–7667.

[6710] Arola E., Strange P. and Gyorffy B.L. (1997). Relativistic theory of magnetic scattering of x-rays: Application to ferromagnetic iron. *Phys. Rev. B*, **55**, 472–484.

[6711] Arp U., Lagutin B.M., Materlik G., Petrov I.D., Sonntag B. and Sukhorukov V.L. (1993). K-absorption spectra of atomic Ca, Cr, Mn and Cu. *J. Phys. B*, **26**, 4381–4398.

[6712] Arratia-Pérez R. (1993a). Spin-orbit effects on RuO_4 and OsO_4. *Chem. Phys. Lett.*, **203**, 409–414.

[6713] Arratia-Pérez R. (1993b). The $M_6S_8L_6$ clusters : an example in cluster and condensed phase chemistry. *Chem. Phys. Lett.*, **213**, 547–553.

[6714] Arratia-Pérez R. and Hernández-Acevedo L. (1993). Spin-orbit effects on heavy metal octahedral clusters. *J. Mol. Str. (Theochem)*, **282**, 131–141.

[6715] Arratia-Pérez R. and Hernández-Acevedo L. (1995). A Dirac molecular orbital study for encapsulated heavy transition metals within yttrium cluster iodides. *Chem. Phys. Lett.*, **247**, 163–167.

[6716] Arratia-Pérez R. and Hernández-Acevedo L. (1996). A Dirac molecular orbital study for tetragonal compression in the $RuY_6I_{12}^{2-}$ cluster. *Chem. Phys. Lett.*, **255**, 217–222.

[6717] Arratia-Pérez R. and Hernández-Acevedo L. (1997). A Dirac molecular orbital study for hexanuclear tungsten cluster structures. *Chem. Phys. Lett.*, **277**, 223–226.

[6718] Arratia-Pérez R. and Hernández-Acevedo L. (1998). Calculated paramagnetic resonance parameters of a gallium arsenide cluster: Ga_2As_3. *J. Chem. Phys.*, **109**, 3497–3500.

[6719] Arratia-Pérez R. and Hernández-Acevedo L. (1999a). The hexanuclear rhenium cluster ions $Re_6S_8X_6^{4-}$ (X=Cl, Br, I): Are these clusters luminescent? *J. Chem. Phys.*, **110**, 2529–2532.

[6720] Arratia-Pérez R. and Hernández-Acevedo L. (1999). Relativistic electronic structure of an icosahedral $Au_{12}Pd$ cluster. *Chem. Phys. Lett.*, **330**, 641–648.

[6721] Arratia-Pérez R. and Hernández-Acevedo L. (1999b). The $Re_6Se_8Cl_6^{4-}$ and $Re_6Se_8I_6^{4-}$ cluster ions: Another example of luminescent clusters ? *J. Chem. Phys.*, **111**, 168–172.

[6722] Arratia-Pérez R., Hernández-Acevedo L. and Alvarez-Thon L. (1998). Calculated paramagnetic hyperfine structure of pentagonal bipyramid Ag_7 cluster. *J. Chem. Phys.*, **108**, 5795–5798.

[6723] Arratia-Pérez R., Hernández-Acevedo L. and Gómez-Jeria J.S. (1995). Calculated paramagnetic properties of matrix isolated Au_3 cluster. *Chem. Phys. Lett.*, **236**, 37–42.

[6724] Arratia-Pérez R., Hernández-Acevedo L. and Weiss-López B. (1999). Calculated paramagnetic properties of the *acute* $GaAs_2$ and *obtuse* Ga_2As clusters. *J. Chem. Phys.*, **110**, 10882–10887.

[6725] Arriola E.R. and Salcedo L.L. (1993). Semiclassical expansion for Dirac Hamiltonians. *Modern Phys. Lett. A*, **8**, 2061–2069.

[6726] Artemyev A.N., Beier T., Plunien G., Shabaev V.M., Soff G. and Yerokhin V.A. (1999). Vacuum-polarization screening corrections to the energy levels of lithiumlike ions. *Phys. Rev. A*, **60**, 45–49.

[6727] Artemyev A.N., Shabaev V.M. and Yerokhin V.A. (1995a). Relativistic nuclear recoil corrections to the energy levels of hydrogenlike and high-Z lithiumlike atoms in all orders in αZ. *Phys. Rev. A*, **52**, 1884–1894.

[6728] Artemyev A.N., Shabaev V.M. and Yerokhin V.A. (1995b). Nuclear recoil corrections to the $2p_{3/2}$ energy of hydrogen-like and high-Z lithium-like atoms in all orders in αZ. *J. Phys. B*, **28**, 5201–5206.

[6729] Artemyev A.N., Shabaev V.M. and Yerokhin V.A. (1997). Vacuum polarization screening corrections to the ground-state energy of two-electron ions. *Phys. Rev. A*, **56**, 3529–3534.

[6730] Artimovich G.K. and Ritus V.I. (1993). Behavior of relativistic particles in the field of a deep well potential. *Zh. Eksp. Teor. Fiz.*, **104**, 2912–2936.

[6731] Arulmozhiraja S., Fujii T. and Tokiwa H. (1999). In^+ cation interactions with some organics: ab initio molecular orbital and density functional theory. *Chem. Phys.*, **250**, 237–242.

[6732] Arvieu R. and Rozmej P. (1994). Spin-orbit pendulum: Oscillations between spin and orbital angular momentum. *Phys. Rev. A*, **50**, 4376–4379.

[6733] Arvieu R. and Rozmej P. (1995). Collapse and revival in the dynamics of a spin with the spin-orbit potential. *Phys. Rev. A*, **51**, 104–119.

[6734] Asaga T., Fujita T. and Hiramoto M. (1998). g factor of tightly bound electron. *Phys. Rev. A*, **57**, 4974–4975.

[6735] Asato M., Settels A., Hoshino T., Asada T., Blügel S., Zeller R. and Dederichs P.H. (1999). Full-potential KKR calculations for metals and semiconductors. *Phys. Rev. B*, **60**, 5202–5210.

[6736] Aschi M., Brönstrup M., Diefenbach M., Harvey J.N., Schröder D. and Schwarz H. (1998). Ein Gasphasenmodell für die Pt^+-katalysierte Kupplung von Methan und Ammoniak. *Angew. Chem.*, **110**, 858–861.

[6737] Aschi M., Brönstrup M., Diefenbach M., Harvey J.N., Schröder D. and Schwarz H. (1998). Pt^+ (in English). *Angew. Chem, Int. Ed. Engl.*, **37**, 829–.

[6738] Asokamani R., Pari G. and Amirthakumari R.M. (1997). A theoretical study on the pressure dependence of the band gap in $A^I B^{III} C_2^{VI}$. *Int. J. Mod. Phys. B*, **11**, 1959–1967.

[6739] Aspromallis G., Sinanis C. and Nicolaides C.A. (1996). The lifetimes of the fine structure levels of the Be^- $1s^2 2s 2p^2$ ^4P metastable state. *J. Phys. B*, **29**, L1–L5.

[6740] Ast H., Keller S., Dreizler R.M., Whelan C.T., Ancarani L.U. and Walters H.R.J. (1996). On the position of the binary peak in relativistic (e,2e) collisions. *J. Phys. B*, **29**, L585–L590.

[6741] Ast H., Keller S., Whelan C.T., Walters H.R.J. and Dreizler R.M. (1994). Electron-impact ionization of the K shell of silver and gold in coplanar asymmetric geometry. *Phys. Rev. A*, **50**, R1–R3.

[6742] Atakishiyev N.M., Jafarov E.I., Nagiyev S.M. and Wolf K.B. (1998). Meixner oscillators. *Rev. Mex. Fis.*, **44**, 235–244.

[6743] Atanasov A.A. and Bankova S.G. (1993). Meson and baryon masses in the local approximation of the Schrödinger equation with relativistic kinematics. *J. Phys. G*, **19**, 827–835.

[6744] Au C.K. and Chu C.S. (1998). Finite mass effect on two-photon processes in hydrogenic systems: effective scalar photon interaction. *Phys. Lett. A*, **244**, 338–348.

[6745] Au C.T., Liao M.S. and Ng C.F. (1997). A theoretical investigation of methane dissociation on Rh(111). *Chem. Phys. Lett.*, **267**, 44–50.

[6746] Aubauer C., Engelhart G., Klapötke T.M. and Schulz A. (1999). 1,1,1,2,2-Pentaiododiphosphanium cations, $P_2I_5^+EI_4^-$ (E= Al, Ga or In): synthesis and characterisation by ^{31}P MAS NMR, IR and Raman spectroscopy. *J. Chem. Soc., Dalton Trans.*, pp. 1729–1733.

[6747] Aucar G.A. (1996). Restrictions on ground state average values imposed by time reversal symmetry. *Chem. Phys. Lett.*, **254**, 13–20.

[6748] Aucar G.A., Botek E., Gómez S., Sproviero E. and Contreras R.H. (1996). RPA AM1 calculations of NMR spin-spin coupling constants: geminal ^{119}Sn-^{119}Sn couplings. *J. Organomet Chem.*, **524**, 1–7.

[6749] Aucar G.A., Jensen H.J.A. and Oddershede J. (1995a). Operator representations in Kramers bases. *Chem. Phys. Lett.*, **232**, 47–53.

[6750] Aucar G.A. and Oddershede J. (1993). Relativistic theory for indirect nuclear spin-spin couplings within the polarization propagator approach. *Int. J. Quantum Chem.*, **47**, 425–435.

[6751] Aucar G.A., Oddershede J. and Sabin J.R. (1995b). Relativistic extension of the Bethe sum rule. *Phys. Rev. A*, **52**, 1054–1059.

[6752] Aucar G.A., Saue T., Visscher L. and Jensen H.J.A. (1999). On the origin and contribution of the diamagnetic term in four-component relativistic calculations of magnetic properties. *J. Chem. Phys.*, **110**, 6208–6218.

[6753] Augenstein B.W. (1995). A lost alternative to Dirac's equation. *Physics Today*, **48(5)**, 86.

[6754] Aullón G., Alemany P. and Alvarez S. (1996). On the existence of a pyramidality effect in d^8...d^8 contacts. Theoretical study and structural correlation. *Inorg. Chem.*, **35**, 5061–5067.

[6755] Aullón G., Ujaque G., Lledós A. and Alvarez S. (1999). Edge-sharing binuclear d^8 complexes with XR bridges: Theoretical and structural database study of their molecular conformation. *Chem. Eur. J.*, **5**, 1391–1410.

[6756] Aullón G., Ujaque G., Lledós A., Alvarez S. and Alemany P. (1998). To bend or not to bend: Dilemma of the edge-sharing binuclear square planar complexes of d^8 transition metal ions. *Inorg. Chem.*, **37**, 804–813.

[6757] Autschbach J. (1999). *Zur Berechnung relativistischer Effekte und zum Verständnis ihrer Trends bei Atomen und Molekülen.* Ph.D. thesis, Universität Siegen, *http://www.ub.uni.siegen.de/epub/diss/autschbach.htm.*

[6758] Avdonina N.B. and Pratt R.H. (1993). Bremsstrahlung spectrum from ions in a model potential approximation. *J. Quant. Spectr. Radiat. Transfer*, **50**, 349–358.

[6759] Avery J., Antonsen F. and Shim I. (1993). 4-currents in relativistic quantum chemistry. *Int. J. Quantum Chem.*, **45**, 573–585.

[6760] Avetissian H.K., Hatsagortsian K.Z., Markossian A.G. and Movsissian S.V. (1999). Generalized eikonal wave function of a Dirac particle interacting with an arbitrary potential and radiation fields. *Phys. Rev. A*, **59**, 549–558.

[6761] Avetissian H.K. and Movsissian S.V. (1996). Theory of elastic scattering of particles in a static potential field. *Phys. Rev. A*, **54**, 3036–3041.

[6762] Avgoustoglou E. and Beck D.R. (1997). All-order relativistic many-body calculations for the electron affinities of Ca^-, Sr^-, Ba^-, and Yb^- negative ions. *Phys. Rev. A*, **55**, 4143–4149.

[6763] Avgoustoglou E. and Beck D.R. (1998). Relativistic many-body calculations for the oscillator strengths of the resonance lines of neon, argon, krypton, and xenon. *Phys. Rev. A*, **57**, 4286–4295.

[6764] Avgoustoglou E., Johnson W.R., Liu Z.W. and Sapirstein J. (1995). Relativistic many-body calculations of $(2p^5 3s)$ excited-state energy levels for neonlike ions. *Phys. Rev. A*, **51**, 1196–1208.

[6765] Avgoustoglou E. and Liu Z.W. (1996). Relativistic many-body calculations of $(2p^5 3d)_{J=1}$ excited-state energy levels for neonlike ions. *Phys. Rev. A*, **54**, 1351–1359.

[6766] Aymar M., Greene C.H. and Luc-Koenig E. (1996). Multichannel Rydberg spectroscopy of complex atoms. *Rev. Mod. Phys.*, **68**, 1015–1123.

[6767] Baba T. and Fukui H. (1999). Calculation of nuclear magnetic shieldings. XIII. Gauge-origin independent relativistic effects. *J. Chem. Phys.*, **110**, 131–137.

[6768] Babb J.F. and Spruch L. (1994). Retardation (or Casimir) potential for the Rydberg hydrogen molecule. *Phys. Rev. A*, **50**, 3845–3855.

[6769] Bacelo D.E., Huang S.D. and Ishikawa Y. (1997). The Au(I)-Au(I) interaction: Hartree-Fock and Møller-Plesset second-order perturbation theory calculations on $[Se_5Au_2]^{2-}$ and $[Se_6Au_2]^{2-}$ complexes. *Chem. Phys. Lett.*, **277**, 215–222.

[6770] Bach V., Barbaroux J.M., Helffer B. and Siedentop H. (1999). On the stability of the electron-positron field. *Comm. Math. Phys.*, **201**, 445–460.

[6771] Bacskay G.B., Bytheway I. and Hush N.S. (1996). H-D spin-spin coupling in stretched molecular hydrogen complexes of osmium(II): density functional studies of J_{HD}. *J. Am. Chem. Soc.*, **118**, 3753–3756.

[6772] Badnell N.R. (1997). On the effects of the two-body non-fine-structure operators of the Breit-Pauli Hamiltonian. *J. Phys. B*, **30**, 1–11.

[6773] Badnell N.R. and Griffin D.C. (1999). Breit-Pauli and intermediate coupling collision strengths for the correlation resonances that arise in the electron-impact excitation of Ni^{4+}. *J. Phys. B*, **32**, 2267–2276.

[6774] Baeck K.K. and Lee Y.S. (1994). Effects of the magnetic part of the Breit term on the $^2\Pi$ states of diatomic hydrides. *J. Chem. Phys.*, **100**, 2888–2895.

[6775] Baerends E.J. and Gritsenko O.V. (1997). A quantum chemical view of density functional theory. *J. Phys. Chem. A*, **101**, 5383–5403.

[6776] Bagatur'yants A.A., Safonov A.A., Stoll H. and Werner H.J. (1998). Ab initio relativistic pseudopotential study of small silver and gold sulfide clusters $(M_2S)_n$, n=1 and 2. *J. Chem. Phys.*, **109**, 3096–3107.

[6777] Bagrov V.G., Belov V.V., Trifonov A.Y. and Yevseyevich A.A. (1994a). Quantization of closed orbits in Dirac theory by Maslov's complex germ method. *J. Phys. A*, **27**, 1021–1043.

[6778] Bagrov V.G., Belov V.V., Trifonov A.Y. and Yevseyevich A.A. (1994b). Quasi-classical spectral series of the Dirac operators corresponding to quantized two-dimensional Langrangian tori. *J. Phys. A*, **27**, 5273–5306.

[6779] Bakalov D. and Korobov V.I. (1998). Hyperfine structure of antiprotonic helium energy levels. *Phys. Rev. A*, **57**, 1662–1667.

[6780] Bakasov A., Ha T.K. and Quack M. (1998). Ab initio calculation of molecular energies including parity violating interactions. *J. Chem. Phys.*, **109**, 7263–7285.

[6781] Baker L.J., Bott R.C., Bowmaker G.A., Healy P.C., Skelton B.W., Schwerdtfeger P. and White A.H. (1995). Structural, far-infrared and ^{31}P nuclear magnetic resonance of two-co-ordinate complexes of tris(2,4,6-trimethoxyphenyl)phosphine with gold(I) halides. *J. Chem. Soc., Dalton Trans.*, pp. 1341–1347.

[6782] Bakhmutov V.I., Bianchini C., Maseras F., Lledos A., Peruzzini M. and Vorontsov E.V. (1999). ^{2}H-T_1 relaxation and deuterium quadrupole coupling constants in transition metal η^2-D_2 complexes. *Chem. Eur. J.*, **5**, 3318–3325.

[6783] Balasubramanian K. (1994). Relativistic effects and electronic structure of lanthanide and actinide molecules. In *Handbook on the Physics and Chemistry of Rare Earths, Vol. 18*, (Edited by Gschneidner Jr. K.A., Eyring L., Choppin G.R. and Lander G.H.), pp. 29–158, Elsevier Science, Amsterdam.

[6784] Balasubramanian K. (1996). Double group of the icosahedral group (I_h) and its applications to fullerenes. *Chem. Phys. Lett.*, **260**, 476–484.

[6785] Balasubramanian K. (1997a). *Relativistic Effects in Chemistry. Part A. Theory and Techniques.* Wiley, New York, 301 p.

[6786] Balasubramanian K. (1997b). *Relativistic Effects in Chemistry. Part B. Applications.* Wiley, New York, 527 p.

[6787] Balasubramanian K. (1998). Relativistic effective core potential techniques for molecules containing very heavy atoms. In *Encyclopedia of Computational Chemistry*, (Edited by von Ragué Schleyer P.), vol. 4, pp. 2471–2480, Wiley, Chichester and New York.

[6788] Balasubramanian K., Chung Y.S. and Glaunsinger W.S. (1993). Geometries and bond energies of PH_n and PH_n^+ (n=1-3). *J. Chem. Phys.*, **98**, 8859–8869.

[6789] Balasubramanian K. and Dai D.G. (1993). Group VI trimers (Se_3, Te_3, and Po_3) electronic states and potential energy surfaces (Erratum: JCP 1997;106:10385). *J. Chem. Phys.*, **99**, 5239–5250.

[6790] Balasubramanian K. and Dai D.G. (1997). Electronic states and potential energy surfaces of the tungsten trimer (W_3). *Chem. Phys. Lett.*, **265**, 538–546.

[6791] Balasubramanian K. and Feng P.Y. (1993). Potential energy curves for electron beam excitation of $RbXe^+$ excimer. *Chem. Phys. Lett.*, **214**, 85–90.

[6792] Balasubramanian K. and Latifzadeh-Masoudipour L. (1999). Spectroscopic properties and potential energy surfaces of electronic states of $SbCl_2$, $SbBr_2$, $SbCl_2^+$ and $SbBr_2^+$. *J. Phys. Chem. A*, **103**, 3044–3053.

[6793] Balasubramanian K. and Xu H. (1995). Spectroscopic properties and potential energy curves of SnF^+. *J. Mol. Spectr.*, **171**, 555–564.

[6794] Balcar E. and Lovesey S.W. (1993). Samarium: magnetic neutron spectroscopic intensities. *J. Phys: CM*, **5**, 7269–7276.

[6795] Baldas J., Heath G.A., Macgregor S.A., Moock K.H., Nissen S.C. and Raptis R.G. (1998). Spectroelectrochemical and computational studies of tetrachloro and tetrabromo oxo- and nitridotechnetium(V) and their Tc^{VI} counterparts. *J. Chem. Soc., Dalton Trans.*, pp. 2303–2314.

[6796] Balinsky A. and Evans W.D. (1999a). On the stability of relativistic one-electron molecules. *J. Phys. A*, **32**, L129–L132.

[6797] Balinsky A. and Evans W.D. (1999b). Stability of one-electron molecules in the Brown-Ravenhall model. *Comm. Math. Phys.*, **202**, 481–500.

[6798] Ballard C.C., Hada M., Kaneko H. and Nakatsuji H. (1996). Relativistic study of nuclear magnetic shielding constants: hydrogen halides. *Chem. Phys. Lett.*, **254**, 170–178.

[6799] Baltz A.J. (1995). Coulomb potential from a particle in uniform ultrarelativistic motion. *Phys. Rev. A*, **52**, 4970–4971.

[6800] Baltz A.J. (1997). Exact Dirac equation calculation of ionization and pair production induced by ultrarelativistic heavy ions. *Phys. Rev. Lett.*, **78**, 1231–1234.

[6801] Baltz A.J., Rhoades-Brown M.J. and Weneser J. (1993). Bound-electron-positron pair production in relativistic heavy-ion collisions. *Phys. Rev. A*, **47**, 3444–3447.

[6802] Baluja K.L. (1995). Relativistic correction to the dipole polarizability of a hydrogenic ion. *Pramana*, **45**, 533–536.

[6803] Baluja K.L. and Gupta A. (1994). Total cross-sections for electron scattering from iron at 10-5000 eV using a model optical potential. *Z. Phys. D*, **31**, 5–11.

[6804] Baluja K.L., Gupta A. and Datta S.M. (1993). Line strengths and transition probabilities for allowed transitions in Fe XXI. *Z. Phys. D*, **28**, 105–108.

[6805] Band I.M., Listengarten M.A. and Trzhaskovskaya M.B. (1992a). Calculation of conversion coefficients for parity doublet in ^{229}Pa. *Izv. Ross. Akad. Nauk. Ser. Fiz.*, **56(11)**, 110–115.

[6806] Band I.M., Listengarten M.A. and Trzhaskovskaya M.B. (1992a). Calculation of conversion coefficients for parity doublet in ^{229}Pa. *Bull. Russian Acad. Sci., Phys.*, **56**, 1749–1755.

[6807] Band I.M. and Trzhaskovskaya M.B. (1993a). Interpretation of the nuclear transition with E_γ=72 keV in $^{187}Re_{75}$ (in Russian). *Yad. Fiz.*, **56(5)**, 1–5.

[6808] Band I.M. and Trzhaskovskaya M.B. (1993a). Interpretation of the nuclear transition with E_γ=72 keV in $^{187}Re_{75}$. *Phys. At. Nucl.*, **56**, 573–575.

[6809] Band I.M. and Trzhaskovskaya M.B. (1993b). Internal conversion coefficients for low-energy nuclear transitions. *At. Data Nucl. Data Tables*, **55**, 43–61.

[6810] Banger K.K., Brisdon A.K., Brain P.T., Parsons S., Rankin D.W.H., Robertson H.E., Smart B.A. and Bühl M. (1999). Experimental and theoretical studies of bis(perfluorovinyl)mercury, $Hg(CF{=}CF_2)_2$: Synthesis, characterization, and structure in the gaseous and crystalline phases. *Inorg. Chem.*, **38**, 5894–5900.

[6811] Bär C. (1990). Das Spektrum von Dirac-Operator. *Bonner Mathematische Schriften*, pp. 217–.

[6812] Barakat T., Odeh M. and Mustafa O. (1998). Perturbed Coulomb potentials in the Klein-Gordon equation via the shifted-l expansion technique. *J. Phys. A*, **31**, 3469–3479.

[6813] Barandiaran Z. and Seijo L. (1994). Quasirelativistic ab initio model potential calculations on the group IV hydrides (XH_2, XH_4; X=Si, Ge, Sn, Pb) and oxides (XO, X=Ge, Sn, Pb). *J. Chem. Phys.*, **101**, 4049–4054.

[6814] Baranger M., Bethe H.A. and Feynman R.P. (1953). Relativistic correction to the Lamb shift. *Phys. Rev.*, **92**, 482–501.

[6815] Barnes L.A., Liu B. and Lindh R. (1993). Bond length dipole moment and harmonic frequency of CO. *J. Chem. Phys.*, **98**, 3972–3977.

[6816] Barone V., Bencini A., Totti F. and Uytterhoeven M.G. (1995). Theoretical study of the electronic structure and the mercury-carbon bonding of methylmercury(II) compounds. *J. Phys. Chem.*, **99**, 12743–12750.

[6817] Barone V., Bencini A., Totti F. and Uytterhoeven M.G. (1996a). Theoretical characterization of the mechanism of Hg-C bond cleavage by halogenic acids. *Organometallics*, **15**, 1465–1469.

[6818] Barone V., Bencini A., Totti F. and Uytterhoeven M.G. (1996b). Comparison between post-Hartree-Fock and DFT methods for the study of strength and mechanism of cleavage of Hg-C bond. *Int. J. Quantum Chem.*, **61**, 361–367.

[6819] Bartlett N. (1998). Relativistic effects and the chemistry of gold. *Gold Bull.*, **31(1)**, 22–25.

[6820] Barton D.H.R., Hall M.B., Lin Z.Y., Parekh S.I. and Reibenspies J. (1993). Unusual attractive interactions between selenium and oxygen in selenoiminoquinones. *J. Am. Chem. Soc.*, **115**, 5056–5059.

[6821] Bartschat K. (1993). Low-energy electron scattering from caesium atoms -comparison of a semirelativistic Breit-Pauli and a full relativistic Dirac treatment. *J. Phys. B*, **26**, 3595–3609.

[6822] Bartschat K. (1994). Close-coupling calculations for low-energy elastic electron scattering from thallium atoms. *Z. Phys. D*, **30**, 85–90.

[6823] Bartschat K., Johnston A.R. and Burrow P.D. (1994). Evidence for the $(6p^2)^3P_2{}^e$ resonance in electron scattering from caesium atoms. *J. Phys. B*, **27**, L231–L234.

[6824] Barut A.O. (1973). Some unusual applications of Lie algebra representations in quantum theory. *SIAM J. Appl. Math.*, **25**, 247–259.

[6825] Barut A.O. (1993). On the 'New constants of the motion of the free Dirac electron'. *Nuovo Cim. A*, **106a**, 583–584.

[6826] Barut A.O. (1994). Localized rotating wavelets with half integer spin. *Phys. Lett. A*, **189**, 277–280.

[6827] Barut A.O. and Bornzin G.L. (1971). SO(4,2)-Formulation of the symmetry breaking in relativistic Kepler problems with or without magnetic charges. *J. Math. Phys.*, **12**, 841–846.

[6828] Barut A.O., Bracken A.J., Komy S. and Unal N. (1993). New approach to the determination of eigenvalues and eigenfunctions for a relativistic two-fermion equation. *J. Math. Phys.*, **34**, 2089–2106.

[6829] Barut A.O. and Cruz M.G. (1993). Classical relativistic spinning particle with anomalous magnetic moment: the precession of spin. *J. Phys. A*, **26**, 6499–6506.

[6830] Barut A.O. and Cruz M.G. (1994). On the Zitterbewegung of the relativistic electron. *Eur. J. Phys.*, **15**, 119–120.

[6831] Barut A.O. and Duru I.H. (1994). Path-integral quantization of the confined solution of a relativistic two-body problem. *Physica. A*, **209**, 249–256.

[6832] Barut A.O. and Pavsic M. (1993). Dirac's shell model of the electron and the general theory of moving relativistic charged membranes. *Phys. Lett. B*, **306**, 49–54.

[6833] Barut A.O. and Unal N. (1993). On Poisson brackets and symplectic structures for the classical and quantum Zitterbewegung. *Found. Phys.*, **23**, 1423–1429.

[6834] Barysz M. (1997). Perturbation theory of relativistic effects: relative inaccuracies of approximate results. *Chem. Phys. Lett.*, **275**, 161–167.

[6835] Barysz M. and Papadopoulos M.G. (1998). On the ground state of NiH_2. *J. Chem. Phys.*, **109**, 3699–3700.

[6836] Barysz M. and Pyykkö P. (1998). Strong chemical bonds to gold. High level correlated relativistic results for diatomic $AuBe^+$, AuC^+, $AuMg^+$, and $AuSi^+$. *Chem. Phys. Lett.*, **285**, 398–403.

[6837] Barysz M. and Sadlej A.J. (1997). Expectation values of operators in approximate two-component relativistic theories. *Theor. Chem. Acc.*, **97**, 260–270.

[6838] Barysz M., Sadlej A.J. and Snijders J.G. (1997). Nonsingular two/one-component relativistic Hamiltonians accurate through arbitrary high order in α^3. *Int. J. Quantum Chem.*, **65**, 225–239.

[6839] Barysz M. and Urban M. (1997). Molecular properties of boron-coinage metal dimers: BCu, BAg, BAu. *Adv. Quantum Chem.*, **28**, 257–272.

[6840] Basavaraju G., Kane P.P., Kissel L.D. and Pratt R.H. (1994). Elastic scattering of 81-keV gamma rays. *Phys. Rev. A*, **49**, 3664–3672.

[6841] Basch H. (1996a). Cation-aromatic bonding in group 14 organometallics. *Inorg. Chim. Acta*, **242**, 191–200.

[6842] Basch H. (1996b). Bond dissociation energies in organometallic compounds. *Inorg. Chim. Acta*, **252**, 265–279.

[6843] Basch H. and Hoz T. (1992). General and theoretical. In *Supplement B: The Chemistry of Acid Derivatives*, (Edited by Patai S.), vol. 2, Wiley, New York.

[6844] Basch H. and Hoz T. (1994). The nature of the triple bond. In *Supplement C2: The Chemistry of Triple-Bonded Functional Groups*, (Edited by Patai S.), pp. 1–108, Wiley, New York.

[6845] Basch H. and Hoz T. (1995). Chapter 1. The nature of the C-M bond (M=Ge,Sn,Pb). In *The Chemistry of Organic Germanium, Tin and Lead Compounds*, (Edited by Patai S.), Wiley, New York.

[6846] Basch H., Musaev D.G. and Morokuma K. (1999a). Why does the reaction of the dihydrogen molecule with $[P_2N_2]Zr(\mu-\eta^2{-}N_2)Zr[P_2N_2]$ produce $[P_2N_2]Zr(\mu-\eta^2\text{-}N_2H)Zr[P_2N_2](\mu\text{-}H)$ but not the thermodynamically more favorable $[P_2N_2]Zr(\mu\text{-}NH)_2Zr[P_2N_2]$? A theoretical study. *J. Am. Chem. Soc.*, **121**, 5754–5761.

[6847] Basch H., Musaev D.G., Morokuma K., Fryzuk M.D., Love J.B., Seidel W.W., Albinati A., Koetzle T.F., Klooster W.T., Mason S.A. and Eckert J. (1999b). Theoretical predictions and single-crystal neutron diffraction and inelastic neutron scattering studies on the reaction of dihydrogen with the dinuclear dinitrogen complex of zirconium $[P_2N_2]Zr(\mu-\eta^2\text{-}N_2)Zr[P_2N_2]$, $P_2N_2{=}PhP(CH_2SiMe_2NSiMe_2CH_2)_2PPh$. *J. Am. Chem. Soc.*, **121**, 523–528.

[6848] Baştuğ T. (1994). *Genaue Berechnung der totalen Energie für kleine Moleküle und Atomcluster mit der Dirac-Fock-Slater Methode*. Ph.D. thesis, University of Kassel.

[6849] Baştuğ T., Fricke B., Finkbeiner M. and Johnson W.R. (1996). The magnetic moment of ^{209}Bi. A molecular determination of the diamagnetic shielding. *Z. Phys. D*, **37**, 281–282.

[6850] Baştuğ T., Heinemann D., Sepp W.D., Kolb D. and Fricke B. (1993). All-electron Dirac-Fock-Slater SCF calculations of the Au_2 molecule. *Chem. Phys. Lett.*, **211**, 119–124.

[6851] Baştuğ T., Kürpick P., Meyer J., Sepp W.J., Fricke B. and Rosén A. (1997). Dirac-Fock-Slater calculations on the geometric and electronic structure of neutral and multiply charged C_{60} fullerenes. *Phys. Rev. B*, **78**, 5015–5020.

[6852] Baştuğ T., Rashid K., Sepp W.D., Kolb D. and Fricke B. (1997). All-electron X_α self-consistent-field calculations of relativistic effects in the molecular properties of Tl_2, Pb_2, and Bi_2 molecules. *Phys. Rev. A*, **55**, 1760–1764.

[6853] Baştuğ T., Sepp W.D., Fricke B., Johnson E. and Barshick C.M. (1995a). All-electron relativistic Dirac-Fock-Slater self-consistent-field calculations of the singly charged diatomic transition-metal- (Fe, Co, Ni, Cu, Zn) argon molecules. *Phys. Rev. A*, **52**, 2734–2736.

[6854] Baştuğ T., Sepp W.D., Kolb D., Fricke B., Baerends E.J. and Te Velde G. (1995). All-electron Dirac-Fock-Slater SCF calculations for electronic and geometric structures of the Hg_2 and Hg_3 molecules. *J. Phys. B*, **28**, 2325–2331.

[6855] Basu C., Roy C.L., Macia E., Dominguez-Adame F. and Sanchez A. (1994). Localization of relativistic electrons in a one-dimensional disordered system. *J. Phys. A*, **27**, 3285–3291.

[6856] Battocletti M., Ebert H. and Akai H. (1996). Influence of gradient corrections to the local-density-approximation on the calculation of hyperfine fields in ferromagnetic Fe, Co, and Ni. *Phys. Rev. B*, **53**, 9776–9783.

[6857] Bauer M., Brooks M.S.S. and Dormann E. (1993). Orbital polarization of the conduction electrons in ferromagnetically ordered $GdAl_2$. *Phys. Rev. B*, **48**, 1014–1021.

[6858] Bauschlicher Jr. C.W. (1999a). The scalar relativistic contribution to gallium halide bond energies. *Theor. Chem. Acc.*, **101**, 421–425.

[6859] Bauschlicher Jr. C.W. (1999b). Correlation consistent basis set for indium. *Chem. Phys. Lett.*, **305**, 446–450.

[6860] Bauschlicher Jr. C.W. (1999c). Accurate indium bonding energies. *J. Phys. Chem. A*, **103**, 6429–6432.

[6861] Bauschlicher Jr C.W. (1999d). TiCl, TiH, and TiH^+ bond energies: a test of a correlation-consistent Ti basis set. *Theor. Chem. Acc.*, **103**, 141–145.

[6862] Bauschlicher Jr C.W. (1999e). Heats of formation for PO_n and PO_nH (n = 1-3). *J. Phys. Chem. A*, **103**, 11126–11129.

[6863] Bauschlicher Jr C.W., Langhoff S.R. and Partridge H. (1994). The low-lying states of AlCu and AlAg. *J. Chem. Phys.*, **100**, 1219–1225.

[6864] Bauschlicher Jr C.W., Langhoff S.R. and Partridge H. (1995). The application of ab initio electronic structure calculations to molecules containing transition metal atoms. In *Modern Electronic Structure Theory*, (Edited by Yarkony D.R.), pp. 1280–1374, World Scientific, Singapore.

[6865] Bauschlicher Jr C.W., Martin J.M.L. and Taylor P.R. (1999). Boron heat of formation revisited: Relativistic effects on the BF_3 atomization energy. *J. Phys. Chem. A*, **103**, 7715–7718.

[6866] Bauschlicher Jr. C.W. and Ricca A. (1998). Atomization energies of SO and SO_2: Basis set extrapolation revisited. *J. Phys. Chem. A*, **102**, 8044–8050.

[6867] Bauschlicher Jr. C.W. and Ricca A. (1999). The heat of formation of C_2F_4. *Chem. Phys. Lett.*, **315**, 449–453.

[6868] Bautista E. (1993). Acceleration through the Dirac-Pauli vacuum and effects of an external field. *Phys. Rev. D*, **48**, 783–789.

[6869] Baylis W.E. and Sienkiewicz J.E. (1995). Polarization trajectories. *J. Phys. B*, **28**, L549–L553.

[6870] Baylis W.E. and Yao Y. (1999). Relativistic dynamics of charges in electromagnetic fields: An eigenspinor approach. *Phys. Rev. A*, **60**, 785–795.

[6871] Bayse C.A. and Hall M.B. (1998). Transition metal polyhydride complexes. 9. The effect of ligand σ- and π-bonding on the H-Ta-H bond angle in six-coordinate tantalum(V) dihydride complexes. *Organometallics*, **17**, 4861–4868.

[6872] Beardmore K.M., Kress J.D., Grønbech-Jensen N. and Bishop A.R. (1998). Determination of the headgroup-gold(111) potential surface for alkanethiol self-assembled monolayers by ab initio calculation. *Chem. Phys. Lett.*, **286**, 40–45.

[6873] Bearpark M.J., Handy N.C., Palmieri P. and Tarroni R. (1993). Spin-orbit interactions from self consistent field wavefunctions. *Mol. Phys.*, **80**, 479–502.

[6874] Bechler A. (1993). Summation formulae for spherical spinors. *J. Phys. A*, **26**, 6039–6042.

[6875] Bechouche P., Mauser N.J. and Poupaud F. (1998). (Semi)-nonrelativistic limits of the Dirac equation with external time-dependent electromagnetic field. *Comm. Math. Phys.*, **197**, 405–425.

[6876] Beck D.R. (1997a). Relativistic configuration interaction results for Xe^{32+}, Ba^{34+}, Nd^{38+}, and Gd^{42+} ^5D J=2 to J=3 energy differences. *Phys. Rev. A*, **56**, 2428–2430.

[6877] Beck D.R. (1997b). Hyperfine structure constants of $(d+s)^3$ states in La I and the Zr II and Hf II isoelectronic sequences. *Int. J. Quantum Chem.*, **65**, 555–564.

[6878] Beck D.R. (1998). Theoretical lifetimes and Landé g values of Cs II $5p^5\,6p$ levels. *Phys. Rev. A*, **57**, 4240–4245.

[6879] Beck D.R. (1999). Energy differences and magnetic dipole ($M1$) decay rates for the W^{52+} and Bi^{61+} members of the nearly-Z-independent $(3d_{3/2})^3 3d_{5/2}$ $J=3--J=2$ transition. *Phys. Rev. A*, **60**, 3304–3305.

[6880] Beck D.R. and Datta D. (1993). Multireference relativistic configuration-interaction calculations for (d+s)(n) transition-metal atomic states: Application to Zr II hyperfine structure. *Phys. Rev. A*, **48**, 182–188.

[6881] Beck D.R. and Datta D. (1995). Theoretical lifetimes of Nb II z $4d^3$ 5p $^5G_3^0$ and $^3D_3^0$ levels. *Phys. Rev. A*, **52**, 2436–2438.

[6882] Becker J.D., Wills J.M., Cox L. and Cooper B.R. (1999). Electronic structure of Pu compounds with group-IIIB metals: Two regimes of behavior. *Phys. Rev. B*, **54**, R17265–R17268.

[6883] Bednar M. and Kolar P. (1993). On the relativistic spin projection operators. *Czech. J. Phys.*, **43**, 777–782.

[6884] Bednyakov I., Labzowsky L. and Soff G. (1995). Parity conserving weak interaction corrections to energy levels of highly charged ions. *J. Phys. B*, **28**, L719–L722.

[6885] Bednyakov I., Labzowsky L., Soff G. and Karasiev V. (1999). The standard model in strong fields: Electroweak radiative corrections for highly charged ions. *Phys. Scripta*, **T80**, 141–142.

[6886] Beideck D.J., Curtis L.J., Irving R.E., Maniak S.T., Hellborg R., Johansson S.G., Martinson I. and Rosberg M. (1993). Lifetimes of the $5d^9 6p$ levels in Au II. *J. Opt. Soc. Am. B*, **10**, 977–981.

[6887] Bei der Kellen S. and Freeman A.J. (1996). Self-consistent relativistic full-potential Korringa-Kohn-Rostoker total-energy method and applications. *Phys. Rev. B*, **54**, 11187–11198.

[6888] Beier T., Artemyev A.N., Eichler J., Shabaev V.M. and Yerokhin V.A. (1999). The influence of QED on the radiative electron capture process in highly charged ions. *Phys. Scripta*, **T80**, 322–323.

[6889] Beier T., Mohr P.J., Persson H., Plunien G., Greiner M. and Soff G. (1997a). Current status of Lamb shift predictions for heavy hydrogen-like ions. *Phys. Lett. A*, **236**, 329–338.

[6890] Beier T., Mohr P.J., Persson H. and Soff G. (1998). Influence of nuclear size on QED corrections in hydrogenlike heavy ions. *Phys. Rev. A*, **58**, 954–963.

[6891] Beier T., Plunien G., Greiner M. and Soff G. (1997b). Two-loop ladder diagram for the vacuum polarization contribution in hydrogen-like ions. *J. Phys. B*, **30**, 2761–2772.

[6892] Beier T., Plunien G. and Soff G. (1997c). The influence of nuclear structure on the Lamb shift in hydrogenlike heavy atoms. *Hyperfine Int.*, **108**, 19–27.

[6893] Beiersdorfer P., Knapp D., Marrs R.E., Elliott S.R. and Chen M.H. (1993). Structure and Lamb shift of $2s_{1/2}$-$2p_{3/2}$ levels in lithiumlike U^{89+} through neonlike U^{82+}. *Phys. Rev. Lett.*, **71**, 3939–3942.

[6894] Beiersdorfer P., Osterheld A.L. and Elliott S.R. (1998a). Measurements and modeling of electric-dipole-forbidden $2p_{1/2} - 2p_{3/2}$ transitions in fluorinelike U^{81+} through berylliumlike U^{88+}. *Phys. Rev. A*, **58**, 1944–1953.

[6895] Beiersdorfer P., Osterheld A.L., Scofield J.H., Crespo López-Urrutia J.R. and Widmann K. (1998b). Measurement of QED and hyperfine splitting in the $2s_{1/2} - 2p_{3/2}$ x-ray transition in Li-like $^{209}Bi^{80+}$. *Phys. Rev. Lett.*, **80**, 3022–3025.

[6896] Belanzoni P., Baerends E.J., van Asselt S. and Langewen P.B. (1995). Density functional study of magnetic coupling parameters. Reconciling theory and experiment for the TiF_3 complex. *J. Phys. Chem.*, **99**, 13094–13102.

[6897] Belanzoni P., Rosi M., Sgamellotti A., Baerends E.J. and Floriani C. (1996). On the electronic structure and bonding of the polynuclear aryl derivatives of the group IB metals $Cu_5(C_6H_5)_5$, $Ag_4(C_6H_5)_4$ and $Au_5(C_6H_5)_5$ by density functional theory. *Chem. Phys. Lett.*, **257**, 41–48.

[6898] Belkhiri L., Benmachiche A. and Boucekkine A. (1997). Influence de la rélativité sur les angles de liaisons et les conformations moléculaires. *J. Soc. Alger. Chim.*, **7**, 289–298.

[6899] Bellemin-Laponnaz S., Le Ny J.P. and Dedieu A. (1999). Mechanism of the allylic rearrangement of allyloxo metal oxo complexes: An ab initio theoretical investigation. *Chem. Eur. J.*, **5**, 57–64.

[6900] Benavides-Garcia M. and Balasubramanian K. (1993a). Spectroscopic constants and potential energy curves of PbBr. *Chem. Phys. Lett.*, **211**, 631–636.

[6901] Benavides-Garcia M. and Balasubramanian K. (1993b). Spectroscopic constants and potential energy curves of PbI. *J. Mol. Spectr.*, **161**, 552–559.

[6902] Bencheikh M. and Schamps J. (1997). A direct relativistic approach to ligand field theory calculations of spin-orbit components of diatomic molecules. *Phys. Scr.*, **56**, 433–435.

[6903] Bencini A., Di Vaira M., Morasi R., Stoppioni P. and Mele F. (1996). Synthesis, x-ray crystal structure and bonding in $[(PPh_3)_2PtSe]_2$. *Polyhedron*, **15**, 2079–2086.

[6904] Bengtsson L.A. and Hoffmann R. (1993). Dilead structural units in lead halide and hydroxo/oxo molecules, clusters, and extended structures found in molten salts, aqueous solutions, and the solid state. *J. Am. Chem. Soc.*, **115**, 2666–2676.

[6905] Bengtsson-Kloo L., Iapalucci C.M., Longoni G. and Ulvenlund S. (1998). Solution structures of mono- and ditriangular Chini clusters of nickel and platinum. An x-ray scattering and quantum chemical study. *Inorg. Chem.*, **37**, 4335–4343.

[6906] Benn I.M. and Tucker R.W. (1988). *An Introduction to Spinors and Geometry with Applications in Physics.*. Institute of Physics Publishing, Bristol, 368 pp.

[6907] Bennett F.R. and Connelly J.P. (1996). Theoretical study of the properties of $InMH_6$ and MBH_6 (M = B, Al, Ga, and In) μ-hydrido-bridged compounds. *J. Phys. Chem.*, **100**, 9308–9313.

[6908] Benson M.T., Bryan J.C., Burrell A.K. and Cundari T.R. (1995). Bonding and structure of heavily π-loaded complexes. *Inorg. Chem.*, **34**, 2348–2355.

[6909] Benson M.T. and Cundari T.R. (1997). Late transition-metal multiple bonding: Platinum phosphinidenes and ruthenium alkylidenes. *Int. J. Quantum Chem.*, **65**, 987–996.

[6910] Benson M.T., Cundari T.R., Li Y.P. and Strohecker L.A. (1994a). Effective core potential study of multiply bonded transition metal complexes of the heavier main group elements. *Int. J. Quantum Chem.*, **S28**, 181–194.

[6911] Benson M.T., Cundari T.R., Lim S.J., Nguyen H.D. and Pierce-Beaver K. (1994b). An effective core potential study of transition-metal chalcogenides. 1. Molecular structure. *J. Am. Chem. Soc.*, **116**, 3955–3966.

[6912] Benvegnù S. (1997). Relativistic point interaction with Coulomb potential in one dimension. *J. Math. Phys.*, **38**, 556–570.

[6913] Benvegnù S. and Dąbrowski L. (1994). Relativistic point interaction. *Lett. Math. Phys.*, **30**, 159–167.

[6914] Berbenni-Bitsh M.E., Meyer S., Schäfer A., Verbaarschot J.J.M. and Wettig T. (1998). Microscopic universality in the spectrum of the lattice Dirac operator. *Phys. Rev. Lett.*, **80**, 1146–1149.

[6915] Bérces A. (1996). Harmonic vibrational frequencies and force constants of $M(CO)_5CX$ (M=Cr, Mo, W; X = O, S, Se). The performance of density functional theory and the influence of relativistic effects. *J. Phys. Chem.*, **100**, 16538–16544.

[6916] Bérces A., Mitchell S.A. and Zgierski M.Z. (1998). Reactions between M_n (M = Nb, Mo and n = 1,2,3, and 4) and N_2. A density functional study. *J. Phys. Chem. A*, **102**, 6340–6347.

[6917] Bergerhoff B. and Soff G. (1999). Scalar potentials in the Dirac equation. *Z. Naturf. A*, **49**, 997–1012.

[6918] Bergner A., Dolg M., Küchle W., Stoll H. and Preuss H. (1993). Ab initio energy-adjusted pseudopotentials for elements of groups 13-17. *Mol. Phys.*, **80**, 1431–1441.

[6919] Bergstrom J., M P., Suric T., Pisk K. and Pratt R.H. (1993). Compton scattering of photons from bound electrons: full relativistic independent-particle-approximation calculations. *Phys. Rev. A*, **48**, 1134–1162.

[6920] Berrington K.A. (1994). A review of electron impact excitation data for the beryllium isoelectronic sequence (Z=4 to 28). *At. Data Nucl. Data Tables*, **57**, 71–95.

[6921] Berny F., Muzet N., Troxler L., Dedieu A. and Wipff G. (1999). Interaction of M^{3+} lanthanide cations with amide, pyridine, and phosphoryl $O=PPh_3$ ligands: A quantum mechanics study. *Inorg. Chem.*, **38**, 1244–1252.

[6922] Berrington K.A., Eissner W.B. and Norrington P.H. (1995). Rmatrix I - Belfast atomic R-matrix codes. *Comput. Phys. Commun.*, **92**, 290–420.

[6923] Berrington K.A., Zeippen C.J., Le Dourneuf M., Eissner W. and Burke P.G. (1991). Electron impact excitation of Fe^{2+}:I. A 17-level fine-structure calculation. *J. Phys. B*, **24**, 3467–3478.

[6924] Berrondo M. and Rivas-Silva J.F. (1995). Stokes shifts in Tl-doped alkali halides. *Int. J. Quantum Chem., Symp*, **29**, 253–256.

[6925] Berry H.G., Dunford R.W. and Livingston A.E. (1993). Comparisons of the QED and relativistic parts of the triplet-state energies in the heliumlike sequence. *Phys. Rev. A*, **47**, 698–701.

[6926] Berseth W.C. and Darewych J.W. (1993). Relativistic corrections to Ps^-. *Phys. Lett. A*, **178**, 347–350.

[6927] Bettega M.H.F., Ferreira L.G. and Lima M.A.P. (1993). Transferability of local-density norm-conserving pseudopotentials to electron-molecule-collision calculations. *Phys. Rev. A*, **47**, 1111–1118.

[6928] Bettega M.H.F., Natalense A.P.P., Lima M.A.P. and Ferreira L.G. (1995). Calculation of elastic scattering cross sections of low-energy electrons by PbH_4 and SnH_4. *J. Chem. Phys.*, **103**, 10566–10570.

[6929] Beyer M., Berg C., Albert G., Achatz U. and Bondybey V.E. (1997). Coordinative saturation of cationic niobium- and rhodium-argon complexes. *Chem. Phys. Lett.*, **280**, 459–463.

[6930] Bharadvaja A. and Baluja K.L. (1996). Systematic trends in the radial matrix elements of E1 transitions in Li, F, Na, and Cu isoelectronic sequences. *Z. Phys. D*, **38**, 1–4.

[6931] Bhatia A.K. and Drachman R.J. (1997). Relativistic, retardation, and radiative corrections in Rydberg states of lithium. *Phys. Rev. A*, **55**, 1842–1845.

[6932] Bhatia A.K. and Drachman R.J. (1998a). Another way to calculate the Lamb shift in two-electron systems. *Phys. Rev. A*, **57**, 4301–4305.

[6933] Bhatia A.K. and Drachman R.J. (1998b). Optical properties of helium including relativistic corrections. *Phys. Rev. A*, **58**, 4470–4472.

[6934] Bhatia A.K. and Drachman R.J. (1998). Relativistic corrections to the binding energy of Ps^-. *Nucl. Instr. Meth. Phys. Res. B*, **143**, 195–198.

[6935] Bhatia A.K. and Drachman R.J. (1999). Energy levels of triply ionized carbon (C IV): Polarization method. *Phys. Rev. A*, **60**, 2848–2852.

[6936] Biedenharn L.C., Mueller B. and Tarlini M. (1993). The Dirac-Coulomb problem for the kappa -Poincare quantum group. *Phys. Lett. B*, **318**, 613–616.

[6937] Bielinska-Waz D., Martin I. and Karwowski J. (1994). Core polarization effects in the relativistic quantum-defect-orbital theory. *Acta Phys. Pol. A*, **85**, 805–812.

[6938] Biémont E. (1997a). Multipole transitions in nickel-like and palladium-like spectra. *J. Phys. B*, **30**, 4207–4222.

[6939] Biémont E. (1997b). On the importance of octupole decays in atomic spectra. *Phys. Scr.*, **T73**, 59–61.

[6940] Biémont E., Fremat Y. and Quinet P. (1999). Ionization potentials of atoms and ions from lithium to tin (Z=50). *At. Data Nucl. Data Tables*, **71**, 117–146.

[6941] Biémont E., Gebarowski R. and Zeippen C.J. (1994a). Resonance transitions in neutral chlorine. *Astronomy Astrophys*, **287**, 290–296.

[6942] Biémont E., Johansson S. and Palmeri P. (1997a). The lowest $5g$ - $6h$ supermultiplet of Fe II. *Phys. Scr.*, **55**, 559–564.

[6943] Biémont E., Marcinek R., Migdalek J. and Quinet P. (1994b). Comment on "Oscillator strengths for some systems with the ns^2np ground-state configuration. II. Gallium isoelectronic sequence". *J. Phys. B*, **27**, 825–828.

[6944] Biémont E., Martin F., Quinet P. and Zeippen C.J. (1994c). The $4s - 4p$ transitions in neutral phosphorus. *Astronomy Astrophys*, **283**, 339–343.

[6945] Biémont E., Pinnington E.H., Kernahan J.A. and Rieger G. (1997b). Beam-laser measurements and relativistic Hartree-Fock calculations of the lifetimes of the $4d^95p$ levels in Ag II. *J. Phys. B*, **30**, 2067–2073.

[6946] Biémont E., Quinet P. and Van Renterghem V. (1998). Theoretical investigation of francium. *J. Phys. B*, **31**, 5301–5314.

[6947] Bieroń J. and Baylis W.E. (1995). Potential energy curves of HgCd and spectroscopic constants of group IIB metal dimers. *Chem. Phys.*, **197**, 129–137.

[6948] Bieroń J., Froese Fischer C. and Grant I.P. (1999a). Large-scale multiconfigurational Dirac-Fock calculations of the hyperfine-structure constants and determination of the nuclear quadrupole moment of ^{49}Ti. *Phys. Rev. A*, **59**, 4295–4299.

[6949] Bieroń J., Froese Fischer C. and Ynnerman A. (1994). Note on MCDF correlation calculations for high-Z ions. *J. Phys. B*, **27**, 4829–4834.

[6950] Bieroń J., Grant I.P. and Froese Fischer C. (1997). Nuclear quadrupole moment of scandium. *Phys. Rev. A*, **56**, 316–321.

[6951] Bieroń J., Grant I.P. and Froese Fischer C. (1998). Multiconfiguration Dirac-Fock calculations of the hyperfine structure constants and determination of the nuclear quadrupole moment of yttrium 90. *Phys. Rev. A*, **58**, 4401–4405.

[6952] Bieroń J., Jönsson P. and Froese Fischer C. (1996a). Large-scale multiconfiguration Dirac-Fock calculations of the hyperfine-structure constants of the $2s^2\ ^2S_{1/2}$, $2p\ ^2P_{1/2}$, and $2p\ ^2P_{3/2}$ states of lithium. *Phys. Rev. A*, **53**, 2181–2188.

[6953] Bieroń J., Jönsson P. and Froese Fischer C. (1999b). Effects of electron correlation, relativity and nuclear structure on hyperfine constants of Be^+ and F^{6+}. *Phys. Rev. A*, **60**, 3547–3557.

[6954] Bieroń J., Parpia F.A., Froese Fischer C. and Jönsson P. (1995). Large-scale multiconfiguration Dirac-Fock calculations of hyperfine interaction constants for nd^2 levels of Sc^+ and Y^+. *Phys. Rev. A*, **51**, 4603–4610.

[6955] Bigeleisen J. (1996). Nuclear size and shape effects in chemical reactions. Isotope chemistry of the heavy elements. *J. Am. Chem. Soc.*, **118**, 3676–3680.

[6956] Bijtebier J. (1997). Bethe-Salpeter equation: 3D reductions, heavy mass limits and abnormal solutions. *Nucl. Phys. A*, **623**, 498–518.

[6957] Bijtebier J. and Broekaert J. (1997). Using Salpeter's propagator to Bethe-Salpeter equation. *Nucl. Phys. A*, **612**, 279–296.

[6958] Birkett B.B., Briand J.P., Charles P., Dietrich D.D., Finlayson K., Indelicato P., Liesen D., Marrus R. and Simionovici A. (1993). Hyperfine quenching and measurement of the $2\,^3P_0$-$2\,^3P_1$ fine-structure splitting in heliumlike silver (Ag^{45+}). *Phys. Rev. A*, **47**, R2454–R2457.

[6959] Biswas B., Sugimoto M. and Sakaki S. (1999). Theoretical study of the structure, bonding nature, and reductive elimination reaction of $Pd(XH_3)(\eta^3$-$C_3H_5)(PH_3)$ (X = C,Si,Ge,Sn). Hypervalent behavior of group 14 elements. *Organometallics*, **18**, 4015–4026.

[6960] Bjerre N., Mitrushenkov A.O., Palmieri P. and Rosmus P. (1998). Spin-orbit and spin-spin couplings in He_2 and He_2^-. *Ther. Chem. Acc.*, **100**, 51–59.

[6961] Blancarte H., Grebert B. and Weder R. (1995). High- and low-energy estimates to the Dirac equation. *J. Math. Phys.*, **36**, 991–1015.

[6962] Blaudeau J.P. (1994). Ab initio calculations of metal-metal complexes using relativistic effective core potentials. *Diss. Abstr.*, **54**, 5682–B.

[6963] Blaudeau J.P. and Curtiss L.A. (1997). Optimized Gaussian basis sets for use with relativistic effective (core) potentials: K, Ca, Ga-Kr. *Int. J. Quantum Chem.*, **61**, 943–952.

[6964] Blaudeau J.P. and Pitzer R.M. (1994). Ab initio studies of Ligand effects on the metal-metal bond in dimolybdenum complexes. *J. Phys. Chem.*, **98**, 4575–4579.

[6965] Blaudeau J.P., Ross R.B., Pitzer R.M., Mougenot P. and Bénard M. (1994). Ab initio calculations of dirhenium complexes using relativistic effective core potentials. *J. Phys. Chem.*, **98**, 7123–7127.

[6966] Blaudeau J.P., Zygmunt S.A., Curtiss L.A., Reed D.T. and Bursten B.E. (1999). Relativistic density functional investigation of $Pu(H_2O)_n^{3+}$ clusters. *Chem. Phys. Lett.*, **310**, 347–354.

[6967] Bleakley K. and Hu P. (1999). A density functional theory study of the interaction between CO and O on a Pt surface: CO/Pt(111), O/Pt(111), and CO/O/Pt(111). *J. Am. Chem. Soc.*, **121**, 7644–7652.

[6968] Bleyer U. (1993). Energy levels of the hydrogen atom due to a generalized Dirac equation. *Found. Phys.*, **23**, 1025–1048.

[6969] Blomberg M.R.A., Karlsson C.A.M. and Siegbahn P.E.M. (1993). Carbonyl insertion into metal-hydrogen and metal-methyl bonds for second-row transition metal atoms. *J. Phys. Chem.*, **97**, 9341–9350.

[6970] Blomberg M.R.A. and Siegbahn P.E.M. (1993). Bridge bonding of N_2 to dinuclear transition metal systems. *J. Am. Chem. Soc.*, **115**, 6908–6915.

[6971] Blomberg M.R.A., Siegbahn P.E.M. and Svensson M. (1994). Reaction of second-row transition-metal cations with methane. *J. Phys. Chem.*, **98**, 2062–2071.

[6972] Blomberg M.R.A., Siegbahn P.E.M. and Svensson M. (1996). Comparisons of results from parametrized configuration interaction (PCI-80) and from hybrid density functional theory with experiments for first row transition metal compounds. *J. Chem. Phys.*, **104**, 9546–9554.

[6973] Blomqvist J. (1972). Vacuum polarization in exotic atoms. *Nucl. Phys. B*, **48**, 95–103.

[6974] Blum H. and Brockmann R. (1999). Causality in relativistic many body theory. *Phys. Rev. C*, **59**, 2546–2557.

[6975] Blundell S.A. (1993a). Calculations of the screened self-energy and vacuum polarization in Li-like, Na-like, and Cu-like ions. *Phys. Rev. A*, **47**, 1790–1803.

[6976] Blundell S.A. (1993b). An initio calculations of QED effects in Li-like, Na-like, and Cu-like ions. *Phys. Scr.*, **T46**, 144–149.

[6977] Blundell S.A. (1994). Screened QED effects in many-electron ions. *Nucl. Instr. Meth. Phys. Res. B*, **87**, 198–203.

[6978] Blundell S.A., Cheng K.T. and Sapirstein J. (1997a). Radiative corrections in atomic physics in the presence of perturbing potentials. *Phys. Rev. A*, **55**, 1857–1865.

[6979] Blundell S.A., Cheng K.T. and Sapirstein J. (1997b). All-order binding corrections to muonium hyperfine splitting. *Phys. Rev. Letters.*, **78**, 4914–4917.

[6980] Blundell S.A., Mohr P.J., Johnson W.R. and Sapirstein J. (1993c). Evaluation of two-photon exchange graphs for highly charged heliumlike ions. *Phys. Rev. A*, **48**, 2615–2626.

[6981] Blundell S.J. (1993). The Dirac comb and the Kronig-Penney model: Comment on"Scattering from a locally periodic potential" by D J Griffiths and N F Taussig [Am. J. Phys. **60**, 883-888, (1992)]. *Am. J. Phys.*, **61**, 1147–1148.

[6982] Bo C., Costas M. and Poblet J.M. (1995). Multiple metal additions to C_{60}. An ab initio study of $(M(PH_3)_2)_nC_{60}$ (M = Pt and Pd; n = 1,2, and 6). *J. Phys. Chem.*, **99**, 5914–5921.

[6983] Boča R. (1994). Tetrahedral gold-phosphine clusters: a relativistic molecular orbital study. *J. Chem. Soc., Dalton Trans.*, pp. 2061–2064.

[6984] Boča R. (1996). Platinum-centered octakis (triphenylphosphino gold) clusters: a relativistic MO study. *Int. J. Quantum Chem.*, **57**, 735–740.

[6985] Boča R., Hvastijová M. and Kohout J. (1994). A molecular orbital approach to coligand isomer formation. *J. Coord. Chem.*, **33**, 137–145.

[6986] Bock H., Hauck T., Näther C., Rösch N., Staufer M. and Häberlen O.D. (1995). The lipophilically wrapped polyion aggregate $((Ba_6Li_3O_2)^{11+}\ (^-OC(CH_3)_3)_{11}(OC_4H_8)_3)$, a face-sharing (octahedron+prismane) $Ba_6Li_3O_2$ polyhedron in a hydrocarbon ellipsoid: preparation, single crystal structure analysis, and density functional calculations. *Angew. Chem. Int. Ed. Engl.*, **34**, 1353–1355.

[6987] Böckmann M., Klessinger M. and Zerner M.C. (1996). Spin-orbit coupling in organic molecules: A semiempirical configuration interaction approach toward triplet state reactivity. *J. Phys. Chem.*, **100**, 10570–10579.

[6988] Bode B.M. and Gordon M.S. (1999). Fast computation of analytical second derivatives with effective core potentials: Application to Si_8C_{12}, Ge_8C_{12}, and Sn_8C_{12}. *J. Chem. Phys.*, **111**, 8778–8784.

[6989] Bodwin G.T. and Yennie D.R. (1988). Some recoil corrections to the hydrogen hyperfine splitting. *Phys. Rev. D*, **37**, 498–523.

[6990] Bodwin G.T., Yennie D.R. and Gregorio M.A. (1985). Recoil effects in the hyperfine structure of QED bound states. *Rev. Mod. Phys.*, **57**, 723–782.

[6991] Boehme C. and Frenking G. (1998). N-Heterocyclic carbene, silylene, and germylene complexes of MCl (M=Cu, Ag, Au). A theoretical study. *Organometallics*, **17**, 5801–5809.

[6992] Boehme C. and Wipff G. (1999). Dithiophosphinate complexes of trivalent lanthanide cations: Consequences of counterions and coordination number for binding energies and selectivity. A theoretical study. *Inorg. Chem.*, **38**, 5734–5741.

[6993] Boero M. and Cortona P. (1994). Transverse exchange energy in relativistic density-functional calculations: An alternative approximation. *Phys. Rev. A*, **49**, 825–828.

[6994] Boettger J.C. (1997). Scalar-relativistic LCGTO DFT calculations for atoms using the Douglas-Kroll transformation. *Int. J. Quantum Chem.*, **65**, 565–574.

[6995] Boettger J.C. (1998a). Scalar-relativistic linear combinations of Gaussian-type-orbitals technique for crystalline solids. *Phys. Rev. B*, **57**, 8743–8746.

[6996] Boettger J.C. (1998b). Relativistic effects on the structural phase stability of molybdenum. *Int. J. Quantum Chem.*, **70**, 825–830.

[6997] Boettger J.C. (1999). Relativistic effects on the structural phase stability of molybdenum. *J. Phys. CM*, **11**, 3237–3246.

[6998] Boettger J.C., Jones M.D. and Albers R.C. (1999). Structural properties of crystalline uranium from linear combination of Gaussian-type orbitals calculations. *Int. J. Quantum Chem.*, **75**, 911–915.

[6999] Boeyens J.C.A. and Kassman R.B. (1996). The Schrödinger equation and spin. *S. Afr. J. Chem.*, **49**, 1–7.

[7000] Bohm A. (1993). Relativistic and nonrelativistic dynamical groups. *Found. Phys.*, **23**, 751–767.

[7001] Bohm A. and Kaldass H. (1999). Relativistic partial-wave analysis using the velocity basis of the Poincaré group. *Phys. Rev. A*, **60**, 4606–4615.

[7002] Bohun C.S. and Cooperstock F.I. (1999). Dirac-Maxwell solitons. *Phys. Rev. A*, **60**, 4291–4300.

[7003] Boisvert G., Lewis L.J., Puska M.J. and Nieminen R.M. (1995). Energetics of diffusion on the (100) and (111) surfaces of Ag, Au, and Ir from first principles. *Phys. Rev. B*, **52**, 9078–9085.

[7004] Bolte J. and Keppeler S. (1998). Semiclassical time evolution and trace formula for relativistic spin-1/2 particles. *Phys. Rev. Lett.*, **81**, 1987–1991.

[7005] Bolte J. and Keppeler S. (1999). A semiclassical approach to the Dirac equation. *Ann. Phys. (NY)*, **274**, 125–162.

[7006] Bonačić-Koutecký V., Češpiva L., Fantucci P. and Koutecký J. (1993a). Effective core potential-configuration interaction study of electronic structure and geometry of small neutral and cationic Ag_n clusters: predictions and interpretation of measured properties. *J. Chem. Phys.*, **98**, 7981–7994.

[7007] Bonačič-Koutecký V., Češpiva L., Fantucci P. and Koutecký J. (1993b). ECP-CI study of electronic structure and geometry of small neutral and charged Ag_n clusters; predictions and interpretation of measured properties. *Z. Phys. D*, **26**, 287–289.

[7008] Bonačic-Koutecký V., Češpiva L., Fantucci P., Pittner J. and Koutecký J. (1994). Effective core potential-configuration interaction study of electronic structure and geometry of small anionic and Ag_n clusters: predictions and interpretation of photo detachment spectra. *J. Chem. Phys.*, **100**, 490–506.

[7009] Bonačic-Koutecký V., Pittner J., Boiron M. and Fantucci P. (1999). An accurate relativistic effective core potential for excited states of Ag atom: An application for studying the absorption spectra of Ag_n and Ag_n^+ clusters. *J. Chem. Phys.*, **110**, 3876–3886.

[7010] Bonča J. and Gubernatis J.E. (1994). Degenerate Anderson impurity model in the presence of spin-orbit and crystal-field splitting. *Phys. Rev. B*, **50**, 10427–10434.

[7011] Bondarev I.V. and Kuten S.A. (1993). Static tensor polarizabilities of hydrogenlike atoms in the $ns_{1/2}$ and $np_{1/2}$ states. *Opt. Spectrosc.*, **75**, 3–4.

[7012] Bondarev I.V. and Kuten S.A. (1993). Static tensor polarizabilities of hydrogenlike atoms in the $ns_{1/2}$ and $np_{1/2}$ states. *Opt. Spektrosk.*, **75**, 6–9.

[7013] Bonesteel N.E. (1993). Spin-orbit scattering and pair breaking in a structurally disordered copper oxide layer. *Phys. Rev. B*, **47**, 9144–9147.

[7014] Bonnelle C., Jonnard P., Barré C., Giorgi G. and Bruneau J. (1997). X-ray emissions in 3d, 4d, and 5d ranges for uranium ions. *Phys. Rev. A*, **55**, 3422–3432.

[7015] Booth H.S. and Radford C.J. (1997). The Dirac-Maxwell equations with cylindrical symmetry. *J. Math. Phys.*, **38**, 1257–1268.

[7016] Borovskii A.V., Zapryagaev S.A., Zatsarinnyi O.I. and Manakov N.L. (1995). *Plasma of Multicharged Ions (in Russian)*. Khimiya, St. Petersburg, 344 p.

[7017] Borrmann H., Campbell J., Dixon D.A., Mercier H.P.A., Pirani A.M. and Schrobilgen G.J. (1998a). $Tl_2Ch_2^{2-}$ (Ch=Se and/or Te) anions: X-ray crystal structures and Raman spectra of (2,2,2-crypt-$K^+)_2Tl_2Se_2^{2-}$ and (2,2,2-crypt-$K^+)_2Tl_2Te_2^{2-}$ and solution ^{77}Se, ^{203}Tl, and ^{205}Tl NMR spectroscopic and theoretical studies of $Tl_2Ch_2^{2-}$, $In_2Se_2^{2-}$, and $In_2Te_2^{2-}$. *Inorg. Chem.*, **37**, 1929–1943.

[7018] Borrmann H., Campbell J., Dixon D.A., Mercier H.P.A., Pirani A.M. and Schrobilgen G.J. (1998b). Trigonal bipyramidal $M_2Ch_3^{-2}$ (M=Sn, Pb; Ch=S, Se, Te) and $TlMTe_3^{3-}$ anions: Multinuclear magnetic resonance, Raman spectroscopic, and theoretical studies, and the x-ray crystal structures of (2,2,2-crypt-$K^+)_3TlPbTe_3^{3-}$·2en and (2,2,2-crypt-$K^+)_2Pb_2Ch_3^{2-}$·0.5en (Ch=S, Se). *Inorg. Chem.*, **37**, 6656–6674.

[7019] Bortman D. and Ron A. (1994). Hysteresis in an atomic system. *Opt. Comm.*, **108**, 253–257.

[7020] Bosanac S.D. (1993). Relativistic dynamics of wave packets for spin-1/2 particles in an electromagnetic field. *J. Phys. A*, **26**, 5523–5540.

[7021] Bose S.K. (1997). Exact bound state of the relativistic Schroedinger equation for the central potential $V(R) = -\alpha/r + \beta/r^{1/2}$. *Nuovo Cim.*, **112B**, 635–638.

[7022] Bose S.K. (1999). Electronic structure of liquid mercury. *J. Phys. CM*, **11**, 4597–4615.

[7023] Bose S.K., Satpathy S. and Jepsen O. (1993). Semiconducting $CsSnBr_3$. *Phys. Rev. B*, **47**, 4276–4280.

[7024] Bosnick K.A., Haslett T.L., Fedrigo S., Moskovits M., Chan W.T. and Fournier R. (1999). Tricapped tetrahedral Ag_7: A structural determination by resonance Raman spectroscopy and density functional theory. *J. Chem. Phys.*, **111**, 8867–8870.

[7025] Bouazza S., Dembczynski J., Stachowska E., Szawiola G. and Ruczkowski J. (1998). Reanalysis and semiempirical predictions of the hyperfine structure of ^{91}Zr I in the model space $(4d+5s)^4$. *Eur. Phys. J. D*, **4**, 39–46.

[7026] Boucher F. and Rousseau R. (1998). Bonding and electronic properties of Cs_3Te_{22}. *Inorg. Chem.*, **37**, 2351–2357.

[7027] Bouchiat M.A. and Bouchiat C. (1997). Parity violation in atoms. *Rep. Prog. Phys.*, **60**, 1351–1396.

[7028] Boudreaux E.A. and Baxter E. (1995). SC-MEH-MO calculations on lanthanide systems. II. $Sm(Cp^*)^+$, $Sm(Cp^*)^{2+}$, and the $(Sm(Cp^*))_4^{8+}$ tetrameter. *Int. J. Quantum Chem.*, **S29**, 605–608.

[7029] Boudreaux E.A. and Baxter E. (1997). QR-SCMEH-MO calculations on lanthanide systems. IV. The $(SmCp^*)_4$ cluster. *Int. J. Quantum Chem.*, **64**, 297–300.

[7030] Boussard P.J.E., Siegbahn P.E.M. and Svensson M. (1994). The interaction of ammonia carbonyl ethylene and water with the copper and silver dimers. *Chem. Phys. Lett.*, **231**, 337–344.

[7031] Boustani I., Rai S.N., Liebermann H.P., Alekseyev A.B., Hirsch G. and Buenker R.J. (1993). Relativistic configuration interaction calculations of potential curves and radiative transition probabilities for the antimony fluoride molecule. *Chem. Phys.*, **177**, 45–52.

[7032] Boutassetta N., Allouche A.R. and Aubert-Frécon M. (1996). Theoretical study of the electronic structure of the Sr_2 molecule. *Phys. Rev. A*, **53**, 3845–3852.

[7033] Boutou V., Allouche A.R., Spiegelmann F., Chevaleyre J. and Aubert-Frécon M. (1998). Predictions of geometrical structures and ionization potentials for small barium clusters Ba_n. *Eur. Phys. J. D*, **2**, 63–73.

[7034] Bowmaker G.A., Schmidbaur H., Krüger S. and Rösch N. (1997). A density functional study of metal-ligand bonding in $((PR_3)_2M)^+$ and (PR_3MCl) (M=Ag, Au; R=H, Me) complexes. *Inorg. Chem.*, **36**, 1754–1757.

[7035] Boya L.J. and Byrd M. (1999). Clifford periodicity from finite groups. *J. Phys. A*, **32**, L201–L205.

[7036] Boykin T.B. (1998). More complete treatment of spin-orbit effects in tight-binding models. *Phys. Rev. B*, **57**, 1620–1625.

[7037] Boyle J.J., Altun Z. and Kelly H.P. (1993). Photoionization cross-section calculation of atomic tungsten. *Phys. Rev. A*, **47**, 4811–4830.

[7038] Bracken A.J. and Melloy G.F. (1999). Localizing the relativistic electron. *J. Phys. A*, **32**, 6127–6139.

[7039] Brage T., Leckrone D.S. and Froese Fischer C. (1996). Core-valence and core-core correlation effects on hyperfine-structure parameters and oscillator strengths in Tl II and Tl III. *Phys. Rev. A*, **53**, 192–200.

[7040] Brage T., Proffitt C.R. and Leckrone D.S. (1999). Relativistic *ab initio* calculations of oscillator strengths and hyperfine structure constants in Tl II. *J. Phys. B*, **32**, 3183–3192.

[7041] Braidwood S., Brunger M. and Weigold E. (1993). Satellite structure of the xenon valence shell by electron-momentum spectroscopy. *Phys. Rev. A*, **47**, 2927–2936.

[7042] Brau F. (1999). Analytical solution of the relativistic Coulomb problem with a hard core interaction for a one-dimensional spinless Salpeter equation [Erratum: Vol. 40, p. 6125]. *J. Math. Phys.*, **40**, 1119–1126.

[7043] Braun J. and Borstel G. (1993). Relativistic photo emission theory applied to GaAs(110). *Phys. Rev. B*, **48**, 14373–14380.

[7044] Braun J.W., Su Q. and Grobe R. (1999). Numerical approach to solve the time-dependent Dirac equation. *Phys. Rev. A*, **59**, 604–612.

[7045] Braun T.P. and Simon A. (1996). Stabilization of octahedral thorium clusters by interstitial hydrogen. *Chem. Eur. J.*, **2**, 511–515.

[7046] Braun T.P., Simon A., Böttcher F. and Ueno F. (1995). $ATh_{12}N_6X_{29}$ (A = Li...Rb; X = Cl, Br): a new type of thorium cluster with a $Th_{12}N_6$ core. *Angew. Chem. Int. Ed. Engl.*, **34**, 567–569.

[7047] Braun T.P., Simon A., Ueno F. and Böttcher F. (1996). The $(Th_{12}N_6X_{41})^{13-}$ cluster: an extension of rare earth metal cluster chemistry. *Eur. J. Solid State Inorg. Chem.*, **33**, 251–264.

[7048] Bravo-Pérez G., Garzón I.L. and Novaro O. (1999a). Non-additive effects in small gold clusters. *Chem. Phys. Lett.*, **313**, 655–664.

[7049] Bravo-Pérez G., Garzón I.L. and Novaro O. (1999b). Ab initio study of small gold clusters. *J. Mol. Str. (Theochem)*, **493**, 225–231.

[7050] Bravo-Vásquez J.P. and Arratia-Pérez R. (1994). Calculated paramagnetic hyperfine structure of the C_{2v} isomers of Ag_3. *J. Phys. Chem.*, **98**, 5627–5631.

[7051] Breidung J. and Thiel W. (1995). The anharmonic force fields of arsine, stibine, and bismutine. *J. Mol. Spectr.*, **169**, 166–180.

[7052] Breidung J., Thiel W. and Demaison J. (1997). Equilibrium structure of PH_2Br. *Chem. Phys. Lett.*, **266**, 515–520.

[7053] Breuer H.P. and Petruccione F. (1998). Relativistic formulation of quantum-state diffusion. *J. Phys. A*, **31**, 33–52.

[7054] Breza M. (1993a). On the axial oxygen position in Y-Ba-Cu-O superconductors. *Solid State Comm.*, **85**, 713–716.

[7055] Breza M. (1993b). CNDO studies of Y-Ba-Cu-O superconductors. III. The axial oxygen position. *Czech. J. Phys.*, **43**, 569–574.

[7056] Breza M. (1994). CNDO studies of Y-Ba-Cu-O superconductors IV The oxygen content effect. *Czech. J. Phys.*, **44**, 695–701.

[7057] Breza M. (1996). Electronic structure of high-temperature superconductors. II. La-Ba-Cu-O systems. *Czech. J. Phys.*, **46**, 503–507.

[7058] Briegel H.J., Englert B.G., Michaelis M. and Süssman G. (1991). Über die Wurzel aus der Klein-Gordon Gleichung als Schrödinger-Gleichung eines relativistischen Spin-0-Teilchens. *Z. Naturf. A*, **46a**, 925–.

[7059] Brinzanescu O., Eichler J., Ichihara A., Shirai T. and Stöhlker T. (1999). Comparison between the nonrelativistic dipole approximation and the exact relativistic theory for radiative recombination. *Phys. Scripta*, **T80**, 324–325.

[7060] Broo A. (1996). Basis set and correlation effects on geometry of octahedral second-row transition-metal complexes. *Int. J. Quantum Chem., Symp.*, **30**, 119–131.

[7061] Brooks M.S.S., Eriksson O. and Johansson B. (1995). From the transition metals to the rare earths - via the actinides. *J. Alloys Compounds*, **223**, 204–210.

[7062] Brooks M.S.S., Eriksson O., Wills J.M. and Johansson B. (1997). Density functional theory of crystal field quasiparticle excitations and the *ab initio* calculation of spin Hamiltonian parameters. Erratum: ibid., 1998, 80, 4108. *Phys. Rev. Lett.*, **79**, 2546–2549.

[7063] Broschag M., Klapötke T.M., Schulz A. and White P.S. (1993). The $ClSeNSeCl^+$ cation. An unusual structurally very flexible ion adopting different solid-state structures as deduced by x-ray and ab initio methods. *Inorg. Chem.*, **32**, 5734–5738.

[7064] Brouder C., Alouani M. and Bennemann K.H. (1996). Multiple scattering theory of x-ray magnetic circular dichroism: Implementation and results for the iron K-edge. *Phys. Rev. B*, **54**, 7334–.

[7065] Brown K.L. and Kaltsoyannis N. (1999). Computational study of the geometric and electronic structures of MN_2 (M = Mo or U). *J. Chem. Soc., Dalton Trans.*, pp. 4425–4430.

[7066] Broyles A.A. (1995). The derivation of the relativistic Hamiltonian for molecules. *Int. J. Quantum Chem.*, **S29**, 257–275.

[7067] Bruce S. and Roa L. (1998). A Dirac particle in a vector-like hydrogenic potential. *Nuovo Cim. A*, **111**, 159–164.

[7068] Bruch R., Safronova U.I., Shlyaptseva A.S., Nilsen J. and Schneider D. (1998a). Theoretical analysis of the doubly excited $3lnl'$ states of sodium-like copper. *Phys. Scr.*, **57**, 334–344.

[7069] Bruch R., Safronova U.I., Shlyaptseva A.S., Nilsen J. and Schneider D. (1998b). New comprehensive theoretical analysis of the doubly excited $3lnl'$ states of sodium-like iron. *J. Quant. Spectr. Rad. Transfer*, **60**, 605–622.

[7070] Bruna P.J. and Grein F. (1998). The electron-spin magnetic moments (g factors) of O_3, O_3Li and O_3Na: An ab initio study. *J. Chem. Phys.*, **109**, 9439–9450.

[7071] Bruna P.J. and Grein F. (1999). Theoretical study of the electron-spin magnetic moments (g factors) of F_2^- and Cl_2^- (X_2^-), as well as MX_2 and $M_2X_2^+$ compounds with M = Li, Na. *Chem. Phys.*, **249**, 169–182.

[7072] Bruss D., Gasenzer T. and Nachtmann O. (1998). New observables for parity violation in atoms: energy shifts in external electric fields. *Phys. Lett. A*, **239**, 81–86.

[7073] Bruss D., Gasenzer T. and Nachtmann O. (1999). Parity violating energy shifts in atoms, I. *Eur. J. Phys. D*, **1**, 1–73.

[7074] Bucher M. (1990). Cohesive properties of silver halides. *J. Imaging Sci.*, **34**, 89–95.

[7075] Buenker R.J., Alekseyev A.B., Liebermann H.P., Lingott R. and Hirsch G. (1998). Comparison of spin-orbit configuration interaction methods employing relativistic effective core potentials for the calculation of zero-field splittings of heavy atoms with a $^2P^o$ ground state. *J. Chem. Phys.*, **108**, 3400–3408.

[7076] Bugacov A. and Shakeshaft R. (1993). Multiphoton transitions in a strong field: inclusion of the photon momentum. *Phys. Rev. A*, **47**, 674–685.

[7077] Bühl M. (1997). NMR chemical shifts of $Zr@C_{28}$. How shielded can ^{91}Zr get? *J. Phys. Chem. A*, **101**, 2514–2517.

[7078] Bühl M. (1999). Density functional calculations of ^{95}Mo NMR chemical shifts: Applications to model catalysts for imine metathesis. *Chem. Eur. J.*, **5**, 3514–3522.

[7079] Bühl M., Kaupp M., Malkina O.L. and Malkin V.G. (1999). The DFT route to NMR chemical shifts. *J. Comp. Chem.*, **20**, 91–105.

[7080] Büker H.H., Maitre P. and Ohanessian G. (1997). Theoretical study of tungsten carbonyl complexes $W(CO)_n{}^+$ (n= 1-6): Structures, binding energies, and implications for gas phase reactivities. *J. Phys. Chem. A*, **101**, 3966–3976.

[7081] Bukowski R. and Jeziorski B. (1993). Nonrelativistic Lamb shift for muonic molecules. *Hyperfine Interactions*, **82**, 179–184.

[7082] Bulgakov E.N., Pichugin K.N., Sadreev A.F., Středa P. and Šeba P. (1999). Hall-like effect induced by spin-orbit interaction. *Phys. Rev. Lett.*, **83**, 376–379.

[7083] Bündgen P., Lushington G.H. and Grein F. (1995). Configuration interaction study of relativistic corrections to the Zeeman effect in diatomic molecules. *Int. J. Quantum Chem.*, **S29**, 283–288.

[7084] Bunge C.F., Jáuregui R. and Ley-Koo E. (1998). Optimal decoupling of positive- and negative-energy orbitals in relativistic electronic structure calculations beyond Hartree-Fock. *Int. J. Quantum Chem.*, **70**, 805–812.

[7085] Burda J.V., Hobza P. and Zahradník R. (1998). $(HX)_2$ species (X = F through At) in the groups of the periodic system: MP2 and CCSD(T) ab initio quantum chemical calculations. *Chem. Phys. Lett.*, **288**, 20–24.

[7086] Burda J.V., Runeberg N. and Pyykkö P. (1998). Chemical bonds between noble metals and noble gases. Ab initio study of the neutral diatomic NiXe, PdXe and PtXe. *Chem. Phys. Lett.*, **288**, 635–641.

[7087] Burda J.V., Šponer J. and Hobza P. (1996a). Ab initio study of the interaction of guanine and adenine with various mono- and bivalent metal cations (Li^+, Na^+, K^+, Rb^+, Cs^+; Cu^+, Ag^+, Au^+; Mg^{2+}, Ca^{2+}, Sr^{2+}, Ba^{2+}; Zn^{2+}, Cd^{2+}, and Hg^{2+}). *J. Phys. Chem.*, **100**, 7250–7255.

[7088] Burda J.V., Zahradník R., Hobza P. and Urban M. (1996b). Dimers of rare gas atoms: CCSD(T), CCSDT and FCI calculations on the $(He)_2$ dimer, CCSD(T) and CCSDT calculations on the $(Ne)_2$ dimer, and CCSD(T) all-electron and pseudopotential calculations on the dimers from $(Ne)_2$ through $(Xe)_2$. *Mol. Phys.*, **89**, 425–432.

[7089] Burdett J.K., Eisenstein O. and Schweizer W.B. (1994). Are strong gold-gold interactions possible in main group $X_nA(AuPR_3)_m$ molecules?. *Inorg. Chem.*, **33**, 3261–3268.

[7090] Burdett J.K. and Sevov S. (1995). Stability of the oxidation states of copper. *J. Am. Chem. Soc.*, **117**, 12788–12792.

[7091] Burdick G.W., Kooy H.J. and Reid M.F. (1993). Correlation contributions to two-photon lanthanide absorption intensities: direct calculations for Eu^{2+} ions. *J. Phys: CM*, **5**, L323–L328.

[7092] Burdick G.W. and Reid M.F. (1993). Many-body perturbation theory calculations of two-photon absorption in lanthanide compounds. *Phys. Rev. Lett.*, **70**, 2491–2494.

[7093] Bürger H., Kuna R., Ma S., Breidung J. and Thiel W. (1994). The vibrational spectra of krypton and xenon difluoride: High-resolution infrared studies and ab initio calculations. *J. Chem. Phys.*, **101**, 1–14.

[7094] Bürger H., Ma S., Breidung J. and Thiel W. (1996). Ab initio computations and high resolution infrared investigation on XeF_4. *J. Chem. Phys.*, **104**, 4945–4953.

[7095] Burgess M. and Jensen B. (1993). Fermions near two-dimensional surfaces. *Phys. Rev. A*, **48**, 1861–1868.

[7096] Bursten B.E., Green J.C. and Kaltsoyannis N. (1994). Theoretical investigation of the effects of spin-orbit coupling on the valence photoelectron spectrum of OsO_4. *Inorg. Chem.*, **33**, 2315–2316.

[7097] Busic O., Grün N. and Scheid W. (1999). e^+e^--Pair production in relativistic atomic heavy ion collisions with capture. *Phys. Scripta*, **T80**, 432–433.

[7098] Buzek P., Klapötke T.M., v R Schleyer P., Tornieporth-Oetting I.C. and White P.S. (1993). Iodazid (in German). *Angew. Chem.*, **105**, 289–290.

[7099] Byrnes T.M.R., Dzuba V.A., Flambaum V.V. and Murray D.W. (1999). Enhancement factor for the electric dipole moment in francium and gold atoms. *Phys. Rev. A*, **59**, 3082–3083.

[7100] Bytheway I., Bacskay G.B. and Hush N.S. (1996). Quantum chemical study of the properties of the molecular hydrogen complexes of osmium(II): a comparison of density functional and conventional ab initio methods. *J. Phys. Chem.*, **100**, 6023–6031.

[7101] Bytheway I., Gillespie R.J., Tang T.H. and Bader R.F.W. (1995). Core distorsions and geometries of the difluorides and dihydrides of Ca, Sr and Ba. *Inorg. Chem.*, **34**, 2407–2414.

[7102] Bytheway I. and Wong M.W. (1998). The prediction of vibrational frequencies of inorganic molecules using density functional theory. *Chem. Phys. Lett.*, **282**, 219–226.

[7103] Byun Y.G., Kan S.Z., Lee S.A., Kim Y.H., Miletic M., Bleil R.E., Kais S. and Freiser B.S. (1996). Experimental and theoretical studies of $Nb_6C_7^{0/+}$. *J. Phys. Chem.*, **100**, 6336–6341.

[7104] Bzowski A., Sham T.K., Watson R.E. and Weinert M. (1995a). Electronic structure of Au and Ag overlayers on Ru(001): the behaviour of the noble-metal *d* bands. *Phys. Rev. B*, **51**, 9979–9984.

[7105] Bzowski A., Yiu Y.M. and Sham T.K. (1995b). Charge redistribution in Au-metalloid intermetallics: a Au $L_{2,3}$-edge x-ray-absorption study. *Phys. Rev. B*, **51**, 9515–9520.

[7106] Cabrol O., Girard B., Spiegelmann F. and Teichteil C. (1996). Relativistic calculation of the electronic structure of the IF molecule. *J. Chem. Phys.*, **105**, 6426–6438.

[7107] Cai W., Fang Q.Y., Zou Y. and Li P. (1997). Excitation-collision parameters and scaled rules of Cu-like ions by electron impact (in Chinese). *High-Power Laser and Particle Beams (Qiangjiguang Yu Lizishu)*, **9**, 199–207.

[7108] Calhorda M.J., Canales F., Gimeno M.C., Jiménez J., Jones P.G., Laguna A. and Veiros L.F. (1997). Gold(I)-gold(III) interactions in polynuclear sulfur-centered complexes. Synthesis and structural characterization of $(S(Au_2\ dppf)(Au(C_6F_5)_3))$ and $((S(Au_2dppf)_2(Au(C_6F_5)_2))OTf(dppf)$ (dppf = 1,1'-bis(diphenylphosphino)ferrocene). *Organometallics*, **16**, 3837–3844.

[7109] Calhorda M.J. and Veiros L.F. (1994). Metal-metal bonds in Au_5 chain and other species. *J. Organomet Chem.*, **478**, 37–44.

[7110] Caliceti E. and Cherubini A.M. (1993). Spectral analysis and stability properties of a relativistic deformation of the harmonic oscillator. *J. Math. Phys.*, **34**, 5451–5467.

[7111] Camanyes S., Maseras F., Moreno M., Lledós A., Lluch J.M. and Bertrán J. (1996). Theoretical study of the hydrogen exchange coupling in the metallocene trihydride complexes $((C_5H_5)_2MH_3)^{n+}$ (M = Mo, W, n=1; M = Nb, Ta, n=0). *J. Am. Chem. Soc.*, **118**, 4617–4621.

[7112] Camanyes S., Maseras F., Moreno M., Lledós A., Lluch J.M. and Bertrán J. (1999). Theoretical study of the effect of Lewis acids on dihydrogen elimination from niobocene trihydrides. *Chem. Eur. J.*, **5**, 1166–1171.

[7113] Camarda H.S. (1996). Determing the eigenvalues of a quaternion matrix with a band structure. *Comp. Phys.*, **10**, 180–185.

[7114] Campbell J., Dixon D.A., Mercier H.P.A. and Schrobilgen G.J. (1995). The nido-Pb_9^{4-} and the Jahn-Teller distorted closo-Pb_9^{3-} Zintl anions: syntheses, x-ray structures, and theoretical studies. *Inorg. Chem.*, **34**, 5798–5809.

[7115] Campbell M.L.F. and Segre G. (1995). Spin-orbit-induced lattice instabilities in the half-filled hopping Hamiltonian. *Phys. Rev. B*, **51**, 6892–6396.

[7116] Campos P.J. and Rodriguez M.A. (1995). On the $C_2H_2I^+$ potential energy hypersurface. An ab initio study. *J. Chem. Soc, Chem. Comm,*, pp. 143–144.

[7117] Campos P.J., Sampedro D. and Rodriguez M.A. (1998). Hydrogen migration and lithium iodide α-elimination in 1-iodo-1-lithioethene. Concerted vs. stepwise mechanism. *Organometallics*, **17**, 5390–5396.

[7118] Cao X.P., Chen X.J. and Li B. (1996). Molecular symmetry and ab initio calculations. II. Symmetry-matrix and symmetry-supermatrix in the Dirac-Fock method. *J. Comp. Chem.*, **17**, 851–863.

[7119] Capelle K. and Gross E.K.U. (1995). Relativistic theory of superconductivity. *Phys. Lett. A*, **198**, 261–266.

[7120] Capelle K. and Gross E.K.U. (1999a). Relativistic framework for microscopic theories of superconductivity. I. The Dirac equation for superconductors. *Phys. Rev. A*, **59**, 7140–7154.

[7121] Capelle K. and Gross E.K.U. (1999b). Relativistic framework for microscopic theories of superconductivity. II. The Pauli equation for superconductors. *Phys. Rev. A*, **59**, 7155–7165.

[7122] Capelle K., Gross E.K.U. and Györffy B.L. (1997). Theory of dichroism in the electromagnetic response of superconductors. *Phys. Rev. Lett.*, **78**, 3753–3756.

[7123] Capelle K., Gross E.K.U. and Györffy B.L. (1998). Analysis of dichronism in the electromagnetic response of superconductors. *Phys. Rev. B*, **58**, 473–489.

[7124] Carlesi S., Franchini A., Bortolani V. and Martinelli S. (1999). Cesium under pressure: First-principles calculation of the bcc-to-fcc phase transition. *Phys. Rev. B*, **59**, 11716–11719.

[7125] Carloni P. and Andreoni W. (1996). Platinum-modified nucleobase pairs in the solid state: A theoretical study. *J. Phys. Chem.*, **100**, 17797–17800.

[7126] Carloni P., Andreoni W., Hutter J., Curioni A., Giannozzi P. and Parrinello M. (1995). Structure and bonding in cisplatin and other Pt(II) complexes. *Chem. Phys. Lett.*, **234**, 50–56.

[7127] Carlson J. and Schiavilla R. (1998). Structure and dynamics of few-nucleon systems. *Rev. Mod. Phys.*, **70**, 743–841.

[7128] Carroll J.J., Haug K.L. and Weisshaar J.C. (1995). Gas phase reactions of second-row transition metal atoms with small hydrocarbons: experiment and theory. *J. Phys. Chem.*, **99**, 13955–13969.

[7129] Casarrubios M. and Seijo L. (1995). Relativistic Wood-Boring ab initio model potential calculations on the platinum atom. *Chem. Phys. Lett.*, **236**, 510–515.

[7130] Casarrubios M. and Seijo L. (1998). The ab initio model potential method. Relativistic Wood-Boring valence spin-orbit potentials and spin-orbit-corrected basis sets from B(Z=5) to Ba(Z=56). *J. Mol. Str. (Theochem)*, **426**, 59–74.

[7131] Casarrubios M. and Seijo L. (1999). The ab initio model potential method: Third-series transition metal elements. *J. Chem. Phys.*, **110**, 784–796.

[7132] Casperson L.W. (1995). Dirac's equation in semiclassical physics. *Phys. Rev. A*, **51**, 1673–1676.

[7133] Castillo S., Bertin V., Solano-Reyes E., Luna-Garcia H., Cruz A. and Poulain E. (1998). Theoretical studies on hydrogen activation by iridium dimers. *Int. J. Quantum Chem.*, **70**, 1029–1035.

[7134] Castillo S., Cruz A., Cuán A., Ramírez-Solís A., Poulain E. and Del Angel G. (1995a). Theoretical study of the rhodium dimer interaction with the hydrogen molecule. *Int. J. Quantum Chem.*, **S29**, 549–557.

[7135] Castillo S., Poulain E., Bertin V. and Cruz A. (1995b). Theoretical studies of the interaction of PtSn systems with H_2. *Int. J. Quantum Chem.*, **S29**, 207–215.

[7136] Castillo S., Cruz A., Bertin V., Poulain E., Arellano J.S. and Del Angel G. (1997). Theoretical study on Pd dimer and trimer interaction with the hydrogen molecule. *Int. J. Quantum Chem.*, **62**, 29–45.

[7137] Cavicchi M. and Vairo A. (1994). A new tool for the Lamb-shift calculation. *Z. Phys. C*, **63**, 455–462.

[7138] Cea P. (1997). Vacuum stability for Dirac fermions in three dimensions. *Phys. Rev. D*, **55**, 7985–7988.

[7139] Celestini F., Ercolessi F. and Tosatti E. (1997). Can liquid metal surfaces have hexatic order? *Phys. Rev. Lett.*, **78**, 3153–3156.

[7140] Cencek W. and Kutzelnigg W. (1996). Accurate relativistic energies of one- and two-electron systems using Gaussian Wave functions. *J. Chem. Phys.*, **105**, 5878–5885.

[7141] Cencek W., Rychlewski J., Jaquet R. and Kutzelnigg W. (1998). Sub-microhartree accuracy potential energy surface for H_2^+ including adiabatic and relativistic effects. I. Calculation of the potential points. *J. Chem. Phys.*, **108**, 2831–2836.

[7142] Centelles M., Vinas X., Barranco M. and Schuck P. (1993). A semiclassical approach to relativistic nuclear mean field theory. *Ann. Phys.*, **221**, 165–204.

[7143] Ceulemans A., Chibotaru L.F. and Fowler P.W. (1998). Molecular anapole moments. *Phys. Rev. Lett.*, **80**, 1861–1864.

[7144] Chakraborty H.S., Gray A., Costello J.T., Desmukh P.C., Haque G.N., Kennedy E.T., Manson S.T. and Mosnier J.P. (1999). Anomalous behavior of the near-threshold photoionization cross section of the neon isoelectronic sequence: A combined experimental and theoretical study. *Phys. Rev. Lett.*, **83**, 2151–2154.

[7145] Chakravorty S.J., Gwaltney S.R., Davidson E.R., Parpia F.A. and Froese Fischer C. (1993). Ground-state correlation energies for atomic ions with 3 to 18 electrons. *Phys. Rev. A*, **47**, 3649–3670.

[7146] Chalker J.T., Daniell G.J., Evangelou S.N. and Nahm I.H. (1993). Eigenfunction fluctuations and correlations at the mobility edge in a two-dimensional system with spin-orbit scattering. *J. Phys:CM*, **5**, 485–490.

[7147] Chamarro M., Dib M., Voliotis V., Filoramo A., Roussignol P., Gacoin T., Boilot J.P., Delerue C., Allan G. and Lannoo M. (1998). Interplay of Coulomb, exchange, and spin-orbit effects in semiconductor nanocrystallites. *Phys. Rev. B*, **57**, 3729–3732.

[7148] Chan W.T. and Fournier R. (1999). Binding of ammonia to small copper and silver clusters. *Chem. Phys. Lett.*, **315**, 257–265.

[7149] Chandra P. and Hess B.A. (1994). A finite-nucleus model for relativistic electronic structure calculations using a Douglas-Kroll-transformed Hamiltonian. *Theor. Chim. Acta*, **88**, 183–199.

[7150] Chang A.H.H. and Yarkony D.R. (1993). On the electronic structure aspects of spin-forbidden processes in N_2O. *J. Chem. Phys.*, **99**, 6824–6831.

[7151] Chang A.H.H., Zhao K., Ermler W.C. and Pitzer R.M. (1994). Electronic structure of actinocenes and actinofullerenes. *J. Alloys Compounds*, **213/214**, 191–195.

[7152] Chang J.J. (1993). General form of the Dirac quantum-defect theory. *Phys. Rev. A*, **48**, 1769–1779.

[7153] Chang J.J. (1998). Similarities between the radial Schrödinger-Coulomb wave function and the radial Dirac-Coulomb wave function and an application to photoionization of hydrogen. *Phys. Rev. A*, **57**, 717–723.

[7154] Chang J.L., Stott M.J. and Alonso J.A. (1996). Theoretical study of gas-phase Na_nPb clusters and implications for liquid Na-Pb alloys. *J. Chem. Phys.*, **104**, 8043–8047.

[7155] Chang T.N. (1993). *Many-body Theory of Atomic Structure and Photoionization (Ed.).* World Scientific, Singapore, 408 p.

[7156] Charpentier T., Zérah G. and Vast N. (1996). Pseudopotentials including semicore states, with an application to barium, α-cerium, and thorium. *Phys. Rev. B*, **54**, 1427–1430.

[7157] Charro E., Martin I. and Lavin C. (1996). Multi-configuration Dirac-Fock and relativistic quantum defect orbital study of triplet-triplet transitions in beryllium-like ions. *J. Quant. Spectr. Rad. Transfer*, **56**, 241–253.

[7158] Chatterjee R. and Buckmaster H.A. (1993a). The relativistic effective operator technique revisited. I. Theory. In *Second International School of Theoretical Physics: Symmetry and Structural Properties of Condensed Matter*, (Edited by Florek W., Lipinski D. and Lulek T.), pp. 421–424, World Scientific, Singapore.

[7159] Chatterjee R. and Buckmaster H.A. (1993b). The relativistic effective operator technique revisited. II. Applications. In *Second International School of Theoretical Physics: Symmetry and Structural Properties of Condensed Matter*, (Edited by Florek W., Lipinski D. and Lulek T.), pp. 425–430, World Scientific, Singapore.

[7160] Chaudhuri R.K., Panda P.K. and Das B.P. (1999a). Hybrid approach to relativistic Gaussian basis functions: Theory and applications. *Phys. Rev. A*, **59**, 1187–1196.

[7161] Chaudhuri R.K., Panda P.K., Das B.P., Mahapatra U.S. and Mukherjee D. (1999b). Relativistic coupled-cluster-based linear response theory for ionization potentials of alkali-metal and alkaline-earth-metal atoms. *Phys. Rev. A*, **60**, 246–252.

[7162] Chauhan H.K. and Raina P.K. (1996). Numerical results for high-energy electron scattering cross sections from two-centre potentials. *J. Phys. B*, **29**, 5889–5899.

[7163] Chen C.Y., Teng Z.X., Yan S.X., Wang Y.S., Yang F.J. and Sun Y.S. (1998). Electron-ion collisional ionization cross sections and rates for the Ne isoelectronic sequence. *At. Data Nucl. Data Tables*, **70**, 255–272.

[7164] Chen G.H. and Raikh M.E. (1999). Exchange-induced enhancement of spin-orbit coupling in two-dimensional electronic systems. *Phys. Rev. B*, **60**, 4826–4833.

[7165] Chen G.X. (1994). Core-polarization on hyperfine interaction constants in the metastable $3d^2$ state of ^{45}Sc II. *Phys. Lett. A*, **193**, 451–456.

[7166] Chen G.X. (1996). Fully relativistic distorted-wave Born procedure for electron-impact excitation. *Phys. Rev. A*, **53**, 3227–3236.

[7167] Chen G.X. and Ong P.P. (1998a). Relativistic calculations for Fe XXIII: Atomic structure. *Phys. Rev. A*, **58**, 1070–1081.

[7168] Chen G.X. and Ong P.P. (1998b). Relativistic calculations for Fe XXIII: Electron-impact excitation. *Phys. Rev. A*, **58**, 1183–1194.

[7169] Chen G.X. and Ong P.P. (1998c). Electron-impact excitation cross sections in F-like selenium. *At. Data Nucl. Data Tables*, **70**, 93–117.

[7170] Chen G.X. and Ong P.P. (1999a). Relativistic distorted-wave excitation cross sections of F-like selenium by electron collision. *J. Phys. B*, **32**, 1121–1137.

[7171] Chen G.X. and Ong P.P. (1999b). RHF-DFT approach for electron affinities of neutral atoms in groups 13 and 14. *J. Phys. B*, **32**, 5351–5356.

[7172] Chen G.X. and Pradhan A.K. (1999). Relativistic and resonance effects in electron impact excitation of Fe^{5+}. *J. Phys. B*, **32**, 1809–1829.

[7173] Chen G.X. and Qiu Y.B. (1997a). Electron-impact excitation for F-like selenium. *Phys. Rev. A*, **56**, 3765–3768.

[7174] Chen G.X. and Qiu Y.B. (1997b). Fully relativistic distorted-wave ellectron impact excitation theory and calculation. *Chin. J. Comp. Phys.*, **14**, 477–479.

[7175] Chen H., Krasowski M. and Fitzgerald G. (1993). Density functional pseudopotential studies of molecular geometries, vibrations and binding energies. *J. Chem. Phys.*, **98**, 8710–8717.

[7176] Chen J., Li K.H. and Wen J.H. (1999). Relativistic high-order harmonics of a hydrogenlike atom in an ultrastrong laser field. *Can. J. Phys.*, **77**, 521–529.

[7177] Chen M.H. (1993a). Relativistic-intermediate-coupling calculations of angular distributions in resonant Auger decay. *Phys. Rev. A*, **47**, 3733–3738.

[7178] Chen M.H. (1993b). Dielectronic recombination coefficients for Ni-like tantalum. *Phys. Rev. A*, **47**, 4775–4778.

[7179] Chen M.H. and Cheng K.T. (1997a). Energy levels of the ground state and the $2s2p$ (J=1) excited states of berylliumlike ions: A large-scale, relativistic configuration-interaction calculation. *Phys. Rev. A*, **55**, 166–174.

[7180] Chen M.H. and Cheng K.T. (1997b). Large-scale relativistic configuration-interaction calculation of the $3s^2\ ^1S_0$ - $3s3p\ ^{1,3}P_1$ transition energies in magnesiumlike ions. *Phys. Rev. A*, **55**, 3440–3446.

[7181] Chen M.H., Cheng K.T. and Johnson W.R. (1993). Relativistic configuration-interaction calculations of $n = 2$ triplet states of heliumlike ions. *Phys. Rev. A*, **47**, 3692–3703.

[7182] Chen M.H., Cheng K.T., Johnson W.R. and Sapirstein J. (1995). Relativistic configuration-interaction calculations for the $n = 2$ states of lithiumlike atoms. *Phys. Rev. A*, **52**, 266–273.

[7183] Chen M.H. and Reed K.J. (1993a). Indirect contributions to electron-impact ionization of Kr^{24+}, Kr^{25+}, and Xe^{43+}. *Phys. Rev. A*, **47**, 1874–1877.

[7184] Chen M.H. and Reed K.J. (1993b). Effects of relativity and M2 transitions on the resonance contributions to electron-impact ionization of highly charged Li-like ions. *Phys. Rev. A*, **48**, 1129–1133.

[7185] Chen M.H. and Reed K.J. (1994). Relativistic effects on angular distribution of Auger electrons emitted from Be-like ions following electron-impact excitation. *Phys. Rev. A*, **50**, 2279–2283.

[7186] Chen M.H., Reed K.J. and Hazi A.U. (1994). Effects of Coster-Kronig transitions on electron-impact excitation rates for fluorinelike ions in their ground states. *Phys. Rev. A*, **49**, 1782–1785.

[7187] Chen M.H., Reed K.J., McWilliams D.M., Guo D.S., Barlow L., Lee M. and Walker W. (1997). K-shell Auger and radiative transitions in the carbon isoelectronic sequence, $6 \leq Z \leq 54$. *At. Data Nucl. Data Tables*, **65**, 289–370.

[7188] Chen M.H. and Scofield J.H. (1995). Relativistic effects on angular distribution and polarization of dielectronic satellite lines of hydrogenlike ions. *Phys. Rev. A*, **52**, 2057–2061.

[7189] Chen M.H. and Scofield J.H. (1998). Erratum: Relativistic effects on angular distribution and polarization of dielectronic satellite lines of hydrogenlike lines [Phys Rev A 52, 2057 (1995)]. *Phys. Rev. A*, **58**, 5011.

[7190] Chen M.K. (1993). Hyperfine splitting for the ground-state muonic ^{3}He atom-corrections up to α^2. *J. Phys. B*, **26**, 2263–2272.

[7191] Chen M.K. (1997). Electric dipole transitions in helium isoelectronic sequences. *Phys. Scr.*, **T73**, 56–58.

[7192] Chen M.K. and Chung K.T. (1994). Absolute term energy of core-excited doublet states of lithium. *Phys. Rev. A*, **49**, 1675–1685.

[7193] Chen X.J., Hua X.L., Hu J.S., Langlois J.M. and Goddard III W.A. (1996). Band structures of II-VI semiconductors using Gaussian basis functions with separable ab initio pseudopotentials: application to prediction of band offsets. *Phys. Rev. B*, **53**, 1377–1387.

[7194] Chen Z., Fonte G. and Goldman S.P. (1994). Upper and lower bounds on the energy eigenvalues of the one-electron Dirac Hamiltonian. *Phys. Rev. A*, **50**, 3838–3844.

[7195] Chen Z. and Goldman S.P. (1993). Relativistic variational calculations with finite nuclear size: application to hydrogenic atoms in strong magnetic fields. *Phys. Rev. A*, **48**, 1107–1113.

[7196] Cheng K.T. and Chen M.H. (1996). Relativistic configuration-interaction calculations for the $2s - 2p_{3/2}$ transition energies of uranium ions. *Phys. Rev. A*, **53**, 2206–2210.

[7197] Cheng K.T., Chen M.H., Johnson W.R. and Sapirstein J. (1994). Relativistic configuration-interaction calculations for the ground state and n=2 singlet states of heliumlike ions. *Phys. Rev. A*, **50**, 247–255.

[7198] Cheng K.T., Johnson W.R. and Sapirstein J. (1993). Lamb-shift calculations for non-Coulomb potentials. *Phys. Rev. A*, **47**, 1817–1823.

[7199] Cheng S., Berry H.G., Dunford R.W., Gemmell D.S., Kanter E.P., Zabransky B.J., Livingston A.E., Curtis L.J., Bailey J. and Nolen Jr J.A. (1993). Branching ratio for the M1 decay of the $2\ ^2S_{1/2}$ state in one-electron krypton. *Phys. Rev. A*, **47**, 903–910.

[7200] Cheng S., Dunford R.M., Liu C.J., Zabransky B.J., Livingston A.E. and Curtis L.J. (1994). M1 decay of the $2\ ^3S_1$ state in heliumlike krypton. *Phys. Rev. A*, **49**, 2374–2353.

[7201] Cheng W.Y. and Huang K.N. (1994a). Polarization correlations of radiation from electron-impact excited atoms. *Chin. J. Phys.*, **32**, 65–86.

[7202] Cheng W.Y. and Huang K.N. (1994b). Relativistic Z-dependence of the oscillator strengths for Be-like ions. *Chin. J. Phys.*, **32**, 361–394.

[7203] Cheng X.L., Zhu Z.H., Gou Q.Q., Peng H.S. and Cheng X.F. (1993). The calculation of transition wavelengths and oscillator strength for Ni-like ions (Ag XX - Xe XXVII). *Yuanzi Yu Fenzi Wuli Xuebao (China)*, **10**, 2989–2994.

[7204] Cheng X.L., Zhu Z.H., Ye A.P., Jiang W.M. and Zhou Y.G. (1994). The calculation of the energy level and transition for the Ni-like ions (Hf^{44+} - Au^{51+}). *Yuanzi Yu Fenzi Wuli Xuebao (China)*, **11**, 29–34.

[7205] Chernysheva L.V. and Yakhontov V.L. (1999). ADJZ. Two-program package to calculate the ground and excited state wave functions in the Hartree-Dirac-Fock approximation. *Comp. Phys. Comm.*, **119**, 232–255.

[7206] Chertihin G.V. and Andrews L. (1995). Reactions of laser-ablated Zr and Hf atoms with hydrogen. Matrix infrared spectra of the MH, MH_2, MH_3, and MH_4 molecules. *J. Phys. Chem.*, **99**, 15004–15010.

[7207] Chertihin G.V. and Andrews L. (1996). Infrared spectra of the reaction products of laser ablated lead atoms and oxygen molecules in condensing argon and nitrogen. *J. Chem. Phys.*, **105**, 2561–2574.

[7208] Chertihin G.V. and Andrews L. (1997). Reactions of laser-ablated Zn and Cd atoms with O_2: Infrared spectra of ZnO, OZnO, CdO, and OCdO in solid argon. *J. Chem. Phys.*, **106**, 3457–3465.

[7209] Chevary J.A. (1993). *Dirac-Hartree-Fock predictions of the ground state electron configurations of atomic negative ions:* Sr^-, Ba^-, Yb^-, Ra^-, La^-, *and* Lu^-. Ph.D. thesis, Univ. of Toronto.

[7210] Chevary J.A. and Vosko S.H. (1994). More theoretical evidence for binding of a $6p$ electron in the lanthanide anions: Tm^- $(\mathrm{Xe})4f^{13}6s^26p^1$. *J. Phys. B*, **27**, 657–665.

[7211] Chi H.C. (1997). Core-polarization effects on near-threshold photoionization of Mg. *Phys. Rev. A*, **56**, 4118–4120.

[7212] Chi H.C. and Huang K.N. (1994). Photoionization of magnesium including double excitations. *Phys. Rev. A*, **50**, 392–398.

[7213] Childs W.J. (1997). Matrix elements of hyperfine structure operators in the SL and jj representations for the s, p^N, and d^N configurations and the $SL-jj$ transformation. *At. Data Nucl. Data Tables*, **67**, 1–70.

[7214] Chisholm M.H., Folting K., Lynn M.L., Tiedtke D.B., Lemoigno F. and Eisenstein O. (1999). Nitrido dimers and trimers of tungsten supported by tBuMe$_2$SiO and CF$_3$Me$_2$CO ligands, respectively. Factors influencing the reductive cleavage of nitriles by tungsten-tungsten triple bonds and an analysis of the structure of the cyclotrimer. *Chem. Eur. J.*, **5**, 2318–2326.

[7215] Choe A.S., Yoo B., Rhee Y. and Lee J. (1993). Comparison of two relativistic B-spline basis sets using a point and a finite-size nucleus for the ions O^{+7} and Hg^{+79}. *J. Kor. Phys. Soc.*, **26**, 597–602.

[7216] Choi S.H., Bytheway I., Lin Z.Y. and Jia G.C. (1998). Understanding the preference for the coplanarity of alkenyl and carbonyl ligands in η^1-alkenyl transition-metal complexes: A simple molecular orbital approach and *ab initio* calculations. *Organometallics*, **17**, 3974–3980.

[7217] Choi S.H., Lin Z.Y. and Xue Z.L. (1999). Theoretical studies of the relative stabilities of transition metal alkylidyne $(CH_3)_2M(\equiv CH)(X)$ and bis(alkylidene)$(CH_3)M(=CH_2)_2(X)$ complexes. *Organometallics*, **18**, 5488–5495.

[7218] Chou H.S., Chang J.Y., Chang Y.H. and Huang K.N. (1996). Energy-level scheme and transition probabilities of S-like ions. *At. Data Nucl. Data Tables*, **62**, 77–145.

[7219] Chou H.S., Chi H.C. and Huang K.N. (1993a). Core-polarization effects for the $(6s^2)\ ^1S_0$ to $(6s6p)\ ^3P_1^o$, $^1P_1^o$ transitions in Hg-like ions. *J. Phys. B*, **26**, 2303–2309.

[7220] Chou H.S., Chi H.C. and Huang K.N. (1993b). Core-polarization effects for the intercombination and resonance transitions in Cd-like ions. *Phys. Rev. A*, **48**, 2453–2456.

[7221] Chou H.S., Chi H.C. and Huang K.N. (1993c). Relativistic excitation energies and oscillator strengths including core-polarization effects for the intercombination and resonance transitions in Mg-like ions. *J. Phys. B*, **26**, 4079–4089.

[7222] Chou H.S., Chi H.C. and Huang K.N. (1993d). Core-polarization effects in the $^3P_1^o$-$^1P_1^o$ separations of Cd- and Hg-like ions. *Phys. Lett. A*, **182**, 302–304.

[7223] Chou H.S., Chi H.C. and Huang K.N. (1994a). Relativistic excitation energies and oscillator strengths for transitions in Zn-like ions including core-polarization effects. *Phys. Rev. A*, **49**, 2394–2398.

[7224] Chou H.S., Chi H.C. and Huang K.N. (1994b). Core polarization effects on the intercombination and resonance transitions in Be-like ions. *Chin. J. Phys.*, **32**, 261–268.

[7225] Chou H.S. and Johnson W.R. (1997). Relativistic many-body perturbation-theory calculations of transition rates for copperlike, silverlike, and goldlike ions. *Phys. Rev. A*, **56**, 2424–2427.

[7226] Christe K.O., Curtis E.C. and Dixon D.A. (1993a). On the structure of IOF_5. *J. Am. Chem. Soc.*, **115**, 9655–9658.

[7227] Christe K.O., Dixon D.A., Mack H.G., Oberhammer H., Pagelot A., Sanders J.C.P. and Schrobilgen G.J. (1993b). Osmium tetrafluoride dioxide cis-OsO_2F_4. *J. Am. Chem. Soc.*, **115**, 11279–11284.

[7228] Christe K.O., Dixon D.A., Mahjoub A.R., Mercier H.P.A., Sanders J.C.P., Seppelt K., Schrobilgen G.J. and Wilson W.W. (1993c). The IOF_6^- anion: the first example of a pentagonal bipyramidal AX_5YZ species. *J. Am. Chem. Soc.*, **115**, 2696–2706.

[7229] Christe K.O., Dixon D.A., Sanders J.C.P., Schrobilgen G.J., Tsai S.S. and Wilson W.W. (1995). On the structure of the $[XeOF_5]^-$ anion and of heptacoordinated complex fluorides containing one or two highly repulsive ligands or sterically active free valence electron pairs. *Inorg. Chem.*, **34**, 1868–1874.

[7230] Christe K.O., Dixon D.A., Sanders J.C.P., Schrobilgen G.J. and Wilson W.W. (1993d). Heptacoordination: pentagonal bipyramidal XeF_7^+ and TeF_7^-ions. *J. Am. Chem. Soc.*, **115**, 9461–9467.

[7231] Christe K.O., Wilson W.W., Drake G.W., Dixon D.A., Boatz J.A. and Gnann R.Z. (1998). Pentagonal planar AX_5 species: synthesis and characterization of the iodine(III) pentafluoride dianion, IF_5^{2-}. *J. Am. Chem. Soc.*, **120**, 4711–4716.

[7232] Christiansen P.A., Moffett T.M. and DiLabio G.A. (1999). Potential curves for the Mg^+Rn complex including charge-transfer states. *J. Phys. Chem. A*, **103**, 8875–8878.

[7233] Christoffersen R.E. and Hall G.G. (1966). Relativistic effects in aromatic molecules. Orbit-orbit coupling in the presence of a magnetic field. Evaluation of integrals involving the interelectronic coordinate. *Theor. Chim. Acta*, **4**, 250–259.

[7234] Chung K.T. (1993). Theory of transition rates in few-electron ions. *AIP Conf. Proc.*, **274**, 381–388.

[7235] Chung K.T. and Zhu X.W. (1993). Energies fine structures and isotope shifts of the $1s^22snl$ excited states of the beryllium atom. *Phys. Rev. A*, **48**, 1944–1954.

[7236] Chung K.T., Zhu X.W. and Wang Z.W. (1993). Ionization potential for ground states of berylliumlike systems. *Phys. Rev. A*, **47**, 1740–1751.

[7237] Chung S.C., Krüger S., Pacchioni G. and Rösch N. (1995). Relativistic effects in the electronic structure of the monoxides and monocarbonyls of Ni, Pd, and Pt: local and gradient-corrected density functional calculations. *J. Chem. Phys.*, **102**, 3695–3702.

[7238] Chung S.C., Krüger S., Ruzankin S.P., Pacchioni G. and Rösch N. (1996a). Effects of relativity on the NiCO, Pd-CO, and Pt-CO bonding mechanism: a constrained space orbital variation analysis of density functional results. *Chem. Phys. Lett.*, **248**, 109–115.

[7239] Chung S.C., Krüger S., Schmidbaur H. and Rösch N. (1996b). A density functional study of trigold oxonium complexes and of their dimerization. *Inorg. Chem.*, **35**, 5387–5392.

[7240] Cioslowski J., Piskorz P. and Rez P. (1997). Accurate analytical representations of the core-electron densities of the elements 3 to 118. *J. Chem. Phys.*, **106**, 3607–3612.

[7241] Citra A. and Andrews L. (1999a). Reactions of laser-ablated iridium atoms with O_2. Infrared spectra and DFT calculations for iridium dioxide and peroxo iridium(VI) dioxide in solid argon. *J. Phys. Chem. A*, **103**, 4182–4190.

[7242] Citra A. and Andrews L. (1999b). Reactions of laser-ablated silver and gold atoms with dioxygen and density functional theory calculations of product molecules. *J. Mol. Struct. (Theochem)*, **489**, 95–108.

[7243] Clarkson E. (1994). Eigenvalues and eigenfunctions of the Dirac operator on spheres and pseudospheres. *J. Math. Phys.*, **35**, 2064–2073.

[7244] Clemence D.P. (1994a). m-function behaviour for a periodic Dirac system. *Proc. Roy. Soc. Edinburgh Sec. A*, **124**, 149–159.

[7245] Clemence D.P. (1994b). Titchmarsh-Weyl theory and Levinson's theorem for Dirac operators. *J. Phys. A*, **27**, 7835–7842.

[7246] Clerk G.J. and McKellar B.H.J. (1993). Relativistic band gaps in one-dimensional disordered system. *Phys. Rev. B*, **47**, 6942–6949.

[7247] Cloke F.G.N., Green J.C. and Jardine C.N. (1999). Electronic structure and photoelectron spectra of bispentalene complexes of thorium and uranium. *Organometallics*, **18**, 1080–1086.

[7248] Cloke F.G.N., Green J.C., Jardine C.N. and Kuchta M.C. (1999). Bonding in bis(pentalene)dimolybdenum: Density functional calculations on $Mo_2(C_8H_6)_2$ and photoelectron spectroscopy of $Mo_2(C_8H_4(1,4\text{-}SiPr^i{}_3)_2)_2$. *Organometallics*, **18**, 1087–1090.

[7249] Clot E. and Eisenstein O. (1998). Influence of ancillary ligands on the kinetics and the thermodynamics of H_2 addition to $IrXH_2(PR_3)_2$ (X = Cl, Br, I and R = H, Me): comparison between density functional theory and perturbation theory. *J. Phys. Chem.*, **102**, 3592–3598.

[7250] Clot E., Leforestier C., Eisenstein O. and Pélissier M. (1995). Dynamics on an ab initio surface for calculating J_{HH} NMR exchange coupling. The case of $OsH_3X(PH_3)_2$. *J. Am. Chem. Soc.*, **117**, 1797–1799.

[7251] Coehoorn R. and Buschow K.H.J. (1993). First principles calculations of the hyperfine field at Gd nuclei in intermetallic compounds. *J. Magnetism Magnetic Materials*, **118**, 175–181.

[7252] Cohen J.M. and Kuharetz B. (1993). Relativistic hydrogen atom: Wave equation in Whittaker form. *J. Math. Phys.*, **34**, 4964–4974.

[7253] Cohen M., Fournier K.B. and Goldstein W.H. (1998). Excitation autoionization rates from ground and excited levels in Li-like Ar^{15+} to S-like Ar^{2+}. *Phys. Rev. A*, **57**, 2651–2671.

[7254] Cohen S.M. and Leung P.T. (1998). General formulation of the semirelativistic approach to atomic sum rules. *Phys. Rev. A*, **57**, 4994–4997.

[7255] Cohen S.M. and Leung P.T. (1999). Comment on "Relativistic correction of the generalized oscillator strength sum rules". *Phys. Rev. A*, **59**, 4847–4848.

[7256] Cole R.J., Brooks N.J., Weightman P. and Matthew J.A.D. (1995). Onset of *d* screening in alkali and alkaline earths. *Phys. Rev. B*, **52**, 2976–2982.

[7257] Collins C.L., Dyall K.G. and Schaefer III H.F. (1995). Relativistic and correlation effects in CuH, AgH, and AuH: comparison of various relativistic methods. *J. Chem. Phys.*, **102**, 2024–2031.

[7258] Collins C.L. and Grev R.S. (1998). Relativistic effects in silicon chemistry: Are the experimental heats of formation of the silicon atom and SiH_4 compatible? *J. Chem. Phys.*, **108**, 5465–5468.

[7259] Comay E. (1993). Remarks on the physical meaning of the Lorentz-Dirac equation. *Found. Phys.*, **23**, 1121–1136.

[7260] Compagno G., Pachucki K. and Salamone G.M. (1993). Non-relativistic radiative corrections to the charge density distribution of the hydrogen atom. *J. Phys. B*, **26**, 1091–1101.

[7261] Compagno G. and Salamone G.M. (1991). N.a. *J. Mod. Opt.*, **37**, 183–.

[7262] Conlon P.C. and Fitzpatrick N.J. (1994). An all-electron and effective core potential study of the effects of electron correlation on the Group 7 complexes $[(C_6H_6)X(CO)_3]^+$ with X=Mn, Tc, Re. *J. Organomet Chem.*, **478**, 173–177.

[7263] Connerade J.P., Grant I.P., Marketos P. and Oberdisse J. (1995). Correlations and level statistics in complex spectra. *J. Phys. B*, **28**, 2539–2551.

[7264] Connerade J.P., Schmidt C. and Warken M. (1993). Strengths of high Rydberg members in alkali spectra within the g-Hartree optimized mean field. *J. Phys. B*, **26**, 3459–3466.

[7265] Continenza A., Monachesi P., Galli M., Marabelli F. and Bauer E. (1996). Theoretical interpretation of optical conductivity of $YbCu_4Ag,Au$. *J. Appl. Phys.*, **79**, 6423–6425.

[7266] Cook D.B. (1995). Effective core potentials and the structures of metallocenes. *Int. J. Quantum Chem.*, **53**, 309–319.

[7267] Cook P.A. (1971). Relativistic harmonic oscillators with intrinsic spin structure. *Lettere Nuovo Cim.*, **1**, 419–426.

[7268] Cooper A.C., Clot E., Huffman J.C., Streib W.E., Maseras F., Eisenstein O. and Caulton K.G. (1999). Computational and experimental test of steric influence on agostic interactions: A homologous series for Ir(III). *J. Am. Chem. Soc.*, **121**, 97–106.

[7269] Cooper B.R., Eriksson O., Hao Y.G. and Fernando G.W. (1992). Surface electronic structure and chemisorption of plutonium and uranium. In *Transuranium Elements A Half Century*, (Edited by Morss L.R. and Fuger J.), pp. 365–377, ACS, Washington DC.

[7270] Cooper F., Khare A. and Sukhatme U. (1995). Supersymmetry and quantum mechanics. *Phys. Rep.*, **251**, 267–385.

[7271] Corà F., Patel A., Harrison N.M., Dovesi R. and Catlow C.R.A. (1996). An ab initio Hartree-Fock study of the cubic and tetragonal phases of bulk tungsten trioxide. *J. Am. Chem. Soc.*, **118**, 12174–12182.

[7272] Corben H.C. (1993). Factors of 2 in magnetic moments, spin-orbit coupling, and Thomas precession. *Am. J. Phys.*, **61**, 551–553.

[7273] Cornehl H.H., Heinemann C., Marçalo J., Pires de Matos A. and Schwarz H. (1996). Das "nackte" uranyl(2+)-Kation UO_2^{2+}. *Angew. Chem.*, **108**, 950–952.

[7274] Cornehl H.H., Heinemann C., Marçalo J., Pires de Matos A. and Schwarz H. (1996). The "bare" uranyl(2+) ion UO_2^{2+}. *Angew. Chem. Int. Ed. Engl.*, **35**, 891–894.

[7275] Cornehl H.H., Wesendrup R., Diefenbach M. and Schwarz H. (1997). A comparative study of oxo-ligand effects in the gas-phase chemistry of atomic lanthanide and actinide cations. *Chem. Eur. J.*, **3**, 1083–1090.

[7276] Cornille M. and Dubau J. (1999). Comparison of the total autoionization rates for $1s2s^22p^2$, $1s2s2p^3$, $1s2p^4$ states of B-like ions ($6 \leq Z \leq 54$). *Phys. Scripta*, **59**, 27–31.

[7277] Cornille M., Dubau J. and Jacquemot S. (1994). Radiative and collisional atomic data for neon-like ions. *At. Data Nucl. Data Tables*, **58**, 1–66.

[7278] Cornille M., Dubau J. and Safronova U.I. (1993). Energies and autoionization rates for Be-like systems Comparison of two methods: AUTOLSJ and MZ. *Phys. Scr.*, **48**, 546–558.

[7279] Cortona P. and Villafiorita Monteleone A. (1996). Ab initio calculations of cohesive and structural properties of the alkali-earth oxides. *J. Phys. C*, **8**, 8983–8994.

[7280] Cory M.G., Kostlmeier S., Kotzian M., Rösch N. and Zerner M.C. (1994). An intermediate neglect of differential overlap technique for actinide compounds. *J. Chem. Phys.*, **100**, 1353–1365.

[7281] Cosentino U., Moro G., Pitea D., Calabi L. and Maiocchi A. (1997). Ab initio effective core potential calculations on lanthanide complexes: basis sets and electron correlation effects in the study of $(Gd\text{-}(H_2O)_9)^{3+}$. *J. Mol. Str. (Theochem)*, **392**, 75–85.

[7282] Cosentino U., Moro G., Pitea D., Villa A., Fantucci P.C., Maiocchi A. and Uggeri F. (1998). Ab initio investigation of gadolinium complexes with polyamino carboxylate ligands and force fields parametrization of metal-ligand interactions. *J. Phys. Chem.*, **102**, 4606–4614.

[7283] Costa Cabral B.J. and Silva Fernandes F.M.S. (1995). Dynamics and structure of molten CsAu. *Conf. Proc. No. 330, E.C.C.C. 1 Comp. Chem.*, **330**, 129–133.

[7284] Costella J.P. and McKellar B.H.J. (1995). The Foldy-Wouthuysen transformation. *Am. J. Phys.*, **63**, 1119–1121.

[7285] Cotaescu I.I. and Dragaescu G. (1997). The operator algebra of the quantum relativistic oscillator. *J. Math. Phys.*, **38**, 5505–5514.

[7286] Cotton F.A., Cowley A.H. and Feng X.J. (1998). The use of density functional theory to understand and predict structures and bonding in main group compounds with multiple bonds. *J. Am. Chem. Soc.*, **120**, 1795–1799.

[7287] Cotton F.A. and Feng X.J. (1997). Density functional theory study of transition-metal compounds containing metal-metal bonds. 1. Molecular structures of dinuclear compounds by complete geometry optimization. *J. Am. Chem. Soc.*, **119**, 7514–7520.

[7288] Cotton F.A., Wilkinson G., Murillo C.A. and Bochmann M. (1999). *Advanced Inorganic Chemistry, 6th Edition.* Wiley, New York, chapter 1-8 'Relativistic Effects'. See also pp. 175, 260, 1084.

[7289] Cotton S.A. (1997). *Chemistry of Precious Metals.* Blackie Academic & Professional, London, 374 p.

[7290] Coudray C. and Coz M. (1971). Construction of relativistic potentials when the energy is fixed. *J. Math. Phys.*, **12**, 1166–1178.

[7291] Coulter B.L. and Adler C.G. (1971). The relativistic one-dimensional square potential. *Am. J. Phys.*, **39**, 305–309.

[7292] Coutinho F.A.B., Nogami Y. and Tomio L. (1999). Two definitions of the electric polarizability of a bound system in relativistic quantum theory. *Am. J. Phys.*, **67**, 735–736.

[7293] Coutinho F.A.B. and Perez J.F. (1993). Boundary conditions in the Aharonov-Bohm scattering of Dirac particles and effect of Coulomb interaction. *Phys. Rev. D*, **48**, 932–939.

[7294] Crater H.W. and Van Alstine P. (1994). Structure of quantum-mechanical relativistic two-body interactions for spinning particles. *Found. Phys.*, **24**, 297–328.

[7295] Crater H.W., Van Alstine P., Becker R.L. and Wong C.Y. (1992). Two-body Dirac equations of constraint dynamics and field theoretic bound states. In *Proc. Int. Symp. on Extended Objects and Bound Systems*, (Edited by Hara O., Ishida S. and Naka S.), pp. 135–160, World Scientific, Singapore.

[7296] Craw J.S., Bacskay G.B. and Hush N.S. (1993). Near doubling of H-H bond length in the "stretched" osmium molecular hydrogen complex $[Os(NH_3)_4OAc(\eta^2\text{-}H_2)]^+$: a theoretical study. *Inorg. Chem.*, **32**, 2230–2231.

[7297] Craw J.S., Vincent M.A., Hillier I.H. and Wallwork A.L. (1995). Ab initio quantum chemical calculations on uranyl UO_2^{2+}, plutonyl PuO_2^{2+}, and their nitrates and sulfates. *J. Phys. Chem.*, **99**, 10181–10185.

[7298] Crawford D.P. and Reiss H.R. (1994). Stabilization in relativistic photoionization with circularly polarized light. *Phys. Rev. A*, **50**, 1844–1850.

[7299] Crawford D.P. and Reiss H.R. (1998). Relativistic ionization of hydrogen by linearly polarized light. *Optics Express*, **2**, 289–297.

[7300] Crawford J.P. (1993). The Dirac oscillator and local automorphism invariance. *J. Math. Phys.*, **34**, 4428–4435.

[7301] Crespo López-Urrutia J.R., Savin D.W., Beiersdorfer P. and Widmann K. (1996). Direct observation of the spontaneous emission of the hyperfine transition F=4 to F=3 in ground state hydrogenlike $^{165}Ho^{66+}$ in an electron beam trap. *Phys. Rev. Lett.*, **77**, 826–829.

[7302] Crespo López-Urrutia J.R., Beiersdorfer P., Widmann K. and Birkett B.B. (1998). Nuclear magnetization distribution radii determined by hyperfine transitions in the 1s level of H-like ions $^{185}Re^{74+}$ and $^{187}Re^{74+}$. *Phys. Rev. A*, **57**, 879–887.

[7303] Crisp M.D. (1996). Relativistic neoclassical radiation theory. *Phys. Rev. A*, **54**, 87–92.

[7304] Cromp B., Carrington Jr T., Salahub D.R., Malkina O.L. and Malkin V.G. (1999). Effect of rotation and vibration on nuclear magnetic resonance chemical shifts: Density functional theory calculations. *J. Chem. Phys.*, **110**, 7153–7159.

[7305] Cruz A., Del Angel G., Poulain E., Martínez-Magadan J.M. and Castro M. (1999). Theoretical study for the Pt_2Au- and $PtAu_2$-ethylene interaction. *Int. J. Quantum Chem.*, **75**, 699–707.

[7306] Cruz A., Poulain E., Del Angel G., Castillo S. and Bertin V. (1998). Theoretical characterization of H_2 adsorption on AuPt clusters. *Int. J. Quantum Chem.*, **67**, 399–409.

[7307] Császár A.G., Kain J.S., Polyansky O.L., Zobov N.F. and Tennyson J. (1998). Relativistic correction to the potential energy surface and vibration-rotation levels of water [Erratum: ibid. 312 (1999) 613-616.]. *Chem. Phys. Lett.*, **293**, 317–323.

[7308] Cui Q., Musaev D.G. and Morokuma K. (1997). Molecular orbital study of the mechanism of platinum(0)-catalyzed alkene and alkyne diboration reactions. *Organometallics*, **16**, 1355–1364.

[7309] Cui Q., Musaev D.G. and Morokuma K. (1998a). Why do $Pt(PR_3)_2$ complexes catalyze the alkyne diboration reaction, but their palladium analogues do not? A density functional study. *Organometallics*, **17**, 742–751.

[7310] Cui Q., Musaev D.G. and Morokuma K. (1998b). Molecular orbital study of H_2 and CH_4 activation on small metal clusters. I. Pt, Pd, Pt_2 and Pd_2. *J. Chem. Phys.*, **108**, 8418–8428.

[7311] Cui Q., Musaev D.G., Svensson M. and Morokuma K. (1996). Analytical second derivatives for effective core potential. Application to transition structures of $Cp_2Ru_2(\mu\text{-}H)_4$ and to the mechanism of reaction Cu + CH_2N_2. *J. Phys. Chem.*, **100**, 10936–10944.

[7312] Cundari T.R. (1993a). Methane adducts of d^0 transition metal imido complexes. *Organometallics*, **12**, 1998–2000.

[7313] Cundari T.R. (1993b). Methane activation by d^0 and d^2 imidos: Effects of d orbital occupation and comparison of [2+2] and oxidative addition. *Organometallics*, **12**, 4971–4978.

[7314] Cundari T.R. (1994a). Methane activation by Group VB bis(imido) complexes. *Organometallics*, **13**, 2987–2994.

[7315] Cundari T.R. (1994b). Calculation of a methane C-H oxidative addition trajectory: Comparison to experiment and methane activation by high-valent complexes. *J. Am. Chem. Soc.*, **116**, 340–347.

[7316] Cundari T.R., Benson M.T., Lutz M.L. and Sommerer S.O. (1996). Effective core potential approaches to the chemistry of the heavier elements. *Rev. Comp. Chem.*, **8**, 145–202.

[7317] Cundari T.R., Conry R.R., Spaltenstein E., Critchlow S.C., Hall K.A., Tahmassebi S.K. and Mayer J.M. (1994). Rhenium-oxo-bis(acetylene) anions: Structure, properties, and electronic structure. Comparison of Re-O bonding with that in other rhenium oxo complexes. *Organometallics*, **13**, 322–331.

[7318] Cundari T.R. and Curtiss S. (1996). Substituent effects on methane activation and elimination by high-valent Zr complexes. *Int. J. Quantum Chem.*, **60**, 779–788.

[7319] Cundari T.R., Fu W.T., Moody E.W., Slavin L.L., Snyder L.A., Sommerer S.O. and Klinckman T.R. (1996). Molecular mechanics force field for platinum coordination complexes. *J. Phys. Chem.*, **100**, 18057–18064.

[7320] Cundari T.R. and Gordon M.S. (1993). Small molecule elimination from group IVB (Ti, Zr, Hf) amido complexes. *J. Am. Chem. Soc.*, **115**, 4210–4217.

[7321] Cundari T.R., Harvey J.N., Klinckman T.R. and Fu W.T. (1999). Multiple bonding involving late transition metals. The case of a silver-oxo complex [Addition: Ibid. 2000, 39, 1336.]. *Inorg. Chem.*, **38**, 5611–5615.

[7322] Cundari T.R. and Klinckman T.R. (1998). Reduced variational space analysis of methane adducts. *Inorg. Chem.*, **37**, 5399–5401.

[7323] Cundari T.R., Kurtz H.A. and Zhou T. (1998). Modeling nonlinear optical properties of transition metal complexes. Basis set, effective core potential, and geometry effects. *J. Phys. Chem. A*, **102**, 2962–2966.

[7324] Cundari T.R. and Li Y.P. (1995). Effective core potential modeling of group IVA - group IVB chemical vapor deposition. *Int. J. Quantum Chem.*, **55**, 315–328.

[7325] Cundari T.R. and Raby P.D. (1997). Theoretical estimation of vibrational frequencies involving transition metal compounds. *J. Phys. Chem. A*, **101**, 5783–5788.

[7326] Cundari T.R., Snyder L.A. and Yoshikawa A. (1998). Ligand and substituent effects in methane activation by mercury(II) complexes. *J. Mol. Str. (Theochem)*, **425**, 13–24.

[7327] Cundari T.R., Sommerer S.O., Strohecker L.A. and Tippett L. (1995). Effective core potential studies of lanthanide complexes. *J. Chem. Phys.*, **103**, 7058–7063.

[7328] Cundari T.R. and Stevens W.J. (1993). Effective core potential methods for the lanthanides. *J. Chem. Phys.*, **98**, 5555–5565.

[7329] Cundari T.R. and Yoshikawa A. (1998). Computational study of methane activation by mercury(II) complexes. *J. Comp. Chem.*, **19**, 902–911.

[7330] Curtis L.J. (1993). A predictive data-based exposition of $nsnp\ ^{1,3}P_1$ lifetimes in the Cd and Hg isoelectronic sequences. *J. Phys. B*, **26**, L589–L594.

[7331] Curtis L.J., Ellis D.G., Matulioniene R. and Brage T. (1997). Relativistic empirical specification of transition probabilities from measured lifetime and energy level data. *Phys. Scr.*, **56**, 240–244.

[7332] Curtis L.J., Ellis D.G., Matulioniene R. and Brage T. (1997). Relativistic empirical specification of transition probabilities from measured lifetime and energy level data. *Phys. Scr.*, **56**, 240–244.

[7333] Cusson R.Y., Reinhard P.G., Molitoris J.J., Stocker H., Strayer M.R. and Greiner W. (1985). Time-dependent Dirac equation with relativistic mean-field dynamics applied to heavy-ion scattering. *Phys. Rev. Lett.*, **55**, 2786–2789.

[7334] Czarnecki A., Melnikov K. and Yelkhovsky A. (1999a). Positronium hyperfine splitting: Analytical value at $O(m\alpha^6)$. *Phys. Rev. Lett.*, **82**, 311–314.

[7335] Czarnecki A., Melnikov K. and Yelkhovsky A. (1999b). α^2 Corrections to parapositronium decay. *Phys. Rev. Lett.*, **83**, 1135–1138.

[7336] Czuchaj E., Rebentrost F., Stoll H. and Preuss H. (1993). Use of non-local l-dependent pseudopotentials in the calculation of the potential energies for the Ba-rare gas systems. *Chem. Phys.*, **177**, 107–117.

[7337] Czuchaj E., Rebentrost F., Stoll H. and Preuss H. (1994). Adiabatic potential curves for the Cd_2 dimer. *Chem. Phys. Lett.*, **225**, 233–239.

[7338] Czuchaj E., Rebentrost F., Stoll H. and Preuss H. (1995). Pseudopotential calculations for the potential energies of LiHe and BaHe. *Chem. Phys.*, **196**, 37–46.

[7339] Czuchaj E., Rebentrost F., Stoll H. and Preuss H. (1996). Potential energy curves for the Zn_2 dimer. *Chem. Phys. Lett.*, **255**, 203–209.

[7340] Czuchaj E., Rebentrost F., Stoll H. and Preuss H. (1997). Calculation of ground- and excited-state potential energy curves for the Hg_2 molecule in a pseudopotential approach. *Chem. Phys.*, **214**, 277–289.

[7341] Czuchaj E., Rebentrost F., Stoll H. and Preuss H. (1998). Calculation of ground- and excited-state potential energy curves for barium - rare gas complexes in a pseudopotential approach. *Theor. Chem. Acc.*, **100**, 117–123.

[7342] Czuchaj E. and Stoll H. (1999). Calculation of ground- and excited-state potential energy curves for the Cd-rare gas complexes. *Chem. Phys.*, **248**, 1–16.

[7343] Dahl J.P. (1997). Physical origin of the Runge-Lenz vector. *J. Phys. A*, **30**, 6831–6840.

[7344] Dahl J.P. and Jørgensen T. (1995). On the Dirac-Kepler problem: the Johnson-Lippmann operator, supersymmetry, and normal-mode representations. *Int. J. Quantum Chem.*, **53**, 161–181.

[7345] Dai D.D. and Li L.M. (1997). Study on europium chalcogenides by means of density functional theory. *J. Mol. Str. (Theochem)*, **417**, 9–17.

[7346] Dai D.D., Li L.M., Ren J. and Whangbo M.H. (1998). Description of ligand field splitting in terms of density functional theory: Calculations of the split levels of the $^2F_{5/2}$ and $^2F_{7/2}$ subterms in CeO and CeF under the weak field coupling scheme. *J. Chem. Phys.*, **108**, 3479–3488.

[7347] Dai D.G., Al-Zahrani M.M. and Balasubramanian K. (1994a). Geometries and energy separations of electronic states of GeF_2, SnF_2, and PbF_2 and their positive ions. *J. Phys. Chem.*, **98**, 9233–9241.

[7348] Dai D.G. and Balasubramanian K. (1993a). Spectroscopic properties and potential energy curves for 30 electronic states of ReH. *J. Mol. Spectr.*, **158**, 455–467.

[7349] Dai D.G. and Balasubramanian K. (1993b). Geometries and potential energies of electronic states of GaX_2 and GaX_3 (X = Cl, Br, and I). *J. Chem. Phys.*, **99**, 293–301.

[7350] Dai D.G. and Balasubramanian K. (1994a). Pt_3Au and PtAu clusters: Electronic states and potential energy surfaces. *J. Chem. Phys.*, **100**, 4401–4407.

[7351] Dai D.G. and Balasubramanian K. (1994b). Potential energy surfaces for Rh-CO, Rh-OC, Ir-CO, and Ir-OC interactions. *J. Chem. Phys.*, **101**, 2148–2156.

[7352] Dai D.G. and Balasubramanian K. (1994c). Spectroscopic constants and potential energy curves of SnF. *Chem. Phys. Lett.*, **224**, 425–431.

[7353] Dai D.G. and Balasubramanian K. (1994d). Twelve electronic states and potential energy surface of Zr_3. *Chem. Phys. Lett.*, **231**, 352–358.

[7354] Dai D.G. and Balasubramanian K. (1995a). Ionization energies of Y_n (n = 1-4). *Chem. Phys. Lett.*, **238**, 203–207.

[7355] Dai D.G. and Balasubramanian K. (1995b). Electronic structures of Pd_4 and Pt_4. *J. Phys. Chem.*, **103**, 648–655.

[7356] Dai D.G. and Balasubramanian K. (1995c). Spectroscopic properties of PtN. *J. Mol. Spectr.*, **172**, 421–429.

[7357] Dai D.G. and Balasubramanian K. (1996a). Geometries and energy separations of 28 electronic states of Ge_5. *J. Chem. Phys.*, **105**, 5901–5906.

[7358] Dai D.G. and Balasubramanian K. (1996b). Electronic states of Rh_4. *Chem. Phys. Lett.*, **263**, 703–709.

[7359] Dai D.G. and Balasubramanian K. (1996c). Geometries and energy separations of 24 electronic states of Sn_5. *J. Phys. Chem.*, **100**, 19321–19325.

[7360] Dai D.G. and Balasubramanian K. (1997). Electronic states of Pb_5: geometries and energy separations. *Chem. Phys. Lett.*, **271**, 118–124.

[7361] Dai D.G. and Balasubramanian K. (1998). Geometries and energy separations of the electronic states of Ge_5^+ and Sn_5^+. *J. Chem. Phys.*, **108**, 4379–4385.

[7362] Dai D.G. and Balasubramanian K. (1999). Geometries and energy separations of electronic states of Pd_5. *Chem. Phys. Lett.*, **310**, 303–312.

[7363] Dai D.G., Liao D.W. and Balasubramanian K. (1995a). Potential energy surfaces for Pt_3+H_2 and Pd_3+H_2 systems. *J. Phys. Chem.*, **102**, 7530–7539.

[7364] Dai D.G., Liao M.Z. and Balasubramanian K. (1996a). Sixteen electronic states of the iridium trimer (Ir_3). *Chem. Phys. Lett.*, **249**, 141–148.

[7365] Dai D.G., Majumder D. and Balasubramanian K. (1998). Theoretical studies of the interaction of butadiene with Rh and Rh_2. *Chem. Phys. Lett.*, **287**, 178–184.

[7366] Dai D.G., Roszak S. and Balasubramanian (1996b). Theoretical study of potential energy surfaces for interactions of Pd_2 with CO. *J. Phys. Chem.*, **104**, 1471–1476.

[7367] Dai D.G., Roszak S. and Balasubramian K. (1999). Electronic states of the hafnium trimer (Hf_3). *Chem. Phys. Lett.*, **308**, 495–502.

[7368] Dai D.G., Xu H. and Balasubramanian K. (1995b). Electronic states of $GeCl^+$. *Chem. Phys. Lett.*, **245**, 503–508.

[7369] Dakternieks D., Henry D.J. and Schiesser C.H. (1998). Equilibria in free-radical chemistry: An ab initio study of hydrogen atom transfer reactions between silyl, germyl, and stannyl radicals and their hydrides. *Organometallics*, **17**, 1079–1084.

[7370] Daniel C., Guillaumont D., Ribbing C. and Minaev B. (1999). Spin-orbit coupling effects on the metal-hydrogen bond homolysis of $M(H)(CO)_3$(H-DAB) (M = Mn,Re; H-DAB = 1,4-diaza-1,3-butadiene). *J. Phys. Chem. A*, **103**, 5766–5772.

[7371] Daniel C., Heitz M.C., Manz J. and Ribbing C. (1995). Spin-orbit induced radiationless transitions in organometallics: Quantum simulation of the $^1E \rightarrow {}^3A_1$ intersystem crossing process in $HCo(CO)_4$. *J. Phys. Chem.*, **102**, 905–912.

[7372] Danilov L.I. (1995). Resolvent estimates and spectrum of the Dirac operator with periodic potential (in Russian). *Teor. Mat. Fiz.*, **103(1)**, 3–22.

[7373] Danilov L.I. (1995). Resolvent estimates and spectrum of the Dirac operator with periodic potential. *Theor. Math. Phys.*, **103**, 349–365.

[7374] Danovich D., Hrušák J. and Shaik S. (1995). Ab initio calculations for small iodo clusters. Good performance of relativistic effective core potentials. *Chem. Phys. Lett.*, **233**, 249–256.

[7375] Danovich D., Marian C.M., Neuheuser T., Peyerimhoff S.D. and Shaik S. (1998). Spin-orbit coupling patterns induced by twist and pyramidalization modes in C_2H_4: A quantitative study and a qualitative analysis. *J. Phys. Chem. A*, **102**, 5923–5936.

[7376] Danovich D. and Shaik S. (1997). Spin-orbit coupling in the oxidative activation of H-H by FeO^+. Selection rules and reactivity effects. *J. Am. Chem. Soc.*, **119**, 1773–1786.

[7377] Daoudi A., Elkhattabi S., Berthier G. and Flament J.P. (1998). On the electronic structure and spectroscopy of the ScN molecule. *Chem. Phys.*, **230**, 31–44.

[7378] Dapprich S. and Frenking G. (1995). Investigation of donor-acceptor interactions: a charge decomposition analysis using fragment molecular orbitals. *J. Phys. Chem.*, **99**, 9352–9362.

[7379] Dapprich S., Ujaque G., Maseras F., Lledós A., Musaev D.G. and Morokuma K. (1996). Theory does not support an osmaoxetane intermediate in the osmium-catalyzed dihydroxylation of olefins. *J. Am. Chem. Soc.*, **118**, 11660–11661.

[7380] Darevich Y.V. (1994). Hamiltonian variational method for relativistic few-particle systems (in Ukrainian). *Ukr. Fiz. Zh.*, **39**, 377–379.

[7381] Darewych J.W. (1999). Integral identities and bounds for scattering calculations and the Dirac formalism. *Phys. Rev. A*, **60**, 290–294.

[7382] Dargel T.K., Hertwig R.H. and Koch W. (1998). Towards an accurate gold carbonyl binding energy in $AuCO^+$: basis set convergence and comparison between functional and conventional methods. *J. Chem. Phys.*, **108**, 3876–3885.

[7383] Dargel T.K., Hertwig R.H. and Koch W. (1999). How do coinage metals bind to benzene? *Mol. Phys.*, **96**, 583–591.

[7384] Dartigeas K., Gonbeau D. and Pfister-Guillouzo G. (1996). Core and valence spectra of TaS_2 and WS_2. Experimental and theoretical studies. *J. Chem. Soc., Faraday Trans.*, **92**, 4561–4566.

[7385] Das B.P. (1997). Computation of correlation effects on the parity-nonconserving electric-dipole transition in atomic ytterbium. *Phys. Rev. A*, **56**, 1635–1637.

[7386] Das J.N. and Dhar S. (1998). Calculation of triple differential cross sections of K-shell ionization of medium-heavy atoms by electrons for symmetric geometry. *Pramana*, **51**, 751–756.

[7387] Das K.K., Alekseyev A.B., Liebermann H.P., Hirsch G. and Buenker R.J. (1995a). Spin-orbit configuration interaction study of the electronic spectrum of antimony iodide. *Chem. Phys.*, **196**, 395–406.

[7388] Das K.K., Alekseyev A.B., Liebermann H.P., Hirsch G. and Buenker R.J. (1995b). Relativistic configuration interaction study of the low-lying electronic states of Bi_2. *J. Chem. Phys.*, **102**, 4518–4530.

[7389] Das K.K., Liebermann H.P., Buenker R.J. and Hirsch G. (1996a). Ab initio configuration interaction calculations of the potential curves and lifetimes of the low-lying electronic states of the lead dimer. *J. Chem. Phys.*, **104**, 6631–6642.

[7390] Das K.K., Liebermann H.P., Hirsch G. and Buenker R.J. (1995c). Use of relativistic core potentials in calculating the electronic spectrum of the antimony dimer. *J. Chem. Phys.*, **102**, 8462–8473.

[7391] Dasgupta A. and Whitney K.G. (1994). Z-scaled data for dielectronic recombination from O-like to F-like ions. *At. Data Nucl. Data Tables*, **58**, 77–99.

[7392] Da Silva A.B.F., Malli G.L. and Ishikawa Y. (1993a). Relativistic universal Gaussian basis set for Dirac-Fock-Coulomb and Dirac-Fock-Breit SCF calculations on heavy atoms. *Chem. Phys. Lett.*, **203**, 201–204.

[7393] Da Silva A.B.F., Malli G.L. and Ishikawa Y. (1993b). Accurate ab initio relativistic Dirac-Fock-Coulomb calculations on heavy atoms using universal Gaussian basis set. *Can. J. Chem.*, **71**, 1713–1715.

[7394] Datta A., Srivastava C.M. and Datta S.N. (1993). Relativistically parametrized extended Huckel calculation of net charges on atoms in $YBa_2Cu_3O_7$. *J. Phys. Chem.*, **97**, 9996–10001.

[7395] Datta D. and Beck D.R. (1993). Possibility of formation of rare-earth negative ions by attachment of f electrons to the atomic ground state. *Phys. Rev. A*, **47**, 5198–5201.

[7396] Datta D. and Beck D.R. (1994). Electron affinities of opposite-parity bound states in Th^-: Relativistic-configuration-interaction studies. *Phys. Rev. A*, **50**, 1107–1111.

[7397] Datta D. and Beck D.R. (1995). Relativistic many-body effects in the fine and hyperfine structure of ^{139}La II $(5d+6s)^2$ J=2 states: the need for second-order electrostatic corrections. *Phys. Rev. A*, **52**, 3622–3627.

[7398] Datta S.N. (1993). Ab-initio nonrelativistic and relativistic Hartree-Fock calculations for both closed- and open-shell molecules using GTO basis. *Pramana J. Phys.*, **41**, 363–370.

[7399] Datta S.N. (1995). Near-Dirac-Hartree-Fock results for first-row atoms calculated with GTO basis sets. *Int. J. Quantum Chem.*, **56**, 91–95.

[7400] Dattoli G., Lorenzutta S., Maino G. and Torre A. (1998). The generating function method and properties of relativistic Hermite polynomials. *Nuovo Cim. B*, **113**, 553–560.

[7401] Datz S., Drake G.W.F., Gallagher T.F., Kleinpoppen H. and zu Putlitz G. (1999). Atomic physics. *Rev. Mod. Phys.*, **71**, S223–S241.

[7402] Daul C., Baerends E.J. and Vernooijs P. (1994). A density functional study of the MLCT states of $[Ru(bpy)_3]^{2+}$ in D_3 symmetry. *Inorg. Chem.*, **33**, 3538–3543.

[7403] Deb A. and Chatterjee A.K. (1998). The electronic structure and chemical bonding mechanism of silver oxide. *J. Phys. CM*, **10**, 11719–11729.

[7404] De Cesare N., Murolo F., Perillo E., Spadaccini G. and Vigilante M. (1994). Systematic study of ^{4}He ion-induced L-shell ionization for $46 \leq Z \leq 70$ elements. *Nucl. Instr. Meth. Phys. Sec. B*, **84**, 295–299.

[7405] Decker S.A. and Klobukowski M. (1998). The first carbonyl bond dissociation energies of $M(CO)_5$ and $M(CO)_4(C_2H_2)$ (M = Fe, Ru, and Os): The role of the acetylene ligand from a density functional perspective. *J. Am. Chem. Soc.*, **120**, 9342–9355.

[7406] Decker S.A., Klobukowski M., Sakai Y. and Miyoshi E. (1998). Calibration and benchmarking of model core potentials: applications to systems containing main-group elements. *J. Mol. Str. (Theochem)*, **451**, 215–226.

[7407] Deco G., Maidagan J., Fojón O. and Rivarola R. (1995). Do symmetric eikonal and continuum distorted wave models satisfy the correct boundary conditions? *Phys. Scr.*, **51**, 334–338.

[7408] Deeth R.J. (1995). Is the ground state of $(RuO_4)^{2-}$ exceptional? *J. Chem. Soc., Dalton Trans.*, pp. 1537–1542.

[7409] Deeth R.J. (1996). Transition state symmetries and theoretical activation enthalpies for chloride exchange at planar $(PdCl_4)^{2-}$. *Chem. Phys. Lett.*, **261**, 45–50.

[7410] Deeth R.J. and Brooke Jenkins H.D. (1997). A density functional and thermochemical study of M-X bond lengths and energies in $(MX_6)^{2-}$ complexes: LDA versus Becke88/Perdew86 gradient-corrected functionals. *J. Phys. Chem. A*, **101**, 4793–4798.

[7411] Dehnen S., Schäfer A., Ahlrichs R. and Fenske D. (1996). An ab initio study of structures and energetics of copper sulfide clusters. *Chem. Eur. J.*, **2**, 429–435.

[7412] Dehnen S., Schäfer A., Fenske D. and Ahlrichs R. (1994). Neue Schwefel- und Selen-verbrückte Kupfercluster; ab-initio-Berechnungen von $[Cu_{2n}Se_n(PH_3)_m]$-Clustern. *Angew. Chem.*, **106**, 786–790.

[7413] Deineka G.V. (1996). Application of a Hermitian basis of B-splines to atomic relativistic calculations (in Russian). *Opt. Spektr.*, **81**, 181–185.

[7414] de Jong W.A. (1998). *Relativistic Quantum Chemistry Applied*. Ph.D. thesis, Groningen, 147 p.

[7415] De Jong W.A. and Nieuwpoort W.C. (1996). Relativity and the chemistry of UF_6: a molecular Dirac-Hartree-Fock-CI study. *Int. J. Quantum Chem.*, **58**, 203–216.

[7416] de Jong W.A., Styszynski J., Visscher L. and Nieuwpoort W.C. (1998). Relativistic and correlation effects on molecular properties: The interhalogens ClF, BrF, BrCl, IF, ICl, and IBr. *J. Chem. Phys.*, **108**, 5177–5184.

[7417] De Jong W.A., Visscher L., Visser O., Aerts P.J.C. and Nieuwpoort W.C. (1994). MOLFDIR: a program package for molecular Dirac-Fock-CI calculations. In *New Challenges in Computational Quantum Chemistry*, (Edited by Broer R., Aerts P.J.C. and Bagus P.S.), pp. 239–248, Univ. of Groningen.

[7418] de Jong W.A., Visscher L. and Nieuwpoort W.C. (1997). Relativistic and correlated calculations on the ground, excited, and ionized states of iodine. *J. Chem. Phys.*, **107**, 9046–9058.

[7419] de Jong W.A., Visscher L. and Nieuwpoort W.C. (1999). On the bonding and the electric field gradient of the uranyl ion. *J. Mol. Str. (Theochem)*, **458**, 41–52.

[7420] Delbourgo R. (1995). Square root of the harmonic oscillator. *Phys. Rev. A*, **52**, 3356–3359.

[7421] De Leo S. (1996). Quaternions and special relativity. *J. Math. Phys.*, **37**, 2955–2968.

[7422] Delin A., Eriksson O., Ahuja R., Johansson B., Brooks M.S.S., Gasche T., Auluck S. and Wills J.M. (1996). Optical properties of the group-IVB refractory metal compounds. *Phys. Rev. B*, **54**, 1673–1681.

[7423] Delin A., Fast L., Eriksson O. and Johansson B. (1998a). Effect of generalized gradient corrections on lanthanide cohesive properties. *J. Alloys Compounds*, **275-277**, 472–475.

[7424] Delin A., Fast L., Johansson B., Eriksson O. and Wills J.M. (1998b). Cohesive properties of the lanthanides: Effect of generalized gradient corrections and crystal structure. *Phys. Rev. B*, **58**, 4345–4351.

[7425] Delin A., Ravindran P., Eriksson O. and Wills J.M. (1998c). Full-potential optical calculations of lead chalcogenides. *Int. J. Quantum Chem.*, **69**, 349–358.

[7426] Delley B. (1998). A scattering theoretic approach to scalar relativistic corrections on bonding. *Int. J. Quantum Chem.*, **69**, 423–433.

[7427] Demachy I., Esteruelas M.A., Jean Y., Lledós A., Maseras F., Oro L.A., Valero C. and Volatron F. (1996). Hydride exchange processes in the coordination sphere of transition metal complexes: The $OsH_3(BH_4)(PR_3)_2$ system. *J. Am. Chem. Soc.*, **118**, 8388–8394.

[7428] Demachy I., Jean Y. and Lledos A. (1999a). Structure and internal rotation in quadruply bonded α-$Mo_2Cl_4(P–P)_2$ complexes: a density functional theory study of the *cis*-$Mo_2Cl_4(PH_3)_4$ complex. *Chem. Phys. Lett.*, **303**, 621–628.

[7429] Demachy I., Lledós A. and Jean Y. (1999b). DFT study of the role of bridging diphosphine ligands in the structure and the internal rotation in quadruply bonded metal dimers of the $Mo_2Cl_4(P-P)_2$ type. *Inorg. Chem.*, **38**, 5443–5448.

[7430] Dembczynski J., Guthohrlein G.H. and Stachowska E. (1993). Sternheimer free determination of the ^{59}Co nuclear quadrupole moment from hyperfine-structure measurements. *Phys. Rev. A*, **48**, 2752–2761.

[7431] Dembczynski J., Stachowska E., Wilson M., Buch P. and Ertmer W. (1994). Measurement and interpretation of the odd-parity levels of Pb I. *Phys. Rev. A*, **49**, 745–754.

[7432] de Mello L.A., Petrilli H.M. and Frota-Pessoa S. (1993). The electric field gradient at the nucleus in HCP Zr and Hf. *J. Phys: CM*, **5**, 8935–8942.

[7433] Dement'ev A.I., Kuznetsov M.L. and Kiselev Y.M. (1997). On extremal oxidation states of heavy 5d elements. *Russian J. Inorg. Chem.*, **42**, 1052–1057.

[7434] Deng L.Q., Branchadell V. and Ziegler T. (1994). Potential energy surfaces of the gas-phase S_N2 reactions $X^- + CH_3X = XCH_3 + X^-$ (X=F, Cl, Br, I): A comparative study by density functional theory and ab initio methods. *J. Am. Chem. Soc.*, **116**, 10645–10656.

[7435] Deng L.Q. and Ziegler T. (1997). Theoretical study of the oxidation of alcohol to aldehyde by d^0 transition-metal - oxo complexes: Combined approach based on density functional theory and the intrinsic reaction coordinate method. *Organometallics*, **16**, 716–724.

[7436] Deng S.Q., Simon A. and Köhler J. (1998). Superconductivity and chemical bonding in mercury. *Ang. Chem. Int. Ed. Engl.*, **37**, 640–643.

[7437] de Oliveira A.E., Guadagnini P.H., Custódio R. and Bruns R.E. (1998). Infrared vibrational intensities, polar tensors, and core electron energies of the group IV hydrides and the fluorosilanes. *J. Phys. Chem.*, **102**, 4615–4622.

[7438] de Oliveira G., Martin J.M.L., de Proft F. and Geerlings P. (1999). Electron affinities of the first- and second-row atoms: Benchmark *ab initio* and density-functional calculations. *Phys. Rev. A*, **60**, 1034–1045.

[7439] Derevianko A., Johnson W.R. and Cheng K.T. (1999c). Non-dipole effects in photoelectron angular distributions for rare gas atoms. *At. Data Nucl. Data Tables*, **73**, 153–211.

[7440] Derevianko A., Johnson W.R., Safronova M.S. and Babb J.F. (1999a). High-precision calculations of dispersion coefficients, static dipole polarizabilities, and atom-wall interaction constants for alkali-metal atoms. *Phys. Rev. Lett.*, **82**, 3589–3592.

[7441] Derevianko A., Safronova M.S. and Johnson W.R. (1999b). *Ab initio* calculations of off-diagonal hyperfine interaction in cesium. *Phys. Rev. A*, **60**, R1741–R1742.

[7442] Derevianko A., Savukov I.M., Johnson W.R. and Plante D.R. (1998). Negative-energy contributions to transition amplitudes in heliumlike ions. *Phys. Rev. A*, **58**, 4453–4461.

[7443] des Cloizeaux J. (1999). Quantum relativistic theory of an electron in terms of local observables. *Eur. Phys. J.*, **8**, 439–443.

[7444] d'Etat B., Briand J.P., Ban G., de Billy L., Briand P., Desclaux J.P., Melin G., Lamy T., Lamboley G., Richard P., Stockli M., Ali R., Renard N., Schneider D., Clark M., Beiersdorfer P. and Decaux V. (1993). X-ray spectroscopy of highly charged ions interacting with surfaces. *AIP Conf. Proc.*, **274**, 592–601.

[7445] Deubel D.V. and Frenking G. (1999). Are there metal oxides that prefer a [2+2] addition over a [3+2] addition to olefins? Theoretical study of the reaction mechanism of $LReO_3$ addition (L = O^-, Cl, Cp) to ethylene. *J. Am. Chem. Soc.*, **121**, 2021–2031.

[7446] Devine N.K. and Wallace S.J. (1995). Instant two-body equation in Breit frame. *Phys. Rev. C*, **51**, 3222–3231.

[7447] Devoto A., Feldman G. and Fulton T. (1993). Simple evaluation of the screened Lamb shift of heavily ionized high-Z atoms. *Phys. Rev. A*, **47**, 1503–1506.

[7448] de Wijs G.A. and de Groot R.A. (1999). Structure and electronic properties of amorphous WO_3. *Phys. Rev. B*, **60**, 16463–16474.

[7449] de Wijs G.A., Pastore G., Selloni A. and van der Lugt W. (1996). First-principles molecular dynamics simulation of liquid Mg_3Bi_2. *J. Phys. C*, **8**, 1879–1896.

[7450] Díaz-Megías S. and Seijo L. (1999). Wood-Boring ab initio model potential relativistic treatment of Ce and CeO. *Chem. Phys. Lett.*, **299**, 613–622.

[7451] Di Bella S., Gulino A., Lanza G., Fragalà I.L. and Marks T.J. (1993a). Photoelectron spectroscopy of f-element organometallic complexes. 10. Investigation of the electronic structure and geometry of bis(η^5-pentamethylcyclopentadienyl)phosphathoracyclobutane by relativistic ab initio, multipolar DV-Xα calculations and gas-phase UV photoelectron spectroscopy. *Organometallics*, **12**, 3326–3332.

[7452] Di Bella S., Gulino A., Lanza G., Fragalà I.L. and Marks T.J. (1993b). An investigation of the electronic structure of some tris(η^5-cyclopentadienyl)thorium(IV) and -uranium(IV) complexes by relativistic effective core potential ab initio calculations and gas-phase UV photoelectron spectroscopy. *J. Phys. Chem.*, **97**, 11673–11676.

[7453] Di Bella S., Gulino A., Lanza G., Fragalà I., Stern D. and Marks T.J. (1994). Photoelectron spectroscopy of f-element organometallic complexes. 12. Comparative investigation of the electronic structure of lanthanide bis(polymethylcyclopentadienyl)hydrocarbyl complexes by relativistic ab initio and DV-Xα calculations and gas-phase UV photoelectron spectroscopy. *Organometallics*, **13**, 3810–3815.

[7454] Di Bella S., Lanza G. and Fragalà I.L. (1993c). Equilibrium geometries and harmonic vibrational frequencies of lanthanum trihalides LaX_3 (X=F, Cl). A relativistic effective core potential ab initio MO study. *Chem. Phys. Lett.*, **214**, 598–602.

[7455] Di Bella S., Lanza G., Fragalà I.L. and Marks T.J. (1996). Electronic structure, molecular geometry, and bonding energetics in zerovalent yttrium and gadolinium bis(arene) sandwich complexes. A theoretical ab initio study. *Organometallics*, **15**, 3985–3989.

[7456] Diefenbach M., Brönstrup M., Aschi M., Schröder D. and Schwarz H. (1999). HCN synthesis from methane and ammonia: Mechanisms of Pt^+-mediated C-N coupling. *J. Am. Chem. Soc.*, **121**, 10614–10625.

[7457] Dieterle M., Harvey J.N., Heinemann C., Schwarz J., Schröder D. and Schwarz H. (1997). Equilibrium studies of weakly bound $Fe(L)^+$ complexes with L = Xe, CO_2, N_2 and CH_4. *Chem. Phys. Lett.*, **277**, 399–405.

[7458] Dietz K. and Pröbsting M. (1998). The structure of the QED vacuum and electron-positron pair production in super-intense, pulsed laser fields. *J. Phys. B*, **31**, L409–L414.

[7459] Diez R.P. (1998). A density functional test on the interaction of H_2 with Pt. *Chem. Phys. Lett.*, **287**, 542–548.

[7460] Dijkstra F., de Jong W.A. and Nieuwpoort W.C. (1995). Electron correlation effects on the f^6-manifold of the Eu^{3+} impurity in Ba_2GdNbO_6. *Int. J. Quantum Chem.*, **S29**, 610–613.

[7461] DiLabio G.A. and Christiansen P.A. (1997). Low-lying 0^+ states of bismuth hydride. *Chem. Phys. Lett.*, **277**, 473–477.

[7462] DiLabio G.A. and Christiansen P.A. (1998). Separability of spin-orbit and correlation energies for the sixth-row main group hydride ground states. *J. Chem. Phys.*, **108**, 7527–7533.

[7463] Dinov K. and Beck D.R. (1995a). Electron affinities of 6*p* electrons in Pr^-. *Phys. Rev. A*, **51**, 1680–1682.

[7464] Dinov K. and Beck D.R. (1995b). Electron affinities and hyperfine structure for U^- and U I obtained from relativistic configuration-interaction calculations. *Phys. Rev. A*, **52**, 2632–2637.

[7465] Dinov K.D. and Beck D.R. (1996). Electron affinity of Pa by 7*p* attachment and hyperfine structure constants for Pa^-. *Phys. Rev. A*, **53**, 4031–4035.

[7466] Dinov K., Beck D.R. and Datta D. (1994). Electron affinities of six bound states of Ce^- formed by attachment of 6p and 5d electrons to Ce. *Phys. Rev. A*, **50**, 1144–1148.

[7467] Dionne G.F. and Allen G.A. (1994). Molecular-orbital analysis of magneto-optical Bi-O-Fe hybrid excited states. *J. Appl. Phys.*, **75**, 6372–6374.

[7468] Dixon D.A., Feller D. and Sandrone G. (1999). Heats of formation of simple perfluorinated carbon compounds. *J. Phys. Chem. A*, **103**, 4744–4751.

[7469] Dmitriev Y.Y. and Fedorova T.A. (1998). Reducible two-photon exchange diagrams and reference state contribution to energy corrections within the modified adiabatic approach. *Phys. Lett. A*, **245**, 555–562.

[7470] Dmitriev Y.Y., Fedorova T.A. and Bogdanov D.M. (1998). A new approach to the direct renormalization of the bound electron self-energy. *Phys. Lett. A*, **241**, 84–89.

[7471] Dolg M. (1996a). Fully relativistic pseudopotentials for alkaline atoms: Dirac-Hartree-Fock and configuration interaction calculations of alkaline monohydrides. *Theor. Chim. Acta*, **93**, 141–156.

[7472] Dolg M. (1996b). On the accuracy of valence correlation energies in pseudopotential calculations. *J. Chem. Phys.*, **104**, 4061–4067.

[7473] Dolg M. (1996c). Accuracy of energy-adjusted quasirelativistic pseudopotentials: a calibration study of XH and X_2 (X=F,Cl,Br,I,At). *Mol. Phys.*, **88**, 1645–1655.

[7474] Dolg F.M. (1997). Quasirelativistische und relativistische energiekonsistente Pseudopotentiale für quantentheoretische Untersuchungen der Chemie schwerer Elemente. *Habilitationsschrift, Universität Stuttgart.*

[7475] Dolg M. (1998). Lanthanides and actinides. In *Encyclopedia of Computational Chemistry*, vol. 2, pp. 1478–1486, Wiley, Chichester and New York.

[7476] Dolg M. and Flad H.J. (1996). Ground state properties of Hg_2. I. A pseudopotential configuration interaction study. *J. Phys. Chem.*, **100**, 6147–6151.

[7477] Dolg M. and Flad H.J. (1997). Size-dependent properties of Hg_n clusters. *Mol. Phys.*, **91**, 815–825.

[7478] Dolg M. and Fulde P. (1998). Relativistic and electron-correlation effects in the ground states of lanthanocenes and actinocenes. *Chem. Eur. J.*, **4**, 200–204.

[7479] Dolg M., Fulde P., Stoll H., Preuss H., Chang A. and Pitzer R.M. (1995). Formally tetravalent cerium and thorium compounds: a configuration interaction study of cerocene $Ce(C_8H_8)_2$ and thorocene $Th(C_8H_8)_2$ using energy-adjusted quasirelativistic ab initio pseudopotentials. *Chem. Phys.*, **195**, 71–82.

[7480] Dolg M., Nicklass A. and Stoll H. (1993a). On the dipole moment of PbO. *J. Chem. Phys.*, **99**, 3614–3616.

[7481] Dolg M., Pyykkö P. and Runeberg N. (1996). Calculated structure and optical properties of $Tl_2Pt(CN)_4$. *Inorg. Chem.*, **35**, 7450–7451.

[7482] Dolg M. and Stoll H. (1996). Electronic structure calculations for molecules containing lanthanide atoms. In *Handbook on the Physics and Chemistry of Rare Earths*, (Edited by Gschneidner Jr. K.A. and Eyring L.), vol. 22, pp. 607–729, Elsevier, Amsterdam.

[7483] Dolg M., Stoll H. and Preuss H. (1993b). A combination of quasirelativistic pesudopotential and ligand field calculations for lanthanoid compounds. *Theor. Chim. Acta*, **85**, 441–450.

[7484] Dolg M., Stoll H., Preuss H. and Pitzer R.M. (1993c). Relativistic and correlation effects for element 105 (Hahnium, Ha). A comparative study of M and MO (M=Nb, Ta, Ha) using energy-adjusted ab initio pseudopotentials. *J. Phys. Chem.*, **97**, 5852–5859.

[7485] Doll K., Pyykkö P. and Stoll H. (1998). Closed-shell interaction in silver and gold chlorides. *J. Chem. Phys.*, **109**, 2339–2345.

[7486] Doll K. and Stoll H. (1998). Ground-state properties of heavy alkali halides. *Phys. Rev. B*, **57**, 4327–4331.

[7487] Dominguez-Adame F. and Rodriguez A. (1995). A one-dimensional relativistic screened Coulomb potential. *Phys. Lett. A*, **198**, 275–278.

[7488] Dong C.Z., Wang J.G. and Qu Y.Z. (1998). Channel dependence of dielectronic recombination rate coefficients of Fe^{25+} ion. *Chin. Phys. Lett.*, **15**, 263–265.

[7489] Dong S.H., Hou X.W. and Ma Z.Q. (1998). Relativistic Levinson theorem in two dimensions. *Phys. Rev. A*, **58**, 2160–2167.

[7490] Dong S.H., Hou X.W. and Ma Z.Q. (1999). Levinson's theorem for the Klein-Gordon equation in two dimensions. *Phys. Rev. A*, **59**, 995–1002.

[7491] Dong Z.C. and Corbett J.D. (1995). Unusual icosahedral cluster compounds: open-shell $Na_4A_6Tl_{13}$ (A = K, Rb, Cs) and the metallic Zintl phase $Na_3K_8Tl_{13}$ (How does chemistry work in solids?). *J. Am. Chem. Soc.*, **117**, 6447–6455.

[7492] Donnelly D., Bell K.L. and Hibbert A. (1997). Breit-Pauli R-matrix calculation of the 3p photoabsorption of singly ionized chromium. *J. Phys. B*, **30**, L285–L291.

[7493] Donnelly D., Bell K.L. and Hibbert A. (1998). Breit-Pauli R-matrix calculation of the 3p photoabsorption of singly ionized manganese. *J. Phys. B*, **31**, L971–L976.

[7494] Donnelly D., Hibbert A. and Bell K.L. (1999). Oscillator strengths for transitions in singly ionized copper. *Phys. Scripta*, **59**, 32–48.

[7495] Doublet M.L., Gallego-Planas N., Philipsen P.H.T., Brec R. and Jobic S. (1998). A new theoretical approach for the electrical properties of TiX_2 (X=S, Se, Te) phases with density functional calculations. *J. Chem. Phys.*, **108**, 649–658.

[7496] Dougherty R.W., Panigrahy S.N., Das T.P. and Andriessen J. (1993). Calculation of the hyperfine fields in the noble-metal atoms. *Phys. Rev. A*, **47**, 2710–2714.

[7497] Dragić A., Marić Z. and Vigier J.P. (1998). The energy spectrum of the hydrogen atom with magnetic spin-orbit and spin-spin interactions. *Phys. Lett. A*, **237**, 349–353.

[7498] Drake G.W., Dixon D.A., Sheehy J.A., Boatz J.A. and Christe K.O. (1998). Seven-coordinated pnicogens. Synthesis and characterization of the SbF_7^{2-} and BiF_7^{2-} dianions and a theoretical study of the AsF_7^{2-} dianion. *J. Am. Chem. Soc.*, **120**, 8392–8400.

[7499] Drake G.W.F. (1988). Theory of transitions and the electron weak interaction. In *The Spectrum of Atomic Hydrogen: Advances*, (Edited by Series G.), World Scientific, Singapore.

[7500] Drake G.W.F. (1993a). High precision calculations and QED effects for two- and three-electron atoms. *Phys. Scr.*, **T46**, 116–124.

[7501] Drake G.W.F. (1993b). QED effects in two- and three-electron ions. *AIP Conf. Proc.*, **275**, 3–19.

[7502] Drake G.W.F. (1993c). Energies and asymptotic analysis for helium Rydberg states. *Adv. At. Mol. Opt. Phys.*, **31**, 1–62.

[7503] Drake G.W.F. (1999). High precision theory of atomic helium. *Phys. Scripta*, **T83**, 83–92.

[7504] Drake G.W.F. and Goldman S.P. (1999). Bethe logarithms for Ps^-, H^-, and heliumlike atoms. *Can. J. Phys.*, **77**, 835–845.

[7505] Drake G.W.F., Khriplovich I.B., Milstein A.I. and Yelkhovsky A.S. (1993). Energy corrections of order $mc^2\alpha^6 \ln\alpha$ in helium. *Phys. Rev. A*, **48**, R15–R17.

[7506] Drchal V., Kudrnovsky J. and Weinberger P. (1994). Relativistic electronic structure of random alloys and their surfaces by linear band-structure methods. *Phys. Rev. B*, **50**, 7903–7914.

[7507] Dreizler R.M. (1993). Density functional theory of relativistic systems. *Phys. Scr.*, **T46**, 167–172.

[7508] Driess M., Janoschek R., Pritzkow H., Rell S. and Winkler U. (1995). Diphosphanyl- and diarsanyl-substituted carbene homologues: germanediyls, stannanediyls, and plumbanediyls with remarkable electronic structures. *Angew. Chem. Int. Ed. Engl.*, **34**, 1614–1616.

[7509] Dronskowski R. (1995). Synthesis, crystal structure, electronic structure, and magnetic properties of In_2ThBr_6. *Inorg. Chem.*, **34**, 4991–4995.

[7510] Dronskowski R., Kohler J. and Long J.R. (1993). SmF_6^{4-} conformation caused by $5d-2p$ orbital interactions. *J. Phys. Chem. Solids*, **54**, 801–808.

[7511] Droz-Vincent P. (1993). The quantum relativistic two-body problem in time-dependent external potentials. *Few-Body Systems.*, **14**, 97–115.

[7512] Du M.L. and Li Z.M. (1994). The contribution from spin-orbit coupling of ligand ions to g-factors in VCl_2 and VBr_2. *J. Phys: CM*, **6**, 6279–6285.

[7513] Dubernet M.L. and Hutson J.M. (1994). Atom-molecule van der Waals complexes containing open-shell atoms. I. General theory and bending levels. *J. Chem. Phys.*, **101**, 1939–1958.

[7514] Dudešek P., Benco L., Daul C. and Schwarz K. (1998). d-to-s bonding in GaN. *J. Phys. C*, **10**, 7155–7162.

[7515] Dufek P., Blaha P. and Schwarz K. (1995). Determination of the nuclear quadrupole moment of ^{57}Fe. *Phys. Rev. Lett.*, **75**, 3545–3548.

[7516] Dunford R.W. (1996). Parity nonconservation in high-Z heliumlike ions. *Phys. Rev. A*, **54**, 3820–3823.

[7517] Dürr D., Goldstein S., Münch-Berndl K. and Zanghi N. (1999). Hypersurface Bohm-Dirac models. *Phys. Rev. A*, **60**, 2729–2736.

[7518] Dushin R.B. and Nekhroshkov S.N. (1996). Relativistic crystal field for Md^{3+}, Er^{3+} and U^{3+} ions (configuration f^3, f^{11}) in octahedral complexes. *Radiokhimiya*, **38**, 210–217.

[7519] Düsterhöft C., Yang L., Heinemann D. and Kolb D. (1994). Solution of the one-electron Dirac equation for the heavy diatomic quasi-molecule $NiPb^{109+}$ by the finite element method. *Chem. Phys. Lett.*, **229**, 667–670.

[7520] Düsterhöft C., Heinemann D. and Kolb D. (1998). Dirac-Fock-Slater calculations for diatomic molecules with a finite element defect correction method (FEM-DKM). *Chem. Phys. Lett.*, **296**, 77–83.

[7521] Dyall K.G. (1993a). All-electron molecular Dirac-Hartree-Fock calculations: properties of the group IV monoxides GeO, SnO, and PbO. *J. Chem. Phys.*, **98**, 2191–2197.

[7522] Dyall K.G. (1993b). Relativistic effects on the bonding and properties of the hydrides of platinum. *J. Chem. Phys.*, **98**, 9678–9686.

[7523] Dyall K.G. (1994a). An exact separation of the spin-free and spin-dependent terms of the Dirac-Coulomb-Breit Hamiltonian. *J. Chem. Phys.*, **100**, 2118–2127.

[7524] Dyall K.G. (1994b). Second-order Møller-Plesset perturbation theory for molecular Dirac-Hartree-Fock wavefunctions. Theory for up to two open-shell electrons. *Chem. Phys. Lett.*, **224**, 186–194.

[7525] Dyall K.G. (1994c). Polyatomic molecular Dirac-Hartree-Fock calculations with Gaussian basis sets: Theory, implementation and applications. In *Relativistic and Electron Correlation Effects in Molecules and Solids*, (Edited by Malli G.L.), pp. 17–58, Plenum Press, New York.

[7526] Dyall K.G. (1997). Interfacing relativistic and non-relativistic methods. I. Normalized elimination of the small component in the modified Dirac equation. *J. Chem. Phys.*, **106**, 9618–9626.

[7527] Dyall K.G. (1998a). Interfacing relativistic and non-relativistic methods. II. Investigation of a low-order approximation. *J. Chem. Phys.*, **109**, 4201–4208.

[7528] Dyall K.G. (1998b). Relativistic and nonrelativistic finite nucleus optimized double zeta basis sets for the $4p$, $5p$ and $6p$ elements. *Theor. Chem. Accounts*, **99**, 366–371.

[7529] Dyall K.G. (1999). Bonding and bending in the actinyls. *Mol. Phys.*, **96**, 511–518.

[7530] Dyall K.G. and Enevoldsen T. (1999). Interfacing relativistic and nonrelativistic methods. III. Atomic 4-spinor expansions and integral approximations. *J. Chem. Phys.*, **111**, 10000–10007.

[7531] Dyall K.G. and Faegri Jr K. (1993). Finite nucleus effects on relativistic energy corrections. *Chem. Phys. Lett.*, **201**, 27–32.

[7532] Dyall K.G. and Faegri Jr K. (1996). Optimization of Gaussian basis sets for Dirac-Hartree-Fock calculations. *Theor. Chim. Acta*, **94**, 39–51.

[7533] Dyall K.G. and Partridge H. (1993). Relativistic corrections to properties of the alkali fluorides. *Chem. Phys. Lett.*, **206**, 565–567.

[7534] Dyall K.G. and van Lenthe E. (1999). Relativistic regular approximations revisited: An infinite-order relativistic approximation. *J. Chem. Phys.*, **111**, 1366–1372.

[7535] Dyke J.M., Haggerston D., Wright A.E., van Lenthe E. and Snijders J.G. (1997). A study of the transition metal tetrafluorides (TiF_4, ZrF_4, HfF_4) using high temperature ultraviolet photoelectron spectroscopy. *J. El. Sp. Rel. Phen.*, **85**, 23–33.

[7536] Dzuba V.A., Flambaum V.V., Gribakin G.F. and Harabati C. (1999a). Calculation of the positron bound state with the copper atom. *Phys. Rev. A*, **60**, 3641–3647.

[7537] Dzuba V.A., Flambaum V.V., Gribakin G.F. and King W.A. (1996a). Many-body calculations of positron scattering and annihilation from noble-gas atoms. *J. Phys. B*, **29**, 3151–3175.

[7538] Dzuba V.A., Flambaum V.V., King W.A., Miller B.N. and Sushkov O.P. (1993). Interaction between slow positrons and atoms. *Phys. Scr.*, **T46**, 248–251.

[7539] Dzuba V.A., Flambaum V.V. and Kozlov M.G. (1994). Calculation of the weak interactions in dysprosium. *Phys. Rev. A*, **50**, 3812–3817.

[7540] Dzuba V.A., Flambaum V.V. and Kozlov M.G. (1996b). Calculations of energy levels for atoms with several valence electrons. *Pis'ma v ZhETF*, **63**, 844–848.

[7541] Dzuba V.A., Flambaum V.V. and Kozlov M.G. (1996c). Combination of the many body perturbation theory with the configuration interaction method. *Phys. Rev. A*, **54**, 3948–3959.

[7542] Dzuba V.A., Flambaum V.V. and Kozlov M.G. (1998). Many body calculations of correlation interaction between core and valence atomic electrons. In *Advances in Quantum Many-Body Theories, Vol. 9*, (Edited by Neilson D. and Bishop R.F.), pp. 167–171, 9th Int. Conf. on Recent Progress in Many-Body Theories, Sydney 1997, World Scientific, Singapore.

[7543] Dzuba V.A., Flambaum V.V. and Sushkov O.P. (1995). Calculation of energy levels; E1 transition amplitudes and parity violation in francium. *Phys. Rev. A*, **51**, 3454–3461.

[7544] Dzuba V.A., Flambaum V.V. and Sushkov O.P. (1997a). Polarizabilities and parity nonconservation in the Cs atom and limits on the deviation from the standard electroweak model. *Phys. Rev. A*, **56**, 4357–4360.

[7545] Dzuba V.A., Flambaum V.V. and Webb J.K. (1999b). Calculations of the relativistic effects in many-electron atoms and space-time variation of fundamental constants. *Phys. Rev. A*, **59**, 230–237.

[7546] Dzuba V.A., Flambaum V.V. and Webb J.K. (1999c). Space-time variation of physical constants and relativistic corrections in atoms. *Phys. Rev. Lett.*, **82**, 888–891.

[7547] Dzuba V.A. and Gribakin G.F. (1994). Correlation-potential method for negative ions and electrons cattering. *Phys. Rev. A*, **49**, 2483–2492.

[7548] Dzuba V.A. and Gribakin G.F. (1997). Fine structure of Ca^-, Sr^-, Ba^-, and Ra^- from the many-body theory calculation. *Phys. Rev. A*, **55**, 2443–2446.

[7549] Dzuba V.A. and Gribakin G.F. (1998). Yb^- $6p_{1/2}$–low-lying shape resonance rather than a bound state. *J. Phys. B*, **31**, L483–L487.

[7550] Dzuba V.A. and Johnson W.R. (1998). Calculation of the energy levels of barium using *B* splines and a combined configuration-interaction and many-body-perturbation-theory method. *Phys. Rev. A*, **57**, 2459–2465.

[7551] Eberhart M.E., Clougherty D.P. and MacLaren J.M. (1993). A theoretical investigation of the mechanisms of fracture in metals and alloys. *J. Am. Chem. Soc.*, **115**, 5762–5767.

[7552] Ebert H. (1996). Magneto-optical effects in transition metal systems. *Rep. Prog. Phys.*, **59**, 1665–1735.

[7553] Ebert H. and Akai H. (1993). Consequences of relativity for the hyperfine interactions - with applications to transition metals. *Hyperfine Interactions*, **78**, 361–375.

[7554] Ebert H., Akai H., Maruyama H., Koizumi A., Yamazaki H. and Schutz G. (1993). Theoretical and experimental investigation of the magnetic X-ray dichroism in diluted and concentrated transition metal alloys. *Int. J. Modern Phys. B*, **7**, 750–755.

[7555] Ebert H., Battocletti M., Deng M., Freyer H. and Voitländer J. (1999a). Fully relativistic description of static magnetic hyperfine interaction in magnetic and nonmagnetic solids. *J. Comp. Chem.*, **20**, 1246–1253.

[7556] Ebert H., Battocletti M. and Gross E.K.U. (1997a). Current density functional theory of spontaneously magnetised solids. *Europhys Lett.*, **40**, 545–550.

[7557] Ebert H., Freyer H. and Deng M. (1997b). Manipulation of the spin-orbit coupling using the Dirac equation for spin-dependent potentials. II. *Phys. Rev. B*, **56**, 9454–9460.

[7558] Ebert H., Freyer H., Vernes A. and Guo G.Y. (1996). Manipulation of the spin-orbit coupling using the Dirac equation for spin-dependent potentials. *Phys. Rev. A*, **53**, 7721–7726.

[7559] Ebert H. and Guo G.Y. (1994). Calculation of magnetic X-ray dichroism (MXD) spectra using the spin polarized relativistic linear-muffin-tin-orbital method of band structure calculation. *Solid State Comm.*, **91**, 85–88.

[7560] Ebert H., Popescu V. and Ahlers D. (1999b). Fully relativistic theory for magnetic EXAFS: Formalism and applications. *J. Comp. Chem.*, **60**, 7156–7165.

[7561] Ebert H. and Schütz G. (1996). Spin-Orbit-Influenced Spectroscopies of Magnetic Solids (Ed.). In *Lecture Notes in Physics*, vol. 466, Springer, Berlin, 287 p.

[7562] Ebert H., Vernes A. and Banhart J. (1997c). Relativistic bandstructure of disordered magnetic alloys. *Solid State Comm.*, **104**, 243–247.

[7563] Ebert H., Vernes A. and Banhart J. (1999d). The influence of spin-orbit coupling and a current dependent potential on the residual resistivity of disordered magnetic alloys. *Solid State Comm.*, **113**, 103–107.

[7564] Eckert J., Jensen C.M., Jones G., Colt E. and Eisenstein O. (1993). An extremely low barrier to rotation of dihydrogen in the complex $IrClH_2(\eta^2\text{-}H_2)(P^iPr_3)_2$. *J. Am. Chem. Soc.*, **115**, 11056–11057.

[7565] Edmonds Jr. J.D. (1997). *Relativistic Reality: A Modern View*. World Scientific, Singapore.

[7566] Edmonds Jr J.D. (1999). Dirac's equation in half of his algebra. *Eur. J. Phys.*, **20**, 461–467.

[7567] Edvardsson D., Lunell S., Rakowitz F., Marian C.M. and Karlsson L. (1998). Calculation of predissociation rates in O_2^{2+} by ab initio MRD-CI methods. *Chem. Phys.*, **229**, 203–216.

[7568] Edvardsson S. and Klintenberg M. (1998b). Role of the electrostatic model in calculating rare-earth crystal-field parameters. *J. Alloys Compounds*, **275-277**, 230–233.

[7569] Ehara M. and Nakatsuji H. (1995). Collision induced absorption spectra and line broadening of CsRg system (Rg = Xe, Kr, Ar, Ne) studied by the symmetry adapted cluster-configuration interaction (SAC-CI) method. *J. Chem. Phys.*, **102**, 6822–6830.

[7570] Ehara M. and Nakatsuji H. (1998). Ionization spectra of $XONO_2$ (X = F,Cl,Br,I) studied by the SAC-CI method. *Chem. Phys.*, **226**, 113–123.

[7571] Ehlers A.W., Böhme M., Dapprich S., Gobbi A., Höllwarth A., Jonas V., Köhler K.F., Stegmann R., Veldkamp A. and Frenking G. (1993). A set of f-polarization functions for pseudo-potential basis sets of the transition metals Sc-Cu, Y-Ag and La-Au. *Chem. Phys. Lett.*, **208**, 111–114.

[7572] Ehlers A.W., Dapprich S., Vyboishchikov S.F. and Frenking G. (1995). Structure and bonding of the transition-metal carbonyl complexes $M(CO)_5L$ (M=Cr,Mo,W) and $M(CO)_3L$ (M=Ni,Pd,Pt); L=CO,SiO, CS, N_2, NO^+, CN^-, NC^-, HCCH, CCH_2, CH_2, CF_2, H_2). *Organometallics*, **15**, 105–117.

[7573] Ehlers A.W. and Frenking G. (1993). Theoretical studies of the M-CO bond lengths and first dissociation energies of the transition metal hexacarbonyls $Cr(CO)_6$, $Mo(CO)_6$ and $W(CO)_6$. *J. Chem. Soc. Chem. Commun.*, pp. 1709–1711.

[7574] Ehlers A.W. and Frenking G. (1994). Structure and bond energies of the transition metal hexacarbonyls $M(CO)_6$ (M = Cr, Mo, W). A theoretical study. *J. Am. Chem. Soc.*, **116**, 1514–1520.

[7575] Ehlers A.W. and Frenking G. (1995). Structures and bond energies of the transition-metal carbonyls $M(CO)_5$ (M = Fe, Ru, Os) and $M(CO)_4$ (M = Ni, Pd, Pt). *Organometallics*, **14**, 423–426.

[7576] Ehlers A.W., Frenking G. and Baerends E.J. (1997a). Structure and bonding of the noble gas-metal carbonyl complexes $M(CO)_5$-Ng (M=Cr,Mo,W and Ng=Ar,Kr,Xe). *Organometallics*, **16**, 4896–4902.

[7577] Ehlers A.W., Lammertsma K. and Baerends E.J. (1998). Phosphinidene complexes $M(CO)_5$-PR: A density functional study on structures and electronic states. *Organometallics*, **17**, 2738–2742.

[7578] Ehlers A.W., Ruiz-Morales Y., Baerends E.J. and Ziegler T. (1997b). Dissociation energies,vibrational frequencies, and ^{13}C NMR chemical shifts of the 18-electron species $[M(CO)_6]^n$ (M=Hf-Ir, Mo, Tc, Ru, Cr, Mn, Fe). A density functional study. *Inorg. Chem.*, **36**, 5031–5036.

[7579] Ehlotzky F., Jaroń A. and Kamiński J.Z. (1997). Electron–atom collisions in a laser field. *Phys. Rep.*, **297**, 63–153.

[7580] Eibler R., Reschbaumer H., Temnitschka C., Podloucky R. and Freeman A.J. (1993). Ab-initio calculation of the electronic structure and energetics of the unreconstructed Au(001) surface. *Surface Sci.*, **280**, 398–414.

[7581] Eichkorn K., Weigend F., Treutler O. and Ahlrichs R. (1997). Auxiliary basis sets for main row atoms and transition metals and their use to approximate Coulomb potentials. *Theor. Chem. Acc.*, **97**, 119–124.

[7582] Eichler J. (1995). Charge transfer from the negative-energy continuum: alternative mechanism for pair production in relativistic atom collisions. *Phys. Rev. Lett.*, **75**, 3653–3656.

[7583] Eichler J. and Belkacem A. (1996). Gauge transformations for coupled-channel calculations applied to pair production in relativistic atomic collisions. *Phys. Rev. A*, **54**, 5427–5430.

[7584] Eichler J., Ichihara A. and Shirai T. (1998). Alignment caused by photoionization and in radiative electron capture into excited states of hydrogenic high-Z ions. *Phys. Rev. A*, **58**, 2128–2135.

[7585] Eichler J. and Meyerhof W.E. (1995). *Relativistic Atomic Collisions*. Acad. Press, New York, 409 pp.

[7586] Eides M.I. (1996). Weak-interaction contributions to hyperfine splitting and Lamb shift. *Phys. Rev. A*, **53**, 2953–2957.

[7587] Eides M.I. and Grotch H. (1995a). Corrections of order α^6 to S levels of two-body systems. *Phys. Rev. A*, **52**, 1757–1760.

[7588] Eides M.I. and Grotch H. (1995b). Corrections of order $\alpha^3(Z\alpha)^4$ and $\alpha^2(Z\alpha)^6$ to the Lamb shift. *Phys. Rev. A*, **52**, 3360–3361.

[7589] Eides M.I. and Grotch H. (1997a). Recoil corrections of order $(Z\alpha)^6(m/M)m$ to the hydrogen energy levels recalculated. *Phys. Rev. A*, **55**, 3351–3360.

[7590] Eides M.I. and Grotch H. (1997b). Radiative correction to the nuclear-size effect and hydrogen-deuterium isotopic shift. *Phys. Rev. A*, **56**, 2507–2509.

[7591] Eides M.I., Grotch H. and Pebler P. (1994). $\alpha^2(Z\alpha)^5m$ contribution to the hydrogen Lamb shift from virtual light-by-light scattering. *Phys. Rev. A*, **50**, 144–170.

[7592] Eides M.I., Grotch H. and Shelyuto V.A. (1997). Analytic contribution of order $\alpha^2(Z\alpha)^5m$ to the Lamb shift. *Phys. Rev. A*, **55**, 2447–2449.

[7593] Eides M.I. and Shelyuto V.A. (1995). Corrections of order $\alpha^2(Z\alpha)^5$ to the hyperfine splitting and the Lamb shift. *Phys. Rev. A*, **52**, 954–961.

[7594] Eissner W. (1991b). Superstructure - an atomic code. *J. Physique IV*, **1(C1)**, 3–13.

[7595] Elander N. and Yarevsky E. (1997). Finite-element calculations of the antiprotonic helium atom including relativistic and QED corrections. Erratum: ibid., 57, 2256. *Phys. Rev. A*, **56**, 1855–1864.

[7596] El-Bahraoui J., Dobado J.A. and Molina J.M. (1999). Metal-metal closed-shell interaction in M_2X_2 (M=Ag,Cu; X=Cl,Br,I) and related compounds $[Ag_2Br_2](PH_3)_3$ and $[Cu_2Cl_2](PH_3)_2$: an RHF, MP2 and DFT study. *J. Mol. Str. (Theochem)*, **493**, 249–257.

[7597] El-Bahraoui J., Molina Molina J. and Portal Olea D. (1998). Theoretical studies of Ag-Ag closed-shell interaction in the silver(I) dimer bis-μ-(5,7-dimethyl[1,2,4]triazolo[1,5-α]. *J. Phys. Chem. A*, **102**, 2443–2448.

[7598] Eliav E., Ishikawa Y., Pyykkö P. and Kaldor U. (1997). Electron affinities of boron, aluminum, gallium, indium, and thallium. *Phys. Rev. A*, **56**, 4532–4536.

[7599] Eliav E. and Kaldor U. (1996a). The relativistic four-component coupled cluster method for molecules: spectroscopic constants of SnH_4. *Chem. Phys. Lett.*, **248**, 405–408.

[7600] Eliav E., Kaldor U. and Hess B.A. (1998a). The relativistic Fock-space coupled-cluster method for molecules: CdH and its ions. *J. Chem. Phys.*, **108**, 3409–3415.

[7601] Eliav E., Kaldor U. and Ishikawa Y. (1994a). Open-shell relativistic coupled-cluster method with Dirac-Fock-Breit wave functions: Energies of the gold atom and its cation. *Phys. Rev. A*, **49**, 1724–1729.

[7602] Eliav E., Kaldor U. and Ishikawa Y. (1994b). Relativistic coupled cluster method based on Dirac-Coulomb-Breit wavefunctions. Ground state energies of atoms with two to five electrons. *Chem. Phys. Lett.*, **222**, 82–87.

[7603] Eliav E., Kaldor U. and Ishikawa Y. (1994c). Ionization potentials and excitation energies of the alkali-metal atoms by the relativistic coupled-cluster method. *Phys. Rev. A*, **50**, 1121–1128.

[7604] Eliav E., Kaldor U. and Ishikawa Y. (1994d). Relativistic coupled cluster theory based on the no-pair Dirac-Coulomb-Breit Hamiltonian: Relativistic pair correlation energies of the Xe atom. *Int. J. Quantum Chem.*, **S28**, 205–214.

[7605] Eliav E., Kaldor U. and Ishikawa Y. (1995a). Ground state electron configuration of rutherfordium: Role of dynamic correlation. *Phys. Rev. Lett.*, **74**, 1079–1082.

[7606] Eliav E., Kaldor U. and Ishikawa Y. (1995b). Relativistic coupled-cluster method: Intrashell excitations in the f^2 shells of Pr^{+3} and U^{+4}. *Phys. Rev. A*, **51**, 225–230.

[7607] Eliav E., Kaldor U. and Ishikawa Y. (1995c). Transition energies of ytterbium, lutetium, and lawrencium by the relativistic coupled-cluster method. *Phys. Rev. A*, **52**, 291–296.

[7608] Eliav E., Kaldor U. and Ishikawa Y. (1995d). Transition energies of mercury and ekamercury (element 112) by the relativistic coupled-cluster method. *Phys. Rev. A*, **52**, 2765–2769.

[7609] Eliav E., Kaldor U. and Ishikawa Y. (1996a). Transition energies of barium and radium by the relativistic coupled-cluster method. *Phys. Rev. A*, **53**, 3050–3056.

[7610] Eliav E., Kaldor U. and Ishikawa Y. (1998b). The relativistic coupled-cluster method: transition energies of bismuth and eka-bismuth. *Mol. Phys.*, **94**, 181–187.

[7611] Eliav E., Kaldor U., Ishikawa Y. and Pyykkö P. (1996b). Element 118: The first rare gas with an electron affinity. *Phys. Rev. Lett.*, **77**, 5350–5352.

[7612] Eliav E., Kaldor U., Ishikawa Y., Seth M. and Pyykkö P. (1996c). Calculated energy levels of thallium and eka-thallium (element 113). *Phys. Rev. A*, **53**, 3926–3933.

[7613] Eliav E., Kaldor U., Schwerdtfeger P., Hess B.A. and Ishikawa Y. (1994e). Ground state electron configuration of element 111. *Phys. Rev. Lett.*, **73**, 3203–3206.

[7614] Eliav E., Shmulyian S., Kaldor U. and Ishikawa Y. (1998c). Transition energies of lanthanum, actinium, and eka-actinium (element 121). *Phys. Rev. A*, **58**, 3954–3958.

[7615] Ellingsen K., Saue T., Aksela H. and Gropen O. (1997). Cl 2p-photoelectron spectrum of HCl studied by fully relativistic, self-consistent-field, and configuration-interaction calculations. *Phys. Rev. A*, **55**, 2743–2747.

[7616] Ellis D.E. and Guenzburger D. (1999). The discrete variational method in density functional theory and its applications to large molecules and solid-state systems. *Adv. Quantum Chem.*, **34**, 51–141.

[7617] El Messaoudi A., Bouchet J. and Joulakian B. (1997). The Breit interaction in relativistic K-shell ionization of heavy atoms by electron impact. *J. Phys. B*, **30**, 4623–4631.

[7618] Endo K., Yamamoto K. and Okada H. (1995). Heavy atom effect on 14 group nuclear shielding constant of SiX_4 and $CH_{4-n}X_n$ (X = Cl, Br, I; n=1, 2, 3, 4). *Bull. Chem. Soc. Jpn.*, **68**, 3341–3345.

[7619] Engel E. (1995). Density functional theory of field theoretical systems. *Int. J. Quantum Chem.*, **56**, 217–223.

[7620] Engel E. and Dreizler R.M. (1996). Relativistic density functional theory. *Topics in Current Chem.*, **181**, 1–80.

[7621] Engel E. and Dreizler R.M. (1999). From explicit to implicit density functionals. *J. Comp. Chem.*, **20**, 31–50.

[7622] Engel E., Facco Bonetti A., Keller S., Andrejkovics I. and Dreizler R.M. (1998). Relativistic optimized-potential method: Exact transverse exchange and Møller-Plesset-based correlation potential. *Phys. Rev. A*, **58**, 964–992.

[7623] Engel E., Keller S. and Dreizler R.M. (1996). Generalized gradient approximation for the relativistic exchange-only energy functional. *Phys. Rev. A*, **53**, 1367–1374.

[7624] Engel E., Keller S. and Dreizler R.M. (1997). Relativistic corrections to the exchange-correlation energy functional. In *Electronic Density Functional Theory, Recent Progress and New Directions*, (Edited by Dobson J.F., Vignale G. and Das M.P.), Plenum, New York.

[7625] Engel E., Keller S., Facco Bonetti A., Müller H. and Dreizler R.M. (1995). Local and nonlocal relativistic exchange-correlation energy functionals: Comparison to relativistic optimized-potential-model results. *Phys. Rev. A*, **52**, 2750–2764.

[7626] Entralgo E., Cabrera B. and Portieles J. (1998). Towards the problem of two point particles with spin. *Nuovo Cim. A*, **111A**, 1185–1195.

[7627] Eriksson L.A., Pettersson L.G.M., Siegbahn P.E.M. and Wahlgren U. (1995). On the accuracy of gradient corrected density functional methods for transition metal complexes. *J. Chem. Phys.*, **102**, 872–878.

[7628] Erkoç Ş., Baştuğ T., Hirata M. and Tachimori S. (1999). Energetics and structural stability of lanthanum microclusters. *Chem. Phys. Lett.*, **314**, 203–209.

[7629] Ermler W.C. and Marino M.M. (1994). Electronic structure of molecules, clusters, and surfaces using ab initio relativistic effective core and core/valence polarization potentials. In *Relativistic and Electron Correlation Effects in Molecules and Solids*, (Edited by Malli G.L.), pp. 71–103, Plenum Press, New York.

[7630] Ermler W.C. and Marino M.M. (1996). Relativistic pseudopotentials and nonlocal effects. In *New Methods in Quantum Theory*, (Edited by Tsipis C.A.), pp. 415–423, Kluwer Acad. Publ., Netherlands.

[7631] Ermolaev A.M. (1998). Atomic states in the relativistic high-frequency approximation of Kristic-Mittleman. *J. Phys. B*, **31**, L65–L74.

[7632] Erokhin [= Yerokhin] V.A. and Shabaev V.M. (1996). Contribution of screened self-energy diagrams to the ground-state Lamb shift of two-electron multicharged ions (in Russian). *Zh. Eksp. Teor. Fiz.*, **110**, 74–94.

[7633] Escalante S., Vargas R. and Vela A. (1999). Structure and energetics of group 14 (IV-A) halides: A comparative density functional-pseudopotential study. *J. Phys. Chem. A*, **103**, 5590–5601.

[7634] Eschrig H. and Servedio V.D.P. (1999). Relativistic density functional approach to open shells. *J. Comp. Chem.*, **20**, 23–30.

[7635] Esquivel R.O., Vivier-Bunge A. and Smith V.H. (1998). Accurate determination of the Fermi-contact interaction in atomic lithium. *J. Mol. Str. (Theochem)*, **433**, 43–50.

[7636] Essén H. (1996). Darwin magnetic interaction energy and its macroscopic consequences. *Phys. Rev. E*, **53**, 5228–5239.

[7637] Esteban M.J. and Séré E. (1999). Solutions of the Dirac-Fock equations for atoms and molecules. *Comm. Math. Phys.*, **203**, 499–530.

[7638] Estiú G.L., Rösch N. and Zerner M.C. (1995). Ground state characteristics and optical spectra of Ce and Eu bis(octaethylporphyrinate) double deckers. *J. Phys. Chem.*, **99**, 13819–13829.

[7639] Evangelou S.N. (1995). Anderson transition, scaling, and level statistics in the presence of spin orbit coupling. *Phys. Rev. Lett.*, **75**, 2550–2553.

[7640] Evans W.D., Lewis R.T., Siedentop H. and Solovej J.P. (1996). Counting eigenvalues using coherent states with an application to Dirac and Schrödinger operators in the semiclassical limit. *Ark. Mat.*, **34**, 265–283.

[7641] Evans W.D., Perry P. and Siedentop H. (1996). The spectrum of relativistic one-electron atoms according to Bethe and Salpeter. *Comm. Math. Phys.*, **178**, 733–746.

[7642] Faas S., Snijders J.G., van Lenthe J.H., van Lenthe E. and Baerends E.J. (1995). The ZORA formalism applied to the Dirac-Fock equation. *Chem. Phys. Lett.*, **246**, 632–640.

[7643] Facco Bonetti A., Engel E., Dreizler R.M., Andrejkovics I. and Müller H. (1998). Relativistic exchange-correlation energy functional: Gauge dependence of the no-pair correlation energy. *Phys. Rev. A*, **58**, 993–1000.

[7644] Faegri K., Martinsen K.G., Strand T.G. and Volden H.V. (1993). The molecular structure of molybdenum pentachloride studied by ab initio molecular orbital calculations and gas electron diffraction. *Acta Chem. Scand.*, **47**, 547–553.

[7645] Fagerli H., Schimmelpfennig B., Gropen O. and Wahlgren U. (1998). Spin-orbit effects in the $PtCH_2^+$ ion. *J. Mol. Str. (Theochem)*, **451**, 227–235.

[7646] Fahmi A. and van Santen R.A. (1996). Density functional study of ethylene adsorption on palladium clusters. *J. Phys. Chem.*, **100**, 5676–5680.

[7647] Faisal F.H.M. and Radozycki T. (1993). Three-dimensional relativistic model of a bound particle in an intense laser pulse II. *Phys. Rev. A*, **48**, 554–557.

[7648] Falomir H., Gamboa Saraví R.E., Muschietti M.A., Santangelo E.M. and Solomin J.E. (1996). Determinants of Dirac operators with local boundary conditions. *J. Math. Phys.*, **37**, 5805–5819.

[7649] Falsaperla P., Fonte G. and Chen J.Z. (1997). Two methods for solving the Dirac equation without variational collapse. *Phys. Rev. A*, **56**, 1240–1248.

[7650] Fan M.F., Lin Z.Y., McGrady J.E. and Mingos D.M.P. (1996). Novel intermolecular C-H$\cdots\pi$ interactions: an ab initio and density functional theory study. *J. Chem. Soc., Perkin Trans.*, pp. 563–568.

[7651] Fanchi J.R. (1993). *Parametrized Relativistic Quantum Theory, Fundamental Theories of Physics, Vol. 56*. Kluwer, Dordrecht.

[7652] Fang B.S., Lo W.S. and Chen H.H. (1993). Spin-orbit effects on electronic structures near E_F for Nb(001). *Phys. Rev. B*, **47**, 10671–10674.

[7653] Fang C.M., Wiegers G.A., Haas C. and de Groot R.A. (1997). Electronic structures of ReS_2, $ReSe_2$ and TcS_2 in the real and the hypothetical undistorted structures. *J. Phys. CM*, **9**, 4411–4424.

[7654] Fang C.M., Wiegers G.A., Meetsma A., de Groot R.A. and Haas C. (1996). Crystal structure and band structure calculations of $Pb_{1/3}TaS_2$ and $Sn_{1/3}NbS_2$. *Physica. B*, **226**, 259–267.

[7655] Fang T.H., McKee M.L. and Worley S.D. (1994). A theoretical study of the interaction of dinitrogen with rhodium. *Can. J. Chem.*, **72**, 519–522.

[7656] Fantacci S., Maseras F. and Lledos A. (1999). Why does {p-Bu^t-calix[4]-$(OMe)_2(O)_2ZrCl_2$} distort away from C_{2v} symmetry? *Chem. Phys. Lett.*, **315**, 145–149.

[7657] Fantacci S., Re N., Rosi M., Sgamellotti A., Guest M.F., Sherwood P. and Floriani C. (1997). Theoretical study of the metathesis-like reaction between ditungsten hexaalkoxides and alkynes. *J. Chem. Soc., Dalton Trans.*, pp. 3845–3852.

[7658] Fantucci P., Polezzo S., Sironi M. and Bencini A. (1995). The nature of the platinum-phosphine bond. An ab initio Hartree-Fock and density functional study. *J. Chem. Soc., Dalton Trans.*, pp. 4121–4126.

[7659] Fantucci P., Lolli S., Pizzotti M. and Ugo R. (1998). An ab initio MO-LCAO investigation of the electronic structure of organic hydroperoxides and platinum(II) hydroperoxo complexes: a contribution to the knowledge of the mechanism of olefin epoxidation. *Inorg. Chim. Acta*, **270**, 479–487.

[7660] Fast L., Wills J.M., Johansson B. and Eriksson O. (1995). Elastic constants of hexagonal transition metals: theory. *Phys. Rev. B*, **51**, 17431–17438.

[7661] Faulkner J.S. (1994). The scattering matrices for a non-spherical scalar-relativistic potential. *Solid State Comm.*, **90**, 791–793.

[7662] Faustov R.N., Karimkhodzhaev A. and Martynenko A.P. (1999). Evaluation of hadronic vacuum polarization contribution to muonium hyperfine splitting. *Phys. Rev. A*, **59**, 2498–2499.

[7663] Fearing H.W., Poulis G.I. and Scherer S. (1994). Effective hamiltonians with relativistic corrections: The Foldy-Wouthuysen transformation versus the direct Pauli reduction. *Nucl. Phys.*, **A570**, 657–685.

[7664] Fedorov V.V., Rodionov A.A., Band I.M. and Trzhaskovskaya M.B. (1995). Hyperfine structure calculations for inner atomic levels. *J. Phys. B*, **28**, 1963–1973.

[7665] Fehrenbach G.M. (1999). *Quasiteilchen und relativistische Korrekturen zur Bandstruktur von Metallen und Halbleitern*. Fakultät für Physik der Ludwig Maximilians Universität, München, habilitationsschrift, 179 S.

[7666] Fehrenbach G.M. and Bross H. (1999). Scalar-relativistic spline augmented plane-wave method using a Douglas Kroll transformation in coordinate representation. *Eur. Phys. J. B*, **9**, 37–48.

[7667] Fehrenbach G.M. and Schmidt G. (1997). Including the relativistic kinetic energy in spline-augmented plane-wave band calculation. *Phys. Rev. B*, **55**, 6666–6669.

[7668] Feiock F.D. and Johnson W.R. (1968). Relativistic evaluation of internal diamagnetic fields for atoms and ions. *Phys. Rev. Lett.*, **21**, 785–786.

[7669] Feldman G. and Fulton T. (1988). Radiative corrections for many-electron atoms in perturbation theory. *Ann. Phys. (NY)*, **184**, 231–253.

[7670] Feldman G. and Fulton T. (1995). Radiative corections in the many-particle and Dirac-Fock approximations: two branches from a common origin. *Ann. Phys. (NY)*, **240**, 315–350.

[7671] Feldmann D.M., Pelzl P.J. and King F.W. (1998). Resolution of some mathematical problems arising in the relativistic treatment of the S states of three-electron systems. *J. Math. Phys.*, **39**, 6262–6275.

[7672] Fell R.N. (1993). Single-transverse-photon contributions of order $\alpha^6 \ln(\alpha)$ to the energy levels of positronium. *Phys. Rev. A*, **48**, 2634–2667.

[7673] Feller D., Glendening E.D. and de Jong W.A. (1999). Structures and binding enthalpies of $M^+(H_2O)_n$ clusters, M=Cu, Ag, Au. *J. Chem. Phys.*, **110**, 1475–1491.

[7674] Feng B., Shen J., Grill V., Evans C. and Cooks R.G. (1998). Cleavage of C-C and C-F bonds by $Xe^{+\cdot}$ and I^+ ions in reactions at a fluorinated self-assembled monolayer surface: Collision energy dependence and mechanisms. *J. Am. Chem. Soc.*, **120**, 8189–8198.

[7675] Feng P.Y. and Balasubramanian K. (1996a). Electronic states of Ga_2P_2. *Chem. Phys. Lett.*, **258**, 387–392.

[7676] Feng P.Y. and Balasubramanian K. (1996b). Geometries and energy separations of the electronic states of In_2P_2. *Chem. Phys. Lett.*, **264**, 449–453.

[7677] Feng P.Y. and Balasubramanian K. (1997a). Electronic states and potential energy curves of Ga_2P, GaP_2, and their ions. *Chem. Phys. Lett.*, **265**, 41–47.

[7678] Feng P.Y. and Balasubramanian K. (1997b). Nearly-degenerate electronic states for Ga_3P_2 and Ga_2P_3. *Chem. Phys. Lett.*, **265**, 547–552.

[7679] Feng P.Y. and Balasubramanian K. (1998a). Electronic states for the In_3P_2 and In_2P_3 clusters. *Chem. Phys. Lett.*, **283**, 167–173.

[7680] Feng P.Y. and Balasubramanian K. (1998b). Electronic states and potential energy curves of In_2P, InP_2, In_2P^+, and InP_2^+. *Chem. Phys. Lett.*, **284**, 313–319.

[7681] Feng P.Y. and Balasubramanian K. (1998c). Electronic states of Ga_3P and GaP_3 clusters. *Chem. Phys. Lett.*, **288**, 1–6.

[7682] Feng P.Y. and Balasubramanian K. (1998d). Electronic states of In_3Sb_2, In_2Sb_3, and their positive ions. *J. Phys. Chem. A*, **102**, 9047–9055.

[7683] Feng P.Y., Liao M.Z. and Balasubramanian K. (1998). Low-lying electronic states of In_2As_2, $In_2As_2^+$ and $In_2As_2^-$. *Chem. Phys. Lett.*, **296**, 283–291.

[7684] Feng X., McEachran R.P. and Stauffer A.D. (1998). Relativistic close-coupling calculations of positron scattering from the heavy alkalis. *Nucl. Instr. Meth. Phys. Res. B*, **143**, 27–31.

[7685] Féret L. and Pascale J. (1998). Configuration-interaction Hartree-Fock calculations for two-electron atoms using a pseudopotential. *Phys. Rev. A*, **58**, 3585–3596.

[7686] Fernández E.J., Gimeno M.C., Jones P.G., Laguna A., Laguna M., López-de-Luzuriaga J.M. and Rodríguez M.A. (1995). Synthesis, structure and reactivity of $((acac)AuCH(PPh_2AuPPh_2)_2$ $CHAu(acac))$, a complex containig the tridentate ligand $(HC(PPh_2)_2)^-$. *Chem. Ber.*, **128**, 121–124.

[7687] Fernández E.J., López-de-Luzuriaga J.M., Monge M., Rodríguez M.A., Crespo O., Gimeno M.C., Laguna A. and Jones P.G. (1998). Theoretical evidence for transannular metal-metal interactions in dinuclear coinage metal complexes. *Inorg. Chem.*, **37**, 6002–6006.

[7688] Fernández Guillermet A., Häglund J. and Grimvall G. (1993). Cohesive properties and electronic structure of $5d$-transition-metal carbides and nitrides in the NaCl structure. *Phys. Rev. B*, **48**, 11673–11684.

[7689] Fernández Pacios L. and Gomez P.C. (1994). Optimized triple-zeta Gaussian basis sets for use with relativistic effective potentials. *Int. J. Quantum Chem.*, **49**, 817–833.

[7690] Ferrari A.M. and Pacchioni G. (1996). Metal deposition on oxide surfaces: a quantum-chemical study of the interaction of Rb, Pd, and Ag atoms with the surface vacancies of MgO. *J. Phys. Chem.*, **100**, 9032–9037.

[7691] Ferrari A.M., Ugliengo P. and Garrone E. (1996). Ab initio study of the adducts of carbon monoxide with alkaline cations. *J. Chem. Phys.*, **105**, 4129–4139.

[7692] Ferrari A.M., Xiao C.Y., Neyman K.M., Pacchioni G. and Rösch N. (1999). Pd and Ag dimers and tetramers adsorbed at the MgO(001) surface: a density functional study. *PCCP*, **1**, 4655–4661.

[7693] Filipponi A. and Di Cicco A. (1995). Atomic background in x-ray absorption spectra of fifth-period elements: evidence for double-electron excitation edges. *Phys. Rev. A*, **52**, 1072–1078.

[7694] Filonov A.B., Migas D.B., Shaposhnikov V.L., Dorozhkin N.N., Borisenko V.E., Heinrich A. and Lange H. (1999). Electronic properties of isostructural ruthenium and osmium silicides and germanides. *Phys. Rev. B*, **60**, 16494–16498.

[7695] Fink R.F., Kivilompolo M. and Aksela H. (1999). Theory and *ab initio* calculations of $2p$ photoabsorption spectra: The lowest Rydberg resonance in HCl. *J. Chem. Phys.*, **111**, 10034–10045.

[7696] Fink R.F., Kivilompolo M., Aksela H. and Aksela S. (1998). Spin-orbit interaction and molecular-field effects in the $L_{2,3}VV$ Auger-electron spectra of HCl. *Phys. Rev. A*, **58**, 1988–2000.

[7697] Finkbeiner M., Fricke B. and Kühl T. (1993). Calculation of the hyperfine structure transition energy and lifetime in the one-electron Bi^{82+} ion. *Phys. Lett. A*, **176**, 113–117.

[7698] Finster F. (1998). Local $U(2,2)$ symmetry in relativistic quantum mechanics. *J. Math. Phys.*, **39**, 6276–6290.

[7699] Fiorentini V., Methfessel M. and Scheffler M. (1993). Reconstruction mechanism of fcc transition metal (001) surfaces. *Phys. Rev. Lett.*, **71**, 1051–1054.

[7700] Flad H.J. and Dolg M. (1996). Ground state properties of Hg_2. 2. A quantum Monte Carlo study. *J. Phys. Chem.*, **100**, 6152–6155.

[7701] Flad H.J. and Dolg M. (1997). Probing the accuracy of pseudopotentials for transition metals in quantum Monte Carlo calculations. *J. Chem. Phys.*, **107**, 7951–7959.

[7702] Flad H.J., Dolg M. and Shukla A. (1997). Spin-orbit coupling in variational Monte Carlo calculations. *Phys. Rev. A*, **55**, 4183–4195.

[7703] Flad H.J., Schautz F., Wang Y.X., Dolg M. and Savin A. (1999). On the bonding of small group 12 clusters. *Eur. Phys. J. D*, **6**, 243–254.

[7704] Flambaum V.V. (1993). Dynamical enhancement of weak interactions and quantum chaos. *Phys. Scr.*, **T46**, 198–207.

[7705] Flambaum V.V. (1999). Enhancement of parity and time invariance violation in the radium atom. *Phys. Rev. A*, **60**, R2611–R2613.

[7706] Flambaum V.V., Gribakina A.A. and Gribakin G.F. (1996). Narrow chaotic compound autoionizing states in atomic spectra. *Phys. Rev. A*, **54**, 2066–2079.

[7707] Flambaum V.V., Gribakina A.A., Gribakin G.F. and Kozlov M.G. (1994). Structure of compound states in the chaotic spectrum of the Ce atom: Localization properties matrix elements and enhancement of weak perturbations. *Phys. Rev. A*, **50**, 267–296.

[7708] Flambaum V.V. and Hanhart C. (1993). Magnetic interaction between relativistic atomic electrons and parity nonconserving nuclear moments. *Phys. Rev. C*, **48**, 1329–1334.

[7709] Flambaum V.V. and Murray D.W. (1997). Limits on the monopole polarization magnetic field from measurements of the electric dipole moments of atoms, molecules, and the neutron. *Phys. Rev. A*, **55**, 1736–1742.

[7710] Flambaum V.V. and Sushkov O.P. (1993). Possibility of observing parity nonconservation by measuring the nuclear anapole moment using the NMR frequency shift in a laser beam. *Phys. Rev. A*, **48**, R751–R754.

[7711] Flament J.P. and Tadjeddine M. (1995). MCSCF calculations on the vibrational properties of CN^- adsorbed on platinum. *Chem. Phys. Lett.*, **238**, 193–198.

[7712] Fleig T. and Marian C.M. (1994). Relativistic all-electron ab initio calculations on the platinum hydride molecule. *Chem. Phys. Lett.*, **222**, 267–273.

[7713] Fleig T. and Marian C. (1996). Ab initio calculation of Ω-splittings and rovibronic states of the PtH and PtD molecules. *J. Mol. Spectr.*, **178**, 1–9.

[7714] Fleig T. and Marian C. (1998). Relativistic ab-initio calculations on PdH and PdD: The rovibronic spectra and rotational splittings. *J. Chem. Phys.*, **108**, 3517–3521.

[7715] Fleig T., Marian C.M. and Olsen J. (1997). Spinor optimization for a relativistic spin-dependent CASSCF program. *Theor. Chem. Acc.*, **97**, 125–135.

[7716] Fleischer H., Stauf S. and Schollmeyer D. (1999). Experimental investigations and ab initio studies of tellurium(II) dithiolates, $Te(SR)_2$. *Inorg. Chem.*, **38**, 3725–3729.

[7717] Fleming J., Vaeck N., Hibbert A., Bell K.L. and Godefroid M.R. (1996). Oscillator strengths for the resonance line of ions in the beryllium isoelectronic sequence. *Phys. Scr.*, **53**, 446–453.

[7718] Fleszar A., Stumpf R. and Eguiluz A.G. (1997). One-electron excitations, correlation effects, and the plasmon in cesium metal. *Phys. Rev. B*, **55**, 2068–2072.

[7719] Fletcher D.A., Dai D., Steimle T.C. and Balasubramanian K. (1993). The permanent electric dipole moment of NbN. *J. Chem. Phys.*, **99**, 9324–9325.

[7720] Flocke N. (1997). Symmetric group approach to relativistic CI. IV. Representations of one-electron spin operators and their products in a symmetric-group-adapted basis of N-electron spin functions. *Int. J. Quantum Chem.*, **61**, 747–757.

[7721] Flocke N., Barysz M., Karwowski J. and Diercksen G.H.F. (1997a). Symmetric group approach to relativistic CI. I. General formalism. *Int. J. Quantum Chem.*, **61**, 1–10.

[7722] Flocke N., Barysz M., Karwowski J. and Diercksen G.H.F. (1997b). Symmetric group approach to relativistic CI. II. Reduction of matrices in the spin space. *Int. J. Quantum Chem.*, **61**, 11–20.

[7723] Flocke N., Barysz M., Karwowski J. and Diercksen G.H.F. (1997c). Symmetric group approach to relativistic CI. III. Matrix elements for spin-dependent operators. *Int. J. Quantum Chem.*, **61**, 21–34.

[7724] Fluchtmann M., Grass M., Braun J. and Borstel G. (1995). Relativistic full-potential photoemission theory for ferromagnetic materials. *Phys. Rev. B*, **52**, 9564–9575.

[7725] Foldy L.L. and Krajcik R.A. (1975). Separable solutions for directly interacting particle systems. *Phys. Rev. D*, **12**, 1700–1710.

[7726] Fontes C.J. (1992). *Inclusion of the generalized Breit interaction in determining relativistic cross sections for excitation of highly charged ions by electron impact*. Ph.D. thesis, Pennsylvania State Univ., University Park, PA, 138 p.

[7727] Fontes C.J. (1998). The role of the $5p^5 5d$ configuration and spin-orbit coupling in the electron-impact excitation of the lowest-lying J=0 and J=2 levels of xenon and krypton. *J. Phys. B*, **31**, 175–181.

[7728] Fontes C.J., Sampson D.H. and Zhang H.L. (1993a). Inclusion of the generalized Breit interaction in excitation of highly charged ions by electron impact. *Phys. Rev. A*, **47**, 1009–1022.

[7729] Fontes C.J., Sampson D.H. and Zhang H.L. (1993b). Method for calculating the electron-impact ionization of ions of any complexity. *Phys. Rev. A*, **48**, 1975–1982.

[7730] Fontes C.J., Sampson D.H. and Zhang H.L. (1994). Use of the factorized form of the collision strength in exploration of the effect of the generalized Breit interaction. *Phys. Rev. A*, **49**, 3704–3711.

[7731] Fontes C.J., Sampson D.H. and Zhang H.L. (1995). Relativistic calculations of cross sections for ionization of U^{90+} and U^{91+} ions by electron impact. *Phys. Rev. A*, **51**, R12–R13.

[7732] Fontes C.J., Sampson D.H. and Zhang H.L. (1999a). Fully relativistic calculations of and fits to $1s$ ionization cross sections. *Phys. Rev. A*, **59**, 1329–1335.

[7733] Fontes C.J., Sampson D.H. and Zhang H.L. (1999b). Fully relativistic distorted-wave cross sections for electron impact ionization from the $2s$, $2p_{1/2}$, and $2p_{3/2}$ subshells in ions with $Z - N \geq 3$, $Z \leq 92$, and $N \leq 12$. *At. Data Nucl. Data Tables*, **72**, 217–238.

[7734] Fontes C.J., Zhang H.L. and Sampson D.H. (1999c). Inclusion of the generalized Breit interaction in electron-impact excitation of ions to specific magnetic sublevels. *Phys. Rev. A*, **59**, 295–299.

[7735] Ford L.H. (1993). Spectrum of the Casimir effect and the Lifshitz theory. *Phys. Rev. A*, **48**, 2962–2967.

[7736] Forest J.L., Pandharipande V.R. and Friar J.L. (1995). Relativistic nuclear Hamiltonians. *Phys. Rev. C*, **52**, 568–575.

[7737] Forstreuter J., Steinbeck L., Richter M. and Eschrig H. (1997). Density-functional calculations for rare-earth atoms and ions. *Phys. Rev. B*, **55**, 9415–9421.

[7738] Forte S. (1994). Relativistic quantum theory with fractional spin and statistics. In *Particles and Fields*, (Edited by Eboli O.J.P. and Rivelles V.O.), pp. 496–572, World Scientific, Singapore.

[7739] Fortunelli A. (1999). Density functional calculations on small platinum clusters: Pt_n^q (n=1-4, q=0,±1). *J. Mol. Str. (Theochem)*, **493**, 233–240.

[7740] Fowler P.W. and Ceulemans A. (1993). Spin-orbit coupling coefficients for icosahedral molecules. *Theor. Chim. Acta*, **86**, 315–342.

[7741] Fowler P.W., Peebles S.A., Legon A.C. and Sadlej A.J. (1996). Electric field gradients and Sternheimer-type properties of the BrCl molecule: correlated, relativistic, ab initio calculations and modelling of nuclear quadrupole coupling constants in complexes B ... BrCl. *Chem. Phys. Lett.*, **257**, 249–256.

[7742] Fraga S. and Karwowski J. (1974). Some forgotten terms of the Dirac-Breit-Pauli equation. *Can. J. Phys.*, **52**, 1045.

[7743] Frankcombe K.E., Cavell K.J., Yates B.F. and Knott R.B. (1995). Large basis set calculations on model zerovalent palladium systems. *J. Phys. Chem.*, **99**, 14316–14322.

[7744] Frankcombe K.E., Cavell K.J., Yates B.F. and Knott R.B. (1996). Ligand substitution: An assessment of the reliability of ab initio calculations. *J. Phys. Chem.*, **100**, 18363–18370.

[7745] Franke R.J. (1994). *Die Berechnung relativistischer Korrekturen von Molekülen mit der direkten Störungstheorie*. Ph.D. thesis, Bochum.

[7746] Franke R. (1997). Numerical study of the one-electron Dirac equation based on 'direct perturbation theory'. *Chem. Phys. Lett.*, **264**, 495–501.

[7747] Franke R. and van Wüllen C. (1998). First-order relativistic corrections to MP2 energy from standard gradient codes: Comparison with results from density functional theory. *J. Comp. Chem.*, **19**, 1596–1603.

[7748] Freindorf M., Marian C.M. and Hess B.A. (1993). Theoretical study of the electronic spectrum of the CoH molecule. *J. Chem. Phys.*, **99**, 1215–1223.

[7749] Frenking G., Antes I., Böhme M., Dapprich S., Ehlers A.W., Jonas V., Neuhaus A., Otto M., Stegmann R., Veldkamp A. and Vyboishchikov S.F. (1996). Pseudopotential calculations of transition metal compounds - scope and limitations. In *Rev. Comp. Chem.*, (Edited by Lipkowitz K.B. and Boyd D.B.), vol. 8, pp. 63–144, VCH, New York.

[7750] Frenking G., Boehme C. and Pidun U. (1999). The electronic structure of transition metal compounds. In *Pauling's Legacy: Modern Modelling of the Chemical Bond*, (Edited by Maksić Z.B. and Orville-Thomas W.J.), pp. 555–570, Elsevier, Amsterdam.

[7751] Frenking G., Fau S., Marchand C.M. and Grützmacher H. (1997). The π-donor ability of the halogens in cations and neutral molecules. A theoretical study of AX_3^+, AH_2X^+, YX_3, and YH_2X (A = C, Si, Ge, Sn, Pb; Y = B, Al, Ga, In, Tl; X = F, Cl, Br, I. *J. Am. Chem. Soc.*, **119**, 6648–6655.

[7752] Frenking G. and Pidun U. (1997). Ab initio studies of transition-metal compounds: the nature of the chemical bond to a transition metal. *J. Chem. Soc., Dalton Trans.*, pp. 1653–1662.

[7753] Friar J.L., Martorell J. and Sprung D.W.L. (1997). Nuclear sizes and the isotope shift. *Phys. Rev. A*, **56**, 4579–4586.

[7754] Friar J.L., Martorell J. and Sprung D.W.L. (1999). Hadronic vacuum polarization and the Lamb shift. *Phys. Rev. A*, **59**, 4061–4063.

[7755] Friar J.L. and Payne G.L. (1997a). Higher-order nuclear-polarizability corrections in atomic hydrogen. *Phys. Rev. C*, **56**, 619–630.

[7756] Friar J.L. and Payne G.L. (1997b). High-order nuclear-size corrections in atomic hydrogen. *Phys. Rev. A*, **56**, 5173–5175.

[7757] Fricke B. (1969). Zur Vakuumpolarisation in Myonenatomen. *Z. Physik*, **218**, 495–508.

[7758] Fricke B., Anton J., Schulze K., Sepp W.D. and Kürpick P. (1998). Relativistic and dynamic contributions in ion-atom and ion-solid collisions. *Adv. Quantum Chem.*, **30**, 273–282.

[7759] Fricke B., Johnson E. and Rivera G.M. (1993). Ionization potentials and radii of atoms and ions of element 105 (unnilpentium) and ions of tantalum derived from multiconfiguration Dirac-Fock calculations. *Radiochim. Acta*, **62**, 17–25.

[7760] Fricke B., Sepp W.D., Bastug T., Varga S., Schulze K., Anton J. and Pershina V. (1997). Use of the DV Xα-method in the field of superheavy atoms. *Adv. Quantum Chem.*, **29**, 109–121.

[7761] Friedberg R., Lee T.D. and Pang Y. (1994). A new lattice formulation of the continuum. *J. Math. Phys.*, **35**, 5600–5629.

[7762] Frisk H. and Guhr T. (1993). Spin-orbit coupling in semiclassical approximation. *Ann. Phys. (NY)*, **221**, 229–257.

[7763] Fritz R., Muther H. and Machleidt R. (1993). Dirac effects in the Hartree-Fock description of finite nuclei employing realistic forces. *Phys. Rev. Lett.*, **71**, 46–49.

[7764] Fritzsche S. (1993). Angular distribution parameters in the resonant xenon $4d^{-1}6p$ Auger spectra. *Phys. Lett. A*, **180**, 262–268.

[7765] Fritzsche S. (1997). CESD97 – A revised version to expand jj-coupled symmetry functions into determinants. *Comp. Phys. Comm.*, **103**, 277–286.

[7766] Fritzsche S., Finkbeiner M., Fricke B. and Sepp W.D. (1995). Level energies and lifetimes in the $3p^43d$ configuration of chlorine-like ions. *Phys. Scr.*, **52**, 258–266.

[7767] Fritzsche S., Froese Fischer C. and Fricke B. (1998). Calculated level energies, transition probabilities, and lifetimes for phosphorus-like ions of the iron group in the $3s3p^4$ and $3s^23p^23d$ configurations. *At. Data Nucl. Data Tables*, **68**, 149–179.

[7768] Fritzsche S. and Grant I.P. (1994a). On the ab initio calculation of the $^5S_2^o$-$^3P_J^e$ intercombination transitions in P II relativity and electron relaxation. *Phys. Lett. A*, **186**, 152–156.

[7769] Fritzsche S. and Grant I.P. (1994b). Ab-initio calculation of the $2s^2\ ^1S_0$-$2s3p\ ^3P_1$ intercombination transition in beryllium-like ions. *Phys. Scr.*, **50**, 473–480.

[7770] Fritzsche S. and Grant I.P. (1995). A program for the expansion of jj-coupled symmetry functions into Slater determinants. *Comp. Phys. Comm.*, **92**, 111–126.

[7771] Fritzsche S., Koike F., Sienkiewicz J.E. and Vaeck N. (1999). Calculation of relativistic atomic transition and ionization properties for highly-charged ions. *Phys. Scripta*, **T80**, 479–481.

[7772] Frochaux E. (1993a). Relativistic corrections to the Schroedinger equation deduced from the quantum field theory. *Helv. Phys. Acta*, **66**, 89–90.

[7773] Frochaux E. (1993b). The bound states of the $P(\phi)_2$ relativistic quantum field models with weak coupling obtained by the variational perturbation method. *Nucl. Phys. B*, **389**, 666–702.

[7774] Frochaux E. (1993c). The bound-states in quantum field theory : review of some analytic problems raised by the variational perturbation method. *Helv. Phys. Acta*, **66**, 567–613.

[7775] Froelich P. and Weyrich W. (1996). On the relativistic corrections to the cross-section for inelastic scattering of photons on atomic electrons. *Z. Physik. D*, **38**, 185–.

[7776] Froese R.D.J., Musaev D.G. and Morokuma K. (1998). Theoretical study of substituent effects in the diimine-M(II) catalyzed ethylene polymerization reaction using the IMOMM method. *J. Am. Chem. Soc.*, **120**, 1581–1587.

[7777] Froese R.D.J., Musaev D.G. and Morokuma K. (1999). Theoretical studies of the Cp_2ZrR^+-catalyzed propylene polymerization reactions and a comparison with ethylene polymerization. *J. Mol. Str. (Theochem)*, **461-462**, 121–135.

[7778] Froese Fischer C. (1994). Allowed transitions and intercombination lines in C III and C II. *Phys. Scr.*, **49**, 323–330.

[7779] Froese Fischer C. (1999). Correlation and relativistic effects on transitions in lighter atoms. *Phys. Scripta*, **T83**, 49–60.

[7780] Froese Fischer C., He X. and Jönsson P. (1998a). The $2p^4\ ^3P_{1,2}$ - $2p^33s\ ^5S_2^o$ and $2p^4\ ^3P_{1,2}$ - $2s2p^5\ ^3P_2^o$ transitions in the oxygen isoelectronic sequence. *Eur. Phys. J. D*, **4**, 285–289.

[7781] Froese Fischer C. and Jönsson P. (1994). MCHF calculations for atomic properties. *Comp. Phys. Comm.*, **84**, 37–58.

[7782] Froese Fischer C. and Parpia F.A. (1993). Accurate spline solutions of the radial Dirac equation. *Phys. Lett. A*, **179**, 198–204.

[7783] Froese Fischer C. and Rubin R.H. (1998). Transition rates for some forbidden lines in Fe IV. *J. Phys. B*, **31**, 1657–1669.

[7784] Froese Fischer C., Saparov M., Gaigalas G. and Godefroid M. (1998b). Breit-Pauli energies, transition probabilities, and lifetimes for $2s, 2p, 3s, 3p, 3d, 4s\ ^2L$ levels of the lithium sequence, Z = 3-8. *At. Data Nucl. Data Tables*, **70**, 119–134.

[7785] Fröhlich N. and Frenking G. (1999). Theoretical models derived from *ab initio* calculations describing the bonding situation in transition metal complexes. In *Solid State Organometallic Chemistry: Methods and Applications*, (Edited by Gielen M., Willem R. and Wrackmeyer B.), pp. 173–226, Wiley.

[7786] Frohn H.J., Klose A., Schroer T., Henkel G., Buss V., Opitz D. and Vahrenhorst R. (1998). Structural, chemical, and theoretical evidence for the electrophilicity of the $[C_6F_5Xe]^+$ cation in $[C_6F_5Xe][AsF_6]$. *Inorg. Chem.*, **37**, 4884–4890.

[7787] Frolov A.M. (1999). Bound-state properties of the positronium negative ion Ps^-. *Phys. Rev. A*, **60**, 2834–2839.

[7788] Fuchs M. and Scheffler M. (1999). Ab initio pseudopotentials for electronic structure calculations of poly-atomic systems using density-functional theory. *Comp. Phys. Comm.*, **119**, 67–98.

[7789] Fuentealba P. and Reyes O. (1993). Polarizabilities and hyperpolarizabilities of the alkali metal atoms. *J. Phys. B*, **26**, 2245–2250.

[7790] Fujimoto S. and Kawakami N. (1993). Persistent currents in mesoscopic Hubbard rings with spin-orbit interaction. *Phys. Rev. B*, **48**, 17406–17412.

[7791] Fuks D., Dorfman F. and Davidov G. (1995). Nonempirical effective potential approach in consistent thermodynamics of solid and liquid tungsten. *Int. J. Quantum Chem.*, **S29**, 675–683.

[7792] Fukui H. and Baba T. (1998). Calculation of nuclear magnetic shieldings. XII. Relativistic no-pair equation. *J. Chem. Phys.*, **108**, 3854–3862.

[7793] Fukui H., Baba T. and Inomata H. (1996). Calculation of nuclear magnetic shieldings. X. Relativistic effects. (Errata, ibid. 1997, 106, 2987). *J. Chem. Phys.*, **105**, 3175–3186.

[7794] Funakubo K., Kakuto A., Otsuki S. and Toyoda F. (1996). Numerical approach to CP-violating Dirac equation. *Progr. Theor. Phys. (Kyoto)*, **95**, 929–941.

[7795] Furst J.E., Wijayaratna W.M.K.P., Madison D.H. and Gay T.J. (1993). Investigation of spin-orbit effects in the excitation of noble gases by spin-polarized electrons. *Phys. Rev. A*, **47**, 3775–3787.

[7796] Gabbaï F.P., Chung S.C., Schier A., Krüger S., Rösch N. and Schmidbaur H. (1997). A novel anionic gold-indium cluster compound: Synthesis and molecular and electronic structure. *Inorg. Chem.*, **36**, 5699–5705.

[7797] Gadea F.X., Savrda J. and Paidarova I. (1994). The structure of Ar_3^+. *Chem. Phys. Lett.*, **223**, 369–376.

[7798] Gagliardi L., Handy N.C., Ioannou A.G., Skylaris C.K., Spencer S., Willetts A. and Simper A.G. (1998a). A two-centre implementation of the Douglas-Kroll transformation in relativistic calculations. *Chem. Phys. Lett.*, **283**, 187–193.

[7799] Gagliardi L., Willetts A., Skylaris C.K., Handy N.C., Spencer S., Ioannou A.G. and Simper A.M. (1998b). A relativistic density functional study on the uranium hexafluoride and plutonium hexafluoride monomer and dimer species. *J. Am. Chem. Soc.*, **120**, 11727–11731.

[7800] Gaigalas G., Kaniauskas J., Kisielius R., Merkelis G. and Vilkas M.J. (1994). Second-order MBPT results for the oxygen isoelectronic sequence. *Phys. Scr.*, **49**, 135–147.

[7801] Gaioli F.H. and Garcia Alvarez E.T. (1995). Some remarks about intrinsic parity in Ryder's derivation of the Dirac equation. *Am. J. Phys.*, **63**, 177–178.

[7802] Gambhir Y.K. and Ring P. (1993). Recipe for the solution of the Dirac equation by the basis set expansion method. *Modern Phys. Lett. A*, **8**, 787–795.

[7803] Gao J.F., Pang W.N., Shang R.C., Gao H. and Long G.L. (1999). Direct and indirect relativistic effect on electron scattering from cesium and gold atoms. *Phys. Rev. A*, **60**, 5108–5110.

[7804] Gao T., Wang H.Y., Yi Y.G., Ran M., Jiang G., Zhu Z.H., Fu Y.B., Sun Y., Tang Y.J. and Wang X.L. (1998). Quantum mechanical calculation on the electronic states $X\ ^1\Sigma^+$ and $H\ ^3\Delta$ of ThO. *Report CNIC-01277*, p. 9 p.

[7805] Gao X.C., Fu J., Li X.H. and Gao J. (1998). Invariant formulation and exact solutions for the relativistic charged Klein-Gordon field in a time-dependent spatially homogeneous electric field. *Phys. Rev. A*, **57**, 753–761.

[7806] Garcia M.E., Pastor G.M. and Bennemann K.H. (1993). Delocalization of a hole in van der Waals cluster: Ionization potential of rare-gas and small Hg_n clusters. *Phys. Rev. B*, **48**, 8388–8397.

[7807] García de la Vega J.M. (1995). Relativistic corrections to the atomic electron affinities. *Phys. Rev. A*, **51**, 2616–2618.

[7808] García de la Vega J.M. and Miguel B. (1994). Roothaan-Hartree-Fock momentum expectation values for heavy elements (Z=55-92). *At. Data Nucl. Data Tables*, **58**, 307–315.

[7809] García-Calderón G., Rubio A. and Villavicencio J. (1999). Low-energy relativistic effects and nonlocality in time-dependent tunneling. *Phys. Rev. A*, **59**, 1758–1761.

[7810] Garzón I.L., Kaplan I.G., Santamaria R. and Novaro O. (1998). Molecular dynamics study of the Ag_6 cluster using an ab initio many-body model potential. *J. Chem. Phys.*, **109**, 2176–2184.

[7811] Garzón I.L., Kaplan I.G., Santamaria R., Vaisberg B.S. and Novaro O. (1997). Ab initio model potential and molecular dynamics simulation of Ag_6 clusters. *Z. Phys. D*, **40**, 202–205.

[7812] Garzón I.L., Michaelian K., Beltrán M.R., Posada-Amarillas A., Ordejón P., Artacho E., Sánchez-Portal D. and Soler J.M. (1998). Lowest energy structures of gold nanoclusters. *Phys. Rev. Lett.*, **81**, 1600–1603.

[7813] Gasche T., Brooks M.S.S. and Johansson B. (1996). Calculated optical properties of thorium, protactinium, and uranium metals. *Phys. Rev. B*, **54**, 2446–2452.

[7814] Gausa M., Kaschner R., Seifert G., Faehmann J.H., Lutz H.O. and Meiwes-Broer K.H. (1996). Photoelectron investigations and density functional calculations of anionic Sb_n^- and Bi_n^- clusters. *J. Chem. Phys.*, **104**, 9719–9728.

[7815] Gayasov R. and Joshi Y.N. (1999). $4d^9-4d^85p$ transitions in La XIII and Ce XIV. *Phys. Scripta*, **60**, 225–227.

[7816] Gdanitz R.J. (1999). Accurately solving the electronic Schrödinger equation of atoms and molecules using explicitly correlated (r_{12}–)MR-CI. The ground state of beryllium dimer (Be_2). *Chem. Phys. Lett.*, **312**, 578–584.

[7817] Ge Q. and King D.A. (1999). Surface diffusion potential energy surfaces from first principles: CO chemisorbed on Pt{110}. *J. Chem. Phys.*, **111**, 9461–9464.

[7818] Gębarowski R., Migdalek J. and Bieroń J.R. (1994). Relativistic multiconfiguration Dirac-Fock study of $3s^23p - 3s^23d$ transition in aluminium isoelectronic sequence. *J. Phys. B*, **27**, 3315–3324.

[7819] Gee C., Boissel P. and Ohanessian G. (1998). Metal-acetylene binding in gaseous $WC_2H_2^+$. *Chem. Phys. Lett.*, **298**, 85–92.

[7820] Geersten J.I. (1969). Evaluation of the Lamb shift for the hydrogen molecule-ion. *J. Chem. Phys.*, **51**, 3181–3185.

[7821] Geetha K.P., Singh A.D., Das B.P. and Unnikrishnan C.S. (1998). Nuclear-spin-dependent parity-nonconserving transitions in Ba^+ and Ra^+. *Phys. Rev. A*, **58**, R16–R18.

[7822] Geipel N.J.M. and Hess B.A. (1997). Scalar-relativistic effects in solids in the framework of a Douglas-Kroll transformed Dirac-Coulomb Hamiltonian. *Chem. Phys. Lett.*, **273**, 62–70.

[7823] Gelabert R., Moreno M., Lluch J.M. and Lledós A. (1997a). Elongated dihydrogen complexes: A combined electronic DFT + nuclear dynamics study of the $[Ru(H...H)(C_5H_5)(H_2PCH_2PH_2)]^+$ complex. *J. Am. Chem. Soc.*, **119**, 9840–9847.

[7824] Gelabert R., Moreno M., Lluch J.M. and Lledós A. (1997b). Structure and dynamics of LRh"H_4" (L = Cp, Tp) systems. A theoretical study. *Organometallics*, **16**, 3805–3814.

[7825] Gel'mukhanov F., Ågren H., Svensson S., Aksela H. and Aksela S. (1996). Theory of Auger spectra for molecular-field-split core levels. *Phys. Rev. A*, **53**, 1379–1387.

[7826] Ghilencea T., Toader O. and Diaconu C. (1994). Numerical evaluation of Coulomb Born approximation for double Bremsstrahlung. *Romanian Rep. Phys.*, **47**, 185–196.

[7827] Gisdakis P., Antonczak S., Köstlmeier S., Herrmann W.A. and Rösch N. (1998). Olefin epoxidation by methyldioxorhenium: A density functional study on energetics and mechanisms. *Angew. Chem. Int. Ed. Engl.*, **37**, 2211–2214.

[7828] Gisdakis P., Antonczak S., Köstlmeier S., Herrmann W.A. and Rösch N. (1998). Olefin epoxidation by methyldioxorhenium: A density functional study on energetics and mechanisms (in German). *Angew. Chem.*, **110**, 2333–2336.

[7829] Gisdakis P., Antonczak S. and Rösch N. (1999). Thermochemistry of oxygen transfer between rhenium and phosphine complexes. A density functional study. *Organometallics*, **18**, 5044–5056.

[7830] Gitman D.M. and Saa A.V. (1993). Pseudoclassical model of spinning particle with anomalous magnetic momentum. *Mod. Phys. Lett. A*, **8**, 463–468.

[7831] Gitman D.M. and Shelepin A.L. (1997). Poincaré group and relativistic wave equations in 2+1 dimensions. *J. Phys. A*, **30**, 6093–6121.

[7832] Glass J.T., McCann J.F. and Crothers D.S.F. (1994a). Relativistic continuum distorted wave theory for electron capture. *J. Phys. B*, **27**, 3445–3460.

[7833] Glass J.T., McCann J.F. and Crothers D.S.F. (1994b). Asymmetric theories of relativistic electron capture. *J. Phys. B*, **27**, 3975–3984.

[7834] Gleichmann M.M. and Hess B.A. (1994a). Relativistic all-electron ab initio calculations for the ground and excited states of the mercury atom. *Chem. Phys. Lett.*, **227**, 229–234.

[7835] Gleichmann M.M. and Hess B.A. (1994b). Relativistic all-electron ab initio calculations of ground and excited states of LiHg including spin-orbit effects. *J. Chem. Phys.*, **101**, 9691–9700.

[7836] Glendening E.D. and Feller D. (1996). Dication-water interactions: $M^{2+}(H_2O)_n$ clusters for alkaline earth metals M = Mg, Ca, Sr, Ba, and Ra. *J. Phys. Chem.*, **100**, 4790–4797.

[7837] Glenzer S., Kunze H.J., Musielok J., Kim Y.K. and Wiese W.L. (1994). Investigation of *LS* coupling in boronlike ions. *Phys. Rev. A*, **49**, 221–227.

[7838] Glukhovtsev M.N. and Bach R.D. (1997). Methylene-iodonium ylide: an isomer of diiodomethane. *Chem. Phys. Lett.*, **269**, 145–150.

[7839] Glukhovtsev M.N., Pross A., McGrath M.P. and Radom L. (1995). Extension of Gaussian-2 (G2) to bromine- and iodine-containing molecules: use of effective core potentials. *J. Chem. Phys.*, **103**, 1878–1885.

[7840] Glukhovtsev M.N., Pross A. and Radom L. (1996). Acidities, proton affinities, and other thermochemical properties of hypohalous acids HOX (X=F-I): a high-level computational study. *J. Phys. Chem.*, **100**, 3498–3503.

[7841] Glushkov A.V., Ambrosov S.V., Orlova V.E. and Orlov S.V. (1996). Calculation of binding energy in negatively charged ions of alkaline earth atoms Ca and Sr. *Izv. VUZov, Fizika.*, **39(9)**, 27–31.

[7842] Glushkov A.V., Efimov V.A., Polevoi A.N., Vitavetskaya L.A., Butenko Y.V., Malinovskii A.V. and Ryazanov E.A. (1998). Calculation of [electron] binding energies for negative ions of heavy elements (in Russian). *Zh. Strukt. Khim.*, **39**, 217–221.

[7843] Godefroid M. and Froese Fischer C. (1999). Isotope shift in the oxygen electron affinity. *Phys. Rev. A*, **60**, R2637–R2640.

[7844] Godefroid M.R., Jönsson P. and Froese Fischer C. (1998). Atomic structure variational calccu-lations in spectroscopy. *Phys. Scripta*, **T78**, 33–46.

[7845] Godfree J.A. and Staunton J.B. (1993). Effect of anisotropy on magnetic interactions: a relativistic electron approach. *J. Appl. Phys.*, **73**, 6557–6559.

[7846] Goidenko I., Labzowsky L., Nefiodov A., Plunien G. and Soff G. (1999). Second-order electron self-energy in hydrogenlike ions. *Phys. Rev. Lett.*, **83**, 2312–2315.

[7847] Goldfuss B. and von Ragué Schleyer P. (1997). Aromaticity in group 14 metalloles: Structural, energetic, and magnetic criteria. *Organometallics*, **16**, 1543–1552.

[7848] Gomes M., Malbouisson J.M.C. and da Silva A.J. (1997). Relativistic corrections to the Aharonov-Bohm scattering. *Phys. Lett. A*, **236**, 373–382.

[7849] Gonzalez A., Loyola G. and Moshinsky M. (1994). Radial equation for the particle-antiparticle system with a Dirac oscillator interaction and a qualitative applications mesons. *Rev. Mexicana Fis.*, **40**, 12–30.

[7850] González J.A., Aucar G.A., Ruiz de Azúa M.C. and Contreras R.H. (1997). CLOPPA RPA-AM1 analysis of the anisotropy of NMR $^1\mathbf{J}$(XY) coupling tensors in Me_3XY compounds (X = ^{13}C, ^{29}Si, ^{119}Sn, ^{207}Pb; Y = ^{19}F, ^{35}Cl). *Int. J. Quantum Chem.*, **61**, 823–833.

[7851] González-Blanco O., Branchadell V., Monteyne K. and Ziegler T. (1998). Nature and strength of metal-chalcogen multiple bonds in high oxidation state complexes. *Inorg. Chem.*, **37**, 1744–1748.

[7852] Gonzalo I. and Santos E. (1997). Radiative corrections in the Zeeman effect of 2 ^3P states of helium. *Phys. Rev. A*, **56**, 3576–3582.

[7853] Gonze X., Stumpf R. and Scheffler M. (1991). Analysis of separable potentials. *Phys. Rev. B*, **44**, 8503–8513.

[7854] Goodman B. and Ignjatović S.R. (1997). A simpler solution of the Dirac equation in a Coulomb potential. *Am. J. Phys.*, **65**, 214–221.

[7855] Goodmanson D.M. (1996). A graphical representation of the Dirac algebra. *Am. J. Phys.*, **64**, 870–880.

[7856] Gorczyca T.W., Felfli Z., Zhou H.L. and Manson S.T. (1998). Spin-orbit effects in the photoionization of neon. *Phys. Rev. A*, **58**, 3661–3672.

[7857] Gordon M.S. and Cundari T.R. (1996). Effective core potential studies of transition metal bonding, structure and reactivity. *Coord. Chem. Rev.*, **147**, 87–115.

[7858] Goreslavsky S.P. and Popruzhenko S.V. (1998). Relativistic deflection of photoelectron trajectories in elliptically polarized laser fields. *Optics Express*, **2**, 271–276.

[7859] Gorshkov V.G. (1961). Contribution to the theory of relativistic Coulomb scattering I. *Zh. Eksp. Teor. Fiz.*, **41**, 977–984.

[7860] Gorshkov V.G. (1962). Contribution to the theory of relativistic Coulomb scattering I. *Sov. Phys. JETP*, **14**, 694–698.

[7861] Gorshkov V.G., Mikhailov A.I. and Polikanov V.S. (1967). Screening in the atomic photoeffect. *Zh. Eksp. Teor. Fiz.*, **52**, 1570–1578.

[7862] Gorshkov V.G., Mikhailov A.I. and Polikanov V.S. (1967). Screening in the atomic photoeffect. *Sov. Phys. JETP*, **25**, 1045–1049.

[7863] Gorvat P.P., Lazur V.Y., Migalina S.I., Shuba I.M. and Yanev R.K. (1996). Term splitting in the quantum chemical two-center problem for the Dirac equation (in Russian). *Teor. Mat. Fiz.*, **109(2)**, 239–249.

[7864] Gosselin P. and Polonyi J. (1998). Path integral for relativistic equations of motion. *Ann. Phys. (NY)*, **268**, 207–224.

[7865] Gotsis H.J. and Strange P. (1994). A first-principles theory of x-ray Faraday effects. *J. Phys:CM*, **6**, 1409–1416.

[7866] Gould M.D. and Battle J.S. (1993). Spin-dependent unitary group approach to the Pauli-Breit Hamiltonian. I. First order energy level shifts due to spin-orbit interactions. *J. Chem. Phys.*, **98**, 8843–8851.

[7867] Gould M.D. and Battle J.S. (1993). Spin-dependent unitary group approach to the Pauli-Breit Hamiltonian. II. First order energy level shifts due to spin-spin interaction. *J. Chem. Phys.*, **99**, 5983–5994.

[7868] Gove S.K., Gropen O., Faegri K., Haaland A., Martinsen K.G., Strand T.G., Volden H.V. and Swang O. (1999). The molecular structure of niobium pentachloride by quantum chemical calculations and gas electron diffraction. *J. Mol. Str.*, **485-486**, 115–119.

[7869] Grabo T., Kreibich T., Kurth S. and Gross E.K.U. (1998). Orbital functionals in density functional theory: the optimized effective potential method. In *Strong Coulomb Correlations in Electronic Structure*, (Edited by Anisimov V.I.), pp. 1–, Gordon and Breach, Tokyo.

[7870] Graham J.P., Wojcicki A. and Bursten B.E. (1999). Molecular orbital description of the bonding and reactivity of the platinum $[\eta^3\text{-CH}_2\text{CCPh})\text{Pt}(\text{PPh}_3)_2]^+$. *Organometallics*, **18**, 837–842.

[7871] Grant I.P. (1994a). Relativistic electronic structure of atoms and molecules. *Adv. At. Mol. Opt. Phys.*, **32**, 169–186.

[7872] Grant I.P. (1994b). Relativistic calculation of atomic properties. *Comp. Phys. Comm.*, **84**, 59–77.

[7873] Grant I.P. (1996). Chapter 22: Relativistic atomic structure. In *Atomic, Molecular and Optical Physics Reference Book*, (Edited by Drake G.W.F.), pp. 258–286, AIP, New York.

[7874] Grass M., Braun J. and Borstel G. (1994a). New developments in the theory of photoemission. *Prog. Surface Sci.*, **46**, 107–122.

[7875] Grass M., Braun J. and Borstel G. (1994b). Relativistic photoemission theory for space-filling potentials. *Phys. Rev. B*, **50**, 14827–14837.

[7876] Green J.C., Guest M.F., Hillier I.H., Jarrett-Sprague S.A., Kaltsoyannis N., MacDonald M.A. and Sze K.H. (1992). Variable photon energy photoelectron spectroscopy of OsO_4 and pseudopotential calculations of the valence ionization energies of OsO_4 and RuO_4. *Inorg. Chem.*, **31**, 1588–159.

[7877] Green J.C., Scurr R.G., Arnold P.L. and Cloke F.G.N. (1997). An experimental and theoretical investigation of the electronic structure of Pd and Pt bis(carbene) complexes. *Chem. Comm.*, pp. 1963–1964.

[7878] Greene T.M., Brown W., Andrews L., Downs A.J., Chertihin G.V., Runeberg N. and Pyykkö P. (1995). Matrix infrared spectroscopic and ab initio studies of ZnH_2, CdH_2 and related metal hydride species. *J. Phys. Chem.*, **99**, 7925–7934.

[7879] Greenwood N.N. and Earnshaw A. (1997). *Chemistry of the Elements*. Butterworth Heinemann, Oxford, see pp. 599, 1180, 1266, 1274.

[7880] Greiner W. (1994). *Relativistic Quantum Mechanics. Wave Equations*. Springer, Berlin, 361 p.

[7881] Greiner W. (1997). *Relativistic Quantum Mechanics. Wave Equations, 2nd Ed.*. Springer, Berlin, 450 p.

[7882] Greiner W. (1998). Correlations in the vacuum. *Adv. Quantum Chem.*, **30**, 195–208.

[7883] Greiner W. and Gupta R.K., (Eds.) (1999). *Heavy Elements and Related New Phenomena*. World Scientific, Singapore, volumes 1-2.

[7884] Gresh N. and Giessner-Prettre C. (1997). Ab initio study of Cu^+, Ag^+, Zn^{++}, and Cd^{++} relaxation by ligands. *New J. Chem.*, **21**, 279–283.

[7885] Greub C., Wyler D., Brodsky S.J. and Munger C.T. (1995). Atomic alchemy: weak decays of muonic and pionic atoms into other atoms. *Phys. Rev. D*, **52**, 4028–4037.

[7886] Gribakina A.A., Flambaum V.V. and Gribakin G.F. (1995). Band structure of the Hamiltonian matrix of a real "chaotic" system: The Ce atom. *Phys. Rev. E*, **52**, 5667–5670.

[7887] Griesemer M. and Tix C. (1999). Instability of a pseudo-relativistic model of matter with self-generated magnetic field. *J. Math. Phys.*, **40**, 1780–1791.

[7888] Griffin D.C., Badnell N.R. and Pindzola M.S. (1998). R-matrix electron-impact excitation cross sections in intermediate coupling: an MQDT transformation approach. *J. Phys. B*, **31**, 3713–3727.

[7889] Griffin D.C., Badnell N.R., Pindzola M.S. and Shaw J.A. (1999). Electron-impact ionization of Mg-like ions. *J. Phys. B*, **32**, 2139–2152.

[7890] Griffin T.R., Cook D.B., Haynes A., Pearson J.M., Monti D. and Morris G.E. (1996). Theoretical and experimental evidence for S_N2 transition states in oxidative addition of methyl iodide to cis-$(M(CO)_2I_2)^-$ (M = Rh, Ir). *J. Am. Chem. Soc.*, **118**, 3029–3030.

[7891] Grigoryan G.V. and Grigoryan R.P. (1993). Canonical quantization of a Dirac spin particle in an external magnetic field (in Russian). *Zh. Eksp. Teor. Fiz.*, **103**, 3–10.

[7892] Grigoryan G.V. and Grigoryan R.P. (1993). Canonical quantization of a Dirac spin particle in an external magnetic field. *J. Exper. Theor. Phys.*, **76**, 1–4.

[7893] Grigoryan G.V. and Grigoryan R.P. (1995). Pseudoclassical Foldy-Wouthuysen transformation and canonical quantization of $D = 2n$-dimensional relativistic spin particles in external electromagnetic field (in Russian). *Teor. Mat. Fiz.*, **102**, 378–383.

[7894] Grimaldi C. and Fulde P. (1997). Theory of screening of the phonon-modulated spin-orbit interaction in metals. *Phys. Rev. B*, **55**, 15523–15530.

[7895] Grin Y., Wedig U. and von Schnering H.G. (1995). Hyperbolic lone pair structure in $RhBi_4$. *Angew. Chem. Int. Ed. Engl.*, **34**, 1204–1206.

[7896] Grin Y., Wedig U. and von Schnering H.G. (1995). Hyperbolic lone pair structure in $RhBi_4$ (in German). *Angew. Chem.*, **107**, 1318–1320.

[7897] Grönbeck H. and Andreoni W. (1997). Theoretical study of $(CO)_n$ chemisorption on Pt and Pt_3: structural, electronic and vibrational properties. *Chem. Phys. Lett.*, **269**, 385–390.

[7898] Grönbeck H., Rosén A. and Andreoni W. (1998). Structural, electronic, and vibrational properties of neutral and charged Nb_n (n=8,9,10) clusters. *Phys. Rev. A*, **58**, 4630–4636.

[7899] Gropen O., Sjøvoll M., Strømsnes H., Karlsen E., Swang O. and Faegri Jr K. (1994). RECP calculations for reactions of H_2 with Pt, Os, Ir, and Re - a systematic comparison. [Erratum: TCA 1995; 91:109-110]. *Theor. Chim. Acta*, **87**, 373–385.

[7900] Grosch G.H. and Range K.J. (1996a). Chemical trends in gold alkali alloys: A density functional study on stability and charge transfer. Part I: Gold alkali alloys of the formula MAu. *J. Alloys Compounds*, **233**, 30–38.

[7901] Grosch G.H. and Range K.J. (1996b). Chemical trends in gold alkali alloys - a DFT-study on stability and charge transfer. Part II: Gold alkali alloys of the formula MAu_5. *J. Alloys Compounds*, **233**, 39–43.

[7902] Gross E.K.U., Dobson J.F. and Petersilka M. (1996). Density functional theory of time-dependent phenomena. *Topics in Current Chem.*, **181**, 81–172.

[7903] Gross F. (1993). *Relativistic Quantum Mechanics and Field Theory*. John. Wiley. & Sons, 629 pp.

[7904] Grosse H., Martin A. and Stubbe J. (1994). Order and spacings of energy levels for the Klein-Gordon equation. *J. Math. Phys.*, **35**, 3805–3816.

[7905] Gruber D., Domiaty U., Li X., Windholz L., Gleichmann M. and Hess B.A. (1995). The NaHg red bands revisited. *J. Chem. Phys.*, **102**, 5174–5180.

[7906] Gruber D., Li X., Windholz L., Gleichmann M.M., Hess B.A., Vezmar I. and Pichler G. (1996). The LiHg($2^2\Pi_{3/2}$–$X^2\Sigma^+_{1/2}$) system. *J. Phys. Chem.*, **100**, 10062–10069.

[7907] Gruber D., Musso M., Windholz L., Gleichmann M., Hess B.A., Fuso F. and Allegrini M. (1994). Study of the LiHg excimer: Blue-green bands. *J. Chem. Phys.*, **101**, 929–936.

[7908] Grypeos M.E., Koutroulos C.G. and Papadopoulos G.J. (1993). A simplified relativistic approach for the study of the single-particle energies in nuclei and hypernuclei. *Acta Phys. Pol. B*, **24**, 607–618.

[7909] Grypeos M.E., Koutroulos C.G. and Papadopoulos G.J. (1994). Approximate treatment of the Dirac equation with scalar and vector potentials of rectangular shapes. *Phys. Rev. A*, **50**, 29–33.

[7910] Guiasu I. and Koniuk R. (1993). Particle interpretation of the Dirac-Coulomb solutions. *Can. J. Phys.*, **71**, 360–364.

[7911] Guil F. and Mañas M. (1996). The Dirac equation and integrable systems of KP (Kadomtsev - Petviashvili) type. *J. Phys. A*, **29**, 641–665.

[7912] Guillaumont D., Wilms M.P., Daniel C. and Stufkens D.J. (1998). Variation in charge-transfer photochemistry clarified by a CASSCF/MR-CCI comparative study of the low-lying excited states of $M(R)(CO)_3$(H-DAB) (M=Mn, R=H, methyl, ethyl), M=Re, R=H, DAB = 1,4-diaza-1,3-butadiene). *Inorg. Chem.*, **37**, 5816–5822.

[7913] Gulino A., Di Bella S., Fragalà I., Casarin M., Seyam A.M. and Marks T.J. (1993). A comparative fully relativistic/nonrelativistic first-principles Xα-DVM and photoelectron spectroscopic investigation of electronic structure in homologous 4f and 5f tris(η^5-cyclopentadienyl)metal(IV) alkoxide complexes. *Inorg. Chem.*, **32**, 3873–3879.

[7914] Günther R., Paulus W., Kratz J.V., Seibert A., Thörle P., Zauner S., Brüchle W., Jäger E., Pershina V., Schädel M., Schausten B., Schumann D., Eichler B., Gäggeler H.W., Jost D.T. and Türler A. (1998). Chromatographic study of Rutherfordium (Element 104) in the system HCl/tributylphosphate (TBP). *Radiochim. Acta*, **80**, 121–128.

[7915] Guo D.S., Gao J. and Chu A.H.M. (1996). Relativistic electron moving in a multimode quantized radiation field. *Phys. Rev. A*, **54**, 1087–1097.

[7916] Guo X.Z., Khakoo M.A., Matthews D.F., Mikaelian G., Crowe A., Kanik I., Trajmar S., Zeman V., Bartschat K. and Fontes C.J. (1999). Differential cross section ratios for low energy electron-impact excitation of the $4p^5 5s$ levels of krypton – sensitive tests of relativistic effects for heavy noble gases. *J. Phys. B*, **32**, L155–L163.

[7917] Gusev D.G., Kuhlman R., Rambo J.R., Berke H., Eisenstein O. and Caulton K.G. (1995). Structural and dynamic properties of $OsH_2X_2L_2$ (X = Cl, Br, I; L = P^iPr_3) complexes: interconversion between remarkable non-octahedral isomers. *J. Am. Chem. Soc.*, **117**, 281–292.

[7918] Gusev D.G., Notheis J.U., Rambo J.R., Hauger B.E., Eisenstein O. and Caulton K.G. (1994). Characterization of $PtH_3(P^tBu_3)_2^+$ as the first dihydrogen complex of d^8 Pt(II). *J. Am. Chem. Soc.*, **116**, 7409–7410.

[7919] Gustavsson M.G.H. and Mårtensson-Pendrill A.M. (1998a). Four decades of hyperfine anomalies. *Adv. Quantum Chem.*, **30**, 343–360.

[7920] Gustavsson M.G.H. and Mårtensson-Pendrill A.M. (1998b). Need for remeasurements of nuclear magnetic dipole moments. *Phys. Rev. A*, **58**, 3611–3618.

[7921] Gutowski M. (1999). Favorable performance of the DFT methods in predicting the minimun-energy structure of the lowest triplet state of WF_4. *Int. J. Quantum Chem.*, **73**, 369–375.

[7922] Gutowski M., Boldyrev I.A., Simons J., Rak J. and Błażejowski J. (1996). Properties of closed-shell, octahedral, multiply-charged hexafluorometallates MF_6^{3-}, M = Sc, Y, La, ZrF_6^{2-}, and TaF_6^-. *J. Am. Chem. Soc.*, **118**, 1173–1180.

[7923] Gutowski M., Rak J., Dokurno P. and Blazejowski J. (1994). Theoretical studies on the structure, thermochemistry, vibrational spectroscopy, and other features of HfX_6^{2-}(X=F, Cl, Br, I). Electrostatic energy in hexahalogenohafnates. *Inorg. Chem.*, **33**, 6187–6193.

[7924] Haaland A., Hammel A., Rypdal K., Swang O., Brunvoll J., Gropen O., Greune M. and Weidlein J. (1993). Molecular structure of pentamethylantimony by gas electron diffraction; structure and bonding in $Sb(CH_3)_5$ and $Bi(CH_3)_5$ studied by ab initio MO calculations. *Acta Chem. Scand.*, **47**, 368–373.

[7925] Haaland A., Martinsen K.J., Swang O., Volden H.V., Booij A.S. and Konings R.J.M. (1995). Molecular structure of monomeric uranium tetrachloride determined by gas electron diffraction at 900 K, gas-phase infrared spectroscopy and quantum-chemical density-functional calculations. *J. Chem. Soc., Dalton Trans.*, pp. 185–190.

[7926] Haaland A., Shorokhov D.J., Volden H.V. and Klinkhammer K.W. (1999). Molecular structure of a monomeric, base-free metal(I) amide, $TlN[Si(CH_3)_3]_2$, by gas electron diffraction and by density functional theory and ab initio MP2 calculations. *Inorg. Chem.*, **38**, 1118–1120.

[7927] Häberlen O.D., Chung S.C. and Rösch N. (1994a). Relativistic density-functional studies of naked and ligated gold clusters. *Int. J. Quantum Chem.*, **S28**, 595–610.

[7928] Häberlen O.D., Chung S.C., Stener M. and Rösch N. (1997). From clusters to bulk: A relativistic density functional investigation on a series of gold clusters Au_n, n= 6, ..., 147. *J. Chem. Phys.*, **106**, 5189–5201.

[7929] Häberlen O.D. and Rösch N. (1993). Effect of phosphine substituents in gold(I) complexes: A theoretical study of $MeAuPR_3$ R=H, Me, Ph. *J. Phys. Chem.*, **97**, 4970–4973.

[7930] Häberlen O.D., Schmidbaur H. and Rösch N. (1994b). Stability of main-group element-centered gold cluster cations. *J. Am. Chem. Soc.*, **116**, 8241–8248.

[7931] Hachem G. (1993). Zeeman effect for Dirac electron (in French). *Ann. Inst. Henri Poincaré Phys. Theor.*, **58**, 105–123.

[7932] Häckl R. and Pilkuhn H. (1996). Transformation of Breit operators into hyperfine-like operators. *J. Phys. B*, **29**, L725–L727.

[7933] Hada M., Ishikawa Y., Nakatani J. and Nakatsuji H. (1999). Dirac-Fock calculations of magnetic shielding constants: hydrogen molecule and hydrogen halides. *Chem. Phys. Lett.*, **310**, 342–346.

[7934] Hada M., Kaneko H. and Nakatsuji H. (1996). Relativistic study of nuclear magnetic shielding constants: tungsten hexahalides and tetraoxide. *Chem. Phys. Lett.*, **261**, 7–12.

[7935] Hada M., Nakatsuji H., Nakai H., Gyobu S. and Miki S. (1993). Theoretical study on the methane activation reactions by Pt, Pt^+, and Pt^- atoms. *J. Mol. Str. (Theochem)*, **281**, 207–212.

[7936] Haftel M.I. and Mandelzweig V.B. (1994a). Nonvariational calculation of the relativistic finite-size and QED corrections for the $2\ ^1S$ excited state of the helium atom. *Phys. Rev. A*, **49**, 3338–3343.

[7937] Haftel M.I. and Mandelzweig V.B. (1994b). Nonvariational calculation of the relativistic and finite-size corrections for the helium ground state. *Phys. Rev. A*, **49**, 3344–3350.

[7938] Hagelberg F., Das T.P. and Weil K.G. (1998). Theoretical investigations of Zintl anions analogous to ozone. *J. Phys. Chem. A*, **102**, 4630–4637.

[7939] Hagelberg F., Srinivas S., Sahoo N., Das T.P. and Weil K.G. (1996). Ab initio investigations on Sb_4 analogous Zintl clusters. *Phys. Rev. A*, **53**, 353–365.

[7940] Hagelin H., Åkermark B. and Norrby P.O. (1999). A solvated transition state for the nucleophilic attack on cationic η^3-allylpalladium complexes. *Chem. Eur. J.*, **5**, 902–909.

[7941] Hagen C.R. and Park D.K. (1996). Relativistic Aharonov-Bohm-Coulomb problem. *Ann. Phys. (NY)*, **251**, 45–63.

[7942] Haire R.G. and Gibson J.K. (1992). The pivotal position of plutonium in the systematics of actinide metals and compounds. In *Transuranium Elements. A Half Century*, (Edited by Morss L.R. and Fuger J.), pp. 426–439, ACS, Washington DC.

[7943] Håkansson H.E.V. (1950). On the Lamb-shift and the ionisation energy of helium-like atoms. *Arkiv Fysik*, **1**, 555–557.

[7944] Häkkinen H., Barnett R.N. and Landman U. (1999). Electronic structure of passivated $Au_{38}(SCH_3)_{24}$ nanocrystal. *Phys. Rev. Lett.*, **82**, 3264–3267.

[7945] Halabuka Z., Perger W. and Trautmann D. (1994). SCA calculations of the inner shell ionization with Dirac-Fock electronic wave functions. *Z. Phys. D*, **29**, 151–158.

[7946] Halabuka Z., Perger W. and Trautmann D. (1995). SCA calculations of the proton induced alignment using relativistic Hartree-Fock wavefunctions. *J. Phys. B*, **28**, 83–89.

[7947] Halasz M.A. and Verbaarschot J.J.M. (1995). Universal fluctuations in spectra of the lattice Dirac operator. *Phys. Rev. Lett.*, **74**, 3920–3923.

[7948] Halilov S.V., Tamura E., Meinert D., Gollisch H. and Feder R. (1995). Relativistic Green function theory of layer densities of states and photoemission from magnetic compounds. *J. Phys:CM.*, **5**, 3859–3870.

[7949] Hall R.L. (1999). Spectral comparison theorem for the Dirac equation. *Phys. Rev. Lett.*, **83**, 468–471.

[7950] Ham F.S. (1965). Dynamical Jahn-Teller effect in paramagnetic resonance spectra: Orbital reduction factors and partial quenching of spin-orbit interaction. *Phys. Rev.*, **138A**, 1727–1740.

[7951] Hamilton J.D. (1996). Relativistic precession. *Am. J. Phys.*, **64**, 1197–1201.

[7952] Hammer B., Morikawa Y. and Nørskov J.K. (1996). CO chemisorption at metal surfaces and overlayers. *Phys. Rev. Lett.*, **76**, 2141–2144.

[7953] Hammer B. and Nørskov J.K. (1995). Why gold is the noblest of all the metals. *Nature*, **376**, 238–240.

[7954] Han Y.K., Bae C. and Lee Y.S. (1999a). Two-component calculations of spin-orbit effects for a van der Waals molecule Rn_2. *Int. J. Quantum Chem.*, **72**, 139–143.

[7955] Han Y.K., Bae C. and Lee Y.S. (1999b). Two-component calculations for the molecules containing superheavy elements: Spin-orbit effects for (117)H, (113)H, and (113)F. *J. Chem. Phys.*, **110**, 8969–8975.

[7956] Han Y.K., Bae C. and Lee Y.S. (1999c). On the consistent definition of spin-orbit effects calculated by relativistic effective core potentials with one-electron spin-orbit operators: Comparison of spin-orbit effects for Tl, TlH, TlH_3, PbH_2, and PbH_4. *J. Chem. Phys.*, **110**, 9353–9359.

[7957] Han Y.K., Bae C., Lee Y.S. and Lee S.Y. (1998). Spin-orbit effects on structures of closed-shell polyatomic molecules containing heavy atoms calculated by two-component Hartree-Fock method. *J. Comp. Chem.*, **19**, 1526–1533.

[7958] Han Y.K. and Lee Y.S. (1999). Structures of RgF_n (Rg = Xe, Rn, and element 118. n = 2,4.) calculated by two-component spin-orbit methods. A spin-orbit induced isomer of $(118)F_4$. *J. Phys. Chem. A*, **103**, 1104–1108.

[7959] Hao Y.G., Eriksson O., Fernando G.W. and Cooper B.R. (1993). Surface electronic structure of γ-uranium. *Phys. Rev. B*, **47**, 6680–6684.

[7960] Haque N., Chakraborty H.S., Desmukh P.C., Manson S.T., Msezane A.Z., Deb N.C., Felfli Z. and Gorczyca T.W. (1999). Relativistic effects in the photoionization of Ne-like iron. *Phys. Rev. A*, **60**, 4577–4581.

[7961] Haque N. and Pradhan A.K. (1999). Photoionization of Fe XV. *Phys. Rev. A*, **60**, R4221–R4224.

[7962] Harada M. and Dexpert H. (1996a). Electronic structure of transition metal clusters from density functional theory: transition monometallic dimers. *C. R. Acad. Sci. Paris. Série II. b*, **322**, 239–246.

[7963] Harada M. and Dexpert H. (1996b). Electronic structure of transition metal clusters from density functional theory. 1. Transition metal dimers. *J. Phys. Chem.*, **100**, 565–572.

[7964] Harra L.K., Boone A.W., Norrington P.H., Keenan F.P. and Kingston A.E. (1993). Energy levels and oscillator strengths for transitions in Helium-like Ni XXVII. *J. Phys. B*, **26**, 2543–2553.

[7965] Hartemann F.V. and Luhmann Jr N.C. (1995). Classical electrodynamical derivation of the radiation damping force. *Phys. Rev. Lett.*, **74**, 1107–1110.

[7966] Hartwigsen C., Goedecker S. and Hutter J. (1998). Relativistic separable dual-space Gaussian pseudopotentials from H to Rn. *Phys. Rev. B*, **58**, 3641–3662.

[7967] Harvey J.N., Diefenbach M., Schröder D. and Schwarz H. (1999). Oxidation properties of the early transition-metal dioxide cations MO_2^+ (M = Ti, V, Zr, Nb) in the gas phase. *Int. J. Mass Spectr.*, **182/183**, 85–97.

[7968] Hascall T., Rabinovich D., Murphy V.J., Beachy M.D., Friesner R.A. and Parkin G. (1999). Mechanistic and theoretical analysis of the oxidative addition of H_2 to six-coordinate molybdenum and tungsten complexes $M(PMe_3)_4X_2$ (M = Mo,W; X = F, Cl, Br, I): An inverse equilibrium isotope effect and an unprecedented halide dependence. *J. Am. Chem. Soc.*, **121**, 11402–11417.

[7969] Hatsugai Y. and Lee P.A. (1993). Numerical study of localization of Dirac fermions on a lattice in two dimensions. *Phys. Rev. B*, **48**, 4204–4207.

[7970] Hättig C. and Hess B.A. (1996). TDMP2 calculation of dynamic multipole polarizabilities and dispersion coefficients of the noble gases Ar, Kr, Xe, and Rn. *J. Phys. Chem.*, **100**, 6243–6248.

[7971] Hättig C. and Hess B.A. (1998). TDMP2 calculation of dynamic multipole polarizabilities and dispersion coefficients of the halogen anions F^-, Cl^-, Br^- and I^-. *J. Chem. Phys.*, **108**, 3863–3870.

[7972] Häussermann U., Dolg M., Stoll H., Preuss H., Schwerdtfeger P. and Pitzer R.M. (1993). Accuracy of energy-adjusted quasirelativistic ab initio pseudopotentials all-electron and pseudopotential benchmark calculations for Hg, HgH and their cations. *Mol. Phys.*, **78**, 1211–1224.

[7973] Havlas Z., Downing J.W. and Michl J. (1998). Spin-orbit coupling in biradicals. 2. Ab initio methodology and applications to 1,1-biradicals: Carbene and silylene. *J. Phys. Chem. A*, **102**, 5681–5692.

[7974] Havlas Z. and Michl J. (1997). Spin-orbit coupling in organic biradicals: zero-field splitting in triplet tetramethyleneethane. *J. Mol. Str. (Theochem)*, **398**, 281–291.

[7975] Havlas Z. and Michl J. (1998). Spin-orbit coupling in biradicals. 3. Heavy atom effects in carbenes. *Coll. Czech. Chem. Comm.*, **63**, 1485–1497.

[7976] Havlas Z. and Michl J. (1999). Ab initio calculation of zero-field splitting and spin-orbit coupling in ground and excited triplets of *m*-xylylene. *J. Chem. Soc., Perkin 2*, pp. 2299–2303.

[7977] Havlin R., McMahon M., Srinivasan R., Le H.B. and Oldfield E. (1997). Solid-state NMR and density functional investigation of carbon-13 shielding tensors in metal-olefin complexes. *J. Phys. Chem. A*, **101**, 8908–8913.

[7978] Hay P.J. and Martin R.L. (1994). Theoretical studies of the structures and electronic properties of $U(NH_2)_3$ and $Np(NH_2)_3$. *J. Alloys Compounds*, **213/214**, 196–198.

[7979] Hay P.J. and Martin R.L. (1998). Theoretical studies of the structures and vibrational frequencies of actinide compounds using relativistic effective core potentials with Hartree-Fock and density functional methods: UF_6, NpF_6, and PuF_6. *J. Chem. Phys.*, **109**, 3875–3881.

[7980] Hayashi Y., Nakai H., Tokita Y. and Nakatsuji H. (1998). A theoretical study of the photochemical reductive elimination and thermal oxidative addition of molecular hydrogen from and to the Ir-complex. *Theor. Chem. Acc.*, **99**, 210–214.

[7981] He G., Seth M., Tokue I. and Macdonald R.G. (1999). Experimental and theoretical determination of the magnetic dipole transition moment for the $Br(4p^5)(^2P_{1/2} \leftarrow {}^2P_{3/2})$ fine-structure transition and the quantum yield of $Br(^2P_{1/2})$ from the 193 nm photolysis of BrCN. *J. Chem. Phys.*, **110**, 7821–7831.

[7982] Hecht L. and Barron L.D. (1993). Time reversal and Hermiticity characteristics of polarizability and optical activity operators. *Mol. Phys.*, **79**, 887–897.

[7983] Heiberg H., Swang O., Ryan O.B. and Gropen O. (1999). C-H activation at a cationic platinum (II) center: A quantum chemical investigation. *J. Phys. Chem. A*, **103**, 10004–10008.

[7984] Hein T.A., Thiel W. and Lee T.J. (1993). Ab initio study of the stability and vibrational spectra of plumbane methylplumbane and homologous compounds. *J. Phys. Chem.*, **97**, 4381–4385.

[7985] Heinemann C. (1995). *Gas-phase ion-molecule chemistry: synergy between experiment and theory.*. Ph.D. thesis, TU Berlin.

[7986] Heinemann C., Cornehl H.H., Schröder D., Dolg M. and Schwarz H. (1996a). The CeO_2^+ cation: gas-phase reactivity and electronic structure. *Inorg. Chem.*, **35**, 2463–2475.

[7987] Heinemann C., Hertwig R.H., Wesendrup R., Koch W. and Schwarz H. (1995a). Relativistic effects on bonding in cationic transition-metal-carbene complexes: a density-functional study. *J. Am. Chem. Soc.*, **117**, 495–500.

[7988] Heinemann C., Koch W., Lindner G.L. and Reinen D. (1995b). Electronic spectrum of S_2^-, the electron affinity of S_2, and the binding energies of neutral and anionic S_3 clusters. *Phys. Rev. A*, **52**, 1024–1038.

[7989] Heinemann C., Koch W., Lindner G.L., Reinen D. and Widmark P.O. (1996b). Ground- and excited-state properties of neutral and anionic selenium dimers and trimers. *Phys. Rev. A*, **54**, 1979–1993.

[7990] Heinemann C., Koch W. and Schwarz H. (1995c). An approximate method for treating spin-orbit effects in platinum. *Chem. Phys. Lett.*, **245**, 509–518.

[7991] Heinemann C. and Schwarz H. (1995). NUO^+, a new species isoelectronic to the uranyl dication UO_2^{2+}. *Chem. Eur. J.*, **1**, 7–11.

[7992] Heinemann C., Schwarz H. and Koch W. (1996c). Ground-state potentials for Co^+/rare-gas interactions. *Mol. Phys.*, **89**, 473–488.

[7993] Heinemann C., Schwarz H., Koch W. and Dyall K.G. (1996d). Relativistic effects in the cationic platinum carbene $PtCH_2^{2+}$. *J. Chem. Phys.*, **104**, 4642–4651.

[7994] Heinemann C., Schwarz J., Koch W. and Schwarz H. (1995d). Interaction of the Fe^+ cation with heavy noble gas atoms. *J. Chem. Phys.*, **103**, 4551–4561.

[7995] Heinemann C., Wesendrup R. and Schwarz H. (1995e). Pt^+-mediated activation of methane: theory and experiment. *Chem. Phys. Lett.*, **239**, 75–83.

[7996] Heinemann M., Terpstra H.J., Haas C. and de Groot R.A. (1995). Electronic structure of β-PbO_2 and its relation with $BaPbO_3$. *Phys. Rev. B*, **52**, 11740–11743.

[7997] Heitz M.C., Ribbing C. and Daniel C. (1997). Spin-orbit induced radiationless transitions in organometallics: Quantum simulation of the intersystem crossing processes in the photodissociation of $HCo(CO)_4$. *J. Chem. Phys.*, **106**, 1421–1428.

[7998] Heiz U., Vayloyan A., Schumacher E., Yeretzian C., Stener M., Gisdakis P. and Rösch N. (1996). Na_xAu and Cs_xAu bimetal clusters: Finite size analogs of sodium-gold and cesium-gold compounds. *J. Chem. Phys.*, **105**, 5574–5585.

[7999] Helffer B. and Parisse B. (1994). Comparison of the decay of eigenfunctions for the Dirac and Klein-Gordon operators. Application to the study of the tunnel effect. *Ann. Inst. Henri Poincaré, Phys. Theor.*, **60**, 147–187.

[8000] Helgaker T., Jaszuński M. and Ruud K. (1999). Ab initio methods for the calculation of NMR shielding and indirect spin-spin coupling constants. *Chem. Rev.*, **99**, 293–352.

[8001] Hemmingsen L., Bauer R., Bjerrum M.J., Schwarz K., Blaha P. and Andersen P. (1999). Structure, chemical bonding, and nuclear quadrupole interactions of β-$Cd(OH)_2$: Experiment and first principles calculations. *Inorg. Chem.*, **38**, 2860–2867.

[8002] Hemmingsen L. and Ryde U. (1996). Ab initio calculations of electric field gradients in cadmium complexes. *J. Phys. Chem.*, **100**, 4803–4809.

[8003] Hemstreet L.A., Fong C.Y. and Nelson J.S. (1993). First-principles calculations of spin-orbit splittings in solids using nonlocal separable pseudopotentials. *Phys. Rev. B*, **47**, 4238–4243.

[8004] Henderson M., Curtis L.J., Matulioniene R., Ellis D.G. and Theodosiu C.E. (1997). Lifetime measurements in Tl III and the determination of the ground-state dipole polarizabilities for Au I - Bi V. *Phys. Rev. A*, **56**, 1872–1878.

[8005] Henderson M., Irving R.E., Curtis L.J., Martinson I., Brage T. and Bengtsson P. (1999). Lifetimes in the $5d^96p$ levels in HgIII. *Phys. Rev. A*, **59**, 4068–4070.

[8006] Hendrickx M., Gong K. and Vanquickenborne L. (1997). Theoretical study of the molecular complexes of Fe^+ with small alkanes. *J. Chem. Phys.*, **107**, 6299–6305.

[8007] Hersbach J.P.T. (1993a). *Relativistic calculations in momentum space: from positronium to quarkonium.*. Ph.D. thesis, Utrecht.

[8008] Hersbach H. (1993b). Relativistic linear potential in momentum space. *Phys. Rev. D*, **47**, 3027–3033.

[8009] Hersbach H. (1994). Relativistic meson spectroscopy in momentum space. *Phys. Rev. D*, **50**, 2562–2575.

[8010] Hertwig R.H., Hrušák J., Schröder D., Koch W. and Schwarz H. (1995). The metal-ligand bond strengths in in cationic gold(I) complexes. Application of approximate density functional theory. *Chem. Phys. Lett.*, **236**, 194–200.

[8011] Hertwig R.H. and Koch W. (1999). A theoretician's view of the C-F bond activation mediated by the lanthanide cations Ce^+ and Ho^+. *Chem. Eur. J.*, **5**, 312–319.

[8012] Hertwig R.H., Koch W., Schröder D., Schwarz H., Hrušák J. and Schwerdtfeger P. (1996). A comparative computational study of cationic coinage metal - ethylene complexes $(C_2H_4)M^+$ (M = Cu, Ag, and Au). *J. Phys. Chem.*, **100**, 12253–12260.

[8013] Hertwig R.H., Koch W. and Yates B.F. (1998). Economical treatments of relativistic effects and electron correlation in WH_6. *J. Comp. Chem.*, **19**, 1604–1611.

[8014] Hess B.A. (1997). Relativistic effects in heavy-element chemistry. *Ber. Bunsenges*, **101**, 1–10.

[8015] Hess B.A. (1998). Relativistic theory and applications. In *Encyclopedia in Computational Chemistry*, (Edited by von Ragué Schleyer P.), vol. 4, pp. 2499–2508, Wiley, Chichester and New York.

[8016] Hess B.A., Marian C.M. and Peyerimhoff S.D. (1995). Ab initio calculation of spin-orbit effects in molecules including electron correlation. In *Modern Electronic Structure Theory*, (Edited by Yarkony D.R.), pp. 152–278, World Scientific, Singapore.

[8017] Hess B.A., Marian C.M., Wahlgren U. and Gropen O. (1996). A mean-field spin-orbit method applicable to correlated wavefunctions. *Chem. Phys. Lett.*, **251**, 365–371.

[8018] Hestenes D. (1993). Zitterbewegung modeling. *Found. Phys.*, **23**, 365–387.

[8019] Hickman A.P., Medikeri-Naphade M., Chapin C.D. and Huestis D.L. (1997). Calculation of fine-structure effects in O^+-O collisions. *Phys. Rev. A*, **56**, 4633–4644.

[8020] Hidaka M., Fujita T., Nakai H. and Nakatsuji H. (1997). Ab initio molecular orbital model of scanning tunneling microscopy. Benzene and benzene adsorbed on a Ag surface . *Chem. Phys. Lett.*, **264**, 371–375.

[8021] Higuchi M. and Hasegawa A. (1997). A relativistic current- and spin-density functional theory and a single-particle equation. *J. Phys. Soc. Japan.*, **66**, 149–157.

[8022] Higuchi M. and Hasegawa A. (1998). Single-particle equation of relativistic current- and spin-density functional theory and its application to the atomic structure of the lanthanide series. *J. Phys. Soc. Japan.*, **67**, 2037–2047.

[8023] Hill R.N. and Krauthauser C. (1994). A solution to the problem of variational collapse for the one-particle Dirac equation. *Phys. Rev. Lett.*, **72**, 2151–2154.

[8024] Himmel H.J., Downs A.J., Greene T.M. and Andrews L. (1999). Methylgallium and methylindium: the first sighting of the simples organic derivatives of Ga(I) and In(I). *Chem. Comm.*, pp. 2243–2244.

[8025] Hirata M. (1996). Relativistic DV-DS MO method for actinide chemistry (in Japanese). *JAERI Res. Rep. 96-033*, 21 p.

[8026] Hirata M., Tachimori S., Sekine R., Onoe J., Takeuchi K., Nakamatsu H. and Mukoyama T. (1998). Electronic structure of actinyl nitrate - triethyl phosphate complexes using the DV-DS method. *J. Alloys Compounds*, **271-273**, 128–132.

[8027] Hjelm A. and Calais J.L. (1993). Self-consistent calculations of the Zeeman splitting in metals. *Phys. Rev. B*, **48**, 8592–8606.

[8028] Hjelm A., Eriksson O. and Johansson B. (1993). Breakdown of Hund's third rule for induced magnetism in U metal. *Phys. Rev. Lett.*, **71**, 1459–1461.

[8029] Hjelm A., Granqvist C.G. and Wills J.M. (1996). Electronic structure and optical properties of WO_3, $LiWO_3$, $NaWO_3$, and HWO_3. *Phys. Rev. B*, **54**, 2436–2445.

[8030] Hnizdo V. (1994). Vacuum-polarization potentials of extended nuclear charges. *Comp. Phys. Comm.*, **83**, 95–106.

[8031] Ho Y.P., Yang Y.C., Klippenstein S.J. and Dunbar R.C. (1997). Binding energies of Ag^+ and Cd^+ complexes from analysis of radiative association kinetics. *J. Phys. Chem. A*, **101**, 3338–3347.

[8032] Hoffman D.C. (1992). Nuclear and chemical properties of elements 103, 104, and 105. In *Transuranium Elements. A Half Century*, (Edited by Morss L.R. and Fuger J.), pp. 104–115, ACS, Washington DC.

[8033] Hoffman D.C. (1994). The heaviest elements. *C. & E. News, May 2*, pp. 24–34.

[8034] Hoffman D.C. (1996). Chemistry of the heaviest elements. *Radiochim. Acta*, **72**, 1–6.

[8035] Hoffman D.C. and Lee D.M. (1999). Chemistry of the heaviest elements - one atom at a time. *J. Chem. Ed.*, **76**, 332–347.

[8036] Hoffman G.J., Swafford L.A. and Cave R.J. (1998). An ab initio study of the mono- and difluorides of krypton. *J. Chem. Phys.*, **109**, 10701–10706.

[8037] Hofmann C., Augustin J., Reinhardt J., Schafer A., Greiner W. and Soff G. (1993). Relativistic Coulomb-distorted plane waves. *Phys. Scr.*, **48**, 257–262.

[8038] Hofmann C.R. and Soff G. (1996). Total and differential conversion coefficients for internal pair creation in extended nuclei. *At. Data Nucl. Data Tables*, **63**, 189–273.

[8039] Hofmann S. (1998). New elements - approaching $Z = 114$. *Rep. Prog. Phys.*, **61**, 639–689.

[8040] Hofstetter S., Hofmann C. and Soff G. (1992). Ionization of hydrogen by a relativistic heavy projectile. *Z. Phys. D*, **23**, 227–232.

[8041] Holas A. and March N.H. (1993). Relativistic density functional theory: reduction of many-electron problem of atoms and molecules to a one-electron Dirac equation. *J. Mol. Str. (Theochem)*, **279**, 273–279.

[8042] Hollemann A.F., Wiberg E. and Wiberg N. (1995). *Lehrbuch der Anorganischen Chemie, 101. Auflage.*. W. de Gruyter, Berlin, kap. 2.1.4. Relativistische Effekte, pp. 338-340.

[8043] Hollwarth A., Böhme M., Dapprich S., Ehlers A.W., Gobbi A., Jonas V., Köhler K.F., Stegmann R., Veldkamp A. and Frenking G. (1993). A set of d-polarization functions for pseudo-potential basis sets of the main group elements Al-Bi and f-type polarization function for Zn, Cd, Hg. *Chem. Phys. Lett.*, **208**, 237–240.

[8044] Holthausen M.C., Heinemann C., Cornehl H.H., Koch W. and Schwarz H. (1995). The performance of density-functional/Hartree-Fock hybrid methods: cationic transition-metal methyl complexes MCH_3^+ (M = Sc-Cu, La, Hf-Au). *J. Chem. Phys.*, **102**, 4931–4941.

[8045] Holzscheiter M.H. and Charlton M. (1999). Ultra-low energy antihydrogen. *Rep. Prog. Phys.*, **62**, 1–60.

[8046] Hong G.Y., Lin X.J., Li L.M. and Xu G.X. (1997). Linkage isomerism and the relativistic effect in interaction of lanthanoid and carbon monoxide. *J. Phys. Chem. A*, **101**, 9314–9317.

[8047] Hong G.Y., Schautz F. and Dolg M. (1999). Ab initio study of metal-ring bonding in the bis(η^6-benzene) lanthanide and -actinide complexes $M(C_6H_6)_2$ (M = La, Ce, Nd, Gd, Tb, Lu, Th, U). *J. Am. Chem. Soc.*, **121**, 1502–1512.

[8048] Honvault P. and Launay J.M. (1999). Effect of spin-orbit corrections on the $F + D_2 \rightarrow DF + D$ reaction. *Chem. Phys. Lett.*, **303**, 657–663.

[8049] Horbatsch M. and Shapoval D.V. (1995a). Analysis of the Klein-Gordon Coulomb problem in the Feshbach-Villars represenation. *Phys. Rev. A*, **51**, 1804–1807.

[8050] Horbatsch M. and Shapoval D.V. (1995b). Relativistic two-particle scattering resonances in the Tamm-Dancoff approximation. *Phys. Rev. D*, **51**, 6008–6016.

[8051] Horbatsch M. and Shapoval D.V. (1995c). Analysis of the Dirac-Coulomb problem in the free-particle representation. *Phys. Rev. A*, **52**, 3348–3351.

[8052] Horodecki P., Kwela J. and Sienkiewicz J.E. (1999). Transition probabilities of forbidden lines in Pb I. *Eur. Phys. J. D*, **6**, 435–440.

[8053] Horwitz L.P. (1993). Dynamical group of the relativistic Kepler problem. *J. Math. Phys.*, **34**, 645–648.

[8054] Hostler L. (1993). Commutator expansion of the self-energy operator for an electron in an external potential. *J. Math. Phys.*, **34**, 5509–5532.

[8055] Hostler L. (1996). Commutator expansion. II. Relativistic reduced Green's functions and the Lamb shift calculation. *J. Math. Phys.*, **37**, 1632–1641.

[8056] Hota R.L., Patnaik R.C., Tripathi G.S. and Misra P.K. (1995). NMR chemical shift in PbTe and spin-orbit effects. *Phys. Rev. B*, **51**, 7291–7294.

[8057] Hou Q., Liu L. and Li J. (1993). Theoretical calculation of differential cross section for e^+-Ar and e^--Ar elastic scattering. *Chin. Phys. Lett.*, **10**, 11–14.

[8058] Howard S.T. (1994). Ab initio effective core potential calculations on HgI_2, PtI_2 and PbI_2. *J. Phys. Chem.*, **98**, 6110–6113.

[8059] Hrušák J. (1997). Relativistic effects in the metal-carbon bond. A comparative ab initio and DFT study of $M{=}CH_2^+$ and $M\text{-}CH_3^+$ (M=Cu, Ag and Au). *S. Afr. J. Chem.*, **50**, 93–101.

[8060] Hrušák J., Hertwig R.H., Schröder D.H., Schwerdtfeger P., Koch W. and Schwarz H. (1995). Relativistic effects in cationic gold(I) complexes: a comparative study of ab initio pseudopotential and density functional methods. *Organometallics*, **14**, 1284–1291.

[8061] Hrušák J., Schröder D. and Schwarz H. (1994). Theoretical prediction of the structure and the bond energy of the gold(I) complex Au^+ (H_2O). *Chem. Phys. Lett.*, **225**, 416–420.

[8062] Hsu J.J., Chung K.T. and Huang K.N. (1994). $^4P_0$ series of lithium. *Phys. Rev. A*, **49**, 4466–4472.

[8063] Hu A., Otto P. and Ladik J. (1998). Relativistic all-electron Hartree-Fock-Dirac calculation of a quasi one-dimensional chain of selenium atoms. *Chem. Phys. Lett.*, **293**, 277–283.

[8064] Hu A., Otto P. and Ladik J. (1999a). Relativistic Gaussian functions for atoms by fitting numerical results with adaptive nonlinear least-square algorithm. *J. Comp. Chem.*, **20**, 655–664.

[8065] Hu A., Otto P. and Ladik J. (1999b). Relativistic all-electron molecular Hartree-Fock-Dirac-(Gaunt) calculations on HgO. *J. Mol. Str. (Theochem)*, **468**, 163–169.

[8066] Hu W.P. and Truhlar D.G. (1994). Structural distortion of CH_3I in an ion-dipole precursor complex. *J. Phys. Chem.*, **98**, 1049–1052.

[8067] Hua G. (1999). Exact solution of the time-dependent Dirac equation. *Wu Li Xue Bao*, **48**, 983–986.

[8068] Huang D.P. and Corbett J.D. (1998). $A_5TaAs_4Tl_2$ (A=Rb,K). Transition-metal Zintl phases with a novel complex ion: Synthesis, structure, and bonding. *Inorg. Chem.*, **37**, 4006–4010.

[8069] Huang W., Zou Y., Tong X.M. and Lim J.M. (1995). Atomic energy levels and Landé g factors: a theoretical study. *Phys. Rev. A*, **52**, 2770–2777.

[8070] Huang Y.S. (1997). Schrödinger-like formalism of relativistic quantum theory for spin-zero particles. *Nuovo Cim. B*, **112**, 75–83.

[8071] Huang Y.S., Yang C.C. and Liaw S.S. (1999). Relativistic solution of hydrogen in a spherical cavity. *Phys. Rev. A*, **60**, 85–90.

[8072] Huber H. and Leeb H. (1998). Spin-orbit potentials from inversion: relativistic versus nonrelativistic schemes. *J. Phys. G*, **24**, 1287–1300.

[8073] Huber H.J.J. (1995). *Inverses Streuproblem mit Spin-Bahn-Term*. Ph.D. thesis, Technische Universität Wien, 158 p.

[8074] Hubert S., Thouvenot P. and Edelstein N. (1993). Spectroscopic studies and crystal-field analyses of Am^{3+} and Eu^{3+} in the cubic-symmetry site of ThO_2. *Phys. Rev. B*, **48**, 5751–5760.

[8075] Hübler K., Hübler U., Roper W.R., Schwerdtfeger P. and Wright L.J. (1997a). The nature of the metal-silicon bond in $[M(SiR_3)H_3(PPh_3)_3]$ (M = Ru,Os) and the crystal structure of $[OsSi(N\text{-pyrrolyl})_3H_3(PPh_3)_3]$. *Chem. Eur. J.*, **3**, 1608–1616.

[8076] Hübler K., Hunt P.A., Maddock S.M., Rickard C.E.F., Roper W.R., Salter D.M., Schwerdtfeger P. and Wright L.J. (1997b). Examination of metal-silicon bonding through structural and theoretical studies of an isostructural set of five-coordinate silyl complexes, $Os(SiR_3)Cl(CO)(PPh_3)_2$ (R = F, Cl, OH, Me). *Organometallics*, **16**, 5076–5083.

[8077] Hughes S.R. and Kaldor U. (1993). The Fock-space coupled-cluster method: electron affinities of the five halogen elements with consideration of triple excitations. *J. Chem. Phys.*, **99**, 6773–6776.

[8078] Hughes V.W. and Kinoshita T. (1999). Anomalous g values of the electron and muon. *Rev. Mod. Phys.*, **71**, S133–S139.

[8079] Huheey J.E., Keiter E.A. and Keiter R.L. (1993). *Inorganic Chemistry. Principles of Structure and Reactivity, 4th Ed.*. HarperCollins CollegePublishers, New York, (See pp. 579, 879-880).

[8080] Huhne T. and Ebert H. (1999). Fully relativistic description of the magneto-optical properties of arbitrary layered systems. *Phys. Rev. B*, **60**, 12982–12989.

[8081] Huhne T., Zecha C., Ebert H., Dederichs P.H. and Zeller R. (1998). Full-potential spin-polarized relativistic Korringa-Kohn-Rostoker method implemented and applied to bcc Fe, fcc Co, and fcc Ni. *Phys. Rev. B*, **58**, 10236–10247.

[8082] Hummer D.G., Eissner B.K.A.W., Pradhan A.K., Saraph H.E. and Tully J.A. (1993). Atomic data from the IRON Project. *Astron. Astrophys*, **279**, 298–309.

[8083] Hunstock E., Mealli C., Calhorda M.J. and Reinhold J. (1999). Molecular structures of $M_2(CO)_9$ and $M_3(CO)_{12}$ (M = Fe,Ru,Os): New theoretical insigts. *Inorg Chem*, **38**, 5053–5060.

[8084] Hunt P. and Schwerdtfeger P. (1996). Are the compounds InH_3 and TlH_3 stable gas phase or solid state species? *Inorg. Chem.*, **35**, 2085–2088.

[8085] Hutschka F., Dedieu A., Troxler L. and Wipff G. (1998). Theoretical studies on the UO_2^{2+} and Sr^{2+} complexation by phosphoryl-containing $O{=}PR_3$ ligands: QM ab initio calculations in the gas phase and MD FEP calculations in aqueous solution. *J. Phys. Chem.*, **102**, 3773–3781.

[8086] Huzinaga S. (1995). Concept of active electrons in chemistry. *Can. J. Chem.*, **73**, 619–628.

[8087] Hwang C., Lee J.W., Kim S.T., Lee D.H. and Onellion M. (1997). Electronic structure of praseodymium. *Solid State Comm.*, **103**, 229–233.

[8088] Hylton D.J. and Snyderman N.J. (1997). Analytic basis set for high-Z atomic QED calculations: Heavy He-like ions. *Phys. Rev. A*, **55**, 2651–2661.

[8089] Iacopini E. (1993). Casimir effect at macroscopic distances. *Phys. Rev. A*, **48**, 129–131.

[8090] Ichihara A., Shirai T. and Eichler J. (1993). Cross sections for electron capture in relativistic atomic collisions. *At. Data Nucl. Data Tables*, **55**, 63–79.

[8091] Ichihara A., Shirai T. and Eichler J. (1996). Radiative electron capture and the photoelectric effect at high energies. *Phys. Rev. A*, **54**, 4954–4959.

[8092] Ichikawa K., Yamanaka T., Takamuku A. and Glaser R. (1997). Neutron diffraction of homopolyatomic bismuth ions in liquid $Bi_5(AlCl_4)_3$ and ab initio study of the structure and bonding of the isolated Bi_5^{3+} ion. *Inorg. Chem.*, **36**, 5284–5290.

[8093] Idlis B.G., Musakhanov M.M. and Usmanov M.S. (1995). Application of supersymmetry and factorization methods to solution of Dirac and Schrödinger equations. *Theor. Math. Phys.*, **101**, 1191–1199.

[8094] Igel-Mann G. (1987). *Semiempirische Pseudopotentiale: Untersuchungen an Hauptgruppenelementen und Nebengruppenelementen mit abgeschlossener d-Schale.*. Ph.D. thesis, Inst. Theor. Chem, Univ. Stuttgart.

[8095] Igel-Mann G., Schlunk R. and Stoll H. (1995). Structure and ionization potentials of clusters containing heavy elements. Part 4: Mixed clusters of alkali and group VI elements. *Mol. Phys.*, **84**, 679–690.

[8096] Igel-Mann G. and Stoll H. (1995). Stucture and ionization potentials of clusters containing heavy elements. Part 3: Mixed clusters of alkali and group V elements. *Mol. Phys.*, **84**, 663–678.

[8097] Ignaczak A. and Gomes J.A.N.F. (1996). Interaction of halide ions with copper: the DFT approach. *Chem. Phys. Lett.*, **257**, 609–615.

[8098] Ihee H., Zewail A.H. and Goddard III W.A. (1999). Conformations and barriers of haloethyl radicals (CH_2XCH_2, X = F,Cl,Br,I): Ab initio studies. *J. Phys. Chem. A*, **103**, 6638–6649.

[8099] Iliaš M., Furdík P. and Urban M. (1998). Comparative study of electron correlation and relativistic effects in CuF, AgF, and AuF. *J. Phys. Chem. A*, **102**, 5263–5268.

[8100] Iliaš M. and Neogrády P. (1999). Ionization potentials of Zn, Cd, Hg and dipole polarizabilities of Zn^+, Cd^+, Hg^+: correlation and relativistic effects. *Chem. Phys. Lett.*, **309**, 441–449.

[8101] Ilyabaev E. and Kaldor U. (1993). Relativistic coupled-cluster calculations for open-shell atoms. *Phys. Rev. A*, **47**, 137–142.

[8102] Indelicato P. (1995). Projection operators in multiconfiguration Dirac-Fock calculations: application to the ground state of heliumlike atoms. *Phys. Rev. A*, **51**, 1132–1145.

[8103] Indelicato P. (1996). Correlation and negative continuum effects for the relativistic M1 transition in two-electron ions using the multiconfiguration Dirac-Fock method. *Phys. Rev. Lett.*, **77**, 3323–3326.

[8104] Indelicato P. (1997). Radiative de-excitation of the $1s^2 2s3p^3$ $^3P_0$ level in beryllium-like ions: a competition between an E2 and a two-electron one-photon E1 transition. *Hyperfine Int.*, **108**, 39–49.

[8105] Indelicato P., Boucard S. and Lindroth E. (1998). Relativistic and many-body effects in K, L, and M shell ionization energy for elements with $10 \leq Z \leq 100$ and the determination of the 1s Lamb shift for heavy elements. *Eur. Phys. J. D*, **3**, 29–41.

[8106] Indelicato P. and Desclaux J.P. (1993). Projection operator in the multiconfiguration Dirac-Fock method. *Phys. Scr.*, **T46**, 110–114.

[8107] Indelicato P. and Mohr P.J. (1995). Asymptotic expansion of the Dirac-Coulomb radial Green's function. *J. Math. Phys.*, **36**, 714–724.

[8108] Indelicato P. and Mohr P.J. (1998a). Coordinate-space approach to the bound-electron self-energy: Coulomb field calculation. *Phys. Rev. A*, **58**, 165–179.

[8109] Indelicato P. and Mohr P.J. (1998b). 6s and 8d state self-energy for hydrogen-like ions and new results on the self-energy screening. *Hyperfine Int.*, **114**, 147–153.

[8110] Ionescu D.C. and Belkacem A. (1999). Relativistic collisions of highly-charged ions. *Phys. Scripta*, **T80**, 128–132.

[8111] Ionescu D.C. and Eichler J. (1996). Bound-free electron-positron pair creation in relativistic heavy-ion collisions as a charge-transfer process. *Phys. Rev. A*, **54**, 4960–4967.

[8112] Ionescu D.C., Greiner W., Muller B. and Soff G. (1993). Collective excitations of the QED vacuum. *Phys. Rev. A*, **47**, 854–867.

[8113] Ionescu D.C., Sørensen A.H. and Belkacem A. (1999). Inner-shell photoionization at relativistic energies. *Phys. Rev. A*, **59**, 3527–3537.

[8114] Ionova G.V., Pershina V.G., Gerasimova G.A., Mikhalko V.K., Kostrubov Y.N. and Suraeva N.I. (1996a). Electronic structure and properties of oxyhalides of group 5 elements (in Russian). *Zh. Neorg. Khim.*, **41**, 632–636.

[8115] Ionova G.V., Pershina V.G., Gerasimova G.A., Mikhalko V.K., Kostrubov Y.N. and Suraeva N.I. (1996a). Electronic structure and properties of oxyhalides of group 5 elements. *Russian J. Inorg. Chem.*, **41**, 610–614.

[8116] Ionova G.V., Pershina V.G., Gerasimova G.A., Mikhalko V.K., Kostrubov Y.N. and Suraeva N.I. (1996b). Relativistic effects on atomic properties of elements 104, 105 and 106 (in Russian). *Zh. Neorg. Khim.*, **41**, 673–678.

[8117] Ionova G.V., Pershina V.G., Gerasimova G.A., Mikhalko V.K., Kostrubov Y.N. and Suraeva N.I. (1996b). Relativistic effects on atomic properties of elements 104, 105 and 106. *Russian J. Inorg. Chem.*, **41**, 651–656.

[8118] Ionova G.V., Pershina V.G., Gerasimova G.A., Mikhalko V.K., Kostrubov Y.N. and Suraeva N.I. (1996c). Electronic structure and properties of halogenides of the groups 4, 5 and 6 (in Russian). *Zh. Neorg. Khim.*, **41**, 1190–1197.

[8119] Ionova G.V., Pershina V.G., Gerasimova G.A., Mikhalko V.K., Kostrubov Y.N. and Suraeva N.I. (1996c). Electronic structure and properties of halides of groups 4-6 elements including transactinides. *Russian J. Inorg. Chem.*, **41**, 1138–1144.

[8120] Ionova G.V., Pershina V.G., Gerasimova G.A., Mikhalko V.K., Kostrubov Y.N. and Suraeva N.I. (1996d). Electronic structure and properties of halogenides of the groups 4, 5 and 6. *Russian J. Inorg. Chem.*, **41**, 775–776.

[8121] Ionova G.V., Pershina V.G., Gerasimova G.A., Mikhalko V.K., Kostrubov Y.N. and Suraeva N.I. (1996d). Electronic structure and properties of Group VI oxohalides (in Russian). *Zh. Neorg. Khim.*, **41**, 802–803.

[8122] Ionova G.V., Pershina V.G., Gerasimova G.A., Mikhalko V.K., Kostrubov Y.N. and Suraeva N.I. (1996e). Fugacity of halides and oxohalides of transition metals and transactinides. *Russian J. Inorg. Chem.*, **41**, 786–789.

[8123] Ionova G.V., Pershina V.G., Zuraeva I.T. and Suraeva N.I. (1996f). Variation law for thermodynamic properties of transition elements in $5d$ series: enthalphy of sublimation. *Russian J. Inorg. Chem.*, **41**, 162–165.

[8124] Irikura K.K. and Goddard III W.A. (1994). Energetics of third-row transition metal methylidene ions MCH_2^+ (M=La, Hf, Ta, W, Re, Os, Ir, Pt, Au). *J. Am. Chem. Soc.*, **116**, 8733–8740.

[8125] Ishii T., Sekine R., Enoki T., Miyazaki E., Miyamae T. and Miyazaki T. (1997). DV-Xα calculation and ultraviolet photoelectron spectra of gold trichloride-graphite intercalation compound ($AuCl_3$-GIC). *J. Phys. Soc. Japan.*, **66**, 3424–3433.

[8126] Ishikawa Y. (1993). Comment on 'Segmented contractions of relativistic Gaussian basis sets'. *Chem. Phys. Lett.*, **212**, 547–550.

[8127] Ishikawa Y. (1998). The use of Gaussian spinors in relativistic electronic structure calculations: the effect of the boundary of the finite nucleus of uniform proton charge distribution. *Chem. Phys.*, **225**, 239–246.

[8128] Ishikawa Y. and Kaldor U. (1996). Relativistic many-body calculations on atoms and molecules. In *Computational Chemistry. Reviews of Current Trends*, (Edited by Leszczynski J.), vol. 1, pp. 1–52, World Scientific, Singapore.

[8129] Ishikawa Y. and Koc K. (1994). Relativistic many-body perturbation theory based on the no-pair Dirac-Coulomb-Breit Hamiltonian: Relativistic correlation energies for the noble-gas sequence through Rn (Z=86), the group-IIB atoms through Hg and the ions of Ne isoelectronic sequence. *Phys. Rev. A*, **50**, 4733–4742.

[8130] Ishikawa Y. and Koc K. (1996). Relativistic many-body perturbation theory for general open-shell multiplet states of atoms. *Phys. Rev. A*, **53**, 3966–3973.

[8131] Ishikawa Y. and Koc K. (1997a). Relativistic many-body perturbation calculations for open-shell systems. *Phys. Rev. A*, **56**, 1295–1304.

[8132] Ishikawa Y. and Koc K. (1997b). Relativistic many-body perturbation calculations for Zn and Cd and their singly ionized ions. *Int. J. Quantum Chem.*, **65**, 545–554.

[8133] Ishikawa Y., Koc K. and Schwarz W.H.E. (1997). The use of Gaussian spinors in relativistic electronic structure calculations: the effect of the boundary of the finite nucleus of uniform proton charge distribution. *Chem. Phys.*, **225**, 239–246.

[8134] Ishikawa Y., Nakajima T., Hada M. and Nakatsuji H. (1998). Relativistic theory of the magnetic shielding constant: a Dirac-Fock finite perturbation study. *Chem. Phys. Lett.*, **283**, 119–124.

[8135] Ishikawa Y. and Quiney H.M. (1993). Relativistic many-body perturbation-theory calculations based on Dirac-Fock-Breit wave functions. *Phys. Rev. A*, **47**, 1732–1739.

[8136] Ismail N., Heully J.L., Saue T., Daudey J.P. and Marsden C.J. (1999). Theoretical studies of the actinides: method calibration for the UO_2^{2+} and PuO_2^{2+} ions. *Chem. Phys. Lett.*, **300**, 296–302.

[8137] Isozaki H. (1997). Inverse scattering theory for Dirac operators. *Ann. Inst. Henri Poincaré Phys. Theor.*, **66(2)**, 237–270.

[8138] Ito H. (1993). On the renormalization condition of the relativistic wave function at the origin. *Progr. Theor. Phys.*, **89**, 763–765.

[8139] Ito H. and Gross F. (1993). Gauge invariance and Compton scattering from relativistic composite systems. *Phys. Rev. C*, **48**, 1948–1972.

[8140] Itoh S., Saito R., Kimura T. and Yabushita S. (1994). Relativistic effect on multiplet terms of rare earth ions. *J. Phys. Soc. Japan.*, **63**, 807–813.

[8141] Ivanov L.N. and Ivanova E.P. (1996). Sturm orbital method for calculating the radiation physics from atoms and ions (in Russian). *Zh. Eksp. Teor. Fiz.*, **110**, 483–498.

[8142] Ivanov L.N., Ivanova E.P. and Knight L.V. (1993a). Energy approach to consistent QED theory for calculation of electron-collision strengths: Ne-like ions. *Phys. Rev. A*, **48**, 4365–4378.

[8143] Ivanov L.N., Ivanova E.P., Knight L.V. and Molchanov A.G. (1993b). Spectrum of plasma containing Ne-like and Na-like ions: Consistent account for Rydberg and autoionizing Rydberg series in balance equations. *Phys. Scr.*, **53**, 653–667.

[8144] Ivanov P.B. and Safronova U.I. (1994). Correlation and relativistic effects for $3l3l'$ autoionization states. *Phys. Scr.*, **49**, 408–416.

[8145] Ivanov V.G. and Karshenbojm S.G. (1997a). Radiative corrections to the level widths in light mesic atoms (in Russian). *Zh. Eksp. Teor. Fiz.*, **112**, 805–817.

[8146] Ivanov V.G. and Karshenbojm S.G. (1997b). Vacuum polarization contribution of order $\alpha(Z\alpha)^6mc^2$ (in Russian). *Yad. Fiz.*, **60**, 333–335.

[8147] Ivanov V.K. (1999). Many-body effects in negative ion photodetachment. *J. Phys. B*, **32**, R67–R101.

[8148] Iversen B.B., Larsen F.K., Pinkerton A.A., Martin A., Darovsky A. and Reynolds P.A. (1998). Characterization of actinide bonding in $Th(S_2PMe_2)_4$ by synchrotron x-ray diffraction. *Inorg. Chem.*, **37**, 4559–4566.

[8149] Jackson P., Diefenbach M., Schröder D. and Schwarz H. (1999). Combined quantum chemical and mass spectrometry study of $[Ge,C,H]^+$ and its neutral counterpart. *Eur. J. Inorg. Chem.*, pp. 1203–1210.

[8150] Jacobsen H. and Berke H. (1997). A density functional study of the systems $(MH_3(PMe_3)_4)^+$ (M=Fe,Ru,Os). *Chem. Ber./Recueil*, **130**, 1273–1277.

[8151] Jacobsen H., Schreckenbach G. and Ziegler T. (1994). The metal carbon double bond in Fischer carbenes: A density functional study of the importance of nonlocal density corrections and relativistic effects. *J. Phys. Chem.*, **98**, 11406–11410.

[8152] Jacobsen H. and Ziegler T. (1994). Nonclassical double bonds in ethylene analogues: Influence of Pauli repulsion on trans bending and π-bond strength. A density functional study. *J. Am. Chem. Soc.*, **116**, 3667–3679.

[8153] Jacobsen H. and Ziegler T. (1996). Transition metal Fischer-type complexes. Density functional analysis of the systems $(CO)_5Cr{=}EH_2$ (E = C, Si, Ge, Sn) and $(CO)_5M{=}CH_2$ (M = Mo, W, Mn^+). *Inorg. Chem.*, **35**, 775–783.

[8154] Jaekel M.T. and Reynaud S. (1997). Movement and fluctuations of the vacuum. *Rep. Prog. Phys.*, **60**, 863–887.

[8155] Jaekel M.T. and Reynaud S. (1999). Observable Dirac electron in accelerated frames. *Phys. Lett. A*, **256**, 95–103.

[8156] Jäger J., Stahl D., Schmidt P.C. and Kniep R. (1993). Ca_3AuN: A calcium auride subnitride (in German). *Angew. Chem.*, **105**, 738.

[8157] Jäger J., Stahl D., Schmidt P.C. and Kniep R. (1993). Ca_3AuN: A calcium auride subnitride. *Angew. Chem. Int. Ed. Engl.*, **32**, 709–710.

[8158] Jakubek Z.J. and Takami M. (1997). Ab initio studies of AgHe exciplex. *Chem. Phys. Lett.*, **265**, 653–659.

[8159] Jallouli H. and Sazdjian H. (1997a). The relativistic two-body potentials of constraint theory from summation of Feynman diagrams. *Ann. Phys. (NY)*, **253**, 376–426.

[8160] Jallouli H. and Sazdjian H. (1997b). Incorporation of anomalous magnetic moments in the two-body relativistic wave equations of constraint theory. *J. Math. Phys.*, **38**, 2181–2196.

[8161] Jallouli H. and Sazdjian H. (1998). Relativistic effects in the pionium lifetime [Erratum: p. 099901]. *Phys. Rev. D*, **58**, 014011, 1–17.

[8162] James J. and Sandars P.G.H. (1999). A parametric approach to nuclear size and shape in atomic parity nonconservation. *J. Phys. B*, **32**, 3295–3307.

[8163] Jamieson M.J., Drake G.W.F. and Dalgarno A. (1995). Retarded dipole-dipole dispersion interaction potential for helium. *Phys. Rev. A*, **51**, 3358–3361.

[8164] Jamorski C., Dargelos A., Teichteil C. and Daudey J.P. (1994). Theoretical determination of spectral lines for the Zn atom and the ZnH molecule. *J. Chem. Phys.*, **100**, 917–925.

[8165] Janiak C. (1997). (Organo)thallium (I) and (II) chemistry: syntheses, structures, properties and applications of subvalent thallium complexes with alkyl, cyclopentadienyl, arene or hydrotris(pyrazolyl) borate ligands. *Coord. Chem. Rev.*, **163**, 107–216, see pp. 188-190 for theory.

[8166] Jaquet R., Cencek W., Kutzelnigg W. and Rychlewski J. (1998). Sub-microhartree accuracy potential energy surface for H_3^+ including adiabatic and relativistic effects. II. Rovibrational analysis for H_3^+ and D_3^+. *J. Chem. Phys.*, **108**, 2837–2846.

[8167] Jarid A., Moreno M., Lledos A., Lluch J.M. and Bertran J. (1993). Ab initio calculations of the quantum mechanical hydrogen exchange coupling in the $[(C_5H_5)Ir(PH_3)H_3]^+$ complex. *J. Am. Chem. Soc.*, **115**, 5861–5862.

[8168] Jáuregui R., Bunge C.F. and Ley-Koo E. (1991). Relativistic CI for atoms. In *Atomic and Molecular Physics*, (Edited by Cisneros C., Alvarez I. and Morgan T.), pp. 276–287, World Scientific, Singapore.

[8169] Jáuregui R., Bunge C.F. and Ley-koo E. (1994). Relativistic atomic CI: A variational principle. In *Proc. 4th US/Mexico Symp. on Atomic and Molecular Physics*, (Edited by Alvarez I., Cisneros C. and Morgan T.J.), pp. 72–79, World Scientific, Singapore.

[8170] Jáuregui R., Bunge C.F. and Ley-Koo E. (1996). Variational principle for atomic relativistic states. In *Proc. 3rd UNAM-CRAY Supercomputing Conf.*, (Edited by Cisneros G., Cogordan J.A., Castro M. and Wang C.M.), pp. 133–145, World Scientific, Singapore.

[8171] Jáuregui R., Bunge C.F. and Ley-Koo E. (1997a). Upper bounds to the eigenvalues of the no-pair Hamiltonian. *Phys. Rev. A*, **55**, 1781–1784.

[8172] Jáuregui R., Bunge C.F. and Ley-Koo E. (1997b). Electrons in atoms: How well can we describe them? *Advances in Applied Clifford Algebras*, **7(S)**, 79–95.

[8173] Jáuregui R., Bunge C.F. and Ley-Koo E. (1997c). Atoms as QED bound atoms. *Rev. Mex. Fis.*, **43**, 673–698.

[8174] Jean Y. and Lledos A. (1998). Phosphines exchange in quadruply bonded metal dimers: theoretical proposal for an alternative to the internal flip mechanism. *Chem. Comm.*, pp. 1443–1444.

[8175] Jefimenko O.D. (1995). Retardation and relativity: The case of a moving line charge. *Am. J. Phys.*, **63**, 454–459.

[8176] Jefimenko O.D. (1996). Derivation of relativistic force transformation equations from Lorentz force law. *Am. J. Phys.*, **64**, 618–620.

[8177] Jemmis E.D. and Giju K.T. (1997). Organometallic analogs of the cyclobutadiene dication: An ab initio MO and density functional study of the symmetrical planar and puckered $(WL_2(\mu\text{-}CR))_2$ complexes (L= H,Me,F,OH; R= H,F,Me). *Organometallics*, **16**, 1425–1429.

[8178] Jemmis E.D., Naga Srinavas G., Leszczynski J., Kapp J., Korkin A.A. and von R Schleyer P. (1995). Group 14 analogs of the cyclopropenium ion: do they favor classical aromatic structures? *J. Am. Chem. Soc.*, **117**, 11361–11362.

[8179] Jemmis E.D., Pavan Kumar P.N.V. and Narahari Sastry G. (1994). Possibility of bond stretch isomerism in $(Cp(CO)_2M)_2(\mu\text{-}E)$ complexes (M=Mn, Re, Cr and W; E=S, Se and Te); a molecular orbital study. *J. Organomet. Chem.*, **478**, 29–36.

[8180] Jenkins A.C. and Strange P. (1994). Relativistic spin-polarized single-site scattering theory. *J. Phys: Conden. Matter.*, **6**, 3499–3517.

[8181] Jenkins A.C. and Strange P. (1995). Electronic structure and x-ray magnetic dichroism in random substational alloys of f-electron elements. *Phys. Rev. B*, **51**, 7279–7282.

[8182] Jenkins A.C. and Strange P. (1998). Applications of the SPRKKR-CPA to the calculation of magnetic x-ray circular dichroism spectra in transition-metal alloys. *J. Magn. Magn. Mat.*, **177-181**, 1039–1041.

[8183] Jennison D.R., Schultz P.A. and Sears M.P. (1997). Ab initio calculations of Ru, Pd, and Ag cluster structure with 55, 135, and 140 atoms. *J. Chem. Phys.*, **106**, 1856–1862.

[8184] Jensen H.J.A., Dyall K.G., Saue T. and Faegri Jr K. (1996). Relativistic four-component multiconfigurational self-consistent-field theory for molecules: Formalism. *J. Chem. Phys.*, **104**, 4083–4097.

[8185] Jentschura U.D., Mohr P.J. and Soff G. (1999). Calculation of the electron self-energy for low nuclear charge. *Phys. Rev. Lett.*, **82**, 53–56.

[8186] Jentschura U. and Pachucki K. (1996). Higher-order binding corrections to the Lamb shift of 2 P states. *Phys. Rev. A*, **54**, 1853–1861.

[8187] Jentschura U.D., Soff G., Ivanov V.G. and Karshenboim S.G. (1997). Bound $\mu^+\mu^-$ system. *Phys. Rev. A*, **56**, 4483–4495.

[8188] Jentschura U.D., Soff G. and Mohr P.J. (1997b). Lamb shift of 3 P and 4 P states and the determination of α. *Phys. Rev. A*, **56**, 1739–1755.

[8189] Jeon Y., Chen J. and Croft M. (1994). X-ray-absorption studies of the d-orbital occupancies of selected $4d/5d$ transition metals compounded with group-III/IV ligands. *Phys. Rev. B*, **50**, 6555–6563.

[8190] Jepson O. and Anderson O.K. (1993). The electronic structure of hcp ytterbium. *Solid State Comm.*, **88**, 871–875.

[8191] Jeung G.H. (1994). Ab initio study of high-spin chemical bonding of MCO and MCS (M=Y-Mo). *Chem. Phys. Lett.*, **221**, 237–240.

[8192] Jeziorski B. and Kołos W. (1969). On the ionization potential of H_2. *Chem. Phys. Lett.*, **3**, 677–678.

[8193] Johansen T.H., Hagen K., Hassler K., Tekautz G. and Stølevik R. (1999). 1,1,2-Triiodosilane (I_2HSi-SiH_2I): molecular structure, internal rotation and vibrational properties determined by gas-phase electron diffraction, infrared and Raman spectroscopy, and ab initio molecular orbital- and density functional calculations. *J. Mol. Str.*, **509**, 237–254.

[8194] Johansson B., Ahuja R., Eriksson O. and Wills J.M. (1995). Anomalous fcc crystal structure of thorium metal (Erratum: p. 3968). *Phys. Rev. Lett.*, **75**, 280–283.

[8195] Johansson B. and Brooks M.S.S. (1993). Theory of cohesion in rare earths and actinides. In *Handbook on the Physics and Chemistry of Rare Earths, Lanthanides/Actinides: Physics - I*, (Edited by Gschneidner K.A., Eyring L., Lander G.H. and Choppin G.R.), vol. 17, pp. 149–243, Elsevier, Amsterdam.

[8196] Johansson M., Hotokka M., Pettersson M. and Räsänen M. (1999). Quantum chemical potential energy surfaces for HXeCl. *Chem. Phys.*, **244**, 25–34.

[8197] Johnson E., Krause M.O. and Fricke B. (1996). Electronic energies of americium from multiconfiguration Dirac-Fock calculations. *Phys. Rev. A*, **54**, 4783–4788.

[8198] Johnson E., Pershina V. and Fricke B. (1999). Ionization potentials of seaborgium. *J. Phys. Chem. A*, **103**, 8458–8462.

[8199] Johnson W.R. (1993). Relativistic many-body theory applied to highly-charged ions. In *Many-Body Theory of Atomic Structure and Photoionization*, (Edited by Chang T.N.), World Scientific, Singapore.

[8200] Johnson W.R. and Cheng K.T. (1996). Relativistic configuration-interaction calculation of the polarizabilities of heliumlike ions. *Phys. Rev. A*, **53**, 1375–1378.

[8201] Johnson W.R., Cheng K.T. and Plante D.R. (1997a). Hyperfine structure of 2 ^3P levels of heliumlike ions. *Phys. Rev. A*, **55**, 2728–2742.

[8202] Johnson W.R. and Guet C. (1994). Elastic scattering of electrons from Xe, Cs^+, and Ba^{2+}. *Phys. Rev. A*, **49**, 1041–1048.

[8203] Johnson W.R., Liu Z.W. and Sapirstein J. (1996). Transition rates for lithium-like ions, sodium-like ions, and neutral alkali-metal ions. *At. Data Nucl. Data Tables*, **64**, 279–300.

[8204] Johnson W.R., Safronova M.S. and Safronova U.I. (1997b). Relativistic many-body calculations of energies of Mg I, Al II, Al I, Hg I, Tl II, Tl I, Pb I, Bi II and Bi I. *Phys. Scr.*, **56**, 252–263.

[8205] Johnson W.R., Safronova M.S. and Safronova U.I. (1997c). Relativistic many-body calculations of energies of $n = 3$ states for boron-like sodium. *Phys. Scr.*, **T73**, 45–47.

[8206] Johnson W.R., Sapirstein J. and Blundell S.A. (1993). Atomic structure calculations associated with PNC experiments in atomic cesium. *Phys. Scr.*, **T46**, 184–192.

[8207] Johnson W.R., Sapirstein J. and Cheng K.T. (1995). Theory of $2s_{1/2}$-$2p_{3/2}$ transitions in highly ionized uranium. *Phys. Rev. A*, **51**, 297–302.

[8208] Jonas V., Boehme C. and Frenking G. (1996). Bent's rule and the structure of transition metal compounds. *Inorg. Chem.*, **35**, 2097–2099.

[8209] Jonas V. and Thiel W. (1995). Theoretical study of the vibrational spectra of the transition metal carbonyls $M(CO)_6$ (M = Cr, Mo, W), $M(CO)_5$ (M = Fe, Ru, Os), and $M(CO)_4$ (M = Ni, Pd, Pt). *J. Chem. Phys.*, **102**, 8474–8484.

[8210] Jonas V. and Thiel W. (1996). Theoretical study of the vibrational spectra of the transition-metal carbonyl hydrides $HM(CO)_5$ (M=Mn,Re), $H_2M(CO)_4$ (M=Fe,Ru,Os), and $HM(CO)_4$ (M=Co,Rh,Ir). *J. Chem. Phys.*, **105**, 3636–3648.

[8211] Jonas V. and Thiel W. (1999). Theoretical study on linear dicyanide and dicarboxyl complexes of the metals Au, Hg, and Tl. On the possible existence of a $[Tl(CO)_2]^{3+}$ cation. *J. Chem. Soc., Dalton Trans.*, pp. 3783–3790.

[8212] Jones B.D. and Perry R.J. (1997). Lamb shift in a light-front Hamiltonian approach. *Phys. Rev. D*, **55**, 7715–7730.

[8213] Jones L.M. (1993). Another Dirac oscillator. *AIP Conf. Proc.*, **272**, 582–584.

[8214] Jönsson P. (1993). Multi-configuration Hartree- and Dirac-Fock calculations of atomic hyperfine structures. *Phys. Scr.*, **48**, 678–687.

[8215] Jönsson P. (1995). *Large-Scale Atomic Calculations using Variational Methods*. Ph.D. thesis, Lund Institute of Technology, 65 p. + 21 original papers.

[8216] Jönsson P. and Froese Fischer C. (1993). Large-scale multiconfiguration Hartree-Fock calculations of hyperfine-interaction constants for low-lying states in beryllium, boron and carbon. *Phys. Rev. A*, **48**, 4113–4123.

[8217] Jönsson P. and Froese Fischer C. (1997a). SMS92: a program for relativistic isotope shift calculations. *Comp. Phys. Comm.*, **100**, 81–92.

[8218] Jönsson P. and Froese Fischer C. (1997b). Accurate multiconfiguration Dirac-Fock calculations of transition probabilities in the Mg isoelectronic sequence. *J. Phys. B*, **30**, 5861–5875.

[8219] Jönsson P. and Froese Fischer C. (1998). Multiconfiguration Dirac-Fock calculations of the $2s^2\,{}^1S_0 - 2s2p\,{}^3P_1$ intercombination transition in C III. *Phys. Rev. A*, **57**, 4967–4970.

[8220] Jönsson P., Froese Fischer C. and Träbert E. (1998). On the status and perspectives of MCDF calculations and measurements of transition data in the Be isoelectronic sequence. *J. Phys. B*, **31**, 3497–3511.

[8221] Jönsson P., Parpia F.A. and Froese Fischer C. (1996). HFS92: A program for relativistic atomic hyperfine structure calculations. *Comp. Phys. Comm.*, **96**, 301–310.

[8222] Jorge F.E., Barreto M.T. and da Silva A.B.F. (1997a). Adapted Gaussian basis sets for closed-shell atoms from samarium to nobelium generated with the generator coordinate Dirac-Fock method. *Chem. Phys.*, **221**, 45–51.

[8223] Jorge F.E., Bobbio T.B. and da Silva A.B.F. (1996). Adapted Gaussian basis sets for the relativistic closed-shell atoms from helium to barium generated with the generator coordinate Dirac-Fock method. *Chem. Phys. Lett.*, **263**, 775–782.

[8224] Jorge F.E. and da Silva A.B.F. (1996a). A generator coordinate version of the closed-shell Dirac-Fock equations. *J. Chem. Phys.*, **104**, 6278–6285.

[8225] Jorge F.E. and da Silva A.B.F. (1996b). On the inclusion of the Breit interaction term in the closed-shell generator coordinate Dirac-Fock formalism. *J. Chem. Phys.*, **105**, 5503–5509.

[8226] Jorge F.E. and da Silva A.B.F. (1997). The generator coordinate Dirac-Fock method applied to helium-like atomic properties. *Z. Phys. D*, **41**, 235–238.

[8227] Jorge F.E. and da Silva A.B.F. (1998). A segmented contraction methodology for Gaussian basis sets to be used in Dirac-Fock atomic and molecular calculations. *Chem. Phys. Lett.*, **289**, 469–472.

[8228] Jorge F.E., de Castro E.V.R. and da Silva A.B.F. (1997b). A universal Gaussian basis set for atoms cerium through lawrencium generated with the generator coordinate Hartree-Fock method. *J. Comp. Chem.*, **18**, 1565–1569.

[8229] Joubert L., Picart G. and Legendre J.J. (1998a). Advantages and drawbacks of the quantum chemistry methodology in predicting the thermochemical data of lanthanide trihalide molecules. *J. Alloys Compounds*, **275-277**, 934–939.

[8230] Joubert L., Picart G. and Legendre J.J. (1998b). Structural and thermochemical ab initio studies of lanthanide trihalide molecules with pseudopotentials. *Inorg. Chem.*, **37**, 1984–1991.

[8231] Joulakian B. and Messaoudi A.E. (1997). Relativistic K-shell ionisation of heavy atoms by electron impact: use of a relativistic helium-like bound state wave function. *Z. Phys. D*, **39**, 85–.

[8232] Judd B.R. (1993a). Spin-dependent contributions to El radiation with application to Ly-α. *J. Phys. B*, **26**, L697–L700.

[8233] Judd B.R. (1993b). Comment on "Many-body perturbation theory calculations of two-photon absorption in lanthanide compound". *Phys. Rev. Lett.*, **71**, 3890.

[8234] Julg A. (1998). A theoretical study of the absorption spectra of Pb^+ and Pb^{3+} in site K^+ of microcline: application to the color of amazonite. *Phys. Chem. Minerals*, **25**, 229–233.

[8235] Jung K.Y., Steimle T.C., Dai D. and Balasubramanian K. (1995). Experimental determination of dipole moments, hyperfine interactions, and ab initio predictions for PtN. *J. Chem. Phys.*, **102**, 643–652.

[8236] Jung W. (1997). Geometrical approach to inverse scattering for the Dirac equation. *J. Math. Phys.*, **38**, 39–48.

[8237] Jursic B.S. (1997). Electron affinities of metals computed by density functional theory and ab initio methods. *Int. J. Quantum Chem.*, **61**, 93–100.

[8238] Kahane S. (1998). Relativistic Dirac-Hartree-Fock photon incoherent scattering functions. *At. Data Nucl. Data Tables*, **68**, 323–347.

[8239] Kaiser B., Bernhardt T.M., Kinne M., Rademann K. and Heidenreich A. (1999). Formation, stability, and structures of antimony oxide cluster ions. *J. Chem. Phys.*, **110**, 1437–1449.

[8240] Kaldor U. (1995). Molecular potentials and relativistic effects. *Few-Body Syst., Suppl.*, **8**, 67–78.

[8241] Kaldor U. (1997). Relativistic coupled cluster: Method and applications. In *Recent Advances in Coupled-Cluster Methods*, (Edited by Bartlett R.J.), World Scientific, Singapore, pp. 125-153.

[8242] Kaldor U. and Eliav E. (1999). High-accuracy calculations for heavy and super-heavy elements. *Adv. Quantum Chem.*, **31**, 313–336.

[8243] Kaldor U. and Hess B.A. (1994). Relativistic all-electron coupled-cluster calculations on the gold atom and gold hydride in the framework of the Douglas-Kroll transformation. *Chem. Phys. Lett.*, **230**, 1–7.

[8244] Kaledin A.L., Heaven M.C. and Morokuma K. (1998). Ab initio potential energy surfaces for the $I(^2P_{3/2}) + O_2(a^1\Delta_g) \Leftrightarrow I(^2P_{1/2}) + O_2(X^3\Sigma_g^-)$. *Chem. Phys. Lett.*, **289**, 110–117.

[8245] Kalkreuter T. (1995). Spectrum of the Dirac operator and multigrid algorithm with dynamical staggered fermions. *Phys. Rev. D*, **51**, 1305–1313.

[8246] Kalkreuter T. and Simma H. (1996). An accelerated conjugate gradient algorithm to compute low-lying eigenvalues - a study for the Dirac operator in SU(2) lattice QCD. *Comp. Phys. Comm.*, **93**, 33–47.

[8247] Kalpana G., Palanivel B. and Rajagopalan M. (1994). Electronic structure and structural phase stability in BaS, BaSe, and BaTe. *Phys. Rev. B*, **50**, 12318–12325.

[8248] Kaltsoyannis N. (1994). A theoretical investigation of the bonding of early transition metals to tellurium. *J. Chem. Soc., Dalton Trans.*, pp. 1391–1400.

[8249] Kaltsoyannis N. (1995). Theoretical investigation of the electronic structures of the mixed-ring sandwich molecules $(M(\eta^7\text{-}C_7H_7)(\eta^5\text{-}C_5H_5))$ (M= Ti, V, Nb or Ta). *J. Chem. Soc., Dalton Trans.*, pp. 3727–3730.

[8250] Kaltsoyannis N. (1997a). Relativistic effects in inorganic and organometallic chemistry. *J. Chem. Soc., Dalton Trans.*, pp. 1–11.

[8251] Kaltsoyannis N. (1997b). Theoretical study of the effects of spin-orbit coupling on the valence photoelectron spectrum of $HfCl_4$. *Chem. Phys. Lett.*, **274**, 405–409.

[8252] Kaltsoyannis N. (1997c). Computational study of the electronic and geometric structures of the dihalogenodimethylselenium compounds, Me_2SeX_2 (X = F, Cl, Br, I or At). *J. Chem. Soc., Dalton Trans.*, pp. 4759–4764.

[8253] Kaltsoyannis N. (1998). Electronic structure of f^1 actinide complexes. Pt. 3. Quasi-relativistic density functional calculations of the optical transition energies of PaX_6^{2-}. *J. Alloys Compounds*, **271-273**, 859–862.

[8254] Kaltsoyannis N. and Bursten B.E. (1995). Electronic structure of f^1 actinide complexes. 1. Non-relativistic and relativistic calculations of the optical transition energies of AnX_6^{q-} complexes. *Inorg. Chem.*, **34**, 2735–2744.

[8255] Kaltsoyannis N. and Bursten B.E. (1997). Electronic structure of f^1 lanthanide and actinide complexes. Part 2. Non-relativistic and relativistic calculations of the ground state electronic structures and optical transition energies of $(Ce(\eta\text{-}C_5H_5)_3)$, $(Th(\eta\text{-}C_5H_5)_3)$ and $(Pa(\eta\text{-}C_8H_8)_2)$. *J. Organomet Chem.*, **528**, 19–33.

[8256] Kaltsoyannis N. and Scott P. (1998). Evidence for actinide metal to ligand π backbonding. Density functional investigations of the electronic structure of $[(NH_2)_3(NH_3)U_2(\mu^2 - \eta^2\text{–}N_2)]$. *Chem. Comm.*, pp. 1665–1666.

[8257] Kaltsoyannis N. and Scott P. (1999). *The f Elements*. Oxford University Press, Oxford, see Ch. 3: 'Relativity, electronic spectroscopy, and magnetism.

[8258] Kane R.S., Cohen R.E. and Silbey R. (1996). Theoretical study of the electronic structure of PbS nanoclusters. *J. Phys. Chem.*, **100**, 7928–7932.

[8259] Kaneko H., Hada M., Nakajima T. and Nakatsuji H. (1996). Spin-orbit effect on the magnetic shielding constant using the ab initio UHF method: tin tetrahalides. *Chem. Phys. Lett.*, **261**, 1–6.

[8260] Kang S.K., Tang H. and Albright T.A. (1993). Structures for d^0 ML_6 and ML_5 complexes. *J. Am. Chem. Soc.*, **115**, 1971–1981.

[8261] Kapala J., Roszak S., Lisek I. and Miller M. (1998). Mass spectrometric and theoretical study of the mixed complex $NaCeCl_4(g)$. *Chem. Phys.*, **238**, 221–229.

[8262] Kapp J., Remko M. and von R Schleyer P. (1996a). H_2XO and $(CH_3)_2XO$ compounds (X=C,Si,Ge,Sn,Pb): Double bonds vs. carbene-like structures – Can the metal compounds exist at all? *J. Am. Chem. Soc.*, **118**, 5745–5751.

[8263] Kapp J., Remko M. and von R Schleyer P. (1997). Reactions of $H_2X{=}XH_2$ and $H_2X{=}O$ double bonds (X = Si,Ge,Sn,Pb): Are 1,3-dioxa-2,4-dimetaletanes unusual molecules? *Inorg. Chem.*, **36**, 4241–4246.

[8264] Kapp J., Schade C., El-Nahasa A.M. and von Ragué Schleyer P. (1996b). Heavy element π donation is not less effective. *Angew. Chem. Int. Ed. (Engl)*, **35**, 2236–2238.

[8265] Kapp J., Schade C., El-Nahasa A.M. and von Ragué Schleyer P. (1996b). Heavy element π donation is not less effective (in German). *Angew. Chem.*, **108**, 2373–2376.

[8266] Kapp J. and von R Schleyer P. (1996). $M(CN)_2$ species (M = Be, Mg, Ca, Sr, Ba): cyanides, nitriles, or neither? *Inorg. Chem.*, **35**, 2247–2252.

[8267] Kapshai V.N. and Alferova T.A. (1999). Relativistic two-particle one-dimensional scattering problem for superposition of δ-potentials. *J. Phys. A*, **32**, 5329–5342.

[8268] Karat E. and Schulz M. (1997). Self-adjoint extensions of the Pauli equation in the presence of a magnetic monopole. *Ann. Phys. (NY)*, **254**, 11–24.

[8269] Karki B.B., Clark S.J., Warren M.C., Hsueh H.C., Ackland G.J. and Crain J. (1997). Ab initio elasticity and lattice dynamics of $AgGaSe_2$. *J. Phys. C*, **9**, 375–380.

[8270] Karpeshin F.F., Wycech S., Band I.M., Trzhaskovskaya M.B., Pfützner M. and Żylicz J. (1998). Rates of transitions between the hyperfine-splitting components of the ground-state and the 3.5 eV isomer in $^{229}Th^{89+}$. *Phys. Rev. C*, **57**, 3085–3088.

[8271] Karshenboim S.G. (1996a). Two-loop logarithmic corrections in the hydrogen Lamb shift. *J. Phys. B*, **29**, L29–L31.

[8272] Karshenboim S.G. (1996b). Leading logarithmic corrections and uncertainty of muonium hyperfine splitting calculations. *Z. Phys. D*, **36**, 11–15.

[8273] Karshenboim S.G. (1997a). Nuclear structure-dependent radiative corrections to the hydrogen hyperfine splitting. *Phys. Lett. A*, **225**, 97–106.

[8274] Karshenboim S.G. (1997b). The Lamb shift of excited S-levels in hydrogen and deuterium atoms. *Z. Phys. D*, **39**, 109–113.

[8275] Karshenboim S.G. (1997c). Radiative corrections to the light muonic atoms decay rate. *Phys. Lett. A*, **235**, 375–378.

[8276] Karshenboim S.G., Ivanov V.G. and Shabaev V.M. (1998a). Some analytic results on the Uehling correction to hyperfine splitting in a muonic atom. *Can. J. Phys.*, **76**, 503–506.

[8277] Karshenboim S.G., Ivanov V.G. and Shabaev V.M. (1999). Analytic results on the VP contribution to the energy in hydrogen-like ions. *Phys. Scripta*, **T80**, 491–492.

[8278] Karshenboim S.G., Jentschura U., Ivanov V. and Soff G. (1998b). Corrections to the wave function and the hyperfine structure in exotic atoms. *Eur. Phys. J. D*, **2**, 209–215.

[8279] Karwowski J. and Fraga S. (1974). Nuclear mass dependence of the Dirac-Breit-Pauli Hamiltonian. *Can. J. Phys.*, **52**, 536–540.

[8280] Karzel H., Potzel W., Köfferlein M., Schiessl W., Steiner M., Hiller U., Kalvius G.M., Mitchell D.W., Das T.P., Blaha P., Schwarz K. and Pasternak M.P. (1996). Lattice dynamics and hyperfine interactions in ZnO and ZnSe at high external pressures. *Phys. Rev. B*, **53**, 11425–11438.

[8281] Kaschner R., Saalmann U., Seifert G. and Gausa M. (1995). Density functional calculations of structures and ionization energies for heavy group V cluster anions. *Int. J. Quantum Chem.*, **56**, 771–777.

[8282] Kaupp M. (1992). *Quantenchemische Untersuchungen zur Chemie schwerer Elemente.*. Ph.D. thesis, Stuttgart.

[8283] Kaupp M. (1996a). NMR chemical-shift anomaly and bonding in piano-stool carbonyl and related complexes - an ab initio ECP/DFT study. *Chem. Eur. J.*, **2**, 348–358.

[8284] Kaupp M. (1996b). Analysis of ^{13}C and ^{17}O chemical shift tensors and an ELF view of bonding in $Fe_2(CO)_9$ and $Rh_6(CO)_{16}$. *Chem. Ber.*, **129**, 527–533.

[8285] Kaupp M. (1996c). Interpretation of ^{31}P-NMR coordination shifts for phosphane ligands. Ab initio ECP/DFT study of chemical shift tensors in $M(CO)_5L$ (M = Cr, Mo, W; L = PH_3, $P(CH_3)_3$, PF_3, PCl_3). *Chem. Ber.*, **129**, 535–544.

[8286] Kaupp M. (1996d). The structure of hexamethyltungsten, $W(CH_3)_6$: Distorted trigonal prismatic with C_3 symmetry. *J. Am. Chem. Soc.*, **118**, 3018–3024.

[8287] Kaupp M. (1996e). Untersuchung der Strukturen, Energien und NMR-Eigenschaften von Uebergangsmetallverbindungen mit Hilfe quantenchemischer Methoden. *Habilitationsschrift, Stuttgart*, p. 188 p.

[8288] Kaupp M. (1996f). ^{13}C NMR chemical-shift tensors of interstitial carbides in transition-metal clusters calculated by density-functional theory. *J. Chem. Soc, Chem. Comm.*, pp. 1141–1142.

[8289] Kaupp M. (1998a). The nonoctahedral structures of d^0, d^1, and d^2 hexamethyl complexes. *Chem. Eur. J.*, **4**, 1678–1686.

[8290] Kaupp M. (1999a). Charting no-man's land in d^0 transition metal six-coordination: Structure predictions for the complexes $[WCl_5CH_3]$, $[WCl_4(CH_3)_2]$, and $[WCl_3(CH_3)_3]$. *Angew. Chem. Int. Ed. Engl.*, **38**, 3034–3037.

[8291] Kaupp M. (1999b). On the relation between π bonding, electronegativity, and bond angles in high-valent transition metal complexes. *Chem. Eur. J.*, **5**, 3631–3643.

[8292] Kaupp M., Aubauer C., Engelhardt G., Klapötke T.M. and Malkina O.L. (1999a). The PI_4^+ cation has an extremely large negative ^{31}P nuclear magnetic resonance chemical shift, due to spin-orbit coupling: A quantum-mechanical prediction and its confirmation by solid-state nuclear magnetic resonance spectroscopy. *J. Chem. Phys.*, **110**, 3897–3902.

[8293] Kaupp M., Malkin V.G. and Malkina O.L. (1998a). NMR of transition metal compounds. In *Encyclopedia of Computational Chemistry*, (Edited by von Ragué Schleyer P.), vol. 3, pp. 1857–1866, Wiley, Chichester and New York.

[8294] Kaupp M., Malkin V.G., Malkina O.L. and Salahub D.R. (1995a). Calculation of ligand NMR chemical shifts in transition-metal complexes using ab initio effective-core potentials and density functional theory. *Chem. Phys. Lett.*, **235**, 382–388.

[8295] Kaupp M., Malkin V.G., Malkina O.L. and Salahub D.R. (1995b). Scalar relativistic effects on ^{17}O NMR chemical shifts in transition-metal oxo complexes. An ab initio ECP/DFT study. *J. Am. Chem. Soc.*, **117**, 1851–1852.

[8296] Kaupp M., Malkin V.G., Malkina O.L. and Salahub D.R. (1996a). Ab initio ECP/DFT calculation and interpretation of carbon and oxygen NMR chemical shift tensors in transition-metal carbonyl complexes. *Chem. Eur. J.*, **2**, 24–30.

[8297] Kaupp M. and Malkina O.L. (1998). Density functional analysis of ^{13}C and ^{1}H chemical shifts and bonding in mercurimethanes and organomercury hydrides: The role of scalar relativistic, spin-orbit, and substituent effects. *J. Chem. Phys.*, **108**, 3648–3659.

[8298] Kaupp M., Malkina O.L. and Malkin V.G. (1997a). Interpretation of ^{13}C NMR chemical shifts in halomethyl cations. On the importance of spin-orbit coupling and electron correlation. *Chem. Phys. Lett.*, **265**, 55–59.

[8299] Kaupp M., Malkina O.L. and Malkin V.G. (1997b). The calculation of ^{17}O chemical shielding in transition metal oxo complexes. I. Comparison of DFT and ab initio approaches, and mechanisms of relativity-induced shielding. *J. Chem. Phys.*, **106**, 9201–9212.

[8300] Kaupp M., Malkina O.L. and Malkin V.G. (1999b). The role of π-type nonbonding orbitals for spin-orbit induced NMR chemical shifts: DFT study of ^{13}C and ^{19}F shifts in the series CF_3IF_n (n=0,2,4,6). *J. Comp. Chem.*, **20**, 1304–1313.

[8301] Kaupp M., Malkina O.L., Malkin V.G. and Pyykkö P. (1998b). How do spin–orbit-induced heavy-atom effects on NMR chemical shifts function? Validation of a simple analogy to spin-spin coupling by density functional theory (DFT) calculations on some iodo compounds. *Chem. Eur. J.*, **4**, 118–126.

[8302] Kaupp M., van Wüllen C., Franke R., Schmitz F. and Kutzelnigg W. (1996b). The structure of XeF_6 and of compounds isoelectronic with it. A challenge to computational chemistry and to the qualitative theory of the chemical bond. *J. Am. Chem. Soc.*, **118**, 11939–11950.

[8303] Kaupp M. and von Ragué Schleyer P. (1993a). Ab initio study of structures and stabilities of substituted lead compounds. Why is inorganic lead chemistry dominated by Pb(II) but organolead chemistry by Pb(IV)? *J. Am. Chem. Soc.*, **115**, 1061–1073.

[8304] Kaupp M. and von Ragué Schleyer P. (1993b). The peculiar coordination of barium: ab initio study of the molecular and electronic structures of the group 2 dihydride dimers M_2H_4 (M=Mg, Ca, Sr, Ba). *J. Am. Chem. Soc.*, **115**, 11202–11208.

[8305] Kaupp M. and von Schnering H.G. (1993). Gaseous mercury(IV) fluoride HgF_4: An ab initio study (in German). *Angew. Chem.*, **105**, 952.

[8306] Kaupp M. and von Schnering H.G. (1993). Gaseous mercury(IV) fluoride HgF_4: An ab initio study. *Angew. Chem. Int. Ed. Engl.*, **32**, 861–863.

[8307] Kaupp M. and von Schnering H.G. (1994a). Dominance of linear 2-coordination in mercury chemistry: Quasirelativistic and nonrelativistic ab initio pseudopotential study of $(HgX_2)_2$ (X=F, Cl, Br, I, H). *Inorg. Chem.*, **33**, 2555–2564.

[8308] Kaupp M. and von Schnering H.G. (1994b). Origin of the unique stability of condensed-phase Hg_2^{2+}. An ab initio investigation of M(I) and M(II) species (M=Zn, Cd, Hg). *Inorg. Chem.*, **33**, 4179–4185.

[8309] Kaupp M. and von Schnering H.G. (1994c). Ab initio comparison of the $(MX_2)_2$ dimers (M=Zn, Cd, Hg; X=F, Cl, H) and study of relativistic effects in crystalline HgF_2. *Inorg. Chem.*, **33**, 4718–4722.

[8310] Kaur S., Srivastava R., McEachran R.P. and Stauffer A.D. (1997). Electron impact excitation of magnesium and zinc atoms in the relativistic distorted-wave approximation. *J. Phys. B*, **30**, 1027–1042.

[8311] Kedziora G.S., Pople J.A., Rassolov V.A., Ratner M.A., Redfern P.C. and Curtiss L.A. (1999). The relativistic Dirac-Fock-Coulomb effect on atomization energies. *J. Chem. Phys.*, **110**, 7123–7126.

[8312] Kegley Jr. D.R., Oberacker V.E., Strayer M.R., Umar A.S. and Wells J.C. (1996). Basis spline collocation method for solving the Schrödinger equation in axillary symmetric systems. *J. Comp. Chem.*, **128**, 197–208.

[8313] Keitel C.H. (1996). Ultra-energetic electron ejection in relativistic atom-laser field interaction. *J. Phys. B*, **29**, L873–L880.

[8314] Keller S., Ast H., Ancarani L.U., Whelan C.T., Walters H.R.J. and Dreizler R.M. (1998). Remark on the theory of three-body effects in relativistic (e,2e) processes. *J. Phys. B*, **31**, L873–L876.

[8315] Keller S. and Dreizler R.M. (1997). Relativistic independent particle approximation study of triply differential cross sections for electron-atom bremsstrahlung. *J. Phys. B*, **30**, 3257–3266.

[8316] Keller S. and Dreizler R.M. (1998). Interpretation of a relativistic $(e, 2e)$ experiment in coplanar asymmetric geometry. *Phys. Rev. A*, **57**, 3652–3659.

[8317] Keller S., Dreizler R.M., Ancarani L.U., Walters H.R.J., Ast H. and Whelan C.T. (1996a). Theoretical analysis of the relativistic first order Born approximation for inner shell (e,2e) processes. *Z. Phys. D*, **37**, 191–196.

[8318] Keller S., Dreizler R.M., Ast H., Whelan C.T. and Walters H.R.J. (1996b). Theory of (e, 2e) processes with spin-polarized relativistic electrons. *Phys. Rev. A*, **53**, 2295–2302.

[8319] Keller S. and Whelan C.T. (1994). On the plane wave Born approximation for relativistic (e, 2e) processes. *J. Phys. B*, **27**, L771–L776.

[8320] Keller S., Whelan C.T., Ast H., Walters H.R.J. and Dreizler R.M. (1994). Relativistic distorted-wave Born calculations for (e, 2e) processes on inner shells of heavy atoms. *Phys. Rev. A*, **50**, 3865–3877.

[8321] Kellö V. and Sadlej A.J. (1993). Quasirelativistic studies of molecular electric properties: Dipole moments of the group IV a oxides and sulfides. *J. Chem. Phys.*, **98**, 1345–1351.

[8322] Kellö V. and Sadlej A.J. (1995a). Polarized basis sets for high-level-correlated calculations of molecular electric properties. *Theor. Chim. Acta*, **91**, 353–371.

[8323] Kellö V. and Sadlej A.J. (1995b). Electron correlation and relativistic effects in the coinage metal compounds. *Theor. Chim. Acta*, **92**, 253–267.

[8324] Kellö V. and Sadlej A.J. (1995c). Electron correlation and relativistic effects in the coinage metal compounds. II. Heteronuclear dimers: CuAg, CuAu, and AgAu. *J. Chem. Phys.*, **103**, 2991–2999.

[8325] Kellö V. and Sadlej A.J. (1996a). Standardized basis sets for high-level-correlated relativistic calculations of atomic and molecular electric properties in the spin-averaged Douglas-Kroll (no-pair) approximation. I. Groups Ib and IIb. *Theor. Chim. Acta*, **94**, 93–104.

[8326] Kellö V. and Sadlej A.J. (1996b). Determination of the quadrupole moment of the halogen nuclei (Cl, Br, I) from molecular data. *Mol. Phys.*, **89**, 127–137.

[8327] Kellö V. and Sadlej A.J. (1998a). Picture change and calculations of expectation values in approximate relativistic theories. *Int. J. Quantum Chem.*, **68**, 159–174.

[8328] Kellö V. and Sadlej A.J. (1998b). The quadrupole moment of the ^{39}K and ^{41}K nuclei from the microwave data for KF and KCl. *Chem. Phys. Lett.*, **292**, 403–410.

[8329] Kellö V. and Sadlej A.J. (1999a). The nuclear quadrupole moment of ^{73}Ge from molecular microwave data. *Mol. Phys.*, **96**, 275–281.

[8330] Kellö V. and Sadlej A.J. (1999b). Nuclear quadrupole moments from molecular microwave data: The quadrupole moment of ^{85}Rb and ^{87}Rb nuclei and survey of molecular data for alkali-metal nuclei. *Phys. Rev. A*, **60**, 3575–3585.

[8331] Kellö V., Sadlej A.J. and Faegri Jr K. (1993). Electron-correlation and relativistic contributions to atomic dipole polarizabilities: Alkali-metal atoms. *Phys. Rev. A*, **47**, 1715–1725.

[8332] Kellö V., Sadlej A.J. and Faegri Jr K. (1998). Electron correlation and relativistic contributions to dipole moments of heavy oxides and sulfides: SnO, PbO, SnS, and PbS. *J. Chem. Phys.*, **108**, 2056–2066.

[8333] Kellö V., Sadlej A.J. and Hess B.A. (1996a). Relativistic effects on electric properties of many-electron systems in spin-averaged Douglas-Kroll and Pauli approximations. *J. Chem. Phys.*, **105**, 1995–2003.

[8334] Kellö V., Sadlej A.J., Pyykkö P., Sundholm D. and Tokman M. (1999). Electric quadrupole moment of the ^{27}Al nucleus: Converging results from the AlF and AlCl molecules and the Al atom. *Chem. Phys. Lett.*, **304**, 414–422.

[8335] Kellö V., Urban M. and Sadlej A.J. (1996b). Electron dipole polarizabilities of negative ions of the coinage metal atoms. *Chem. Phys. Lett.*, **253**, 383–389.

[8336] Kenny S.D., Rajagopal G. and Needs R.J. (1995). Relativistic corrections to atomic energies from quantum Monte Carlo calculations. *Phys. Rev. A*, **51**, 1898–1904.

[8337] Kenny S.D., Rajagopal G., Needs R.J., Leung W.K., Godfrey M.J., Williamson A.J. and Foulkes W.M.C. (1996). Quantum Monte Carlo calculations of the energy of the relativistic homogeneous electron gas. *Phys. Rev. Lett.*, **77**, 1099–1102.

[8338] Kessler J. (1985). *Polarized Electrons, 2nd Ed.*. Springer-Verlag, Berlin.

[8339] Kettle S.F.A. (1998). Double groups revisited and their extension to triple and quadruple groups. *Spectrochim. Acta A*, **54**, 1633–1638.

[8340] Khabibullaev P.K., Matveev V.I. and Matrasulov D.U. (1998). Inelastic collisions of relativistic highly charged ions with heavy atoms and K-vacancy production. *J. Phys. B*, **31**, L607–L611.

[8341] Khandogin Y., Alekseyev A.B., Liebermann H.P., Hirsch G. and Buenker R.J. (1997). Ab initio relativistic CI calculations of the spectroscopic constants and transition probabilities for the low-lying states of the BiOH/HBiO isomers. *J. Mol. Spectr.*, **186**, 22–33.

[8342] Khare S.P. and Wadehra J.M. (1995). K-shell ionization of atoms by electron impact. *Phys. Lett. A*, **198**, 212–216.

[8343] Khare S.P. and Raj D. (1993). The spin polarization of electrons elastically scattered by argon atoms. *J. Phys. B*, **26**, 4807–4814.

[8344] Khare S.P., Sinha P. and Wadehra J.M. (1994). L3-shell ionization of xenon and gold by electron and positron impacts. *Phys. Lett. A*, **184**, 204–208.

[8345] Khasanov A.B. (1995). On eigenvalues of the Dirac operator located on the continuous spectrum (in Russian). *Teor. Mat. Fiz.*, **99(1)**, 20–26.

[8346] Khasanov A.B. (1995). On eigenvalues of the Dirac operator located in the continuous spectrum. *Theor. Math. Phys.*, **99**, 396–401.

[8347] Khelashvili A. and Kiknadze N. (1997). Bound states in continuum induced by relativity. *Phys. Rev. A*, **55**, 2557–2563.

[8348] Kholomaj B.V. (1995). Approximate solution of the Dirac equation in arbitrary periodic electrostatic field. *Izv. Vyssh. Uchebn. Zaved. Fizika*, **38(7)**, 45–47.

[8349] Kholomaj B.V. (1998). Approximate solution of Dirac equation in transversal and longitudinal focusing periodical fields of crystal (in Russian). *Izv. Vyssh. Uchebn. Zaved. Fizika*, **41(11)**, 66–69.

[8350] Khriplovich I.B. (1993). On the checks of fundamental symmetries in atomic physics. *Phys. Scr.*, **T46**, 90–91.

[8351] Khriplovich I.B. and Lamoreaux S.K. (1997). *CP Violation without Strangeness. Electric Dipole Moments of Particles, Atoms, and Molecules*. Springer, Berlin, 230 pp.

[8352] Khriplovich I.B., Milstein A.I. and Sen'kov R.A. (1996). Nature of the Darwin term and $(Z\alpha)^4 m^3/M^2$ contribution to the Lamb shift for an arbitrary spin of the nucleus. *Phys. Lett. A*, **221**, 370–374.

[8353] Khriplovich I.B., Milstein A.I. and Yelkhovsky A.S. (1993). Logarithmic corrections in the two-body QED problem. *Phys. Scr.*, **T46**, 252–260.

[8354] Khriplovich I.B., Milstein A.I. and Yelkhovsky A.S. (1994). Infrared divergence, Thomson scattering, and Lamb shift. *Am. J. Phys.*, **62**, 70–73.

[8355] Khriplovich I.B., Milstein A.I. and Yelkhovsky A.S. (1998). Comment on "Corrections to $O(\alpha^7(\ln\alpha)mc^2)$ fine-structure splittings and $O(\alpha^6(\ln\alpha)mc^2)$ energy levels in helium". *Phys. Rev. A*, **57**, 1462–1463.

[8356] Khriplovich I.B. and Sen'kov R.A. (1998). Nucleon polarizability contribution to the hydrogen Lamb shift and hydrogen-deuterium isotope shift. *Phys. Lett. A*, **249**, 474–476.

[8357] Kickelbick G. and Schubert U. (1997). $ClMPH_3$ and $(MPH_3)^+$ (M = Cu, Ag, Au); a density functional study. *Inorg. Chim. Acta*, **262**, 61–64.

[8358] Kijowski J. (1994). On electrodynamical self-interaction. *Acta Phys. Pol. A*, **85**, 771–787.

[8359] Kiljunen T., Eloranta J. and Kunttu H. (1999). Ab initio and molecular dynamics studies on rare gas hydrides: Potential-energy curves, isotropic hyperfine properties, and matrix cage trapping of atomic hydrogen. *J. Chem. Phys.*, **110**, 11814–11822.

[8360] Kim E. and Cox D.L. (1998). Knight-shift anomalies in heavy-electron materials. *Phys. Rev. B*, **58**, 3313–3340.

[8361] Kim D.H., Lee P.A. and Wen X. (1997). Massless Dirac fermions, gauge fields, and underdoped cuprates. *Phys. Rev. Lett.*, **79**, 2109–2112.

[8362] Kim J.H., Lee S.W., Maassen H. and Lee H.W. (1996). Relativistic oscillator of constant period. *Phys. Rev. A*, **53**, 2991–2997.

[8363] Kim M.C., Lee H.S., Lee Y.S. and Lee S.Y. (1998). Low-lying states of Tl_2 calculated by the configuration interaction methods based upon relativistic effective core potentials and two-component spinors. *J. Chem. Phys.*, **109**, 9384–9390.

[8364] Kim M.C., Lee S.Y. and Lee Y.S. (1996). Spin-orbit effects calculated by a configuration interaction method using determinants of two-component molecular spinors: test calculations on Rn and TlH. *Chem. Phys. Lett.*, **253**, 216–222.

[8365] Kim S.H. (1993). Quantum mechanical theory of Thomson scattering by a relativistic electron. *J. Phys. Soc. Japan*, **62**, 1–5.

[8366] Kim W.S. and Kaltsoyannis N. (1998). Theoretical study of the geometric and electronic structures and spectra of *trans*-$ME_2(PH_3)_4$ complexes (M=Mo, W; E=S, Se, Te). *Inorg. Chem.*, **37**, 674–678.

[8367] Kim Y.K. (1993a). Unsettled questions in relativistic atomic structure theory. *Comm. At. Mol. Phys.*, **28**, 201–210.

[8368] Kim Y.K. (1993b). What's new in relativistic atomic structure theory. *Phys. Scr.*, **T47**, 54–58.

[8369] Kim Y.K. (1997). Strengths and weaknesses of relativistic atomic structure calculations. *Phys. Scr.*, **T73**, 19–24.

[8370] Kim Y.K., Parente F., Marques J.P., Indelicato P. and Desclaux J.P. (1998). Failure of multiconfiguration Dirac-Fock wave functions in the nonrelativistic limit. *Phys. Rev. A*, **58**, 1885–1888.

[8371] Kim Y.S., Lee S.Y., Oh W.S., Park B.H., Han Y.K., Park S.J. and Lee Y.S. (1998). Kramers' unrestricted Hartree-Fock and second-order Møller-Plesset perturbation methods using relativistic effective core potentials with spin-orbit operators: Test calculations for HI and CH_3I. *Int. J. Quantum Chem.*, **66**, 91–98.

[8372] King F.W. (1997). Progress on high precision calculations for the ground state of atomic lithium. *J. Mol. Str. (Theochem)*, **400**, 7–56.

[8373] King F.W. (1999). High-precision calculations for the ground and excited states of the lithium atom. *Adv. At. Mol. Opt. Phys.*, **40**, 57–112.

[8374] King F.W., Ballegeer D.G., Larson D.J., Pelzl P.J., Nelson S.A., Prosa T.J. and Hinaus B.M. (1998). Hylleraas-type calculations of the relativistic corrections for the ground state of the lithium atom. *Phys. Rev. A*, **58**, 3597–3603.

[8375] King W.A., Di Bella S., Lanza G., Khan K., Duncalf D.J., Cloke F.G.N., Fragalà I.L. and Marks T.J. (1996). Metal-ligand bonding and bonding energetics in zerovalent lanthanide, group 3, group 4, and group 6 bis(arene) sandwich complexes. A combined solution thermochemical and ab initio quantum chemical investigation. *J. Am. Chem. Soc.*, **118**, 627–635.

[8376] Kinne M., Heidenreich A. and Rademann K. (1998). Reactions of selected bismuth oxide cluster cations with propene. *Angew. Chem. Int. Ed. Engl.*, **37**, 2509–2511.

[8377] Kinoshita T. (1995). New value of the α^3 electron anomalous magnetic moment. *Phys. Rev. Lett.*, **75**, 4728–4731.

[8378] Kinoshita T. (1996). The fine structure constant. *Rep. Prog. Phys.*, **59**, 1459–1492.

[8379] Kinoshita T. and Nio M. (1994). Improved theory of the muonium hyperfine structure. *Phys. Rev. Lett.*, **72**, 3803–3806.

[8380] Kinoshita T. and Nio M. (1999). Sixth-order vacuum-polarization contribution to the Lamb shift of muonic hydrogen. *Phys. Rev. Lett.*, **82**, 3240–3243.

[8381] Kirchhoff F., Binggeli N., Galli G. and Massidda S. (1994a). Structural and bonding properties of solid tellurium from first-principles calculations. *Phys. Rev. B*, **50**, 9063–9071.

[8382] Kirchhoff F., Holender J.M. and Gillan M.J. (1994b). Energetics and electronic structure of silver chloride. *Phys. Rev. B*, **49**, 17420–17423.

[8383] Kirchner E.J.J. (1994). *Interactions at surfaces. Quantum chemical studies*. Ph.D. thesis, Vrije Universiteit, Amsterdam, 146 p.

[8384] Kirpekar S., Aagaard Jensen H.J. and Oddershede J. (1997). Spin-orbit corrections to the indirect nuclear spin-spin coupling constants in XH_4 (X=C, Si, Ge, and Sn). *Theor. Chim. Acta*, **95**, 35–47.

[8385] Kiselev A.A., Ivchenko E.L. and Rössler U. (1998). Electron g factor in one- and zero-dimensional semiconductor nanostructures. *Phys. Rev. B*, **58**, 16353–16359.

[8386] Kisielius R., Berrington K.A. and Norrington P.H. (1995). Relativistic study of electron-impact excitation of hydrogen-like ions. *J. Phys. B*, **28**, 2459–2471.

[8387] Kivimäki A., Aksela H., Jauhiainen J., Naves de Brito A., Sairanen O.P., Aksela S., Aus-mees A., Osborne S.J. and Svensson S. (1994). Kr $M_{4,5}N_{2,3}$-$N_{2,3}N_{2,3}N_{2,3}N_{2,3}$ and Xe $N_{4,5}O_{2,3}$-$O_{2,3}O_{2,3}O_{2,3}$ satellite Auger spectra following direct double ionization. *Phys. Rev. A*, **49**, 5124–5127.

[8388] Kiyonaga H., Morihashi K. and Kikuchi O. (1998). Calculation of contributions of one- and two-electron spin-orbit coupling terms to the parity-violating energy shifts for amino acids and helical alkanes. *J. Chem. Phys.*, **108**, 2041–2043.

[8389] Klapötke T.M. and Schulz A. (1996). *Quantum Chemical Methods in Main-Group Chemistry*. Wiley, Chichester, 262 p. For relativistic effects, see pp. 76-82, 160-162.

[8390] Klapötke T.M. and Schulz A. (1997). The structure of astatine azide, AtN_3 – A theoretical study. *Structural Chemistry*, **8**, 421–423.

[8391] Klapötke T.M. and Tornieporth-Oetting I.C. (1994). *Nichtmetallchemie*. VCH, Weinheim, kap. 1.9. Einfluss relativistischer Effekte, pp. 71-75.

[8392] Klarsfeld S. (1977). Analytical expressions for the evaluation of vacuum polarization potentials in muonic atoms. *Phys. Lett. B*, **66**, 86–88.

[8393] Kleppner D. (1999). A short history of atomic physics in the twentieth century. *Rev. Mod. Phys.*, **71**, S78–S84.

[8394] Kless A., Börner A., Heller D. and Selke R. (1997). Ab initio studies of rhodium(I)-N-alkenylamide complexes with cis- and trans-coordinating phosphines: Relevance for the mechanism of catalytic asymmetric hydrogenation of prochiral dehydroamino acids. *Organometallics*, **16**, 2096–2100.

[8395] Klink W.H. (1998a). Point form relativistic quantum mechanics and electromagnetic form factors. *Phys. Rev. C*, **58**, 3587–3604.

[8396] Klink W.H. (1998b). Relativistic simultaneously coupled multiparticle states. *Phys. Rev. C*, **58**, 3617–3626.

[8397] Klinkhammer K.W., Fässler T.F. and Grützmacher H. (1998). The formation of heteroleptic carbene homologues by ligand exchange – Synthesis of the first plumbanediyl dimer. *Angew. Chem. Int. Ed. Engl.*, **37**, 124–126.

[8398] Klinkhammer K.W. and Pyykkö P. (1995). Ab initio interpretation of the closed-shell intermolecular E...E attraction in dipnicogen $(H_2E\text{-}EH_2)_2$ and dichalcogen $(HE\text{-}EH)_2$ hydride model dimers. *Inorg. Chem.*, **34**, 4134–4138.

[8399] Klobukowski M. (1993a). Ab initio SCF and Møller-Plesset studies on hexafluorides of selenium and tellurium. *J. Comp. Chem.*, **14**, 1234–1239.

[8400] Klobukowski M. (1993b). Ab initio SCF and Møller-Plesset studies on hexachlorides of selenium and tellurium and their dianions. *Can. J. Chem.*, **71**, 141–146.

[8401] Klopper W. (1997). Simple recipe for implementing computation of first-order relativistic corrections to electron correlation energies in framework of direct perturbation theory. *J. Comp. Chem.*, **18**, 20–27.

[8402] Klotz R. (1985). *Ab-Initio-Berechnung von spinverbotenen elektronischen Uebergängen*. Ph.D. thesis, Bonn.

[8403] Knöpfle K., Sandratskii L.M. and Kübler J. (1997). Symmetry properties of intra-atomic spin and angular momentum densities: application to U_3Sb_4. *J. Phys. CM*, **9**, 7095–7104.

[8404] Knuts S., Minaev B.F., Vahtras O. and Ågren H. (1995). Spin-orbit coupling in the intersystem crossing of the ring-opened oxirane biradical. *Int. J. Quantum Chem.*, **55**, 23–34.

[8405] Kobayashi K., Miura H. and Nagase S. (1994). The heavier group 15 analogues of benzene cyclobutadiene and their valence isomers (M_6 and M_4 M=P, As, Sb and Bi). *J. Mol. Str. (Theochem)*, **311**, 69–77.

[8406] Kobayashi K. and Nagase S. (1996). Structures and electronic states of endohedral dimetallofullerenes: $M_2@C_{80}$ (M=Sc, Y, La, Ce, Pr, Eu, Gd, Yb and Lu) . *Chem. Phys. Lett.*, **262**, 227–232.

[8407] Kobayashi K. and Nagase S. (1998). Structures and electronic states of $M@C_{82}$ (M=Sc, Y, La and lanthanides). *Chem. Phys. Lett.*, **282**, 325–329.

[8408] Kobayashi K., Nagase S. and Akasaka T. (1995). A theoretical study of C_{80} and $La_2@C_{80}$. *Chem. Phys. Lett.*, **245**, 230–236.

[8409] Koc K. and Ishikawa Y. (1994). Single-Fock-operator method for matrix Dirac-Fock self-consistent-field calculations on open-shell atoms. *Phys. Rev. A*, **49**, 794–798.

[8410] Koc K., Ishikawa Y. and Kagawa T. (1994). Relativistic configuration interaction calculations for open-shell atomic systems. *Chem. Phys. Lett.*, **231**, 407–413.

[8411] Koc K., Ishikawa Y., Kagawa T. and Kim Y.K. (1996). Relativistic modification of asymptotic configuration interaction in the carbon isoelectronic sequence. *Chem. Phys. Lett.*, **263**, 338–344.

[8412] Kocinski J. (1999). A five-dimensional form of the Dirac equation. *J. Phys. A*, **4257-4277**.

[8413] Koga N. (1999). Ab initio MO study of ethylene insertion into the Sm-C bond of $H_2SiCp_2SmCH_3$. *Theor. Chem. Acc.*, **102**, 285–292.

[8414] Koga N. and Morokuma K. (1993a). Ab initio MO calculation of $(\eta^2\text{-}C_{60})Pt(PH_3)_2$. Electronic structure and interaction between C_{60} and Pt. *Chem. Phys. Lett.*, **202**, 330–334.

[8415] Koga N. and Morokuma K. (1993b). SiH, SiSi, and CH bond activation by coordinatively unsaturated $RhCl(PH_3)_2$. Ab initio molecular orbital study. *J. Am. Chem. Soc.*, **115**, 6883–6892.

[8416] Koga T., Aoki H., Garcia de la Vega J.M. and Tatewaki H. (1997). Atomic ionization potentials and electron affinities with relativistic and mass corrections. *Theor. Chem. Acc.*, **96**, 248–255.

[8417] Köhler K., Silverio S.J., Hyla-Kryspin I., Gleiter R., Zsolnai L., Driess A., Huttner G. and Lang H. (1997). Trigonal-planar-coordinated organogold(I) complexes stabilized by organometallic 1,4-diynes: Reaction behavior, structure, and bonding. *Organometallics*, **16**, 4970–4979.

[8418] Kohstall C., Fritzsche S., Fricke B. and Sepp W.D. (1998). Calculated level energies, transition probabilities, and lifetimes of silicon-like ions. *At. Data Nucl. Data Tables*, **70**, 63–92.

[8419] Kohstall C., Fritzsche S., Fricke B., Sepp W.D. and Träbert E. (1999). Comment on the lifetimes of the $3s3p^6\ {}^2S_{1/2}$ level for chlorine-like ions. *Phys. Scripta*, **T80**, 482–484.

[8420] Koike F. (1993). Theory of inelastic scattering of low energy electrons by lanthanide and actinide atoms. *Phys. Lett. A*, **176**, 439–442.

[8421] Koizumi H. and Sugano S. (1995). Geometric phase in two Kramers doublets molecular systems. *J. Chem. Phys.*, **102**, 4472–4481.

[8422] Kokalj A., Lesar A. and Hodoščček M. (1997). Interaction of oxygen with the Pt(111) surface: a cluster model study. *Chem. Phys. Lett.*, **268**, 43–49.

[8423] Kokko K. and Das M.P. (1998). Some ground-state properties of $3d$ and $4d$ metals studied using the generalized gradient approximation. *J. Phys. CM*, **10**, 1285–1291.

[8424] Kolakowska A. (1996). Application of the minimax principle to the Dirac-Coulomb problem. *J. Phys. B*, **29**, 4515–4527.

[8425] Kolakowska A. (1997). Explicitly correlated trial functions in relativistic variational calculations. *J. Phys. B*, **30**, 2773–2779.

[8426] Kolakowska A., Talman J.D. and Aashamar K. (1996). Minimax variational approach to the relativistic two-electron problem. *Phys. Rev. A*, **53**, 168–177.

[8427] Kollár J., Vitos L. and Skriver H.L. (1997). Anomalous atomic volume of α-Pu. *Phys. Rev. B*, **55**, 15353–15355.

[8428] Komninos Y., Aspromallis G. and Nicolaides C.A. (1995). Theory and computation of perturbed spectra: applications to the Al ${}^2D$ relativistic spectrum. *J. Phys. B*, **28**, 2049–2067.

[8429] Kónya B. and Papp Z. (1999). On the Coulomb Sturmian matrix elements of relativistic Coulomb Green's operators. *J. Math. Phys.*, **40**, 2307–2310.

[8430] Kornienko A.A., Dunina E.B. and Yankevich V.L. (1995). Influence of configuration mixing on the spin-orbit interaction parameter of rare-earth ions. *Opt. Spectr.*, **79**, 700–705.

[8431] Kornienko A.A., Dunina E.B. and Yankevich V.L. (1995). Influence of configuration mixing on the spin-orbit interaction parameter of rare-earth ions (in Russian). *Opt. Spektr.*, **79**, 761–766.

[8432] Korobov V.I. and Bakalov D.D. (1997). Energies and relativistic corrections for the metastable states of antiprotonic helium atoms. *Phys. Rev. Lett.*, **79**, 3379–3382.

[8433] Korzhavyi P.A., Abrikosov I.A., Johansson B., Ruban A.V. and Skriver H.L. (1999). First-priciples calculations of the vacancy formation energy in transition and noble metals. *Phys. Rev. B*, **59**, 11693–11703.

[8434] Koseki S., Gordon M.S., Schmidt M.W. and Matsunaga N. (1995). Main group effective nuclear charges for spin-orbit calculations. *J. Phys. Chem.*, **99**, 12764–12772.

[8435] Koshibae W., Ohta Y. and Maekawa S. (1993). Electronic and magnetic structures of cuprates with spin-orbit interaction. *Phys. Rev. B*, **47**, 3391–3400.

[8436] Koskinen J.T. and Cooks R.G. (1999). Novel rare gas ions BXe^+, BKr^+, and BAr^+ formed in a halogen/rare gas exchange reaction. *J. Phys. Chem. A*, **103**, 9565–9568.

[8437] Köstlmeier S., Häberlen O.D., Rösch N., Herrmann W.A., Solouki B. and Bock H. (1996). Density functional study on the electronic structure of trioxorhenium organyls. *Organometallics*, **15**, 1872–1878.

[8438] Köstlmeier S., Nasluzov V.A., Herrmann W.A. and Rösch N. (1997). Lewis acidity and reactivity of transition metal oxo complexes. A comparative density functional study of CH_3ReO_3, CH_3TcO_3, and their base adducts. *Organometallics*, **16**, 1786–1792.

[8439] Köstlmeier S., Pacchioni G., Herrmann W.A. and Rösch N. (1996). Structure and properties of dimer, trimer, and tetramer aggregates of methyltrioxorhenium (MTO): an ab initio study. *J. Organomet. Chem.*, **514**, 111–117.

[8440] Kotochigova S. and Lambropoulos P. (1993). An approach to multiphoton ionization and autoionization of Si I. *J. Phys. B*, **26**, L731–L736.

[8441] Kotochigova S. and Lambropoulos P. (1994). Theory of multi-photon ionization and autoionization of Si. *Z. Phys. D*, **31**, 41–48.

[8442] Kotochigova S., Levine Z.H., Shirley E.L., Stiles M.D. and Clark C.W. (1996). Reference data for electronic structure calculations. http://math.nist.gov/ DFTdata/. *Bull. Am. Phys. Soc.*, **41**, 65.

[8443] Kotochigova S., Levine Z.H., Shirley E.L., Stiles M.D. and Clark C.W. (1997). Local-density-functional calculations of the energy of atoms. *Phys. Rev. A*, **55**, 191–199.

[8444] Kotochigova S., Levine H. and Tupitsyn I. (1997). Correlated relativistic calculation of the giant resonance in the Gd^{3+} absorption spectrum. *Int. J. Quantum Chem.*, **65**, 575–584.

[8445] Kotochigova S. and Tupitsyn I. (1995). Electronic structure of molecules by the numerical generalized-valence-bond wave functions. *Int. J. Quantum Chem.*, **S29**, 307–312.

[8446] Koures V.G. and Harris F.E. (1995). Light cone Hamiltonian in quantum chemistry: Gaussian basis representation for quantum electrodynamics. *Int. J. Quantum Chem.*, **S29**, 277–282.

[8447] Koutroulos C.G. and Papadopoulos J. (1993). Root mean square radii of the Λ-particle orbits in hypernuclei using rectangular shape potentials in a relativistic treatment. *Prog. Theor. Phys.*, **90**, 1039–1048.

[8448] Kovács A. and Csonka G.I. (1997). Vibrational analysis of $TeCl_4$. II. A Hartree-Fock, MP2, and density functional study. *Int. J. Quantum Chem.*, **65**, 817–826.

[8449] Kovács A., Csonka G.I. and Keserű G.M. (1998). Comparison of *ab initio* and density functional methods for vibrational analysis of $TeCl_4$. *J. Comp. Chem.*, **19**, 308–318.

[8450] Kovács A. and Frenking G. (1999). Bonding interactions of a molecular pair of tweezers with transition metals. Theoretical study of bis(η^2-alkyne) complexes of copper(I), silver(I), and gold(I). *Organometallics*, **18**, 887–894.

[8451] Kovács A., Konings R.J.M. and Booij A.S. (1997). High-temperature infrared spectra of $LaCl_3$, $LaBr_3$, and LaI_3. *Chem. Phys. Lett.*, **268**, 207–212.

[8452] Kovalik A. (1997). Effects of relativity on the KL_1L_2 ($^3P_0$) transition rate of $^{159}_{65}Tb$. *J. El. Sp. Rel. Phen.*, **87**, 1–7.

[8453] Kowalski F.V. (1993). Relativistic nature of the Schrödinger equation. *Phys. Lett. A*, **182**, 23–27.

[8454] Kozelka J., Savinelli R., Berthier G., Flament J.P. and Lavery R. (1993). Force field for platinum binding to adenine. *J. Comp. Chem.*, **14**, 45–53.

[8455] Kozin I.N., Jensen P., Li Y., Buenker R.J., Hirsch G. and Klee S. (1996). An ab initio calculation of the dipole moment surfaces and the vibrational transition moments of the H_2Te molecule. *J. Mol. Spectr.*, **181**, 108–118.

[8456] Kozlov M.G. (1997). Enhancement of the electric dipole moment of the electron in the YbF molecule. *J. Phys. B*, **30**, L607–L612.

[8457] Kozlov M.G. and Ezhov V.F. (1994). Enhancement of the electric dipole moment of the electron in the YbF molecule. *Phys. Rev. A*, **49**, 4502–4507.

[8458] Kozlov M.G. and Labzowsky L.N. (1995). Parity violation effects in diatomics. *J. Phys. B*, **28**, 1933–1961.

[8459] Kozlov M.G. and Porsev S.G. (1999). Polarizabilities and hyperfine structure constants of the low-lying levels of barium. *Eur. Phys. J. D*, **5**, 59–63.

[8460] Kozlov M.G., Porsev S.G. and Flambaum V.V. (1996). Manifestation of the nuclear anapole moment in the M1 transitions in bismuth. *J. Phys. B*, **29**, 689–697.

[8461] Kozlov M.G., Titov A.V., Mosyagin N.S. and Souchko P.V. (1997). Enhancement of the electric dipole moment of the electron in the BaF molecule. *Phys. Rev. A*, **56**, 3326–3329.

[8462] Kozlov M.G. and Yashchuk V.V. (1996). Estimate of $P-$ and $P,T-$odd effects in diatomic van der Waals molecules. *JETP Letters*, **64**, 709–713.

[8463] Kraft T., Oppeneer P.M., Antonov V.N. and Eschrig H. (1995). Relativistic calculations of the magneto-optical Kerr spectra in (001) and (111) US, USe, and UTe. *Phys. Rev. B*, **52**, 3561–3570.

[8464] Kragten D.D., van Santen R.A. and Lerou J.J. (1999a). Density functional study of the palladium acetate catalyzed Wacker reaction in acetic acid. *J. Phys. Chem. A*, **103**, 80–88.

[8465] Kragten D.D., van Santen R.A., Neurock M. and Lerou J.J. (1999b). A density functional study of acetoxylation of ethylene to vinyl acetate catalyzed by palladium acetate. *J. Phys. Chem. A*, **103**, 2756–2765.

[8466] Krainov V.P. (1998). Energy distribution of relativistic electrons in the tunneling ionization of atoms by super-intense laser radiation. *Optics Express*, **2**, 268–270.

[8467] Krainov V.P. (1999). Energy and angular distribution of relativistic electrons in the tunnelling ionization of atoms by circularly polarized light. *J. Phys. B*, **32**, 1607–1614.

[8468] Kramkowski P., Baum G., Radius U., Kaupp M. and Scheer M. (1999). Novel complexes with a short tungsten-phosphorus triple bond. *Chem. Eur. J.*, **5**, 2890–2898.

[8469] Krasovska O.V., Krasovskii E.E. and Antonov V.N. (1995). Ab initio calculation of the optical and photoelectron properties of RuO_2. *Phys. Rev. B*, **52**, 11825–11829.

[8470] Krasovska O.V., Winkler B., Krasovskii E.E., Antonov V.N. and Yavorsky B.Y. (1998). The colour of sulphur. *J. Phys. CM*, **10**, 4093–4100.

[8471] Krasovskii A.E. (1999). Improved approach to spin-polarized relativistic LMTO formalism: Application to the electronic structure of Fe-Ni compounds at the Earth's core conditions. *Phys. Rev. B*, **60**, 12788–12798.

[8472] Kratz J.V. (1995). Chemie der schwersten Elemente. *Chemie in unserer Zeit*, **29**, 194–206.

[8473] Kratz J.V. (1999). Chemical properties of the transactinide elements. In *Heavy Elements and Related New Phenomena, Vol. 1*, (Edited by Greiner W. and Gupta R.K.), pp. 129–193, World Scientific, Singapore.

[8474] Kreibich T., Gross E.K.U. and Engel E. (1998). Approximate relativistic optimized potential method. *Phys. Rev. A*, **57**, 138–148.

[8475] Kresse G. and Hafner J. (1994). Norm-conserving and ultrasoft pseudopotentials for first-row and transition elements. *J. Phys. CM*, **6**, 8245–8257.

[8476] Kresse G. and Hafner J. (1997). Ab initio simulation of the metal/nonmetal transition in expanded fluid mercury. *Phys. Rev. B*, **55**, 7539–7548.

[8477] Kretzschmar I., Fiedler A., Harvey J.N., Schröder D. and Schwarz H. (1997). Effects of sequential ligation of molybdenum cation by chalcogenides on electronic structure and gas-phase reactivity. *J. Phys. Chem. A*, **101**, 6252–6264.

[8478] Kroeger H. and Scheu N. (1998). Structure functions from Breit-frame regularized Hamiltonian. In *Advances in Quantum Many-Body Theories, Vol. 9*, (Edited by Neilson D. and Bishop R.F.), pp. 179–182, World Scientific, Singapore, proc. 9th Int. Conf. on Recent Progress in Many-Body Theories, Sydney 1997.

[8479] Kroger S. and Kroger M. (1995). A program to compute the angular coefficients of the relativistic one-electron hyperfine structure parameters. *Comp. Phys. Comm.*, **90**, 381–387.

[8480] Kronfeldt H.D., Ashkenasi D., Basar G., Neale L. and Wilson M. (1993a). Isotope shifts for the $5d^5 6s7s$ and $5d^5 6s6d$ configurations of Re I. *Z. Phys. D*, **25**, 185–189.

[8481] Kronfeldt H.D., Ashkenasi D., Kroger S. and Wyart J.F. (1993b). Fine and hyperfine structure for $4f^{11} 5d6s6p$ in Er I. *Phys. Scr.*, **48**, 688–698.

[8482] Krüger S., Vent S. and Rösch N. (1997). Size dependence of bond length and binding energy in palladium and gold clusters. *Ber. Bunsenges*, **101**, 1640–1643.

[8483] Ktitarev D.V. and Yegikian R.S. (1993). Feynman path integral for Dirac system with analytic potential. *J. Math. Phys.*, **34**, 2821–2826.

[8484] Kuch W., Zharnikov M., Dittschar A., Meinel K., Schneider C.M., Kirschner J., Henk J. and Feder R. (1996). Magnetic dichroism study of the relativistic electronic structure of perpendicularly magnetized Ni/Cu(001). *J. Appl. Phys.*, **79**, 6426–6428.

[8485] Küchle W., Dolg M. and Stoll H. (1997). Ab initio study of the lanthanide and actinide contraction. *J. Phys. Chem. A*, **101**, 7128–7133.

[8486] Küchle W., Dolg M., Stoll H. and Preuss H. (1994). Energy-adjusted pseudopotentials for the actinides Parameter sets and test calculations for thorium and thorium monoxide. *J. Chem. Phys.*, **100**, 7535–7542.

[8487] Kudo T. and Nagase S. (1993). Cations of strained polycyclic compounds with Si, Ge, Sn, and Pb skeletons: Theoretical study of structures and properties. *Rev. Heteroatom Chem.*, **8**, 122–142.

[8488] Kukla K.W., Livingston A.E., Suleiman J., Berry H.G., Dunford R.W., Gemmell D.S., Kanter E.P., Cheng S. and Curtis L.J. (1995). Fine-structure energies for the $1s2s\ ^3S$ - $1s2p\ ^3P$ transition in heliumlike Ar^{16+}. *Phys. Rev. A*, **51**, 1905.

[8489] Kulakowski K., Maksymowicz A.Z. and Magdon M. (1994). Magnetic anisotropy and magnetostriction of atom pairs. *Acta Phys. Pol. A*, **85**, 869–873.

[8490] Kulkarni S.A. and Koga N. (1999). Ab initio mechanistic investigation of samarium(III)-catalyzed olefin hydroboration reaction. *J. Mol. Str. (Theochem)*, **461-462**, 297–310.

[8491] Kulkarni S., Warke C.S. and Gambhir Y.K. (1995). Relativistic mean-field approach to the anapole moment: Atomic parity-violating hyperfine transitions. *Phys. Rev. C*, **52**, 1047–1060.

[8492] Kulkarni S., Warke C.S. and Gambhir Y.K. (1998). Relativistic mean field calculation of parity-violating observables in francium. *Phys. Rev. C*, **57**, 1485–1491.

[8493] Kullie O., Düsterhöft C. and Kolb D. (1999). Dirac-Fock finite element method (FEM) calculations for some diatomic molecules. *Chem. Phys. Lett.*, **314**, 307–310.

[8494] Kumar P., Jain A.K., Tripathi A.N. and Nahar S.N. (1994a). Spin-polarization parameters and cross sections for electron scattering from zinc and lead atoms. *Phys. Rev. A*, **49**, 899–907.

[8495] Kumar P., Jain A.K., Tripathi A.N. and Nahar S.N. (1994b). Spin polarization and cross sections of electrons elastically scattered from heavy alkaline-earth atoms. *Z. Phys. D*, **30**, 149–153.

[8496] Kunz C.F., Burghardt I. and Heß B.A. (1998). *Ab initio* relativistic all-electron calculation of the Ar–I_2 ground state potential. *J. Chem. Phys.*, **109**, 359–366.

[8497] Kunz C.F., Hättig C. and Hess B.A. (1996). Ab initio study of the individual interaction energy components in the ground state of the mercury dimer. *Mol. Phys.*, **89**, 139–156.

[8498] Kuo T.Y., Chen C.M.J., Hsu S.W. and Huang K.N. (1993). Relativistic cross sections of positron-impact ionization of hydrogenic ions. *Phys. Rev. A*, **48**, 357–363.

[8499] Kurihara M., Hirata M., Sekine R., Onoe J., Nakamatsu H., Mukoyama T. and Adachi H. (1999). Discrete-variational Dirac-Slater calculation on valence band XPS for UC. *J. Alloys Compounds*, **283**, 128–132.

[8500] Kürpick P. (1996). Influence of atomic excitation processes on nuclear resonance yield curves: an ab initio study. *J. Phys. B*, **29**, L169–L172.

[8501] Kürpick P., Auer U., Sepp W.D. and Fricke B. (1993). Relativistic ab initio calculations for ion-atom collisions. *AIP Conf. Proc.*, **274**, 298–302.

[8502] Kürpick P., Baştuğ T., Fricke B., Sepp W.D., Warczak A., Jaeger M., Ullrich J., Kandler T., Schulz M., Demian A., Damrau M., Braeuning H. and Schmidt-Boecking H. (1995a). Full scale relativistic ab initio time dependent calculations for the L-K vacancy transfer in 208 MeV Ni^{23+} on a Ge solid target. *Phys. Lett. A*, **207**, 199–202.

[8503] Kürpick P., Baştuğ T., Sepp W.D. and Fricke B. (1995b). Relativistic ab initio interpretation of L-K vacancy sharing in ion-solid-target collisions. *Phys. Rev. A*, **52**, 2132–2135.

[8504] Kürpick P., Sepp W.D. and Fricke B. (1995c). Relativistic ab initio description of the K-vacancy production in heavy-ion-atom collision systems with solid targets. *Phys. Rev. A*, **51**, 3693–3697.

[8505] Kushto G.P., Souter P.F. and Andrews L. (1998). An infrared spectroscopic and quasirelativistic theoretical study of the coordination and activation of dinitrogen by thorium and uranium atoms. *J. Chem. Phys.*, **108**, 7121–7130.

[8506] Kushto G.P., Souter P.F., Andrews L. and Neurock M. (1997). A matrix isolation FT-IR and quasirelativistic density functional theory investigation of the reaction products of laser-ablated uranium atoms with NO, NO_2 and N_2O. *J. Chem. Phys.*, **106**, 5894–5903.

[8507] Kushto G.P., Souter P.F., Chertihin G.V. and Andrews L. (1999). An infrared spectroscopic and density functional investigation of dinitrogen activation by group IV metal atoms. *J. Chem. Phys.*, **110**, 9020–9031.

[8508] Kutzelnigg W. (1996). Stationary direct perturbation theory of relativistic corrections. *Phys. Rev. A*, **54**, 1183–1198.

[8509] Kutzelnigg W. (1997). Relativistic one-electron Hamiltonians 'for electrons only' and the variational treatment of the Dirac equation. *Chem. Phys.*, **225**, 203–222.

[8510] Kutzelnigg W. (1999a). Effective Hamiltonians for degenerate and quasidegenerate direct perturbation theory of relativistic effects. *J. Chem. Phys.*, **110**, 8283–8294.

[8511] Kutzelnigg W. (1999b). Relativistic corrections to magnetic properties. *J. Comp. Chem.*, **20**, 1199–1219.

[8512] Kutzelnigg W., Fleischer U. and van Wüllen C. (1996). IGLO method for chemical shift tensor calculations. In *Encyclopedia of NMR*.

[8513] Kutzelnigg W., Franke R., Ottschofski E. and Klopper W. (1994). Relativistic Hartree-Fock based on direct perturbation theory. In *New Challenges in Computational Quantum Chemistry*, (Edited by Aerts P.J.C., Bagus P.S. and Broer R.), pp. 112–133.

[8514] Kutzelnigg W., Ottschoffski E. and Franke R. (1995). Relativistic Hartree-Fock by means of stationary direct perturbation theory. I. General theory. *J. Chem. Phys.*, **102**, 1740–1751.

[8515] Kutzner M. and Radojevic V. (1994). Effects of relaxation and interchannel coupling in inner-shell photoionization of atomic ytterbium. *Phys. Rev. A*, **49**, 2574–2579.

[8516] Kutzner M., Tidwell C., Vance S.E. and Radojevic V. (1994). Inner-shell photoionization of group-IIB atoms. *Phys. Rev. A*, **49**, 300–309.

[8517] Kutzner M. and Vance S.E. (1994). Photoionization of palladium including relaxation effects. *Phys. Rev. A*, **50**, 4836–4841.

[8518] Kutzner M., Winn D. and Mattingly S. (1993). Inner-shell photoionization of alkaline-earth-metal atoms. *Phys. Rev. A*, **48**, 404–413.

[8519] Kuznetsov A.M. (1994). Comparative pseudopotential and all-electron ab initio studies of mono-, hexa- and octahydrate halogenide complexes. *Chem. Phys.*, **179**, 47–53.

[8520] Kwato Njock M.G., Nana Engo S.G., Owono Owono L.C., Lagmago G. and Oumarou B. (1994a). Supersymmetry-based quantum-defect theory of the Dirac equation in the central-potential approximation. *Phys. Lett. A*, **187**, 191–196.

[8521] Kwato Njock M.G., Oumarou B., Owono L.C., Kenmogne J.D., Nana S.G. and Onana Boyomo M. (1994b). Use of Dirac functions in the limit of exact quantum-defect theory for the calculation of oscillator strengths for lithium-like ions. *Phys. Lett. A*, **184**, 352–359.

[8522] Kwato Njock M.G., Oumarou B., Waha Ndeuna L. and Nana Engo S.G. (1993). Relativistic electric dipole radial integrals between Rydberg atomic states in the semiclassical Coulomb approximation. *Phys. Lett. A*, **180**, 124–131.

[8523] Kwon I., Collins L.A., Kress J.D. and Troullier N. (1995). First-principle study of solid Ar and Kr under high compression. *Phys. Rev. B*, **52**, 15165–15169.

[8524] Kylstra N.J., Ermolaev A.M. and Joachain C.J. (1997). Relativistic effects in the time evolution of a one-dimensional model atom in an intense laser field. *J. Phys. B*, **30**, L449–L460.

[8525] Labelle P., Lepage G.P. and Magnea U. (1994). Order $m\alpha^8$ contributions to the decay rate of orthopositronium. *Phys. Rev. Lett.*, **72**, 2006–2008.

[8526] Labelle P. and Zebarjad S.M. (1999). Derivation of the Lamb shift using an effective field theory. *Can. J. Phys.*, **77**, 267–278.

[8527] Labzowsky L.N. (1993). Adiabatic S-matrix approach in QED theory of highly charged two-electron ions. *J. Phys. B*, **26**, 1039–1069.

[8528] Labzowsky L.N. (1996). *Atomic Theory. The Quantum Electrodynamics of Electronic Shells and Radiation Processes (in Russian).* Nauka - Fizmatlit, Moscow,, 304 p.

[8529] Labzowsky L.N. and Goidenko I.A. (1997). Multiple commutator expansion for the Lamb shift in a strong Coulomb field. *J. Phys. B*, **30**, 177–187.

[8530] Labzowsky L., Goidenko I., Gorshtein M., Soff G. and Pyykkö P. (1997a). Hyperfine structure of the $2p_{3/2}$ state of highly charged $^{209}_{83}$Bi ions. *J. Phys. B*, **30**, 1427–1435.

[8531] Labzowsky L.N., Goidenko I.A. and Liesen D. (1997b). The non-resonant corrections to the process of the radiative electron capture of highly charged heavy ions. *Phys. Scr.*, **56**, 271–274.

[8532] Labzowsky L.N., Goidenko I.A. and Nefiodov A.V. (1998a). Electron self-energy calculations for tightly bound electrons in atoms. *J. Phys. B*, **31**, L477–L482.

[8533] Labzowsky L., Goidenko I. and Pyykkö P. (1999a). Estimates of the bound-state QED contributions to the g-factor of valence ns electrons in alkali metal atoms. *Phys. Lett. A*, **258**, 31–37.

[8534] Labzowsky L., Goidenko I., Tokman M. and Pyykkö P. (1999b). Calculated self-energy contributions for an ns valence electron using the multiple-commutator method. *Phys. Rev. A*, **59**, 2707–2711.

[8535] Labzowsky L.N., Johnson W.R., Soff G. and Schneider S.M. (1995). Dynamic proton model for the hyperfine sructure of the hydrogenlike ion $^{209}_{83}Bi^{82+}$. *Phys. Rev. A*, **51**, 4597–4602.

[8536] Labzowsky L., Karasiev V. and Goidenko I. (1994). Importance of the non-resonant corrections for the modern Lamb shift measurements in the multicharged hydrogen-like ions. *J. Phys. B*, **27**, L439–L445.

[8537] Labzowsky L., Karasiev V., Lindgren I., Persson H. and Salomonson S. (1993a). Higher-order QED corrections for multi-charged ions. *Phys. Scr.*, **T46**, 150–156.

[8538] Labzowsky L.N., Klimchitskaya G.L. and Dmitriev Y.Y. (1993b). *Relativistic Effects in the Spectra of Atomic Systems*. Institute of Physics Publishing, Bristol, 340 pp.

[8539] Labzowsky L.N. and Mitrushenkov A.O. (1996). Renormalization of the second-order electron self-energy for a tightly bound atomic electron: a detailed derivation. *Phys. Rev. A*, **53**, 3029–3043.

[8540] Labzowsky L., Mitrushenkov A., Shelyuto V. and Soff G. (1998b). The second-order electron self-energy counterterms in bound state QED. *Phys. Lett. A*, **240**, 225–234.

[8541] Labzowsky L., Mitrushenkov A., Shelyuto V. and Soff G. (1998c). Counterterms for the second-order electron self-energy in bound-state QED. *Phys. Rev. A*, **57**, 4038–4040.

[8542] Labzowsky L.N. and Nefiodov A.V. (1994a). Radiative interference effects in the dielectronic recombination process of an electron with hydrogenlike uranium. *Phys. Rev. A*, **49**, 236–239.

[8543] Labzowsky L.N. and Nefiodov A.V. (1994b). Analytic evaluation of the nuclear polarization contribution to the energy shift in heavy ions. *Phys. Lett. A*, **188**, 371–375.

[8544] Labzowsky L.N., Nefiodov A.V., Plunien G., Beier T. and Soff G. (1996). Vacuum polarization - nuclear polarization corrections to the Lamb shift in heavy atoms. *J. Phys. B*, **29**, 3841–3854.

[8545] Labzowsky L., Nefiodov A., Plunien G., Soff G. and Pyykkö P. (1997c). Vacuum-polarization corrections to the hyperfine-structure splitting of highly charged $^{209}_{83}$Bi ions. *Phys. Rev. A*, **56**, 4508–4516.

[8546] Labzowsky L.N. and Tokman M.A. (1995). Reference state contributions to the two-photon interaction corrections for the energy shifts in multicharged few-electron ions. *J. Phys. B*, **28**, 3717–3727.

[8547] Labzowsky L.N. and Tokman M.A. (1998). The reference state Coulomb-Breit QED corrections for the few-electron highly charged ions. *Adv. Quantum Chem.*, **30**, 393–413.

[8548] La Cour Jansen T., Rettrup S., Sharma C.R., Snijders J.G. and Palmieri P. (1999). On the evaluation of spin-orbit matrix elements in a spin-adapted basis. *Int. J. Quantum Chem.*, **73**, 23–27.

[8549] Ladik J.J. (1997). Four-component Dirac-Hartree-Fock equations for solids; generalization of the relativistic Hartree-Fock equations. *J. Mol. Str. (Theochem)*, **391**, 1–14.

[8550] Laerdahl J.K., Faegri Jr K., Visscher L. and Saue T. (1998). A fully relativistic Dirac-Hartree-Fock and second-order Møller-Plesset study of the lanthanide and actinide contraction. *J. Chem. Phys.*, **109**, 10806–10817.

[8551] Laerdahl J.K., Saue T. and Faegri Jr K. (1997a). Direct relativistic MP2: properties of ground state CuF, AgF and AuF. *Theor. Chem. Acc.*, **97**, 177–184.

[8552] Laerdahl J.K., Saue T., Faegri Jr. K. and Quiney H.M. (1997b). Ab initio study of PT-odd interactions in thallium fluoride . *Phys. Rev. Lett.*, **79**, 1642–1645.

[8553] Laerdahl J.K. and Schwerdtfeger P. (1999). Fully relativistic *ab initio* calculations of the energies of chiral molecules including parity-violating weak interactions. *Phys. Rev. A*, **60**, 4439–4453.

[8554] Lagmago Kamta G., Nana Engo S.G., Kwato Njock M.G. and Oumarou B. (1998). Consistent description of Klein-Gordon dipole matrix elements. *J. Phys. B*, **31**, 963–997.

[8555] Lagutin B.M., Demekhin F.V., Petrov I.D., Sukhorukov V.L., Eresman A., Vollweiler F., Smolander K. and Schartner K.H. (1998). Relativistic effects on the photoabsorption of outer atomic shells (in Russian). *Zh. Strukt. Khim.*, **39**, 992–1000.

[8556] Laidler K.J. and Meiser J.H. (1999). *Physical Chemistry, Third Edition.* Houghton Mifflin Company, Boston and New York, see pp. 543-544, 549.

[8557] LaJohn L. and Luke T.M. (1993). On the construction of an efficient basis for CI calculations: application to a spin forbidden transition in P II. *J. Phys. B*, **26**, 863–871.

[8558] LaJohn L. and Talman J.D. (1998). Variational solution of the single-particle Dirac equation in the field of two nuclei using relativistically adapted Slater basis functions. *Theor. Chem. Acc.*, **99**, 351–356.

[8559] Lalić M.V., Popović Z.V. and Vukajlović F.R. (1999). Electronic structure and electric field gradient calculations for Hf_2Fe intermetallic compounds. *J. Phys. CM*, **11**, 2513–2522.

[8560] Lamare L. and Michel-Calendini F. (1997). LDA electronic structure calculations on Au_{13} cluster. *Int. J. Quantum Chem.*, **61**, 635–639.

[8561] Lambert C., Kaupp M. and von Ragué Schleyer P. (1993). "Inverted" sodium-lithium electronegativity: Polarity and metalation energies of organic and inorganic alkali-metal compounds. *Organometallics*, **12**, 853–859.

[8562] Lämmerzahl C. (1993). The pseudodifferential operator square root of the Klein-Gordon equation. *J. Math. Phys.*, **34**, 3918–3932.

[8563] Lämmerzahl C. (1994). Addendum to "The pseudopotential operator square root of the Klein-Gordon equation" [J. Math. Phys. **34**, 3918, (1993)]. *J. Math. Phys.*, **35**, 3769.

[8564] Land M.C., Arshansky R.I. and Horwitz L.P. (1994). Selection rules for dipole radiation from a relativistic bound state. *Found. Phys.*, **24**, 563–578.

[8565] Land M.C. and Horwitz L.P. (1995). The Zeeman effect for the relativistic bound state. *J. Phys. A*, **28**, 3289–3304.

[8566] Landis C.R., Firman T.K., Root D.M. and Cleveland T. (1998). A valence bond perspective on the molecular shapes of simple metal alkyls and hydrides. *J. Am. Chem. Soc.*, **120**, 1842–1854.

[8567] Landrum G.A., Dronskowski R., Niewa R. and DiSalvo F.J. (1999). Electronic structure and bonding in cerium (nitride) compounds: trivalent versus tetravalent cerium. *Chem. Eur. J.*, **5**, 515–522.

[8568] Lanza G. and Fragalà I.L. (1996). A relativistic effective core potential ab initio study of molecular geometries and vibrational frequencies of lanthanide trihalides LnX_3 (Ln = Gd, Lu; X = F, Cl). *Chem. Phys. Lett.*, **255**, 341–346.

[8569] Lanza G. and Fragalà I.L. (1998). Theoretical study of the molecular properties of cerium trihalides and tetrahalides CeX_n (n = 3,4; X = F,Cl). *J. Phys. Chem. A*, **102**, 7990–7995.

[8570] Lanzisera D.V. and Andrews L. (1997). Reactions of laser-ablated Mg, Ca, Sr, and Ba atoms with hydrogen cyanide in excess argon. Matrix infrared spectra and density functional calculations on novel isocyanide products. *J. Phys. Chem. A*, **101**, 9666–9672.

[8571] Larson L.J., McCauley E.M., Weissbart B. and Tinti D.S. (1995). Luminescent gold(I) complexes. Optical and ODMR studies of mononuclear halo(triphenylphosphine)- and halo(triphenylarsine)gold(I) complexes. *J. Phys. Chem.*, **99**, 7218–7226.

[8572] Latifzadeh L. and Balasubramanian K. (1996a). Electronic states of the diatomic antimony fluoride (SbF) . *Chem. Phys. Lett.*, **257**, 257–264.

[8573] Latifzadeh-Masoudipur L. and Balasubramanian K. (1996b). Electronic states and potential energy curves of SbF_2 and SbF_2^+ . *Chem. Phys. Lett.*, **262**, 553–558.

[8574] Latifzadeh-Masoudipur L. and Balasubramanian K. (1997a). Geometries and energy separations of electronic states of $AsCl_2$, $AsBr_2$, $AsCl_2{}^+$, and $AsBr_2{}^+$. *J. Chem. Phys.*, **106**, 2695–2701.

[8575] Latifzadeh-Masoudipur L. and Balasubramanian K. (1997b). Electronic states and potential energy curves of PI_2 and $PI_2{}^+$. *Chem. Phys. Lett.*, **267**, 545–550.

[8576] Launila O., Wahlgren U., Schimmelpfennig B., Taklif A.G., Fagerli H. and Gropen O. (1997). Spectroscopy of NbO: Characterization of the doublet manifold. *J. Mol. Spectr.*, **186**, 131–143.

[8577] Lavaty T.G., Wikrent P., Drouin P.J. and Kukolich S.G. (1998). Microwave measurements and calculations on the molecular structure of tetracarbonyldihydroruthenium. *J. Chem. Phys.*, **109**, 9473–9478.

[8578] Lavin C., Alvarez A.B. and Martin I. (1997). Systematic trends in the relativistic oscillator strengths for fine-structure transitions in the aluminum isoelectronic sequence. *J. Quant. Spectr. Rad. Transfer*, **57**, 831–845.

[8579] Lavin C., Barrientos C. and Martin I. (1994). Relativistic quantum defect calculations on the copper isoelectronic sequence. *Int. J. Quantum Chem.*, **50**, 411–428.

[8580] Lavin C. and Martin I. (1993). Oscillator strengths in the boron isoelectronic sequence. *J. Quant. Spectr. Rad. Transfer*, **50**, 611–619.

[8581] Lavin C. and Martin I. (1994). Relativistic quantum defect orbital calculations on triplet-triplet transitions in cadmium-like ions. *J. Quant. Spectr. Rad. Transfer*, **52**, 21–29.

[8582] Lavín C., Martin P., Martin I. and Karwowski J. (1993). Relativistic quantum defect orbital calculations of singlet-singlet transitions in the zinc and cadmium isoelectronic sequences. *Int. J. Quantum Chem.*, **S27**, 385–397.

[8583] Lawrence W.G. and Apkarian V.A. (1994). Many-body potentials for an open shell atom: Spectroscopy of spin-orbit transitions of iodine in crystalline Xe and Kr. *J. Chem. Phys.*, **101**, 1820–1831.

[8584] Lazur V.Y., Gorvat P.P., Migalina S.Y., Shuba J.M. and Yanev R.K. (1996a). Asymptotic approach to the energy splitting in the relativistic two-centre problem. 1. The wave function of the Dirac electron in the internuclear region (in Ukrainian). *Ukr. Fiz. Zh.*, **41**, 605–611.

[8585] Lazur V.Y., Gorvat P.P., Migalina S.Y., Shuba J.M. and Yanev R.K. (1996b). Asymptotic approach to the energy splitting in the relativistic two-centre problem. 2. The wave function of the Dirac electron in the region of another ion (in Ukrainian). *Ukr. Fiz. Zh.*, **41**, 612–619.

[8586] Lazur V.Y., Gorvat P.P., Migalina S.Y., Shuba J.M. and Yanev R.K. (1996c). Asymptotic approach to the energy splitting in the relativistic two-centre problem. 3. Potential of the one-electron exchange interaction (in Ukrainian). *Ukr. Fiz. Zh.*, **41**, 620–626.

[8587] Lazzeretti P. and Zanasi R. (1997). On the calculation of parity-violating energies in hydrogen peroxide and hydrogen disulphide molecules within the random-phase approximation. *Chem. Phys. Lett.*, **279**, 349–354.

[8588] Lazzeretti P., Zanasi R. and Faglioni F. (1999). Energetic stabilization of *d*-camphor via neutral currents. *Phys. Rev. E*, **60**, 871–874.

[8589] Le A.T., Le V.H., Komarov L.I. and Romanova T.S. (1994). Operator representation of the Dirac Coulomb Green function and relativistic polarizability of hydrogen-like atoms. *J. Phys. B*, **27**, 4083–4094.

[8590] Le A.T., Le V.H., Komarov L.I. and Romanova T.S. (1996). Relativistic dynamical polarizability of hydrogen-like atoms. *J. Phys. B*, **29**, 2897–2906.

[8591] Le V.H. and Nguyen T.G. (1993). On the Green function for a hydrogen-like atom in the Dirac monopole field plus the Aharonov-Bohm field. *J. Phys. A*, **26**, 3333–3338.

[8592] Lee E.P.F., Soldán P. and Wright T.G. (1998). Geometries and binding energies of $Rg{\cdot}NO^+$ cationic complexes (Rg = He,Ne, Ar, Kr, and Xe). *J. Phys. Chem. A*, **102**, 6858–6864.

[8593] Lee E.P.F. and Wright T.G. (1997). Boron dibromide and boron diiodide ground state neutral and cation. Use of effective core potentials combined with ab initio and density functional theory. *J. Phys. Chem. A*, **101**, 1374–1377.

[8594] Lee E.P.F. and Wright T.G. (1999). Interaction energy of the radon-water ($Rn{\cdot}H_2O$) complex. *J. Phys. Chem. A*, **103**, 7843–7847.

[8595] Lee H.S., Han Y.K., Kim M.C., Bae C. and Lee Y.S. (1998). Spin-orbit effects calculated by two-component coupled-cluster methods: test calculations on AuH, Au_2, TlH and Tl_2. *Chem. Phys. Lett.*, **293**, 97–102.

[8596] Lee R.N. and Milstein A.I. (1995). Quasiclassical Green function and Delbrueck scattering in a screened Coulomb field. *Phys. Lett. A*, **198**, 217–224.

[8597] Lee S.H. and Liu K.P. (1999). Exploring the spin-orbit reactivity in the simplest chlorine atom reaction. *J. Chem. Phys.*, **111**, 6253–6259.

[8598] Leininger T., Berning A., Nicklass A., Stoll H., Werner H.J. and Flad H.J. (1997). Spin-orbit interaction in heavy group 13 atoms and TlAr. *Chem. Phys.*, **217**, 19–27.

[8599] Leininger T., Nicklass A., Küchle W., Stoll H., Dolg M. and Bergner A. (1996a). The accuracy of the pseudopotential approximation: non-frozen-core effects for spectroscopic constants of alkali fluorides XF (X = K, Rb, Cs). *Chem. Phys. Lett.*, **255**, 274–280.

[8600] Leininger T., Nicklass A., Stoll H., Dolg M. and Schwerdtfeger P. (1996b). The accuracy of the pseudopotential approximation. II. A comparison of various core sizes for indium pseudopotentials in calculations for spectroscopic constants of InH, InF, and InCl. *J. Chem. Phys.*, **105**, 1052–1059.

[8601] Leininger T., Riehl J.F., Jeung G.H. and Pélissier M. (1993). Comparison of the widely used HF pseudo-potentials: MH^+ (M=Fe, Ru, Os). *Chem. Phys. Lett.*, **205**, 301–305.

[8602] Lerner L. (1996). Derivation of the Dirac equation from a relativistic representation of spin. *Eur. J. Phys.*, **17**, 172–175.

[8603] Lesar A., Muri G. and Hodoščček M. (1998). Ab initio studies on the structures and vibrational frequencies of rare earth fluorides LnF_n (Ln = Er, Tm; n = 1, 2, 3) and their positive ions and an assessment of their ionization and dissociation energies. *J. Phys. Chem. A*, **102**, 1170–1176.

[8604] Lévai G. and Del Sol Mesa A. (1996). On some solutions of the Dirac equation. *J. Phys. A*, **29**, 2827–2832.

[8605] Levin F.S. and Micha D.A. (1993). *Long-Range Casimir Forces – Theory and Recent Experiments on Atomic Systems (Ed.)*. Plenum Press, New York and London, 357 pp.

[8606] Lewis R.R. and Blinder S.M. (1995). Stark-induced anapole magnetic fields in atoms. *Phys. Rev. A*, **52**, 4439–4446.

[8607] Lewis R.T., Siedentop H. and Vugalter S. (1997). The essential spectrum of relativistic multi-particle operators. *Ann. Inst. Henri Poincaré Phys. Theor.*, **67**, 1–28.

[8608] Le Yaouanc A., Oliver L. and Raynal J. (1995). High-order expansion of the energy eigenvalues of a relativistic Coulomb equation. *Ann. Phys. (NY)*, **239**, 243–271.

[8609] Ley-Koo E., Bunge C.F. and Jáuregui R. (1997). Evaluation of relativistic atomic integrals using perimetric coordinates. *Int. J. Quantum Chem.*, **63**, 93–97.

[8610] Ley-Koo E., Jáuregui R., Góngora-T A. and Bunge C.F. (1993). General method to evaluate two-body integrals for relativistic atomic calculations. *Phys. Rev. A*, **47**, 1761–1770.

[8611] Li B.L., Cao Y. and Feng J.W. (1997). An ab initio study of the structure of the MoC_4 cluster. *J. Mol. Str.*, **407**, 149–153.

[8612] Li G., Samuel M.A. and Eides M.I. (1993). One-logarithmic recoil correction in muonium hyperfine splitting. *Phys. Rev. A*, **47**, 876–878.

[8613] Li J., Dickson R.M. and Ziegler T. (1995a). Dihydrogen versus dihydride: relativistic effects on the relative stabilities of nonclassical and classical isomers of $M(PH_3)_3H_4$ (M = Fe, Ru, Os). *J. Am. Chem. Soc.*, **117**, 11482–11487.

[8614] Li J., Fisher C.L., Chen J.L., Bashford D. and Noodleman L. (1996). Calculation of redox potentials and pK_a values of hydrated transition metal cations by a combined density functional and continuum dielectric theory. *Inorg. Chem.*, **35**, 4694–4702.

[8615] Li J. and Pyykkö P. (1993). Structure of $E(AuPH_3)_4^+$ E=N, P, As: T_d or C_{4v}. *Inorg. Chem.*, **32**, 2630–2634.

[8616] Li J., Schreckenbach G. and Ziegler T. (1994a). First bond dissociation energy of $M(CO)_6$ (M=Cr, Mo, W) revisited: The performance of density functional theory and influence of relativistic effects. *J. Phys. Chem.*, **98**, 4838–4841.

[8617] Li J., Schreckenbach G. and Ziegler T. (1995b). A reassessment of the first metal-carbonyl disssociation energy in $M(CO)_4$ (M = Ni, Pd, Pt), $M(CO)_5$ (M = Fe, Ru, Os), and $M(CO)_6$ (M = Cr, Mo, W) by a quasirelativistic density functional method. *J. Am. Chem. Soc.*, **117**, 486–494.

[8618] Li J., Schreckenbach G. and Ziegler T. (1995c). Relativistic effects on metal-ligand bond strengths in π-complexes: a quasi-relativistic density functional study of $M(PH_3)_2X_2$ (M = Ni,Pd, Pt; X_2 = O_2, C_2H_2, C_2H_4) and $M(CO)_4(C_2H_4)$ (M = Fe, Ru, Os). *Inorg. Chem.*, **34**, 3245–3252.

[8619] Li J. and Bursten B.E. (1997). Electronic structure of cycloheptatrienyl sandwich compounds of actinides: $An(\eta^7\text{-}C_7H_7)_2$ (An = Th, Pa, U, Np, Pu, Am). *J. Am. Chem. Soc.*, **119**, 9021–9032.

[8620] Li J. and Bursten B.E. (1999). Bis(arene) actinide sandwich complexes, $(\eta^6\text{-}C_6H_3R_3)_2An$: Linear or bent? *J. Am. Chem. Soc.*, **121**, 10243–10244.

[8621] Li J., Irle S. and Schwarz W.H.E. (1996b). Electronic structure and properties of trihalogen catons X_3^+ and XY_2^+ (X, Y = F, Cl, Br, I). *Inorg. Chem.*, **35**, 100–109.

[8622] Li J., Liu C.W. and Lu J.X. (1994b). Electronic structures and d-pπ bonding of some $M_3X_4^{4+}$ cluster compounds. *Polyhedron*, **13**, 1841–1851.

[8623] Li J., Liu C.W. and Lu J.X. (1994c). Ab initio studies of electronic structures and quasi-aromaticity in $M_3S_{4-n}O_n^{4+}$ (M=Mo, W; n=0-4) clusters. *J. Chem. Soc., Faraday Trans.*, **90**, 39–45.

[8624] Li L., Hackett P.A. and Rayner D.M. (1993). Relativistic effects in reactions of the coinage metal dimers in the gas phase. *J. Chem. Phys.*, **99**, 2583–2590.

[8625] Li R.N. and Mil'shtejn A.I. (1995). Quasiclassical Green function and Delbrueck scattering in a screened Coulomb field (in Russian). *Zh. Eksp. Teor. Fiz.*, **107**, 1393–1402.

[8626] Li S.H. and Hall M.B. (1999). Transition metal polyhydride complexes. 11. Mechanistic studies of the cis to trans isomerization of the iridium(III) dihydride $Ir(H)_2(CO)L$ ($L=C_6H_3(CH_2P(H)_2)_2$). *Organometallics*, **18**, 5682–5687.

[8627] Li W.J., Hong M.C., Cao R., Kang B.S. and Liu H.Q. (1999a). g-matrix based on configuration interaction and Stone's perturbation theory. *J. Magn. Res.*, **138**, 74–79.

[8628] Li W.J., Hong M.C., Cao R., Kang B.S. and Liu H.Q. (1999b). The g-values of some halomolybdenyl, -vanadyl, and -chromyl complexes. *J. Magn. Res.*, **138**, 80–88.

[8629] Li Y., Liebermann H.P., Hirsch G. and Buenker R.J. (1994). Relativistic configuration interaction study of the electronic spectrum of thallium chloride. *J. Mol. Spectr.*, **165**, 219–232.

[8630] Li Z.S., Norin J., Persson A., Wahlström C.G., Svanberg S., Doidge P.S. and Biémont E. (1999). Radiative properties of neutral germanium obtained from excited-state lifetime and branching-ratio measurements and comparison with theoretical calculations. *Phys. Rev. A*, **60**, 198–208.

[8631] Liao D.W. and Balasubramanian K. (1993). Germanium monochloride (GeCl). Spectroscopic constants and potential energy curves. *Chem. Phys. Lett.*, **213**, 174–180.

[8632] Liao D.W. and Balasubramanian K. (1994). Spectroscopic constants and potential energy curves for GeF. *J. Mol. Spectr.*, **163**, 284–290.

[8633] Liao M.S. (1993). *Relativistische Dichte-Funktional-Berechnungen an schwermetall-haltigen Festkörpern.*. Ph.D. thesis, Universität-Gesamthochschule-Siegen, 209 p.

[8634] Liao M.S., Au C.T. and Ng C.F. (1997). Methane dissociation on Ni, Pd, Pt and Cu metal (111) surfaces - a theoretical comparative study. *Chem. Phys. Lett.*, **272**, 445–452.

[8635] Liao M.S., Lü X. and Zhang Q.E. (1998). Cyanide adsorbed on coinage metal electrodes: A relativistic density functional investigation. *Int. J. Quantum Chem.*, **67**, 175–185.

[8636] Liao M.S. and Schwarz W.H.E. (1994). Effective radii of the monovalent coin metals. *Acta Cryst.*, **B50**, 9–12.

[8637] Liao M.S. and Schwarz W.H.E. (1997a). Bonding and geometry of $(X\text{-}M\text{-}X)^{3-}$ groups in crystalline A_3MX_2 compounds (A = alkali, M = coinage metal, X = chalcogen). *J. Alloys Compounds*, **246**, 2–17.

[8638] Liao M.S. and Schwarz W.H.E. (1997b). On the structural data of Hg(I) halides. *J. Alloys Compounds*, **246**, 124–130.

[8639] Liao M.S. and Zhang Q.E. (1995). Hg-Hg bonding in mercurous $Hg(I)_2L_2$ compounds: the influence of ligand electronegativity. *J. Mol. Str.*, **358**, 195–203.

[8640] Liao M.S. and Zhang Q.E. (1997). A theoretical study of complexes MH_x^{2-} and MCl_y^{2-} in crystalline A_2MH_x and A_2MCl_y compounds (A = alkali, alkaline earth; M = Ni, Pd, Pt; $x =$ 2,4,6; $y =$ 4,6. *Inorg. Chem.*, **36**, 396–405.

[8641] Liao M.S. and Zhang Q.E. (1998a). Chemical bonding in XeF_2, XeF_4, KrF_2, KrF_4, RnF_2, $XeCl_2$ and $XeBr_2$: from the gas-phase to the solid state. *J. Phys. Chem.*, **102**, 10647–10654.

[8642] Liao M.S. and Zhang Q.E. (1998b). Dissociation of methane on different transition metals. *J. Mol. Catal. A. - Chemical*, **136**, 185–194.

[8643] Liao M.S. and Zhang Q.E. (1999). Application of an improved point-charge model to study the crystal Hg_2F_2. *J. Solid State Chem.*, **146**, 239–244.

[8644] Liao M.S., Zhang Q.E. and Schwarz W.H.E. (1995). Properties and stabilities of MX, MX_2, and M_2X_2 compounds (M = Zn, Cd, Hg; X = F, Cl, Br, I). *Inorg. Chem.*, **34**, 5597–5605.

[8645] Liao M.S., Zhang Q.E. and Schwarz W.H.E. (1998). On bonding, structure, and stability of ternary hydrides A_2MH_2 (A=Li,Na; M=Pd,Pt). *Z. anorg. allg. Chem.*, **624**, 1419–1428.

[8646] Liao M.Z., Dai D. and Balasubramanian K. (1995). Electronic states of the Ga_3As_2 and Ga_2As_3 clusters. *Chem. Phys. Lett.*, **239**, 124–130.

[8647] Liaw S.S. (1992b). Dipole oscillator strengths for the alkali-like ions. *Can. J. Phys.*, **70**, 1279–1282.

[8648] Liaw S.S. (1993a). Perturbative calculation of transition amplitudes for cesium. *Phys. Rev. A*, **47**, 1726–1731.

[8649] Liaw S.S. (1993b). Energy levels and transition amplitudes for alkali-metal atoms in the Brueckner approximation. *Phys. Rev. A*, **48**, 3555–3560.

[8650] Liaw S.S. (1995). Bound states of Ca^-. *Phys. Rev. A*, **52**, 1754–1756.

[8651] Lichnerowicz A. (1998). First eigenvalue of the Dirac operator for a Kähler manifold of even complex dimension: The limiting case. *Lett. Math. Phys.*, **46**, 71–80.

[8652] Lieb E., Loss M. and Siedentop H. (1996). Stability of relativistic matter via Thomas–Fermi theory. *Helv. Phys. Acta*, **69**, 974–984.

[8653] Lieb E.H., Siedentop H. and Solovej J. (1997a). Stability and instability of relativistic electrons in classical electromagnetic fields. *J. Statist. Phys.*, **89**, 37–59.

[8654] Lieb E.H., Siedentop H. and Solovej J. (1997b). Stability of relativistic matter with magnetic fields. *Phys. Rev. Lett.*, **79**, 1785–1788.

[8655] Liebermann H.P., Boustani I., Rai S.N., Alekseyev A.B., Hirsch G. and Buenker R.J. (1993). Use of relativistic core potentials to compute potential curves and lifetimes of low-lying states of arsenic fluoride. *Chem. Phys. Lett.*, **214**, 381–390.

[8656] Lim I.S., Pernpointner M., Seth M., Laerdahl J.K., Schwerdtfeger P., Neogrady P. and Urban M. (1999). Relativistic coupled-cluster static dipole polarizabilities of the alkali metals from Li to element 119. *Phys. Rev. A*, **60**, 2822–2828.

[8657] Lin C.D. (1993). Classifications and properties of doubly excited states of atoms. In *Fundamental Processes and Applications of Atoms and Ions*, (Edited by Lin C.D.), World Scientific, Singapore.

[8658] Lin D.H. (1997). Relativistic fixed-energy amplitudes of the step and square well potential problems. *J. Phys. A*, **30**, 4365–4372.

[8659] Lin D.H. (1999). Path integral for the relativistic three-dimensional Aharonov-Bohm-Coulomb system. *J. Math. Phys.*, **40**, 1246–1254.

[8660] Lin Q.G. (1998). Levinson theorem for Dirac particles in two dimensions. *Phys. Rev. A*, **57**, 3478–3488.

[8661] Lin Q.G. (1999). Levinson theorem for Dirac particles in one dimension. *Eur. Phys. J. D*, **7**, 515–524.

[8662] Lin Z.Y. and Hall M.B. (1993). Theoretical studies on the stability of M_8C_{12} clusters. *J. Am. Chem. Soc.*, **115**, 11165–11168.

[8663] Lin Z.Y., Hall M.B., Guest M.F. and Sherwood P. (1994). Theoretical studies of inorganic and organometallic reaction mechanisms. 8. Hydrogen exchange in the β-agostic ethylene complex of cyclopentadienyl rhodium. *J. Organomet. Chem.*, **478**, 197–203.

[8664] Lindgren I. (1992). Many-body problems in atomic physics. In *Recent Progress in Many-Body Theories*, (Edited by Ainsworth T.L.), vol. 3, pp. 245–276, Plenum Press, New York.

[8665] Lindgren I. (1994). Relativistic many-body calculations on atomic systems. *Phys. Reports*, **242**, 269–284.

[8666] Lindgren I. (1996). Relativistic many-body and QED calculations on atomic systems. *Int. J. Quantum Chem.*, **57**, 683–695.

[8667] Lindgren I. (1998). Electron correlation and quantum electrodynamics. *Mol. Phys.*, **94**, 19–28.

[8668] Lindgren I. (1999). QED effects in strong nuclear fields. *Phys. Scr.*, **T80**, 133–140.

[8669] Lindgren I., Martinson I. and Schuch R. (1993a). Heavy-Ion Spectroscopy and QED Effects in Atomic Systems (Ed.). *Phys. Scr.*, **T46**, proc. **85**th Nobel Symposium, 269 p.

[8670] Lindgren I. and Morrison J. (1986). *Atomic Many-Body Theory, 2nd Edition*. Springer-Verlag, Berlin, 466 p. Chapter 14.7. 'Relativistic Effects'.

[8671] Lindgren I., Persson H., Salomonson S., Karasiev V., Labzowsky L., Mitrushenkov A. and Tokman M. (1993b). Second-order QED corrections for few-electron heavy ions: reducible Breit-Coulomb correction and mixed self-energy–vacuum polarization correction. *J. Phys. B*, **26**, L503–L509.

[8672] Lindgren I., Persson H., Salomonson S. and Labzowsky L. (1995a). Full QED calculations of the two-photon exchange for helium-like systems. *Phys. Rev. A*, **51**, 1167–1195.

[8673] Lindgren I., Persson H., Salomonson S. and Sunnergren P. (1995b). Few-body problems in atomic physics. *Few-Body Syst., Suppl.*, **8**, 60–66.

[8674] Lindgren I., Persson H., Salomonson S. and Sunnergren P. (1998). Analysis of the electron-self-energy for tightly bound electrons. *Phys. Rev. A*, **58**, 1001–1015.

[8675] Lindgren I., Persson H., Salomonson S. and Ynnerman A. (1993c). Bound-state self-energy calculation using partial-wave renormalization. *Phys. Rev. A*, **47**, R4555–R4558.

[8676] Lindhard J. and Sørensen A.H. (1996). Relativistic theory of stopping for heavy ions. *Phys. Rev. A*, **53**, 2443–2456.

[8677] Lindroth E. (1994). Calculation of doubly excited states of helium with a finite discrete spectrum. *Phys. Rev. A*, **49**, 4473–4480.

[8678] Lindroth E., Bürgers A. and Brandefelt N. (1998). Relativistic effects on the H^- resonances converging to the H(n=2) threshold. *Phys. Rev. A*, **57**, R685–R688.

[8679] Lindroth E. and Indelicato P. (1993). Inner shell transitions in heavy atoms. *Phys. Scr.*, **T46**, 139–143.

[8680] Lindroth E. and Indelicato P. (1994). High precision calculations of inner shell transitions in heavy elements. *Nucl. Instr. Meth. Phys. Res. B*, **87**, 222–226.

[8681] Lindroth E. and Ynnerman A. (1993). Ab initio calculations of g_j factors for Li, Be^+, and Ba^+. *Phys. Rev. A*, **47**, 961–970.

[8682] Lingott R.M., Liebermann H.P., Alekseyev A.B. and Buenker R.J. (1999). Electronic states and transitions of bismuth sulfide. *J. Chem. Phys.*, **110**, 11294–11302.

[8683] Liu J.C., Liu H.K., Wang Z.L., Cheng Y.S. and Yang X.D. (1998). Spin - orbit components of resonant satellite photoionization of Ca^+. *J. Phys. B*, **31**, L533–L537.

[8684] Liu L. and Li J.M. (1991b). Amplitudes of electronic wavefunctions at nuclei for atoms: Pt. 2. Amplitudes of electronic wavefunctions at nuclei for atomic ions. *Acta Phys. Sinica*, **40**, 1929–1933.

[8685] Liu L., Li J.M. and Pratt R.H. (1994). Amplitudes at nuclei of electronic wave functions for atomic ions. *Phys. Rev. A*, **49**, 3770–3775.

[8686] Liu W.J. and Dolg M. (1998). Benchmark calculations for lanthanide atoms: Calibration of ab initio and density-functional methods. *Phys. Rev. A*, **57**, 1721–1728.

[8687] Liu W.J., Dolg M. and Fulde P. (1997a). Low-lying electronic states of lanthanocenes and actinocenes $M(C_8H_8)_2$ (M=Nd,Tb, Yb, U). *J. Chem. Phys.*, **107**, 3584–3591.

[8688] Liu W.J., Dolg M. and Fulde P. (1998a). Calculated properties of lanthanocene anions and the unusual electronic structure of their neutral counterparts. *Inorg. Chem.*, **37**, 1067–1072.

[8689] Liu W.J., Dolg M. and Li L.M. (1998b). Fully relativistic density functional calculations of the ground and excited states of Yb, YbH, YbF, and YbO. *J. Chem. Phys.*, **108**, 2886–2895.

[8690] Liu W.J., Franke R. and Dolg M. (1999). Relativistic ab initio and density functional theory calculations on the mercury fluorides: Is HgF_4 thermodynamically stable? *Chem. Phys. Lett.*, **302**, 231–239.

[8691] Liu W.J., Hong G.Y., Dai D.D., Li L.M. and Dolg M. (1997b). The Beijing four-component density functional program package (BDF) and its applications to EuO, EuS, YbO and YbS. *Theor. Chem. Acc.*, **96**, 75–83.

[8692] Liu W.J., Küchle W. and Dolg M. (1998c). Ab initio pseudopotential and density-functional all-electron study of ionization and excitation energies of actinide atoms. *Phys. Rev. A*, **58**, 1103–1110.

[8693] Liu W.J. and Li L.M. (1994). New advances in relativistic quantum chemistry (in Chinese). *Chemistry (China)*, **11**, 1–5.

[8694] Liu W.J. and Li L.M. (1995). Studies on the electronic structure of MX_4 (M = Ti, Zr, Hf; X = Cl, Br) (in Chinese). *Acta Chim. Sinica*, **53**, 431–437.

[8695] Liu W.J. and van Wüllen C. (1999). Spectroscopic constants of gold and eka-gold (element 111) diatomic compounds: The importance of spin-orbit coupling. *J. Chem. Phys.*, **110**, 3730–3735.

[8696] Liu W.Y., Gu S.H. and Li B.W. (1996). Fine structure in a strong magnetic field: Paschen-Back effect reconsidered in Rydberg atoms. *Phys. Rev. A*, **53**, 3044–3049.

[8697] Liu X.Y., Palacios A.A., Novoa J.J. and Alvarez S. (1998). Framework bonding and coordination sphere rearrangement in the M_2X_2 cores of synthetic analogues of oxyhemocyanin and related Cu and Pt complexes. *Inorg. Chem.*, **37**, 1202–1212.

[8698] Liu Y. and Allen R.E. (1995). Electronic structure of the semimetals Bi and Sb . *Phys. Rev. B*, **52**, 1566–1577.

[8699] Liu Y., Xu Z. and Johnson P.D. (1995). Spin-orbit coupling, exchange interaction, and hybridization in the photoexcitation of the Ni $3p$ core level. *Phys. Rev. B*, **52**, R8593–R8596.

[8700] Liu Z.W. and Kelly H.P. (1993). Relativistic calculation of multiphoton ionization in strong laser fields. *Phys. Rev. A*, **47**, 1460–1466.

[8701] Liubich V., Fuks D. and Dorfman S. (1999). Interstitial boron in tungsten: Electronic structure, ordering tendencies, and total energy calculations. *Int. J. Quantum Chem.*, **75**, 917–926.

[8702] Lledos A. and Jean Y. (1998). A density functional study of the internal rotation in the quadruply bonded $Mo_2Cl_4(PH_3)_4$ complex. *Chem. Phys. Lett.*, **287**, 243–249.

[8703] Llusar R., Casarrubios M., Barandiarán Z. and Seijo L. (1996). Ab initio model potential calculations on the electronic spectrum of Ni^{2+}-doped MgO including correlation, spin-orbit and embedding effects. *J. Chem. Phys.*, **105**, 5321–5330.

[8704] Lobanov A.E. (1997). On the problem of electron polarization in a pulsed electromagnetic field (in Russian). *Vestn. Mosk. Univ, Ser. 3. Fiz. Astr.*, **(2)**, 59–60.

[8705] Lobayan R.M. and Aucar G.A. (1998a). NMR-K reduced coupling constant calculations within the CLOPPA-PM3 approach: I. General results. *J. Mol. Str. (Theochem)*, **452**, 1–11.

[8706] Lobayan R.M. and Aucar G.A. (1998b). NMR-K reduced coupling constant calculations within the CLOPPA-PM3 approach: II: shortcomings and how to overcome them. *J. Mol. Str. (Theochem)*, **452**, 13–23.

[8707] Loginov A.V. (1995). Estimation of the contribution from the configuration superposition to the semiempirical hyperfine structure parameters of atomic spectra . *Opt. Spektr.*, **79**, 5–11.

[8708] Loginov A.V. (1995). Estimation of the contribution from the configuration superposition to the semiempirical hyperfine structure parameters of atomic spectra. *Opt. Spectr.*, **79**, 1–7.

[8709] Lohmann B. (1993). Correlation effects on relative intensities, angular distribution and spin polarization of Auger electrons from mercury atoms. *J. Phys. B*, **26**, 1623–1629.

[8710] Lohmann B. and Fritzsche S. (1994). Absolute Auger rates relative intensities and angular distribution of the KLL spectra of Auger electrons from alkali atoms. *J. Phys. B*, **27**, 2919–2941.

[8711] Lohr L.L. and Ashe III A.J. (1993). Electronic structure of organobismuth compounds: Effective core potential and semiempirical calculations. *Organometallics*, **12**, 343–346.

[8712] Löken S., Felser C. and Tremel W. (1998). $Tl_2Au_4S_3$: $x = 4/3$ member of the series $A_{2-x}Au_xQ$. Preparation and an analysis of its gold-gold bonding. *Chem. Comm.*, pp. 1603–1604.

[8713] Long P. and Crater H.W. (1998). Two-body Dirac equations for general covariant interactions and their coupled Schrödinger-like forms. *J. Math. Phys.*, **39**, 124–126.

[8714] Loock H.P., Simard B., Wallin S. and Linton C. (1998). Ionization potentials and bond energies of TiO, ZrO, NbO and MoO. *J. Chem. Phys.*, **109**, 8980–8992.

[8715] López N. and Illas F. (1997). The nature of metal-oxide chemical bond: Electronic structure of PdMgO and PdOMg molecules. *J. Chem. Phys.*, **107**, 7345–7349.

[8716] Lorenzana H.A., Klepeis J.E., Lipp M.J., Evans W.J., Radousky H.B. and van Schilfgaarde M. (1997). High-pressure phases of PbF_2: A joint experimental and theoretical study. *Phys. Rev. B*, **56**, 543–551.

[8717] Lounesto P. (1993). Clifford algebras and Hestenes spinors. *Found. Phys.*, **23**, 1203–1237.

[8718] Lovatt S.C., Gyorffy B.L. and Guo G.Y. (1993). Relativistic spin-polarized scattering theory for space-filling potentials. *J. Phys: CM*, **5**, 8005–8030.

[8719] Low F.E. (1998). Run-away electrons in relativistic spin $\frac{1}{2}$ quantum electrodynamics. *Ann. Phys. (NY)*, **266**, 274–292.

[8720] Lowther J.E., Dewhurst J.K., Leger J.M. and Haines J. (1999). Relative stability of ZrO_2 and HfO_2 structural phases. *Phys. Rev. B*, **60**, 14485–14488.

[8721] Lu H.G. and Li L.M. (1999). Density functional study on zerovalent lanthanide bis(arene)-sandwich complexes. *Theor. Chem. Acc.*, **102**, 121–126.

[8722] Lu X., Xu X., Wang N.Q. and Zhang Q.E. (1999). Bonding of NO_2 to the Au atom and Au(111) surface: A quantum chemical study. *J. Phys. Chem. A*, **103**, 10969–10974.

[8723] Lu Z.W., Klein B.M. and Zunger A. (1995). Spin-polarization-induced structural selectivity in Pd_3X and Pt_3X (X = 3d) compounds. *Phys. Rev. Lett.*, **75**, 1320–1323.

[8724] Lu Z.W., Wei S.H. and Zunger A. (1991b). Long-range order in binary late-transition-metal alloys. *Phys. Rev. Lett.*, **66**, 1753–1756.

[8725] Lucenti A., Lusanna L. and Pauri M. (1998). Dirac observables and spin bases for N relativistic particles. *J. Phys. A*, **31**, 1633–1656.

[8726] Lucha W. and Schöberl F.F. (1994). Variational approach to the spinless relativistic Coulomb problem. *Phys. Rev. D*, **50**, 5443–5445.

[8727] Lucha W. and Schöberl F.F. (1995). Semirelativistic Hamiltonians of apparently nonrelativistic form. *Phys. Rev. A*, **51**, 4419–4426.

[8728] Lucha W. and Schöberl F.F. (1996). Relativistic Coulomb problem: Analytic upper bounds on energy levels. *Phys. Rev. A*, **54**, 3790–3794.

[8729] Luc-Koenig E., Aymar M., Lecomte J.M. and Lyras A. (1999). R-matrix calculation of Raman couplings and dynamical Stark shifts in heavy alkaline-earth atoms. *Eur. Phys. J. D*, **7**, 487–497.

[8730] Luke T.M. (1996). Relativistic transition rates for 3 quartet multiplets in neutral vanadium. *Phys. Scr.*, **54**, 346–351.

[8731] Lulek T. (1993). Spin-orbit coupling for f-electrons in a crystalline field. *Acta Phys. Pol. A*, **84**, 867–873.

[8732] Lun D.R. and Buckman S.J. (1997). Extraction of spin-orbit interactions from phase shifts via inversion. *Phys. Rev. Lett.*, **79**, 541–544.

[8733] Lun D.R., Eberspächer M., Amos K., Scheid W. and Buckman S.J. (1998). Improved spin-orbit inversion method. *Phys. Rev. A*, **58**, 4993–4996.

[8734] Lundell J., Berski S. and Latajka Z. (1999). Density functional study of the $Xe_2H_3^+$ cation. *Chem. Phys.*, **247**, 215–224.

[8735] Lundell J., Nieminen J. and Kunttu H. (1993). All-electron and effective core potential studies on ground state ArH^+, KrH^+ and XeH^+ ions. *Chem. Phys. Lett.*, **208**, 247–255.

[8736] Lundell J. and Pettersson M. (1999a). The dihydrogen-bonded complex XeH_2–H_2O. *PCCP*, **1**, 1691–1692.

[8737] Lundell J. and Pettersson M. (1999b). Molecular properties of $Xe_2H_3^+$. *J. Mol. Str.*, **509**, 49–54.

[8738] Lundell J., Pettersson M. and Räsänen M. (1999). The proton-bound rare gas compounds $(RgHRg')^+$ (Rg = Ar, Kr, Xe) – a computational approach. *PCCP*, **1**, 4151–4155.

[8739] Lundell J., Räsänen M. and Kunttu H. (1995). Predicted structure, spectra and stability of $ArHX^+$, $KrHX^+$ and $XeXH^+$ (X = Cl, Br or I). *J. Mol. Str.*, **358**, 159–165.

[8740] Lundin U., Fast L., Nordstrm L., Johansson B., Wills J.M. and Eriksson O. (1998). Transition-metal dioxides with a bulk modulus comparable to diamond. *Phys. Rev. B*, **57**, 4979–4982.

[8741] Luo F., Kim G., McBane G.C., Giese C.F. and Gentry W.R. (1993). Influence of retardation on the vibrational wave function and binding energy of the helium dimer. *J. Chem. Phys.*, **98**, 9687–9690.

[8742] Luo J., Trammell G.T. and Hannon J.P. (1993). Scattering operator for elastic and inelastic resonant X-ray scattering. *Phys. Rev. Lett.*, **71**, 287–290.

[8743] Luo L.B., Lanza G., Fragalá I.L., Stern C.L. and Marks T.J. (1998). Energetics of metal-ligand multiple bonds. A combined solution thermochemical and ab initio quantum chemical study of M=O bonding in group 6 metallocene oxo complexes. *J. Am. Chem. Soc.*, **120**, 3111–3122.

[8744] Luo Z.F., Xu Z.Z. and Qiu X.J. (1994). Energy-dependent potential and hyperfine mass splitting of quarkonium. *Comm. Theor. Phys.*, **21**, 217–222.

[8745] Lupinetti A.J., Fau S., Frenking G. and Strauss S.H. (1997). Theoretical analysis of the bonding between CO and positively charged atoms. *J. Phys. Chem. A*, **101**, 9551–9559.

[8746] Lupinetti A.J., Frenking G. and Strauss S.H. (1998). Nonclassical metal carbonyls: Appropriate definitions with a theoretical justification. *Angew. Chem. Int. Ed*, **37**, 2113–2116.

[8747] Lupinetti A.J., Frenking G. and Strauss S.H. (1998). Nonclassical metal carbonyls: Appropriate definitions with a theoretical justification (in German). *Angew. Chem.*, **110**, 2229–2232.

[8748] Lupinetti A.J., Jonas V., Thiel W., Strauss S.H. and Frenking G. (1999). Trends in molecular geometries and bond strengths of the homoleptic d^{10} metal carbonyl cations $[M(CO)_n]^{x+}$ (M^{x+} = Cu^+, Ag^+, Au^+, Zn^{2+}, Cd^{2+}, Hg^{2+}; n = 1-6): A theoretical study. *Chem. Eur. J.*, **5**, 2573–2583.

[8749] Lushington G.H., Buendgen P. and Grein F. (1995). Ab initio study of molecular g-tensors. *Int. J. Quantum Chem.*, **55**, 377–392.

[8750] Lushington G.H. and Grein F. (1996). Complete to second-order ab initio level calculations of electronic g-tensors. *Theor. Chim. Acta*, **93**, 259–267.

[8751] L'vov A.I. and Milstein A.I. (1994). Relativistic oscillator model and Delbrück scattering. *Phys. Lett. A*, **192**, 185–191.

[8752] Lyanda-Geller Y. (1998). Quantum interference and electron-electron interactions at strong spin-orbit coupling in disordered systems. *Phys. Rev. Lett.*, **80**, 4273–4276.

[8753] Ma N.L. (1998). How strong is the Ag^+-ligand bond? *Chem. Phys. Lett.*, **297**, 230–238.

[8754] Ma N.L., Ng K.M. and Tsang C.W. (1997). Theoretical studies of the structure and stability of $Ag(C_6H_6)_n^+$ (n = 1 and 2). *Chem. Phys. Lett.*, **277**, 306–310.

[8755] Macdonald C.L.B. and Cowley A.H. (1999). A theoretical study of free and $Fe(CO)_4$-complexed borylenes (boranediyls) and heavier congeners: The nature of the iron–Group 13 element bonding. *J. Am. Chem. Soc.*, **121**, 12113–12126.

[8756] Macedo A.M.S. (1994). Intensity correlations in electronic-wave propagation in a disordered medium: The influence of spin-orbit scattering. *Phys. Rev. B*, **49**, 11736–11741.

[8757] Macgregor S.A. and Moock K.H. (1998). Stabilization of high oxidation states in transition metals. 2. WCl_6 oxidizes $[WF_6]^-$, but would $PtCl_6$ oxidize $[PtF_6]^-$? An electrochemical and computational study of 5d transition metal halides: $[MF_6]^z$ versus $[MFCl_6]^z$ (M = Ta to Pt; z = 0,1-,2-). *Inorg. Chem.*, **37**, 3284–3292.

[8758] Mackay K.M., Mackay R.A. and Henderson W. (1996). *Introduction to Modern Inorganic Chemistry, 5th Edition.* Stanley Thornes (Publishers) Ltd., Cheltenham, UK, chapter 16.13. 'Relativistic effects'.

[8759] Maehira T., Higuchi M., Nakamura M. and Hasegawa A. (1998). The Fermi surface of 5f-electron compounds. *J. Phys. CM*, **10**, 11565–11570.

[8760] Mahé L. and Barthelat J.C. (1995). Is phosphorus able to form double bonds with arsenic, antimony, or bismuth? An ab initio study of the PXH_2 potential energy surfaces. *J. Phys. Chem.*, **99**, 6819–6827.

[8761] Mahé L. and Barthelat J.C. (1997). Pseudopotentials including part of the atomic correlation energy. *J. Mol. Str. (Theochem)*, **401**, 93–105.

[8762] Mahé L., Boughdiri S.F. and Barthelat J.C. (1997). Electronic structure and energetics in the CuX and Cu_2X series (X = O, S, Se, Te, Po). *J. Phys. Chem.*, **101**, 4224–4230.

[8763] Maier T.J., Dreizler R.M. and Ixaru L.G. (1993). Nonperturbative pair production by a Dirac square well with time-dependent depth. *Phys. Rev. A*, **48**, 2031–2039.

[8764] Majumdar D. and Balasubramanian K. (1996). A theoretical study of the curves for Ta + CO interaction. *Chem. Phys. Lett.*, **262**, 263–268.

[8765] Majumdar D. and Balasubramanian K. (1997a). Theoretical study of the electronic states of Rh_5. *J. Chem. Phys.*, **106**, 4053–4060.

[8766] Majumdar D. and Balasubramanian K. (1997b). Theoretical studies of CO interaction on Rh_3 cluster. *J. Chem. Phys.*, **106**, 7215–7222.

[8767] Majumdar D. and Balasubramanian K. (1997c). Theoretical study of the electronic states of Zr_5. *Chem. Phys. Lett.*, **279**, 403–410.

[8768] Majumdar D. and Balasubramanian K. (1997d). Electronic states of Ta_2C^+. *Chem. Phys. Lett.*, **280**, 212–218.

[8769] Majumdar D. and Balasubramanian K. (1998a). Theoretical study of the electronic states of Rh_n^+ (n=3–5). *J. Chem. Phys.*, **108**, 2495–2503.

[8770] Majumdar D. and Balasubramanian K. (1998b). A theoretical study of the potential energy curves and spectroscopic constants of TaC and TaC^+. *Chem. Phys. Lett.*, **284**, 273–280.

[8771] Majumdar D., Roszak S. and Balasubramanian K. (1997). Theoretical study of the interaction of benzene with Rh^+ and Rh_2^+ cations. *J. Chem. Phys.*, **107**, 408–414.

[8772] Maksudov F.G. and Allakhverdiev B.P. (1996). On the spectral theory of non-self-adjoint Dirac operators. *Dokl. Akad. Nauk.*, **348**, 305–306.

[8773] Malhotra S., Singh A.D. and Das B.P. (1995). Relativistic configuration-interaction analysis of parity nonconservation in Ba^+. *Phys. Rev. A*, **51**, R2665–R2667.

[8774] Malkin V.G., Malkina O.L. and Salahub D.R. (1996). Spin-orbit correction to NMR shielding constants from density functional theory. *Chem. Phys. Lett.*, **261**, 335–345.

[8775] Malkina O.L., Schimmelpfennig B., Kaupp M., Hess B.A., Chandra P., Wahlgren U. and Malkin V.G. (1998). Spin-orbit corrections to NMR shielding constants from density functional theory. How important are the two-electron terms? *Chem. Phys. Lett.*, **296**, 93–104.

[8776] Mallampalli S. and Sapirstein J. (1996). Fourth-order vacuum-polarization contribution to the Lamb shift. *Phys. Rev. A*, **54**, 2714–2717.

[8777] Mallampalli S. and Sapirstein J. (1998a). Fourth-order self-energy contribution to the Lamb shift. *Phys. Rev. A*, **57**, 1548–1564.

[8778] Mallampalli S. and Sapirstein J. (1998b). Perturbed orbital contribution to the two-loop Lamb shift in hydrogen. *Phys. Rev. Lett.*, **80**, 5297–5300.

[8779] Mallampalli S. and Sapirstein J. (1998c). Finite basis sets in momentum space. *J. Phys. B*, **31**, 3779–3787.

[8780] Malli G.L. (1994a). *Relativistic and Electron Correlation Effects in Molecules and Solids (Ed.)*. Plenum Press, New York, 478 p.

[8781] Malli G.L. (1994b). Relativistic and electron correlation effects in molecules of heavy elements. In *Relativistic and Electron Correlation Effects in Molecules and Solids*, (Edited by Malli G.L.), pp. 1–15, Plenum Press, New York.

[8782] Malli G.L., Da Silva A.B.F. and Ishikawa Y. (1993a). Universal Gaussian basis set for relativistic calculations on atoms and molecules. *Chem. Phys. Lett.*, **201**, 37–40.

[8783] Malli G.L., Da Silva A.B.F. and Ishikawa Y. (1993b). Universal Gaussian basis set for accurate ab initio relativistic Dirac-Fock calculations. *Phys. Rev. A*, **47**, 143–146.

[8784] Malli G.L., Da Silva A.B.F. and Ishikawa Y. (1994). Highly accurate relativistic universal Gaussian basis set: Dirac-Fock-Coulomb calculations for atomic systems up to nobelium. *J. Chem. Phys.*, **101**, 6829–6833.

[8785] Malli G.L. and Ishikawa Y. (1998). The generator coordinate Dirac-Fock method for open-shell atomic systems. *J. Chem. Phys.*, **109**, 8759–8763.

[8786] Malli G.L. and Styszynski J. (1994). Ab initio all-electron Dirac-Fock-Breit calculations for ThF_4 using relativistic universal Gaussian basis set. *J. Chem. Phys.*, **101**, 10736–10745.

[8787] Malli G.L. and Styszynski J. (1996). Ab initio all-electron Dirac-Fock-Breit calculations for UF_6. *J. Chem. Phys.*, **104**, 1012–1017.

[8788] Malli G.L. and Styszynski J. (1998). Ab initio all-electron fully relativistic Dirac-Fock-Breit calculations for molecules of the superheavy transactinide elements: Rutherfordium tetrachloride. *J. Chem. Phys.*, **109**, 4448–4455.

[8789] Malvetti M. and Pilkuhn H. (1994). Equal-time relativistic two-body equations. *Phys. Reports C*, **248**, 1–60.

[8790] Manakov N.L. and Zapriagaev S.A. (1997). Solution of the Dirac-Coulomb problem by the second-order Dirac equation approach. *Phys. Scr.*, **T37**, 36–37.

[8791] Mäntykenttä A. (1993). Channel interaction and relaxation effects in Xe $N_{45}OO$ Auger transitions. *Phys. Rev. A*, **47**, 3961–3965.

[8792] Mäntykenttä A., Aksela H., Aksela S., Tulkki J. and Åberg T. (1993). Electron correction in the 4d hole state of Ba studied by Auger and photoelectron spectroscopy. *Phys. Rev. A*, **47**, 4865–4873.

[8793] March N.H. (1993). Completely local relativistic density-functional theory: The role of the virial. *Phys. Rev. A*, **48**, 4778–4779.

[8794] March N.H. (1997). Density-functional approach to relativistic charge expansion theory. *Phys. Rev. A*, **55**, 3935–3936.

[8795] Marcinek R. and Migdalek J. (1993a). Oscillator strengths for some systems with the ns^{2}np ground-state configuration. I. Aluminium isoelectronic sequence. *J. Phys. B*, **26**, 1391–1402.

[8796] Marcinek R. and Migdalek J. (1993b). Oscillator strengths for some systems with the ns^{2}np ground-state configuration. II. Gallium isoelectronic sequence. *J. Phys. B*, **26**, 1403–1414.

[8797] Margl P.M., Woo T.K. and Ziegler T. (1998). Potential catalyst deactivation reaction in homogeneous Ziegler-Natta polymerization of olefins: Formation of an allyl intermediate. *Organometallics*, **17**, 4997–5002.

[8798] Marian C.M. (1993). Stability and the CO stretching vibrational frequency of molecular AgCO. *Chem. Phys. Lett.*, **215**, 582–586.

[8799] Marian C.M. (1994). Relativistic calculations on transition-metal compounds. In *New Challenges in Computational Quantum Chemistry*, (Edited by Broer R., Aerts P.J.C. and Bagus P.S.), pp. 145–164, Univ. Groningen.

[8800] Marian C.M. and Perić M. (1996). Ab initio calculation of the potential energy surface for the large-amplitude bending and symmetric stretching vibration in the electronic ground state of XeF_2. *Z. Phys. D*, **36**, 285–291.

[8801] Marian C.M. and Wahlgren U. (1996). A new mean-field and ECP-based spin-orbit method. Applications to Pt and PtH. *Chem. Phys. Lett.*, **251**, 357–364.

[8802] Marian T.A. (1996). Higher-order multipole expansion in the Dirac equation. *Phys. Rev. A*, **53**, 1992–1999.

[8803] Marinescu M., Babb J.F. and Dalgarno A. (1994a). Long-range potentials including retardation for the interaction of two alkali-metal atoms. *Phys. Rev. A*, **50**, 3096–3104.

[8804] Marinescu M., Florescu V. and Dalgarno A. (1994b). Two-photon excitation of the 5 ^2D states of rubidium. *Phys. Rev. A*, **49**, 2714–2718.

[8805] Marinescu M., Vrinceanu D. and Sadeghpour H.R. (1998). Radiative transitions and van der Waals coefficients for francium. *Phys. Rev. A*, **58**, R4259–R4262.

[8806] Marino M.M. and Ermler W.C. (1993). Ab initio rep-based relativistic core/valence polarization operator. *Chem. Phys. Lett.*, **206**, 271–277.

[8807] Markendorf R., Schober C. and John W. (1994). Analysis of nuclear spin-lattice relaxation in HCP transition metals based on Dirac theory. *J. Phys:CM*, **6**, 3965–3986.

[8808] Marketos P. (1993). Relativistic corrections to alkali atom g-factors. *Z. Phys. D*, **27**, 219–222.

[8809] Marketos P. (1994). Forbidden electric quadrupole transitions between even O III levels. *Z. Phys. D*, **29**, 247–251.

[8810] Marketos P. and Nandi T. (1997). Theoretical lifetimes for certain O II levels. *Z. Phys. D*, **42**, 237–242.

[8811] Marketos P. and Zambetaki I. (1993). A relativistic CI calculation for a two-photon non-resonant transition on O III. *J. Phys. B*, **26**, L249–L254.

[8812] Marketos P., Zambetaki I. and Kleidis M. (1993). A relativistic CI study on OIII: wavefunctions, excitation energies and transition probabilities. *Z. Phys. D*, **27**, 17–27.

[8813] Markham G.D., Bock C.L., Trachtman M. and Bock C.W. (1999). Intramolecular non-bonded interactions between oxygen and group VIA elements. An ab initio molecular orbital and density functional theory investigation of the structures of HX-CH_2-CHO (X = S, Se and Te). *J. Mol. Str. (Theochem)*, **459**, 187–199.

[8814] Maron L. and Teichteil C. (1999). Erratum to 'On the accuracy of averaged relativistic shape-consistent pseudopotentials' [Chem. Phys. 237 (1998) 105-122]. *Chem. Phys.*, **242**, 293.

[8815] Maron L., Leininger T., Schimmelpfennig B., Vallet V., Heully J.L., Teichteil C., Gropen O. and Wahlgren U. (1999). Investigation of the low-lying states of PuO_2^{2+}. *Chem. Phys.*, **244**, 195–201.

[8816] Marques J.P., Parente F. and Indelicato P. (1993a). Hyperfine quenching of the $1s^22s2p\ ^3P_0$ level in berylliumlike ions. *Phys. Rev. A*, **47**, 929–935.

[8817] Marques J.P., Parente F. and Indelicato P. (1993b). Hyperfine quenching of the $1s^22s^22p^63s3p\ ^3P_0$ level in magnesium-like ions. *At. Data Nucl. Data Tables*, **55**, 157–170.

[8818] Marquez A., Anguiano J., Gonzalez G. and Sanz J.F. (1995). A theoretical approach to the molecular structure of vinylstannane and some structural isomers. *J. Organomet. Chem.*, **486**, 45–50.

[8819] Marquez A., Capitan M.J., Odriozola J.A. and Sanz J.F. (1994). Spectroscopic properties and potential energy curves some low-lying electronic states of AlO, AlO^+, LaO, and LaO^+: An ab initio CASSCF study. *Int. J. Quantum Chem.*, **52**, 1329–1338.

[8820] Marrocco M., Weidinger M., Sang R.T. and Walther H. (1998). Quantum electrodynamic shifts of Rydberg energy levels between parallel metal plates. *Phys. Rev. Lett.*, **81**, 5784–5787.

[8821] Marsden C.J. (1995). Structures and binding energies of EM_6 clusters (E = C, Si, Ge, Sn, Pb; M = Li, Na, K, Rb, Cs): $PbNa_6$ is not unique. *Chem. Phys. Lett.*, **245**, 475–483.

[8822] Mårtensson-Pendrill A.M. (1993a). Many-body perturbation theory in atomic structure calculations. *Phys. Scr.*, **T46**, 102–109.

[8823] Mårtensson-Pendrill A.M. (1993b). Parity non-conserving effects in atomic systems. *Phys. Scr.*, **T46**, 182–183.

[8824] Mårtensson-Pendrill A.M. (1995). Magnetic moment distribution in Tl nuclei. *Phys. Rev. Lett.*, **74**, 2184–2187.

[8825] Mårtensson-Pendrill A.M., Gough D.S. and Hannaford P. (1994). Isotope shifts and hyperfine structure in the 369.4-nm $6s - 6p_{1/2}$ resonance line of singly ionized ytterbium. *Phys. Rev. A*, **49**, 3351–3365.

[8826] Mårtensson-Pendrill A.M., Lindgren I., Lindroth E., Salomonson S. and Staudte D.S. (1995). Convergence of relativistic perturbation theory for the $1s2p$ states in low-Z heliumlike systems. *Phys. Rev. A*, **51**, 3630–3635.

[8827] Martin A. and Stubbe J. (1995). Bargmann- and Calogero-type bounds for the Dirac equation. *J. Math. Phys.*, **36**, 4680–4690.

[8828] Martin I. (1998). The relativistic quantum defect orbital method and some of its applications. In *Quantum Systems in Chemistry and Physics*, (Edited by McWeeny R.), Kluwer, Dordrecht.

[8829] Martin I., Almaraz M.A. and Lavin C. (1995). Relativistic oscillator strengths for transitions in the principal spectral series of the silver isoelectronic sequence. *Z. Phys. D*, **34**, 239–246.

[8830] Martin I., Karwowski J., Diercksen G.H.F. and Barrientos C. (1993). Transition probabilities in the lithium sequence. *Astronomy & Astrophys, Suppl. Ser.*, **100**, 595–605.

[8831] Martin J.M.L. (1999). The ground-state spectroscopic constants of Be_2 revisited. *Chem. Phys. Lett.*, **303**, 399–407.

[8832] Martin J.M.L. and de Oliveira G. (1999). Towards standard methods for benchmark quality *ab initio* thermochemistry–W1 and W2 theory. *J. Chem. Phys.*, **111**, 1843–1856.

[8833] Martin J.M.L. and Taylor P.R. (1994). Basis set convergence for geometry and harmonic frequencies. Are h functions enough?. *Chem. Phys. Lett.*, **225**, 473–479.

[8834] Martin J.M.L. and Taylor P.R. (1999). A definitive heat of vaporization of silicon through benchmark *ab initio* calculations on SiF_4. *J. Phys. Chem. A*, **103**, 4427–4431.

[8835] Martin P., Lavín C. and Martin I. (1994). Triplet-triplet transitions in zinc-like ions. *Z. Phys. D*, **30**, 279–284.

[8836] Martin W.C. and Sugar J. (1996). Designations of ds^2p energy levels in neutral zirconium, hafnium, and rutherfordium ($Z = 104$). *Phys. Rev. A*, **53**, 1911–1914.

[8837] Martínez-y-Romero R.P., Salas-Brito A.L. and Saldaña-Vega J. (1999). Nonunitary representations of the SU(2) algebra in the Dirac equation with a Coulomb potential. *J. Math. Phys.*, **40**, 2324–2336.

[8838] Martínez-y-Romero R.P., Saldaña-Vega J. and Salas-Brito A.L. (1998). New non-unitary representations in a Dirac hydrogen atom. *J. Phys. A*, **31**, L157–L161.

[8839] Martynenko A.P. and Faustov R.N. (1997). Relativistic corrections to fine structure of positronium (in Russian). *Yad. Fiz.*, **60**, 1407–1417.

[8840] Maseras F. and Eisenstein O. (1998). Opposing steric and electronic contributions in $OsCl_2H_2(PPr^i_3)_2$. A theoretical study of an unusual structure. *New J. Chem.*, **22**, 5–9.

[8841] Maseras F., Koga N. and Morokuma K. (1993a). Ab initio molecular orbital characterization of the $(Os(PR_3)_3"H_5")^+$ complex. *J. Am. Chem. Soc.*, **115**, 8313–8320.

[8842] Maseras F., Koga N. and Morokuma K. (1994). Ab initio MO and MM study on the nature of $(Ru(P\text{-}P)_2"H_3")^+$ (P-P = dppb, diop, dpmb, dppe) complexes. *Organometallics*, **13**, 4008–4016.

[8843] Maseras F., Li X.K., Koga N. and Morokuma K. (1993b). Ab initio molecular orbital study of the $(Os(PR_3)_3H_4)$ system. Peeking into the peculiarities of seven-coordination. *J. Am. Chem. Soc.*, **115**, 10974–10980.

[8844] Maseras F., Lledós A., Costas M. and Poblet J.M. (1996). Bonding in elongated dihydrogen complexes. Theoretical analysis of the electron density in ML_n(H...H) species. *Organometallics*, **15**, 2947–2953.

[8845] Maseras F., Lockwood M.A., Eisenstein O. and Rothwell I.P. (1998). Four-electron reduction of diazo compounds at a single tungsten metal center: A theoretical study of the mechanism. *J. Am. Chem. Soc.*, **120**, 6598–6602.

[8846] Maslen P., Faeder J. and Parson R. (1996). Ab initio calculations of the ground and excited states of I_2^- and ICl^- . *Chem. Phys. Lett.*, **263**, 63–72.

[8847] Massidda S., Posternak M. and Baldereschi A. (1993). Hartree-Fock LAPW approach to the electronic properties of periodic systems. *Phys. Rev. B*, **48**, 5058–5068.

[8848] Matrasulov D.U. (1998). Chaotic ionization of relativistic atom. I. (in Russian). *Uzb. Fiz. Zh.*, **4**, 19–23.

[8849] Matrasulov D.U. (1999). Chaotic ionization of relativistic atom. II. (in Russian). *Uzb. Fiz. Zh.*, **5**, 25–30.

[8850] Matrasulov D.U., Matveev V.I. and Musakhanov M.M. (1993). Near continuum states of a relativistic electron in the field of a finite electric dipole. *Doga Turkish J. Phys.*, **17**, 743–749.

[8851] Matrasulov D.U., Matveev V.I. and Musakhanov M.M. (1999). Eigenvalue problem for the relativistic electric-dipole system. *Phys. Rev. A*, **60**, 4140–4143.

[8852] Matsubara T., Koga N., Musaev D.G. and Morokuma K. (1998). Density functional study on activation of *ortho*-CH bond in aromatic ketone by Ru complex. Role of unusual five-coordinated d^6 metallacycle intermediate with agostic interaction. *J. Am. Chem. Soc.*, **120**, 12692–12693.

[8853] Matsubara T., Maseras F., Koga N. and Morokuma K. (1996). Application of the new "integrated MO + M" (IMOMM) method to the organometallic reaction $Pt(PR_3)_2 + H_2$ (R = H, Me, t-Bu, and Ph). *J. Phys. Chem.*, **100**, 2573–2580.

[8854] Matsunaga N., Koseki S. and Gordon M.S. (1996). Relativistic potential energy surfaces of XH_2 (X=C,Si,Ge,Sn, and Pb) molecules: Coupling of $^1A_1$ and $^3B_1$ states. *J. Chem. Phys.*, **104**, 7988–7996.

[8855] Matsuoka O., Pisani L. and Clementi E. (1993). All-electron Dirac-Fock-Roothaan calculations on lead oxide. *Chem. Phys. Lett.*, **202**, 13–17.

[8856] Matsuzawa N., Seto J. and Dixon D.A. (1997). Density functional theory predictions of second-order hyperpolarizabilities of metallocenes. *J. Phys. Chem. A*, **101**, 9391–9398.

[8857] Mattes M. and Sorg M. (1994). Relativistic Schrödinger equations and the Bohm-Aharonov effect. *J. Phys. Soc. Japan*, **63**, 2532–2537.

[8858] Mattes M. and Sorg M. (1999a). Second-order mixtures in relativistic Schrödinger theory. *J. Math. Phys.*, **40**, 71–92.

[8859] Mattes M. and Sorg M. (1999b). Two-particle systems in relativistic Schrödinger theory. *J. Phys. A*, **32**, 4761–4786.

[8860] Matveev A.V., Neyman K.M., Pacchioni G. and Rösch N. (1999). Density functional study of M_4 clusters (M=Cu, Ag, Ni, Pd) deposited on the regular MgO(001) surface. *Chem. Phys. Lett.*, **299**, 603–612.

[8861] Matveev V.I. and Matrasulov D.U. (1999). Single and double K-vacancy production in the collision of relativistic hghly charged ions with heavy atoms. *Phys. Scripta*, **T80**, 429.

[8862] Matveev V.I., Matrasulov D.U. and Rakhimov K.Y. (1997). Two-centre problem for the Dirac equation (in Russian). *Uzbek. Fiz. Zh.*, **2**, 15–20.

[8863] Matveev V.I. and Musakhanov M.M. (1994). Inelastic processes in collisions of relativistic multicharged ions with atoms. *J. Exp. Theor. Phys.*, **78**, 149–152.

[8864] Matveev V.I. and Musakhanov M.M. (1994). Inelastic processes in collisions of relativistic multicharged ions with atoms. *Zh. Eksp. Teor. Fiz.*, **105**, 280–287.

[8865] Maul M., Schäfer A., Greiner W. and Indelicato P. (1996). Prospects for parity non-conservation experiments with highly charged heavy ions. *Phys. Rev. A*, **53**, 3915–3925.

[8866] Maul M., Schäfer A. and Indelicato P. (1998). Stark quenching for the $1s^22s2p\ ^3P_0$ level in beryllium-like ions and parity-violating effects. *J. Phys. B*, **31**, 2725–2734.

[8867] Maung K.M., Kahana D.E. and Norbury J.W. (1993). Solution of two-body relativistic bound-state equations with confining plus Coulomb interactions. *Phys. Rev. D*, **47**, 1182–1189.

[8868] Mayer M., Häberlen O.D. and Rösch N. (1996). Relevance of relativistic exchange-correlation functionals and of finite nuclei in molecular density-functional calculations. *Phys. Rev. A*, **54**, 4775–4782.

[8869] Mayor-López M.J. and Weber J. (1997). DFT calculations of the binding energy of metallocenes. *Chem. Phys. Lett.*, **281**, 226–232.

[8870] Mayor-López M.J., Weber J., Mannfors B. and Cunningham Jr. A.F. (1998). Density functional study of protonated, acetylated, and mercurated derivatives of ferrocene: Mechanism of the electrophilic substitution reaction. *Organometallics*, **17**, 4983–4991.

[8871] Mazevet S., McCarthy I.E., Madison D. and Weigold E. (1998). Semirelativistic DWBA for the ionization of closed shell atoms at intermediate energies. *J. Phys. B*, **31**, 2187–2202.

[8872] McCann J.F., Glass J.T. and Crothers D.S.F. (1996). The energy dependence of relativistic nonradiative electron capture. *J. Phys. B*, **29**, 6155–6164.

[8873] McCullough Jr E.A., Aprà E. and Nichols J. (1997). Comparison of the Becke-Lee-Yang-Parr and Becke-Perdew-Wang exchange-correlation functionals for geometries of cyclopentadienyl-transition metal complexes. *J. Phys. Chem. A*, **101**, 2502–2508.

[8874] McKee M.L. and Worley S.D. (1997). Ab initio study of the interaction of rhodium with dinitrogen and carbon monoxide. *J. Phys. Chem. A*, **101**, 5600–5603.

[8875] McQuarrie B.R. and Vrscay E.R. (1993). Rayleigh-Schrödinger perturbation theory at large order for radial Klein-Gordon equations. *Phys. Rev. A*, **47**, 868–875.

[8876] Mealli C., Ienco A., Galindo A. and Perez Carreño E. (1999). Theoretical overview of Pd(I) and Pt(I) dimers with bridging phosphido ligands. *Inorg. Chem.*, **38**, 4620–4625.

[8877] Mebel A.M., Lin H.L. and Lin S.H. (1999). Ab initio molecular orbital and density functional study of the $C_6H_6{\cdot}I_2$ complex in the ground and excited electronic states. *Int. J. Quantum Chem.*, **72**, 307–318.

[8878] Mebel A.M., Morokuma K. and Isobe K. (1995). A theoretical study of rectangular tetrasulfur in a gas phase and in the tetranuclear $[\{Rh_2(\eta^5\text{-}C_5Me_5)_2(\mu\text{-}CH_2)_2\}(\mu\text{-}S_4)]^{2+}$ complex. *Inorg. Chem.*, **34**, 1208–1211.

[8879] Medvedeva N.I., Gubanov V.A., Novikov D.I. and Klein B.M. (1995). Oxygen defects ordering in δ-Bi_2O_3: LMTO-ASA and FPLMTO calculations. *Int. J. Quantum Chem.*, **S29**, 541–547.

[8880] Medvedeva N.I., Zhukov V.P., Novikov D.L. and Gubanov V.A. (1996). Electronic structure and chemical bonding of δ-Bi_2O_3 (in Russian). *Zh. Strukt. Khim.*, **37(1)**, 48–58.

[8881] Mehl M.J. (1996). First-principles study of the structure of mercury. In *Materials Theory, Simulations and Parallel Algorithms*, (Edited by Kaxiras E.), pp. 383–388, MRS, Pittsburgh, PA.

[8882] Meir Y. and Wingreen N.S. (1994). Spin-orbit scattering and the Kondo effect. *Phys. Rev. B*, **50**, 4947–4950.

[8883] Mejías J.A. (1996). Theoretical study of adsorption of Cu, Ag, and Au on the NaCl(100) surface. *Phys. Rev. B*, **53**, 10281–10288.

[8884] Melić B. (1994). Hydrogenic atoms with spinless nucleus in a magnetic field. *J. Phys. B*, **27**, 3849–3862.

[8885] Menchi M. and Bosin A. (1997). DFT-LDA pseudopotentials in quantum Monte Carlo . *Int. J. Quantum Chem.*, **61**, 295–302.

[8886] Mendez B., Dominguez-Adame F. and Macia E. (1993). A transfer matrix method for the determination of one-dimensional band structures. *J. Phys. A*, **26**, 171–177.

[8887] Mendizabal F. (1999). Theoretical study of the Au-ethylene interaction. *Int. J. Quantum Chem.*, **73**, 317–324.

[8888] Mendoza C., Eissner W., Le Dourneuf M. and Zeippen C.J. (1995). Atomic data for opacity calculations XXIII - the aluminium isoelectronic sequence. *J. Phys. B*, **28**, 3485–3504.

[8889] Mercero J.M., Lopez X., Fowler J.E. and Ugalde J.M. (1997). Ab-initio studies of alternant X_2Y_2 rings (X = N, P, As, and Sb and Y = O, S, Se, and Te). Planar versus butterfly structures. *J. Phys. Chem. A*, **101**, 5574–5579.

[8890] Merchan M., Pou-Amérigo R. and Roos B.O. (1996). A theoretical study of the dissociation energy of Ni_2^+. A case of broken symmetry. *Chem. Phys. Lett.*, **252**, 405–414.

[8891] Merenga H. (1997). *Electronic Structure Calculations on Cerium-Containing Crystals. Towards a Better Understanding of Scintillation in Ionic Crystals.*. Ph.D. thesis, Delft Univ. Press, 169 p.

[8892] Merenga H. and Andriessen J. (1994). Four-component Hartree-Fock-Dirac calculations with the MOLFDIR program package: the challenge to improve efficiency. In *New Challenges in Computational Quantum Chemistry*, (Edited by Broer R., Aerts P.J.C. and Bagus P.S.), pp. 249–254, Univ. Groningen.

[8893] Merkelis G., Martinson I. and Vilkas M.J. (1999). Ab initio calculation of electric quadrupole and magnetic dipole transitions in ions of the N I isoelectronic sequence. *Phys. Scr.*, **59**, 122–132.

[8894] Meservey R. and Tedrow P.M. (1994). Spin-polarized electron tunneling. *Phys. Reports*, **238**, 173–243.

[8895] Mestres J., Duran M., Martin-Zarza P., Medina de la Rosa E. and Gili P. (1993). Ab initio theoretical study on geometries, chemical bonding, and infrared and electronic spectra of the $M_2O_7^{2-}$ (M=Cr, Mo, W) anions. *Inorg. Chem.*, **32**, 4708–4713.

[8896] Meyer J. (1994). Erratum: Construction of linearly independent relativistic symmetry orbitals for finite double-point groups including time reversal symmetry. *Int. J. Quantum Chem.*, **52**, 1369–1372.

[8897] Meyer J. (1997). Addendum to construction of linearly independent relativistic symmetry orbitals for finite double-point groups including time-reversal symmetry. *Int. J. Quantum Chem.*, **61**, 929–933.

[8898] Meyer J., Sepp W.D., Fricke B. and Rosén A. (1996). A new version of the program TSYM generating relativistic molecular symmetry orbitals for finite double point groups. *Comp. Phys. Comm.*, **96**, 263–287.

[8899] Meyerhofer D.D., Knauer J.P., McNaught S.J. and Moore C.I. (1996). Observation of the relativistic mass shift effects during high-intensity laser-electron interactions. *J. Opt. Soc. Am. B*, **13**, 113–117.

[8900] Miadoková I., Kellö V. and Sadlej A.J. (1997). Standardized basis sets for high-level-correlated relativistic calculations of atomic and molecular electric properties in the Douglas-Kroll approximation. *Theor. Chem. Acc.*, **96**, 166–175.

[8901] Miadoková I., Kellö V. and Sadlej A.J. (1999). Electron correlation and relativistic effects in electric properties of the alkali metal fluorides. *Mol. Phys.*, **96**, 179–187.

[8902] Mian M., Harrison N.M., Saunders V.M. and Flavell W.R. (1996). An ab initio Hartree-Fock investigation of galena (PbS). *Chem. Phys. Lett.*, **257**, 627–632.

[8903] Michl J. (1996). Spin-orbit coupling in biradicals. 1. The 2-electrons-in-2-orbitals model revisited. *J. Am. Chem. Soc.*, **118**, 3568–3579.

[8904] Mickelsson J. (1998). Vacuum polarization and the geometric phase: Gauge invariance. *J. Math. Phys.*, **39**, 831–837.

[8905] Miecznik G. and Greene C.H. (1995). Spin-orbit effects in aluminum photoionization. *Phys. Rev. A*, **51**, 513–527.

[8906] Migdalek J. and Stanek M. (1993). The spin-allowed and spin-forbidden $5s^2\ {}^1S_0$ - $5s5p\ {}^1P_1, {}^3P_1$ transitions in strontium isoelectronic sequence. *Z. Phys. D*, **27**, 9–15.

[8907] Mijatović M., Ivanovski G. and Veljanoski B. (1994). Scattering and bound states of a relativistic neutral spin-1/2 particle in a magnetic field. *Z. Phys. A*, **348**, 139–146.

[8908] Mikhailov A.I. and Polikanov V.S. (1968). Perturbation theory for the Dirac equation. *Zh. Eksp. Teor. Fiz.*, **54**, 175–182.

[8909] Mikhailov A.I. and Polikanov V.S. (1968). Perturbation theory for the Dirac equation. *Soviet Phys. JETP*, **27**, 95–98.

[8910] Mikheev N.B. and Rumer I.A. (1999). Stabilization of the divalent state for the lanthanides and actinides in solutions, melts and clusters. *Radiochim. Acta*, **85**, 49–55.

[8911] Milet A. and Dedieu A. (1995). Theoretical study of the protonation of square-planar palladium(II) complexes. Assessment of basis set and correlation effects. *Theor. Chim. Acta*, **92**, 361–367.

[8912] Milhorat J. (1998). Spectrum of the Dirac operator on $\mathrm{Gr}_2\ (C^{m+2})$. *J. Math. Phys.*, **39**, 594–609.

[8913] Miller III T.F. and Hall M.B. (1999). Structural and bonding trends in platinum-carbon clusters. *J. Am. Chem. Soc.*, **121**, 7389–7396.

[8914] Milletti M.C. (1993). Theoretical study of the influence of the halide ligands on the metal-metal quadruple bond in the $M_2X_4(PH_3)_4$(M=Mo, W; X=Br, Cl, I) series of complexes. *Polyhedron*, **12**, 401–405.

[8915] Milonni P.W., Schaden M. and Spruch L. (1999). Lamb shift of an atom in a dielectric medium. *Phys. Rev. A*, **59**, 4259–4263.

[8916] Milonni P.W. and Shih M.L. (1992b). Casimir forces. *Contemporary Phys.*, **33**, 313–322.

[8917] Mil'shtein A.I. and Khriplovich I.B. (1994). Large relativistic corrections to the positronium decay probability (in Russian). *Zh. Eksp. Teor. Fiz.*, **106**, 689–697.

[8918] Mil'shtein A.I. and Khriplovich I.B. (1994). Large relativistic corrections to the positronium decay probability. *J. Exp. Theor. Phys.*, **79**, 379–383.

[8919] Minaev B.F. (1983). *Theoretical analysis and prediction of spin-orbit interaction effects in molecular spectroscopy and chemical reactions (in Russian).*. Ph.D. thesis, Avtoreferat dissertatsii, Inst. Khim. Fiz. AN SSSR, 52 p.

[8920] Minaev B.F. (1996). Paramagnetic spin catalysis of a radical recombination reaction. *Mol. Engineering*, **6**, 261–279.

[8921] Minaev B.F. and Ågren H. (1995). Spin-orbit coupling induced chemical reactivity and spin-catalysis phenomena. *Collect. Czech. Chem. Comm.*, **60**, 339–371.

[8922] Minaev B.F. and Ågren H. (1996). Spin-catalysis phenomena. *Int. J. Quantum Chem.*, **57**, 519–532.

[8923] Minaev B.F. and Ågren H. (1998). Spin-orbit coupling in oxygen containing diradicals. *J. Mol. Str. (Theochem)*, **434**, 193–206.

[8924] Minaev B. and Ågren H. (1999). Spin uncoupling in ethylene activation by palladium and platinum atoms. *Int. J. Quantum Chem.*, **72**, 581–596.

[8925] Minaev B.F., Knuts S. and Ågren H. (1994). On the interpretation of the external heavy-atom effect on singlet-triplet transitions. *Chem. Phys.*, **181**, 15–28.

[8926] Minaev B.F. and Lunell S. (1993). Classification of spin-orbit effects in organic chemical reactions. *Z. Phys. Chem.*, **182**, 263–284.

[8927] Minaev B.F., Lunell S. and Kobzev G.I. (1993). The influence of intermolecular interaction on the forbidden near-IR transitions in molecular oxygen. *THEOCHEM*, **103**, 1–9.

[8928] Minaev B.F., Norman P., Jonsson J. and Ågren H. (1995). Response theory calculations of singlet-triplet transitions in molecular nitrogen. *Chem. Phys.*, **190**, 11–29.

[8929] Minaev B., Vaara J., Ruud K., Vahtras O. and Ågren H. (1998). Internuclear distance dependence of the spin-orbit coupling contributions to proton NMR chemical shifts. *Chem. Phys. Lett.*, **295**, 455–461.

[8930] Minaev B., Vahtras O. and Ågren H. (1996). Magnetic phosphorescence of molecular oxygen. A study of the b $^1\Sigma_g^+$ - X $^3\Sigma_g^-$ transition probability using multiconfiguration response theory. *Chem. Phys.*, **208**, 299–311.

[8931] Minami T. and Matsuoka O. (1995). Relativistic Gaussian basis sets for radon through plutonium. *Theor. Chim. Acta*, **90**, 27–39.

[8932] Mingos D.M.P. (1998). *Essential trends in inorganic chemistry*. Oxford Univ. Press, Oxford, pp. 26, 367.

[8933] Mishra S.K., Satpathy S. and Jepsen O. (1997). Electronic structure and thermoelectric properties of bismuth telluride and bismuth selenide. *J. Phys. C*, **9**, 461–470.

[8934] Mitas L. (1994). Quantum Monte Carlo calculation of the Fe atom. *Phys. Rev. A*, **49**, 4411–4414.

[8935] Mitrushenkov A., Labzowsky L., Lindgren I., Persson H. and Salomonson S. (1995). Second order loop after loop self-energy correction for few-electron multicharged ions. *Phys. Lett. A*, **200**, 51–55.

[8936] Miyagi H., Yamaguchi K., Matsuo H. and Mukose K. (1998). First-principles study of solid iodine and bromine under high pressure. *J. Phys. CM*, **10**, 11203–11213.

[8937] Miyake K. and Sakai Y. (1994). Model potential method in molecular calculation: application to SrX_2 and BaX_2 (X=F, Cl, Br, and I). *J. Mol. Str. (Theochem)*, **311**, 123–135.

[8938] Miyazaki T. and Ohno T. (1999). First-principles study of the electronic structure of the organic solids $(CH_3)_4N[M(\mathrm{dmit})_2]_2$ (M=Ni and Pd): Role of dimerization and the stability of the formation of a dimer. *Phys. Rev. B*, **59**, R5269–R5272.

[8939] Miyazaki T., Terakura K., Morikawa Y. and Yamasaki T. (1995). First-principles theoretical study of metallic states of DCNQI-(Cu,Ag) systems: simplicity and variety in complex systems. *Phys. Rev. Lett.*, **74**, 5104–5107.

[8940] Miyoshi E., Sakai Y., Tanaka K. and Masamura M. (1998). Relativistic *dsp*-model core potentials for main group elements in the fourth, fifth and sixth row and their applications. *J. Mol. Str. (Theochem)*, **451**, 73–79.

[8941] Moc J. (1999). An ab initio QCISD study of the periodic trends in structures and reactivities of the XH_5^- anions (X = Si, Ge, Sn, Pb). *J. Mol. Str. (Theochem)*, **461-462**, 249–259.

[8942] Moc J. and Morokuma K. (1994). Ab initio MO study on the periodic trends in structures and energies of hypervalent compounds: Four-coordinated XH_4^- and XF_4^- anions containing a group 15 central atom (X=P, As, Sb, Bi). *Inorg. Chem.*, **33**, 551–560.

[8943] Moc J. and Morokuma K. (1995). Ab initio molecular orbital study on the periodic trends in structures and energies of hypervalent compounds: Five-coordinated XH_5 species containing a group 15 central atom (X = P, As, Sb, and Bi). *J. Am. Chem. Soc.*, **117**, 11790–11797.

[8944] Moc J. and Morokuma K. (1997). Ab initio MO study of the periodic trends in structures and energies of hypervalent compounds: five-, six-, and seven-coordinated XF_5, XH_6^-, XF_6^-, XH_7^{2-} and XF_7^{2-} species containing a group 15 central atom (where X is P, As, Sb, Bi). *J. Mol. Str.*, **436-437**, 401–418.

[8945] Mocanu C.I., Morega M.L. and Morega A.M. (1993). On the Thomas rotation paradox . *Hadronic J.*, **16**, 295–329.

[8946] Mohamed A., Parisse B. and Outassourt A. (1993). Asymptotic band spectrum of the Dirac operator with a periodic potential. *Helv. Phys. Acta*, **66**, 192–215.

[8947] Mohan M., Eissner W., Hibbert A. and Burke P.G. (1995). Electron-impact fine-structure transitions in Cu XX from its ground state. *J. Phys. B*, **28**, 2249–2256.

[8948] Mohan M., Le Dourneuf M., Hibbert A. and Burke P.G. (1998). Relativistic calculation for photoionization of the ground state of neonlike Fe XVII. *Phys. Rev. A*, **57**, 3489–3492.

[8949] Mohanty A.K. and Parpia F.A. (1996). Fully relativistic calculations for the ground state of the AgH molecule. *Phys. Rev. A*, **54**, 2863–2867.

[8950] Mohr P.J. (1993). Quantum electrodynamics calculations in few-electron systems. *Phys. Scr.*, **T46**, 44–51.

[8951] Mohr P.J. (1994). QED effects in high-Z few-electron atoms. *Nucl. Instr. Meth. Phys. Res. B*, **87**, 232–236.

[8952] Mohr P.J. (1996). Tests of fundamental physics. In *Atomic, Molecular and Optical Physics Handbook*, (Edited by Drake G.W.F.), AIP, Woodbury, NY, pp. 341-351.

[8953] Mohr P.J. (1997). QED corrections in heavy atoms. *Phys. Rep.*, **293**, 227–369.

[8954] Mohr P.J. (1998). Quantum electrodynamics and the fundamental constants. *Adv. Quantum Chem.*, **30**, 77–97.

[8955] Mohr P.J., Plunien G. and Soff G. (1998). QED corrections in heavy atoms. *Phys. Rep.*, **293**, 227–369.

[8956] Mohr P.J. and Soff G. (1993). Nuclear size correction to the electron self-energy. *Phys. Rev. Lett.*, **70**, 158–161.

[8957] Moiseiwitsch B.L. (1997). The virial theorem for electron capture. *J. Phys. B*, **30**, 1097–1104.

[8958] Molina L.M., Alonso J.A. and Stott M.J. (1999a). Assembling alkali-lead solid compounds from clusters. *J. Chem. Phys.*, **111**, 7053–7061.

[8959] Molina L.M., López M.J., Rubio A., Alonso J.A. and Stott M.J. (1998). Mixed lead-alkali clusters in the gas phase and in liquid alloys. *Int. J. Quantum Chem.*, **69**, 341–348.

[8960] Molina L.M., López M.J., Rubio A., Balbás L.C. and Alonso J.A. (1999b). Pure and mixed Pb clusters of interest for liquid ionic alloys. *Adv. Quantum Chem.*, **33**, 329–348.

[8961] Möller H., Niu J.E., Lutz H.D. and Schwarz W.H.E. (1997). Structural, spectroscopic and electronic properties of hydrogen-bonded water molecules in crystals. Ab initio calculations and experimental data of $MCl_2{\cdot}n(H,D)_2O$, M = Sr or Ba. *J. Mol. Str.*, **436-437**, 233–245.

[8962] Molzberger K. and Schwarz W.H.E. (1996). Numerical investigations of different orders of relativistic effects in atomic shells. *Theor. Chim. Acta*, **94**, 213–222.

[8963] Momberger K., Belkacem K. and Sørensen A.H. (1995). Non-perturbative momentum space approach to relativistic heavy-ion collisions. *Europhys Lett.*, **32**, 401–406.

[8964] Momberger K., Grün N. and Scheid W. (1991). Coupled channel analysis of electron-positron pair production in relativistic heavy ion collisions. *Z. Phys. D*, **18**, 133–137.

[8965] Momberger K., Grün N. and Scheid W. (1993). Non-perturbative character of bound-free electron-positron pair production and the role of phase-distortion effects. *J. Phys. B*, **26**, 1851–1862.

[8966] Monahan A.H. and McMillan M. (1998). Faddeev equations for a relativistic two- and three-body system. *Phys. Rev. A*, **58**, 4226–4228.

[8967] Moore E.A. (1999). Relativistic chemical shielding: formally exact solutions for one-electron atoms of maximum total angular momentum for any principal quantum number. *Mol. Phys.*, **97**, 375–380.

[8968] Moores D.L. and Reed K.J. (1995a). Electron collisions with very highly charged ions - relativistic calculations. *Nucl. Instr. Meth. Phys. Res.*, **98**, 122–124.

[8969] Moores D.L. and Reed K.J. (1995b). Electron impact ionization of highly-charged uranium ions. *J. Phys. B*, **28**, 4861–4865.

[8970] Moores D.L. and Reed K.J. (1995c). Effect of the Møller interaction on electron-impact ionization of high-Z hydrogenlike ions. *Phys. Rev. A*, **51**, R9–R11.

[8971] Moreau P., Brun N., Walsh C.A., Colliex C. and Howie A. (1997). Relativistic effects in electron-energy-loss-spectroscopy observations of the Si/SiO_2 interface plasmon peak. *Phys. Rev. B*, **56**, 6774–6781.

[8972] Moreau W., Easther R. and Neutze R. (1994). Relativistic (an)harmonic oscillator. *Am. J. Phys.*, **62**, 531–535.

[8973] Morgon N.H. (1995). A theoretical study of the structure and stability of $C_6H_4X^-$ (X = F, Cl, Br, and I) ions. *J. Phys. Chem.*, **99**, 17832–17837.

[8974] Moroz A.V. and Barnes C.H.W. (1999). Effect of the spin-orbit interaction on the band structure and conductance of quasi-one-dimensional systems. *Phys. Rev. B*, **60**, 14272–14285.

[8975] Moruzzi V.L. and Marcus P.M. (1993). Trends in bulk moduli from first-principles total-energy calculations. *Phys. Rev. B*, **48**, 7665–7667.

[8976] Moshinsky M., del Sol Mesa A. and Smirnov Y.F. (1995). Symmetry Lie algebra of the two body system with a Dirac oscillator interaction. *Rev. Mex. Fis.*, **41**, 322–341.

[8977] Moshinsky M. and Loyola G. (1993). Barut equation for the particle-antiparticle system with a Dirac oscillator interaction. *Found. Phys.*, **23**, 197–210.

[8978] Moss R.E. (1993a). Calculations for vibration-rotation levels of HD^+ in particular for high N. *Mol. Phys.*, **78**, 371–405.

[8979] Moss R.E. (1993b). The 2pσ_u-1sσ_g electronic spectrum of D_2^+. *J. Chem. Soc., Faraday Trans.*, **89**, 3851–3855.

[8980] Moss R.E. (1993c). Calculations for the vibration-rotation levels of H_2^+ in its ground and first excited electronic states. *Mol. Phys.*, **80**, 1541–1554.

[8981] Moss R.E. and Sadler I.A. (1986). The electric quadrupole moments of one-electron atoms. *J. Phys. B*, **19**, L503–L506.

[8982] Mosyagin N.S., Kozlov M.G. and Titov A.V. (1998). Electric dipole moment of the electron in the YbF molecule. *J. Phys. B*, **31**, L763–L767.

[8983] Mosyagin N.S., Titov A.V. and Latajka Z. (1997). Generalized relativistic effective core potential: Gaussian expansions of potentials and pseudospinors for atoms Hg through Rn. *Int. J. Quantum Chem.*, **63**, 1107–1122.

[8984] Mota F., Novoa J.J., Losada J., Alvarez S., Hoffmann R. and Silvestre J. (1993). Pyramidality and metal-metal multiple bonding: structural correlations and theoretical study. *J. Am. Chem. Soc.*, **115**, 6216–6229.

[8985] Mourad J. and Sazdjian H. (1994). The two-fermion relativistic wave equations of constraint theory in the Pauli-Schrödinger form. *J. Math. Phys.*, **35**, 6379–6406.

[8986] Moussa M.H.Y. and Baseia B. (1998). Nonlocality of a single particle: From the Fock space to cavity QED. *Phys. Lett. A*, **245**, 335–338.

[8987] Mudry C., Simons B.D. and Altland A. (1998). Random Dirac fermions and non-Hermitian quantum mechanics. *Phys. Rev. Lett.*, **80**, 4257–4260.

[8988] Muguruma C., Koga N., Kitaura K. and Morokuma K. (1995). The potential energy function for a ligand substitution reaction of square-planar platinum (II) complex in water: the important role of three-body effect. *J. Chem. Phys.*, **103**, 9274–9291.

[8989] Muinasmaa U., Burk P. and Pentchuk J. (1997). Complexes between divalent metals and carboxylic acids: Semiempirical study. *Int. J. Quantum Chem.*, **62**, 653–658.

[8990] Mukoyama T., Nakamatsu H. and Adachi H. (1993). Relativistic calculations of photoelectron and conversion-electron spectra for UF_4. *J. El. Spectr. Rel. Phen.*, **63**, 409–417.

[8991] Müller A., Wittneben V., Diemann E., Hormes J. and Kuetgens U. (1994). Electronic structure of thiometalates $(MS_4)^{n-}$ (M=Mo, W, Re) XANES spectra and SCF-Xα-SW calculations. *Chem. Phys. Lett.*, **225**, 359–363.

[8992] Müller C. (1994). Finite Elemente zur Lösung der zeitabhängigen Dirac-Gleichung. *Diplomarbeit, Giessen.*

[8993] Müller C., Grün N. and Scheid W. (1998). Finite element formulation of the Dirac equation and the problem of fermion doubling. *Phys. Lett. A*, **242**, 245–250.

[8994] Müller H., Franke R., Vogtner S., Jaquet R. and Kutzelnigg W. (1998). Toward spectroscopic accuracy of ab initio calculations of vibrational frequencies and related quantities: a case study of the HF molecule. *Theor. Chem. Acc.*, **100**, 85–102.

[8995] Müller-Nehler U. and Soff G. (1994). Electron excitations in superheavy quasimolecules. *Phys. Rep.*, **246**, 101–250.

[8996] Munzar D. and Christensen N.E. (1994). Electronic structure of Sn/Ge superlattices. *Phys. Rev. B*, **49**, 11238–11247.

[8997] Münzenberg G. (1995). Discovery, synthesis and nuclear properties of the heaviest elements. *Radiochim. Acta*, **70/71**, 193–206.

[8998] Münzenberg G. (1999). Discoveries of the heaviest elements. *J. Phys. G*, **25**, 717–725.

[8999] Musaev D.G., Froese R.D.J. and Morokuma K. (1997). Transition-metal catalyzed olefin polymerization reactions: A theoretical comparison of mechanisms for diimine-M(II) (M=Ni, Pd, and Pt) and zirconocene catalysts. *New J. Chem.*, **21**, 1269–1282.

[9000] Musaev D.G. and Morokuma K. (1995). Does the tetrahydroborate species $AuBH_4$ exist? Ab initio MO study of the structure and stability of $CuBH_4$, $AgBH_4$, and $AuBH_4$. *Organometallics*, **14**, 3327–3334.

[9001] Musaev D.G. and Morokuma K. (1996). Potential energy surfaces of transition-metal-catalyzed chemical reactions. *Adv. Chem. Phys.*, **95**, 61–128.

[9002] Musaev D.G. and Morokuma K. (1999). Theoretical studies of the mechanism of ethylene polymerization reaction catalyzed by diimine-M(II) (M=Ni, Pd and Pt) and Ti- and Zr-chelating alkoxides. *Topics in Catal.*, **7**, 107–123.

[9003] Musakhanov Y. and Matveev V.I. (1993). Inelastic collisions of relativistic multiply charged ions with atoms. *Doga Turkish J. Phys.*, **17**, 756–762.

[9004] Mustafa O. and Odeh M. (1999). Quasi-relativistic harmonic oscillator bound states. *J. Phys. A*, **32**, 6653–6662.

[9005] Mustafa O. and Sever R. (1993). Approach to the shifted $1/N$ expansion for spin-1/2 relativistic particle. *J. Quant. Spectr. Radiat. Transfer*, **49**, 65–69.

[9006] Nag N. and Roychoudhury R. (1994). Exact solutions of two body Dirac equations. *Z. Phys. A*, **49**, 983–986.

[9007] Nag N. and Roychoudhury R. (1995). Algebraic approach to the fixed point structure of the quantum mechanical Dirac-Coulomb system. *Z. Naturf.*, **50a**, 995–997.

[9008] Nagase S. (1991). Interesting properties of the heavier group 14 analogues of aromatic and polycyclic carbon compounds. A theoretical study. *Polyhedron*, **10**, 1299–1309.

[9009] Nagase S. (1993). Theoretical study of heteroatom-containing compounds. From aromatic and polycyclic molecules to hollow cage clusters. *Pure. & Appl. Chem.*, **65**, 675–682.

[9010] Nagase S., Kobayashi K. and Kudo T. (1994). Theoretical study of the aromatic and polyhedral compounds with Ge, Sn and Pb skeletons. *Main Group Metal Chem.*, **17**, 171–181.

[9011] Nagel B. (1994). The relativistic Hermite polynomial is a Gegenbauer polynomial. *J. Math. Phys.*, **35**, 1549–1554.

[9012] Nagy A. (1994). Relativistic density-functional theory for ensembles of excited states. *Phys. Rev. A*, **49**, 3074–3076.

[9013] Nahar S.N. (1999). Oscillator strengths for dipole-allowed fine-structure transitions in FeXIII. *At. Data Nucl. Data Tables*, **72**, 129–151.

[9014] Naito T., Nagase S. and Yamataka H. (1994). Theoretical study of the structure and reactivity of ylides of N, P, As, Sb, and Bi. *J. Am. Chem. Soc.*, **116**, 10080–10088.

[9015] Nakajima T. and Hirao K. (1999). A new relativistic theory: a relativistic scheme by eliminating small components (RESC). *Chem. Phys. Lett.*, **302**, 383–391.

[9016] Nakajima T., Suzumura T. and Hirao K. (1999). A new relativistic scheme in Dirac-Kohn-Sham theory. *Chem. Phys. Lett.*, **304**, 271–277.

[9017] Nakamatsu H., Mukoyama T. and Adachi H. (1995). Ionic and covalent bonds in CeO_2 crystal. *Chem. Phys. Lett.*, **247**, 168–172.

[9018] Nakamura E., Yu Y., Mori S. and Yamago S. (1997). Unusually stable organomercury hydrides and radicals. *Angew. Chem. Int. Ed. Engl.*, **36**, 374–376.

[9019] Nakao T., Dixon D.A. and Chen H. (1993). Electronic structure of palladium dimer from density functional theory. *J. Phys. Chem.*, **97**, 12665–12667.

[9020] Nakatsuji H. and Ehara M. (1994). Symmetry adapted cluster-configuration interaction study on the excited and ionized states of $TiBr_4$ and TiI_4. *J. Chem. Phys.*, **101**, 7658–7671.

[9021] Nakatsuji H., Hada M., Kaneko H. and Ballard C.C. (1996a). Relativistic study of nuclear magnetic shielding constants: mercury dihalides. *Chem. Phys. Lett.*, **255**, 195–202.

[9022] Nakatsuji H., Hada M., Tejima T., Nakajima T. and Sugimoto M. (1996b). Spin-orbit effect on the magnetic shielding constant using the ab initio UHF method. Electronic mechanism in the aluminium compounds, AlX_4^- (X = H, F, Cl, Br and I). *Chem. Phys. Lett.*, **249**, 284–289.

[9023] Nakatsuji H., Hu Z.M. and Nakai H. (1997a). Theoretical studies of the catalytic activity of Ag surface for the oxidation of olefins. *Int. J. Quantum Chem.*, **65**, 839–855.

[9024] Nakatsuji H., Hu Z.M. and Nakajima T. (1997b). Spin-orbit effect on the magnetic shielding constant: niobium hexahalides and titanium tetrahalides. *Chem. Phys. Lett.*, **275**, 429–436.

[9025] Nakatsuji H., Nakajima T., Hada M., Takashima H. and Tanaka S. (1995a). Spin-orbit effect on the magnetic shielding constant using the ab initio UHF method: silicon tetrahalides. *Chem. Phys. Lett.*, **247**, 418–424.

[9026] Nakatsuji H., Takashima H. and Hada M. (1995b). Spin-orbit effect on the magnetic shielding constant using the ab initio UHF method. *Chem. Phys. Lett.*, **233**, 95–101.

[9027] Nana Engo S.G., Kwato Njock M.G., Owono Owono L.C., Oumarou B., Lagmago Kamta G. and Motapon O. (1997). Comparison of phenomenological and supersymmetry-inspired quantum-defect methods in their relativistic and quasirelativistic formulations. *Phys. Rev. A*, **56**, 2624–2647.

[9028] Nana Engo S.G., Owono Owono L.C., Dada J.P., Waha Ndeuna L., Kwato Njock M.G., Oumarou B. and Motapon O. (1995). Relativistic semiclassical description of dipole matrix elements for arbitrary $nlj \rightarrow n'l'j'$ transitions in non-hydrogenic ions. *J. Phys. B*, **28**, 2333–2353.

[9029] Narayan Vaidya A. and Barbosa da Silva Filho P. (1999). Green function for a charged spin-$\frac{1}{2}$ particle with anomalous magnetic moment in a plane-wave external electromagnetic field. *J. Phys. A*, **32**, 6605–6612.

[9030] Nash C.S. and Bursten B.E. (1995). Comparisons among transition metal, actinide and trans-actinide complexes: the relativistic electronic structures of $Cr(CO)_6$, $W(CO)_6$, $U(CO)_6$ and $Sg(CO)_6$. *New J. Chem.*, **19**, 669–675.

[9031] Nash C.S. and Bursten B.E. (1999a). Spin-orbit effects, VSEPR theory, and the electronic structures of heavy and superheavy group IVA hydrides and group VIIIA tetrafluorides. A partial role reversal for elements 114 and 118. *J. Phys. Chem. A*, **103**, 402–410.

[9032] Nash C.S. and Bursten B.E. (1999b). Spin-orbit coupling versus the VSEPR method: On the possibility of a nonplanar structure for the super-heavy noble gas tetrafluoride $(118)F_4$. *Angew. Chem. Int. Ed. Engl.*, **38**, 151–153.

[9033] Nash C.S. and Bursten B.E. (1999c). Spin-orbit coupling versus the VSEPR method: On the possibility of a nonplanar structure for the super-heavy noble gas tetrafluoride $(118)F_4$ (in German). *Angew. Chem.*, **111**, 115–117.

[9034] Nash C.S. and Bursten B.E. (1999d). Spin-orbit effects on the electronic structure of heavy and superheavy hydrogen halides: Prediction of an anomalously strong bond in H[117]. *J. Phys. Chem. A*, **103**, 632–636.

[9035] Nash C.S. and Bursten B.E. (1999e). Prediction of the bond lengths, vibrational frequencies, and bond dissociation energy of octahedral seaborgium hexacarbonyl, $Sg(CO)_6$. *J. Am. Chem. Soc.*, **121**, 10830–10831.

[9036] Nash C.S., Bursten B.E. and Ermler W.C. (1997). Ab initio relativistic potentials with spin-orbit operators. VII. Am through element 118. [Erratum: JCP 111 (1999) 2347]. *J. Chem. Phys.*, **106**, 5133–5142.

[9037] Nasluzov V.A. and Rösch N. (1996). Density functional based structure optimization for molecules containing heavy elements: analytical energy gradients for the Douglas-Kroll-Hess scalar relativistic approach to the LCGTO-DF method. *Chem. Phys. Lett.*, **210**, 413–425.

[9038] Naumkin F.Y. (1998). Single- versus multi-state DIM model for Rg X_2 systems: On the influence of spin-orbit coupling on Ar - I_2 potentials. *Chem. Phys.*, **226**, 319–335.

[9039] Navarro J.A.R., Romero M.A., Salas J.M., Quirós M., El Bahraoui J. and Molina J. (1996). Binuclear platinum(II) triazolopyrimidine bridged complexes. Preparation, crystal structure, NMR spectroscopy, and ab initio MO investigation on the bonding nature of the Pt(II)...Pt(II) interaction in the model compound $Pt_2[NHCHN(C(CH_2)(CH_3))]_4$. *Inorg. Chem.*, **35**, 7829–7835.

[9040] Neale L. and Wilson M. (1995). Core polarization in Kr VIII. *Phys. Rev. A*, **51**, 4272–4275.

[9041] Nedjadi Y., Ait-Tahar S. and Barrett R.C. (1998). An extended relativistic quantum oscillator for S=1 particles. *J. Phys. A*, **31**, 3867–3874.

[9042] Neese F. and Solomon E.I. (1998). Calculation of zero-field splittings, g-values, and the relativistic nephelauxetic effect in transition metal complexes. Application to high-spin ferric complexes. *Inorg. Chem.*, **37**, 6568–6582.

[9043] Nefiodov A.V., Labzowsky L.N. and Goidenko I.A. (1999a). A new approach to the electron self-energy calculations. *Phys. Scripta*, **T80**, 498–499.

[9044] Nefiodov A.V., Labzowsky L.N. and Moores D.L. (1999b). Overlapping identical resonances and radiative interference effects in recombination of heavy multicharged ions. *Phys. Rev. A*, **60**, 2069–2075.

[9045] Nefiodov A.V., Labzowsky L.N., Plunien G. and Soff G. (1996). Nuclear polarization effects in spectra of multicharged ions. *Phys. Lett. A*, **222**, 227–232.

[9046] Nefiodov A.V., Karasiev V.V. and Yerokhin V.A. (1994). Interference effects in the recombination process of hydrogenlike lead. *Phys. Rev. A*, **50**, 4975–4978.

[9047] Nemoshkalenko V.V. and Antonov V.N. (1998). *Computational Methods in Solid State Physics.* Gordon and Breach, 312 pp.

[9048] Nemukhin A.V., Ermilov A.Y., Petrukhina M.A., Klotzbücher W.E. and Smets J. (1997). Predicting lanthanide cluster properties: A comparison with the observed optical spectra of Ho_2. *Spectrokhim. Acta*, **A53**, 1803–.

[9049] Nemukhin A.V., Togonidze V.V., Kovba V.V. and Orlov R.Y. (1998). Estimated solvent shifts of the vibrational spectra of gold thiocomplexes on the basis of nonempirical calculations (in Russian). *Zh. Neorg. Khim.*, **39**, 460–463.

[9050] Nenciu G. and Purice R. (1996). One dimensional periodic Dirac Hamiltonians: Semiclassical and high-energy asymptotics for the gaps. *J. Math. Phys.*, **37**, 3153–3167.

[9051] Neogrády P., Kellö V., Urban M. and Sadlej A.J. (1996). Polarized basis sets for high-level-correlated calculations of molecular electric properties. VII. Elements of the group Ib: Cu, Ag, Au. *Theor. Chim. Acta*, **93**, 101–129.

[9052] Neogrády P., Kellö V., Urban M. and Sadlej A.J. (1997). Ionization potentials and electron affinities of Cu, Ag, and Au: Electron correlation and relativistic effects. *Int. J. Quantum Chem.*, **63**, 557–565.

[9053] Neuhaus A., Veldkamp A. and Frenking G. (1994). Oxo and nitrido complexes of molybdenum, tungsten, rhenium, and osmium. A theoretical study. *Inorg. Chem.*, **33**, 5278–5286.

[9054] Newton R.G. (1994). Comment on"Normalization of scattering states scattering phase shifts and Levinson theorem" by N. Poliatzky (Helv Phys Acta 66, 241-263, 1993). *Helv. Phys. Acta*, **67**, 20–21.

[9055] Ng S.M., Lau C.P., Fan M.F. and Lin Z.Y. (1999). Experimental and theoretical studies of highly fluxional $TpRu(PPh_3)$'H_2SiR_3' complexes (Tp = hydridotris(pyrazolyl)borate. *Organometallics*, **18**, 2484–2490.

[9056] Nicklass A. and Stoll H. (1995). On the importance of core polarization in heavy post-d elements: a pseudopotential calibration study for X_2H_6 (X = Si, Ge, Sn, Pb). *Mol. Phys.*, **86**, 317–326.

[9057] Nicklass A., Dolg M., Stoll H. and Preuss H. (1995). Ab initio energy-adjusted pseudopotentials for noble gases Ne through Xe: Calculation of atomic dipole and quadropole polarizabilities. *J. Chem. Phys.*, **102**, 8942–8952.

[9058] Nieminen J., Kauppi E., Lundell J. and Kunttu H. (1993). Potential energy surface and vibrational analysis along the stretching vibrations of $XeHXe^+$ ion. *J. Chem. Phys.*, **98**, 8698–8703.

[9059] Nieuwpoort W.C., Aerts P.J.C. and Visscher L. (1994). Molecular electronic structure calculations based on the Dirac-Coulomb-(Breit) Hamiltonian. In *Relativistic and Electron Correlation Effects in Molecules and Solids*, (Edited by Malli G.L.), pp. 59–70, Plenum Press, New York.

[9060] Nikitin A.G. (1998). On exact Foldy-Wouthuysen transformation. *J. Phys. A*, **31**, 3297–3300.

[9061] Nio M. and Kinoshita T. (1997). Radiative corrections to the muonium hyperfine structure. II. The $\alpha(Z\alpha)^2$ correction. *Phys. Rev. D*, **55**, 7267–7290.

[9062] Nitta H., Kudo T. and Minowa H. (1999). Motion of a wave packet in the Klein paradox. *Am. J. Phys.*, **67**, 966–971.

[9063] Nogami Y. and Toyama F.M. (1993). Supersymmetry aspects of the Dirac equation in one dimension with a Lorentz scalar potential. *Phys. Rev. A*, **47**, 1708–1714.

[9064] Nogami Y. and Toyama F.M. (1996). Coherent state of the Dirac oscillator. *Can. J. Phys.*, **74**, 114–121.

[9065] Nogami Y. and Toyama F.M. (1998). Reflectionless potentials for the one-dimensional Dirac equation: Pseudoscalar potentials. *Phys. Rev. A*, **57**, 93–97.

[9066] Nogueira F., Fiolhais C., He J.S., Perdew J.P. and Rubio A. (1996). Transferability of a local pseudopotential based on solid-state electron density. *J. Phys. CM*, **8**, 287–302.

[9067] Nogueira S.R. and Guenzburger D. (1996). Relativistic effects on the electronic structure and bonding of $[Ir(CN)_5]^{3-}$. *Int. J. Quantum Chem.*, **57**, 471–479.

[9068] Norbury J.W., Maung K.M. and Kahana D.E. (1994). Exact numerical solution of the spinless Salpeter equation for the Coulomb potential in momentum space. *Phys. Rev. A*, **50**, 3609–3613.

[9069] Norman M.R. and Koelling D.D. (1993). Electronic structure, Fermi surfaces, and superconductivity in f electron metals. In *Handbook on the Physics and Chemistry of Rare Earths*, (Edited by Gschneidner Jr. K.A., Eyring L., Lander G.H. and Choppin G.R.), vol. 17, pp. 1–85, Elsevier Science.

[9070] Norman N.C. (1994). *Periodicity and the p-Block Elements*. Oxford Univ. Press, see p. 31.

[9071] Norman N.C. (1997). *Periodicity and the s- and p-Block Elements*. Oxford Univ. Press, see p. 30.

[9072] Norquist P.L., Beck D.R., Bilodeau R.C., Scheer M., Srawley R.A. and Haugen H.K. (1999). Theoretical and experimental binding energies for the $d^7s^2\,^4F$ levels in Ru^-, including calculated hyperfine structure and $M1$ decay rates. *Phys. Rev. A*, **59**, 1896–1902.

[9073] Novoa J.J., Aullón G., Alemany P. and Alvarez S. (1995). On the bonding nature of the M...M interactions in dimers of square-planar Pt(II) and Rh(I) complexes. *J. Am. Chem. Soc.*, **117**, 7169–7171.

[9074] Nowakowski M. (1999). The quantum mechanical current of the Pauli equation. *Am. J. Phys.*, **67**, 916–919.

[9075] Noyes H.P. and Jones E.D. (1999). Solution of a relativistic three-body problem. *Few-Body Systems*, **27**, 123–139.

[9076] Nunes G.S., Allen P.B. and Martins J.L. (1998a). Electronic structure of silver halides. *Solid State Comm.*, **230**, 377–380.

[9077] Nunes G.S., Allen P.B. and Martins J.L. (1998b). Pressure-induced phase transitions in silver halides. *Phys. Rev. B*, **57**, 5098–5105.

[9078] Nunzi F., Sgamellotti A., Re N. and Floriani C. (1999). A density functional study of $[M(PH_3)_2(\eta^2\text{-}C_2X_4)]$ alkene compexes for the Group 10 metals Ni, Pd, Pt: the effects of electron-attracting substituents. *Dalton Trans.*, pp. 3487–3492.

[9079] Nygren M.A. and Pettersson L.G.M. (1996). H_2O interaction with the polar $Cu_2O(100)$ surface: A theoretical study. *J. Phys. Chem.*, **100**, 1874–1878.

[9080] Oberle C. and Eysel H.H. (1993). Ab initio calculations for some oxo-anions of chlorine, bromine and iodine. *J. Mol. Str. (Theochem)*, **280**, 107–115.

[9081] Ochs U. and Sorg M. (1995). Relativistic Schroedinger equations and the particle-wave duality. *J. Phys. Soc. Japan.*, **64**, 1120–1131.

[9082] Oda Y., Nakamura Y. and Adachi H. (1997). Discrete-variational Dirac-Slater calculation of uranyl(VI) nitrate complexes. *J. Alloys Compounds*, **255**, 24–30.

[9083] Ogurisu O. (1996). Essential spectrum of the Dirac Hamiltonian for a spin 1/2 neutral particle with an anomalous magnetic moment in an asymptotically constant magnetic field. *J. Math. Phys.*, **37**, 1234–1243.

[9084] Ohlmann D., Marchand C.M., Grützmacher H., Chen G.S., Farmer D., Glaser R., Currao A., Nesper R. and Pritzkow H. (1996). Tris(chalcogenato)carbenium ions $(C(XR_3))^{3+}$ (X = O, S, Se, Te): an experimental and quantum-chemical comparison. *Angew. Chem. Int. Ed. Engl.*, **35**, 300–303.

[9085] Ohno M. and von Niessen W. (1995). Green's-function calculations of valence photoemission spectra Pd_2CO and Pt_2CO. *Phys. Rev. B*, **51**, 13547–13553.

[9086] Oja A.S. and Lounasmaa O.V. (1997). Nuclear magnetic ordering in simple metals at positive and negative nanokelvin temperatures. *Rev. Mod. Phys.*, **69**, 1–136.

[9087] Okamoto M. and Takayanagi K. (1999). Structure and conductance of a gold atomic chain. *Phys. Rev. B*, **60**, 7808–7811.

[9088] Okolowski J.A. and Slomiana M. (1993). Comment on "Relativistic aspects of nonrelativistic quantum mechanics", by Dennis Dieks and and Gerard Nienhuis (Am J Phys 1990;58:650-655). *Am. J. Phys.*, **61**, 376–377.

[9089] Olenev A.V., Shevelkov A.V. and Popovkin B.A. (1999). Crystal structure of Hg_2PCl_2 and electronic structure of its main "building unit" – the (P_2Hg_6) octahedron. *J. Solid State Chem.*, **142**, 14–18.

[9090] Olivera P.P., Shustorovich E., Patrito E.M. and Sellers H. (1997). Thermodynamic and ab initio calculations of properties of chemisorbed ions and application to adsorbed sulfur oxide anions. *J. Mol. Catal. A*, **119**, 275–287.

[9091] Olsen H.A. and Kunashenko Y. (1997). Dirac states of relativistic electrons channeled in a crystal and high-energy channeling electron-positron pair production by photons. *Phys. Rev. A*, **56**, 527–537.

[9092] Olsen J., Minaev B., Vahtras O., Ågren H., Jørgensen P., Jensen H.J.A. and Helgaker T. (1994). The Vegard-Kaplan band and the phosphorescent decay of N_2. *Chem. Phys. Lett.*, **231**, 387–394.

[9093] O'Malley S.M. and Beck D.R. (1996). Relativistic configuration-interaction results for the hyperfine-structure constants of ^{133}Cs II and ^{137}Ba III $5p^5(5d+6s+6p)$ levels. *Phys. Rev. A*, **54**, 3894–3902.

[9094] O'Malley S.M. and Beck D.R. (1998). Electron affinities, magnetic dipole decay rates, and hyperfine structure for the excited states of $5p^3$ Sn^-. *Phys. Rev. A*, **57**, 1743–1746.

[9095] O'Malley S.M. and Beck D.R. (1999). Electron affinities and $E1$ f values for 11 bound states of La^- formed by $6p$ and $5d$ attachment. *Phys. Rev. A*, **60**, 2558–2561.

[9096] Omary M.A. and Patterson H.H. (1998). Temperature-dependent photoluminescence properties of $Tl[Ag(CN)_2]$: Formation of luminescent metal-metal-bonded inorganic exciplexes in the solid state. *Inorg. Chem.*, **37**, 1060–1066.

[9097] Omary M.A. and Patterson H.H. (1998). Luminescent homoatomic exciplexes in dicyanoargentate(I) ions doped in alkali halide crystals. 1. "Exciplex tuning" by site-selective excitation. *J. Am. Chem. Soc.*, **120**, 7696–7705.

[9098] Omnès R. (1997). Localization of relativistic particles. *J. Math. Phys.*, **38**, 708–715.

[9099] Onoe J. (1997). Relativistic effects on covalent bonding: Role of individual valence atomic orbitals. *J. Phys. Soc. Japan*, **66**, 2328–2336.

[9100] Onoe J., Nakamatsu H., Mukoyama T., Sekine R., Adachi H. and Takeuchi K. (1997). Structure and bond nature of the UF_5 monomer. *Inorg. Chem.*, **36**, 1934–1938.

[9101] Onoe J., Nakamatsu H., Mukoyama T., Sekine R., Adachi H. and Takeuchi K. (1996). Relativistic change in bond overlap population for the study of relativistic effects on bond length of diatomic molecules. *J. Phys. Soc. Japan*, **65**, 2459–2462.

[9102] Onoe J., Nakamatsu H., Sekine R., Mukoyama T., Adachi H. and Takeuchi K. (1994a). Note on the contribution of relativity to Cu_2 bonding. *J. Phys. Soc. Japan*, **63**, 3992–3995.

[9103] Onoe J., Sekine R., Takeuchi K., Nakamatsu H., Mukoyama T. and Adachi H. (1994b). Atomic-number dependence of relativistic effects on chemical bonding using the non-relativistic discrete-variational Xα methods. *Chem. Phys. Lett.*, **217**, 61–64.

[9104] Onoe J., Takeuchi K., Nakamatsu H., Mukoyama T., Sekine R., Kim B.I. and Adachi H. (1993). Relativistic effects on the electronic structure and chemical bonding of UF_6. *J. Chem. Phys.*, **99**, 6810–6817.

[9105] Ord G.N. (1996). The Schrödinger and Dirac free particle equations without quantum mechanics. *Ann. Phys. (NY)*, **250**, 51–62.

[9106] Ord G.N. and McKeon D.G.C. (1993). On the Dirac equation in 3+1 dimensions. *Ann. Phys. (NY)*, **222**, 244–253.

[9107] Orlova G. and Goddard J.D. (1999). Density functional study of tetra-atomic clusters and complexes of the group 16 elements: Trends in structure and bonding. *J. Phys. Chem. A*, **103**, 6825–6834.

[9108] Osaka N., Akita M., Fujii S. and Itoh K. (1996). Ab initio molecular orbital normal frequency calculation of 1,3-butadiene-silver ion complexes as models for adsorbates on coldly evaporated silver films. *J. Phys. Chem.*, **100**, 17606–17612.

[9109] Ostanin S.A. and Shirokovskii V.P. (1994). A new method of ab initio calculation of the magnetocrystalline anisotropy energy in relativistic ferromagnets. *J. Magn. Magn. Mater.*, **135**, 135–140.

[9110] Ottschofski E. and Kutzelnigg W. (1995). Relativistic Hartree-Fock by means of stationary direct perturbation theory. II. Ground states of rare gas atoms. *J. Chem. Phys.*, **102**, 1752–1757.

[9111] Ottschofski E. and Kutzelnigg W. (1997). Direct perturbation theory of relativistic effects for explicitly correlated wave functions: The He isoelectronic sequence. *J. Chem. Phys.*, **106**, 6634–6646.

[9112] Oudet X. (1996). Atomic magnetic moments and spin notion. *J. Appl. Phys.*, **79**, 5416–5418.

[9113] Ouyang P., Mohta V. and Jaffe R.L. (1999). Dirac particles in twisted tubes. *Ann. Phys. (NY)*, **275**, 297–313.

[9114] Ovcharov I.N. and Fedosov N.I. (1993). Compact (correlated) coherent states of an electron moving in the field of a plane electromagnetic wave (in Russian). *Izv. Vyssh. Uche. Zav. Fiz.*, **36(2)**, 45–49.

[9115] Ovcharov I.N. and Fedosov N.I. (1993). Compact (correlated) coherent states of an electron moving in the field of a plane electromagnetic wave. *Russian Phys. J.*, **36**, 137–140.

[9116] Owusu A., Dougherty R.W., Gowri G., Das T.P. and Andriessen J. (1997a). Relativistic many-body investigation of hyperfine interactions in excited S states of alkali metals: Francium and potassium. *Phys. Rev. A*, **56**, 305–309.

[9117] Owusu A., Yuan X., Panigrahy S.N., Dougherty R.W., Das T.P. and Andriessen J. (1997b). Theory of hyperfine interactions in potassium and Sc^{2+} ion: Trends in systems isoelectronic with potassium. *Phys. Rev. A*, **55**, 2644–2650.

[9118] Ozdemir L. and Karal H. (1999). Energy and oscillator strength calculations using H-F with relativistic corrections. *J. Quant. Spectr. Radiat. Transfer*, **62**, 655–663.

[9119] Ozoliņš V. and Körling M. (1993). Full-potential calculations using the generalized gradient approximation: Structural properties of transition metals. *Phys. Rev. B*, **48**, 18304–18307.

[9120] Ozoliņš V., Wolverton C. and Zunger A. (1998). Cu-Au, Ag-Au, Cu-Ag, and Ni-Au intermetallics: First-principles study of phase diagrams and structures. *Phys. Rev. B*, **57**, 6427–6443.

[9121] Pacchioni G., Chung S.C., Krüger S. and Rösch N. (1997). Is CO chemisorbed on Pt anomalous compared with Ni and Pd? An example of surface chemistry dominated by relativistic effects. *Surf. Sci.*, **392**, 173–184.

[9122] Pacchioni G., Mayer M., Krüger S. and Rösch N. (1999). Bonding and adsorbate core level shifts of transition metal atoms on the Al(100) surface from density functional calculations. *Chem. Phys. Lett.*, **299**, 137–144.

[9123] Pachucki K. (1993a). Radiative correction to the electron charge density in the hydrogen. *Phys. Rev. A*, **48**, 120–128.

[9124] Pachucki K. (1993b). Contributions to the binding two-loop correction to the Lamb shift. *Phys. Rev. A*, **48**, 2609–2614.

[9125] Pachucki K. (1993c). Higher-order binding corrections to the Lamb shift. *Ann. Phys. (NY)*, **226**, 1–87.

[9126] Pachucki K. (1995). Radiative recoil correction to the Lamb shift. *Phys. Rev. A*, **52**, 1079–1085.

[9127] Pachucki K. (1996a). Theory of the Lamb shift in muonic hydrogen. *Phys. Rev. A*, **53**, 2092–2100.

[9128] Pachucki K. (1996b). $\alpha(Z\alpha)^2 E_F$ correction to hyperfine splitting in hydrogenic atoms. *Phys. Rev. A*, **54**, 1994–1998.

[9129] Pachucki K. (1997). Effective Hamiltonian approach to the bound state: Positronium hyperfine state. *Phys. Rev. A*, **56**, 297–304.

[9130] Pachucki K. (1998a). Effective Hamiltonian approach to the bound state: energy of helium $n\ {}^3S_1$ states to order $m\alpha^6$. *J. Phys. B*, **31**, 2489–2499.

[9131] Pachucki K. (1998b). Quantum electrodynamics effects on singlet S-states of helium of order $m\alpha^6$. *J. Phys. B*, **31**, 3547–3556.

[9132] Pachucki K. (1998c). Simple derivation of helium Lamb shift. *J. Phys. B*, **31**, 5123–5133.

[9133] Pachucki K. (1999a). Quantum electrodynamics effects on helium fine structure. *J. Phys. B*, **32**, 137–152.

[9134] Pachucki K. (1999b). Proton structure effects in muonic hydrogen. *Phys. Rev. A*, **60**, 3593–3598.

[9135] Pachucki K. and Karshenboim S.G. (1995). Nuclear-spin-dependent recoil correction to the Lamb shift. *J. Phys. B*, **28**, L221–L224.

[9136] Pachucki K. and Karshenboim S.G. (1998). Complete results for positronium energy levels at order $m\alpha^6$. *Phys. Rev. Lett.*, **80**, 2101–2104.

[9137] Pachucki K. and Karshenboim S.G. (1999). Higher-order recoil corrections to energy levels of two-body systems. *Phys. Rev. A*, **60**, 2792–2798.

[9138] Pachucki K., Leibfried D. and Hänsch T.W. (1993). Nuclear-structure correction to the Lamb shift. *Phys. Rev. A*, **48**, R1–R4.

[9139] Pachucki K., Leibfried D., Weitz M., Huber A., König W. and Hänsch T.W. (1996). Theory of the energy levels and precise two-photon spectroscopy of atomic hydrogen and deuterium. *J. Phys. B*, **29**, 177–195.

[9140] Pacios L.F. and Gómez P.C. (1998). Ab initio study of BrO_3 isomers. *Chem. Phys. Lett.*, **289**, 412–418.

[9141] Padden W.E.P. (1994). Coulomb interaction correction to the vacuum polarization tensor in a super-strong magnetic field. *J. Phys. B*, **27**, 5419–5452.

[9142] Padma R. and Deshmukh P.C. (1994). Near-threshold behaviour of generalised oscillator strengths for $2p$ ionisation. *Australian J. Phys.*, **47**, 271–278.

[9143] Pak M.V., Tulub A.V. and Brattsev V.F. (1996). Spectral D-line of Na-like multicharged ions and phenomenological inclusion of the Lamb shift in many-electron systems (in Russian). *Opt. Spektr.*, **80**, 570–576.

[9144] Pak M.V., Tulub A.V. and Brattsev V.F. (1996). Spectral D-line of Na-like multicharged ions and phenomenological inclusion of the Lamb shift in many-electron systems. *Opt. Spectr.*, **80**, 507–513.

[9145] Pal'chikov V.G. (1998). Relativistic transition probabilities and oscillator strengths in hydrogen-like atoms. *Phys. Scr.*, **57**, 581–593.

[9146] Pal'chikov V.G., Sokolov Y.L. and Yakovlev V.P. (1997). On the accuracy of Lamb shift measurements in hydrogen. *Phys. Scr.*, **55**, 33–40.

[9147] Pal'chikov V.G. and von Oppen G. (1995). The spin-spin mixing of $1s3s\ ^3S_1$ and $1s3d\ ^3D_1$ states in the helium isoelectronic sequence. *Phys. Scr.*, **52**, 366–371.

[9148] Palmieri P., Tarroni R. and Amos R.D. (1996). Fine structure of the lowest vibronic transition of NH^+. *J. Phys. Chem.*, **100**, 6958–6965.

[9149] Panat P.V. and Paranjape V.V. (1999). Dipolar and quadrupolar contributions to self-energy of a hydrogenic atom placed between two metallic slabs. *Solid State Comm.*, **110**, 443–446.

[9150] Panchanan S., Roy B. and Roychoudhury R. (1995). Group theoretic approach for a Dirac particle in Coulomb-like potentials. *J. Phys. A*, **28**, 6467–6477.

[9151] Panek P., Kamiński J.Z. and Ehlotzky F. (1999). Angular and polarization effects in relativistic potential scattering of electrons in a powerful laser field. *Can. J. Phys.*, **77**, 591–602.

[9152] Panina N.S. and Kukushkin Y.N. (1998). *Ab initio* calculations of the electronic structure and normal vibration frequencies of acetonitrile (CH_3CN). *Russian J. Inorg. Chem.*, **43**, 405–409.

[9153] Papakondylis A. and Sautet P. (1996). Ab initio study of the structure of the α-MoO_3 solid and study of the absorption of H_2O and CO molecules on its (100) surfaces. *J. Phys. Chem.*, **100**, 10681–10683.

[9154] Papanikolaou N., Stefanou N., Zeller R. and Dederichs P.H. (1993). Can $5d$ and sp impurities be magnetic?. *Phys. Rev. Lett.*, **71**, 629–632.

[9155] Papp E. (1993). Energy and β-function solutions to relativistic Hamiltonians with Coulombic and linear potentials. *Phys. Rev. A*, **48**, 4091–4096.

[9156] Papp T., Campbell J.L. and Raman S. (1993). Experimental test of Dirac-Fock versus Dirac-Hartree-Slater L X-ray intensity rations. *J. Phys. B*, **26**, 4007–4017.

[9157] Parente F., Marques J.P. and Indelicato P. (1994). Hyperfine quenching of the $3s^23p^63d^4$ J=4 level in titaniumlike ions. *Europhys. Lett.*, **26**, 437–442.

[9158] Park C.Y. and Almlöf J.E. (1994). Two-electron relativistic effects in molecules. *Chem. Phys. Lett.*, **231**, 269–276.

[9159] Park K.T., Novikov D.L., Gubanov V.A. and Freeman A.J. (1994). Electronic structure of noble-metal monoxides: PdO, PtO, and AgO. *Phys. Rev. B*, **49**, 4425–4431.

[9160] Park S.J., Kim M.C., Lee Y.S. and Jeung G.H. (1997). Ab initio calculations on the electronic states of GaAr and $GaAr^+$. *J. Chem. Phys.*, **107**, 2481–2487.

[9161] Parpia F.A. (1997). Electric-dipole hyperfine matrix elements of the ground state of the TlF molecule in the Dirac-Fock approximation. *J. Phys. B*, **30**, 3983–4001.

[9162] Parpia F.A. (1998). Ab initio calculation of the enhancement of the electric dipole moment of an electron in the YbF molecule. *J. Phys. B*, **31**, 1409–1430.

[9163] Parpia F.A., Froese Fischer C. and Grant I.P. (1996). GRASP92: A package for large-scale relativistic atomic structure calculations. *Comp. Phys. Comm.*, **94**, 249–271.

[9164] Parpia F.A. and Mohanty A.K. (1995a). Numerical study of the convergence of the linear expansion method for the one-electron Dirac equation. *Chem. Phys. Lett.*, **238**, 209–214.

[9165] Parpia F.A. and Mohanty A.K. (1995b). Dirac-Fock calculations for the ground states for some small molecules. *Phys. Rev. A*, **52**, 962–968.

[9166] Parpia F.A., Wijesundera W.P. and Grant I.P. (1993). A program for generating complete active spaces for relativistic atomic structure calculations. *Comp. Phys. Comm.*, **76**, 127–139.

[9167] Parrot R. and Boulanger D. (1993). Molecular spin-orbit interaction for d^5 ions in covalent crystals: spin-lattice coupling coefficients of Mn^{2+} in II-VI compounds. *Phys. Rev. B*, **47**, 1849–1857.

[9168] Parrot R. and Boulanger D. (1997). Molecular spin-orbit interaction for d^5 ions in crystals. Spin-lattice coupling constants and first-order spin-orbit interaction for the fluorescent levels. *phys. stat. sol. (b)*, **201**, 405–412.

[9169] Parsons I.W. and Till S.J. (1993). Quasi-relativistic MSXα calculations of a_{1g} potential-energy curves for MoF_6, WF_6 and UF_6. *J. Chem. Soc., Faraday Trans.*, **89**, 25–28.

[9170] Partridge H., Bauschlicher Jr C.W. and Visscher L. (1995). The dissociation energies of AlH_2 and AlAr. *Chem. Phys. Lett.*, **246**, 33–39.

[9171] Pascual J.L., Seijo L. and Barandiarán Z. (1993). Ab initio model potential study of environmental effects on the Jahn-Teller parameters of Cu^{2+} and Ag^{2+} impurities in MgO, CaO, and SrO hosts. *J. Chem. Phys.*, **98**, 9715–9724.

[9172] Passante R. (1998). Radiative level shifts of an accelerated hydrogen atom and the Unruh effect in quantum electrodynamics. *Phys. Rev. A*, **57**, 1590–1594.

[9173] Patchkovskii S. and Ziegler T. (1999). Prediction of electron paramagnetic resonance g-tensors of transition metal complexes using density functional theory: First applications to some axial d^1 MEX_4 systems. *J. Chem. Phys.*, **111**, 5730–5740.

[9174] Patnaik R.C., Hota R.L. and Tripathi G.S. (1998). Indirect nuclear spin-spin interactions in PbTe. *Phys. Rev. B*, **58**, 3924–3931.

[9175] Paulus B., Fulde P. and Stoll H. (1995). Electron correlations for ground-state properties of group-IV semiconductors. *Phys. Rev. B*, **51**, 10572–10578.

[9176] Paulus B., Fulde P. and Stoll H. (1996). Cohesive energies of cubic III-V semiconductors. *Phys. Rev. B*, **54**, 2556–2560.

[9177] Paulus W., Kratz J.V., Strub E., Zauner S., Brüchle W., Pershina V., Schädel M., Schausten B., Adams J.L., Gregorich K.E., Hoffman D.C., Lane M.R., Laue C., Lee D.M., McGrath C.A., Shaughnessy D.K., Strellis D.A. and Sylwester E.R. (1999). Chemical properties of element 105 in aqueous solution: Extraction of the fluoride-, chloride-, and bromide complexes of the Group-5 elements into an aliphatic amine. *Radiochim. Acta*, **84**, 69–77.

[9178] Pauri M. (1993). On the constants of the motion for the free Dirac particle. *Nuovo Cim. A*, **106**, 427–429.

[9179] Pavlov M., Blomberg M.R.A., Siegbahn P.E.M., Wesendrup R., Heinemann C. and Schwarz H. (1997). Pt^+-catalyzed oxidation of methane: Theory and experiment. *J. Phys. Chem. A*, **101**, 1567–1579.

[9180] Pavone P., Baroni S. and de Gironcoli S. (1998). $\alpha \leftrightarrow \beta$ phase transition in tin: A theoretical study based on density-functional perturbation theory. *Phys. Rev. B*, **57**, 10421–10423.

[9181] Pavšič M., Recami E., Rodrigues Jr W.A., Maccarrone G.D., Raciti F. and Salesi G. (1993). Spin and electron structure. *Phys. Lett. B*, **318**, 481–488.

[9182] Pegarkov A. (1993). Adiabatic states of diatomic molecules in the presence of strong spin-orbit interaction. *Opt. Spectr.*, **75**, 425–431.

[9183] Pegarkov A. (1993). Adiabatic states of diatomic molecules in the presence of strong spin-orbit interaction. *Opt. Spektr.*, **75**, 717–729.

[9184] Pelc O. and Horwitz L.P. (1997). Construction of a complete set of states in relativistic scattering theory. *J. Math. Phys.*, **38**, 115–138.

[9185] Pénicaud M. (1997). Electron localization in the series of actinide metals. The cases of δ-Pu and Es. *J. Phys. CM*, **9**, 6341–6349.

[9186] Perera S.A. and Bartlett R.J. (1993). Relativistic effects at the correlated level. An application to interhalogens. *Chem. Phys. Lett.*, **216**, 606–612.

[9187] Perger W.F., Halabuka Z. and Trautmann D. (1993). Continuum wavefunction solver for GRASP. *Comp. Phys. Comm.*, **76**, 250–262.

[9188] Peric M., Marian C.M. and Peyerimhoff S.D. (1994). Ab initio investigation of the structure of the X ^{2}A', A ^{2}A"(1 $^2\Pi$) spectral system of HCO: theoretical treatment of the vibronic and spin-orbit coupling. *J. Mol. Spectr.*, **166**, 406–422.

[9189] Pernpointner M. and Schwerdtfeger P. (1998). Accurate nuclear quadrupole moments for the gallium isotopes ^{69}Ga and ^{71}Ga within the PCNQM model. *Chem. Phys. Lett.*, **295**, 347–353.

[9190] Pernpointner M., Schwerdtfeger P. and Hess B.A. (1998a). The nuclear quadrupole moment of ^{133}Cs: accurate relativistic coupled cluster calculations for CsF within the point-charge model for nuclear quadrupole moments. *J. Chem. Phys.*, **108**, 6739–6747.

[9191] Pernpointner M., Seth M. and Schwerdtfeger P. (1998b). A point-charge model for the nuclear quadrupole moment: coupled-cluster, Dirac-Fock, Douglas-Kroll, and nonrelativistic Hartree-Fock calculations for the Cu and F electric field gradients in CuF. *J. Chem. Phys.*, **108**, 6722–6738.

[9192] Perry J.K., Ohanessian G. and Goddard III W.A. (1994). Mechanism and energetics for dehydrogenation of methane by gaseous iridium ions. *Organometallics*, **13**, 1870–1877.

[9193] Pershina V.G. (1996). Electronic structure and properties of the transactinides and their compounds. *Chem. Rev.*, **96**, 1977–2010.

[9194] Pershina V. (1998a). Solution chemistry of element 105 part I: hydrolysis of group 5 cations: Nb, Ta, Ha and Pa. *Radiochim. Acta*, **80**, 65–73.

[9195] Pershina V. (1998b). Solution chemistry of element 105 part II: hydrolysis and complex formation of Nb, Ta, Ha and Pa in HCl solutions. *Radiochim. Acta*, **80**, 75–84.

[9196] Pershina V. and Bastug T. (1999). Solution chemistry of element 105. Part III: Hydrolysis and complex formation of Nb, Ta, Db and Pa in HF and HBr solutions. *Radiochim. Acta*, **84**, 79–84.

[9197] Pershina V., Bastug T. and Fricke B. (1998). Recent progress in theoretical investigations of the electronic structure of the transactinides. *J. Alloys Compounds*, **271-273**, 283–286.

[9198] Pershina V. and Fricke B. (1993). Relativistic effects in physics and chemistry of element 105. IV. Their influence on the electronic structure and related properties. *J. Chem. Phys.*, **99**, 9720–9729.

[9199] Pershina V. and Fricke B. (1994a). Electronic structure and properties of the group 4, 5, and 6 highest chlorides including elements 104, 105, and 106. *J. Phys. Chem.*, **98**, 6468–6473.

[9200] Pershina V. and Fricke B. (1994b). The electronic structure of the group 6 oxyanions $(MO_4)^{2-}$where M = Cr, Mo, W, and element 106. *Radiochim. Acta*, **65**, 13–17.

[9201] Pershina V. and Fricke B. (1995). Group 6 oxychlorides $MOCl_4$, where M = Mo, W, and element 106 (Sg): Electronic structure and thermochemical stability. *J. Phys. Chem.*, **99**, 144–147.

[9202] Pershina V. and Fricke B. (1996). Group 6 dioxychlorides MO_2Cl_2 (M = Cr, Mo, W, and element 106, Sg): The electronic structure and thermochemical stability. *J. Phys. Chem.*, **100**, 8748–8751.

[9203] Pershina V. and Fricke B. (1999). Electronic structure and chemistry of the heaviest elements. In *Heavy Elements and Related New Phenomena, Vol. 1*, (Edited by Greiner W. and Gupta R.K.), pp. 194–262, World Scientific, Singapore.

[9204] Pershina V., Fricke B. and Ionova G.V. (1994a). Theoretical study of the physicochemical properties of the light transactinides. *J. Alloys Compounds*, **213/214**, 33–37.

[9205] Pershina V., Fricke B., Ionova G.V. and Johnson E. (1994b). Thermodynamic functions of element 105 in neutral and ionized states. *J. Phys. Chem.*, **98**, 1482–1486.

[9206] Pershina V., Fricke B., Kratz J.V. and Ionova G.V. (1994c). The electronic structure of anionic halide complexes of element 105 in aqueous solutions and their extraction by aliphatic amines. *Radiochim. Acta*, **64**, 37–48.

[9207] Pershina V., Johnson E. and Fricke B. (1999). Theoretical estimates of redox potentials for Group 6 elements, including Element 106, seaborgium, in acid solutions. *J. Phys. Chem. A*, **103**, 8463–8470.

[9208] Persson B.J., Roos B.O. and Pierloot K. (1994). A theoretical study of the chemical bonding in $M(CO)_x$ (M=Cr, Fe, and Ni). *J. Chem. Phys.*, **101**, 6810–6821.

[9209] Persson C. and Lindefelt U. (1997). Relativistic band structure calculation of cubic and hexagonal SiC polytypes. *J. Appl. Phys.*, **82**, 5496–5508.

[9210] Persson H. (1993). *QED effects in highly charged ions.*. Ph.D. thesis, Department of Physics, University of Göteborg.

[9211] Persson H., Lindgren I., Labzowsky L., Plunien G., Beier T. and Soff G. (1996a). Second-order self-energy– vacuum-polarization contributions to the Lamb shift in highly charged few-electron ions. *Phys. Rev. A*, **54**, 2805–2813.

[9212] Persson H., Lindgren I. and Salomonson S. (1993a). A new approach to the electron self energy calculation. *Phys. Scr.*, **T46**, 125–131.

[9213] Persson H., Lindgren I., Salomonson S. and Sunnergren P. (1993b). Accurate vacuum-polarization calculations. *Phys. Rev. A*, **48**, 2772–2778.

[9214] Persson H., Salomonson S. and Sunnergren P. (1998). Regularization corrections to the partial-wave renormalization procedure. *Adv. Quantum Chem.*, **30**, 379–392.

[9215] Persson H., Salomonson S., Sunnergren P. and Lindgren I. (1996b). Two-electron Lamb-shift calculations on heliumlike ions. *Phys. Rev. Lett.*, **76**, 204–207.

[9216] Persson H., Salomonson S., Sunnergren P. and Lindgren I. (1997a). Radiative corrections to the electron g-factor in H-like ions. *Phys. Rev. A*, **56**, 2499–2502.

[9217] Persson H., Salomonson S., Sunnergren P., Lindgren I. and Gustavsson M.G.H. (1997b). A theoretical survey of QED tests in highly charged ions. *Hyperfine Int.*, **108**, 3–17.

[9218] Persson H., Schneider S.M., Greiner W., Soff G. and Lindgren I. (1996c). Self-energy correction to the hyperfine structure splitting of hydrogenlike atoms. *Phys. Rev. Lett.*, **76**, 1433–1436.

[9219] Persson J. (1993). Investigation of the Zeeman effect in the np^2-configuration elements. *Z. Phys. D*, **28**, 23–26.

[9220] Persson J.L., Hui Q., Jakubek Z.J., Nakamura M. and Takami M. (1996). Formation of $AgHe_2$ exciplex in liquid helium. *Phys. Rev. Lett.*, **76**, 1501–1504.

[9221] Persson J.R. (1997). On the hyperfine structure in the yttrium ion. *Z. Phys. D*, **42**, 259–262.

[9222] Petit A. (1999). Fine structure parametric analysis of the $f^3ds^2 + F^3d^2s$ configurations in UI. *Eur. Phys. J. D*, **6**, 157–170.

[9223] Petráš M. (1995). The SO(3,3) group as a common basis for Dirac's and Proca's equations. *Czech. J. Phys.*, **45**, 455–464.

[9224] Petrilli H.M., Blöchl P.E., Blaha P. and Schwarz K. (1998). Electric-field-gradient calculations using the projector augmented wave method. *Phys. Rev. B*, **57**, 14690–14697.

[9225] Petrucci R.H. (1989). *General Chemistry*. Macmillan Publishing Company, New York, p. 285.

[9226] Petrucci R.H. and Harwood W.S. (1993). *General Chemistry, 6th Ed.*. Macmillan, New York, see pp. 346-7, 864.

[9227] Pettersson M., Khriachtchev L., Lundell J. and Räsänen M. (1999a). A chemical compound formed from water and xenon: HXeOH. *J. Am. Chem. Soc.*, **121**, 11904–11905.

[9228] Pettersson M., Lundell J., Khriachtchev L. and Räsänen M. (1998a). Neutral rare-gas containing charge-transfer molecules in solid matrices. III. HXeCN, HXeNC, and HKrCN in Kr and Xe. *J. Chem. Phys.*, **109**, 618–625.

[9229] Pettersson M., Lundell J., Khriachtchev L., Isoniemi E. and Räsänen M. (1998b). HXeSH, the first example of a xenon-sulfur bond. *J. Am. Chem. Soc.*, **120**, 7979–7980.

[9230] Pettersson M., Lundell J. and Räsänen M. (1995a). Neutral rare-gas containing charge-transfer molecules in solid matrices I: HXeCl. HXeBr, HXeI and HKrCl in Kr and Xe. *J. Chem. Phys.*, **102**, 6423–6431.

[9231] Pettersson M., Lundell J. and Räsänen M. (1995b). Neutral rare-gas containing charge-transfer molecules in solid matrices. II. HXeH, HXeD, and DXeD in Xe. *J. Chem. Phys.*, **103**, 205–210.

[9232] Pettersson M., Lundell J. and Räsänen M. (1999b). New rare-gas-containing neutral molecules. *Eur. J. Inorg. Chem.*, **1**, 729–737.

[9233] Phatak S.C., Pal S. and Biswas S. (1995). Semiclassical features in the quantum description of a Dirac particle in a cavity. *Phys. Rev. E*, **52**, 1333–1344.

[9234] Philipsen P.H.T., van Lenthe E., Snijders J.G. and Baerends E.J. (1997). Relativistic calculations on the absorption of CO on the (111) surfaces of Ni, Pd, and Pt within the zeroth-order regular approximation. *Phys. Rev. B*, **56**, 13556–13562.

[9235] Phillips D.R. and Wallace S.J. (1996). Relativistic bound-state equations in three dimensions. *Phys. Rev. C*, **54**, 507–522.

[9236] Philpott R.J. (1996). Thomas precession and the Liénard-Wiechert field. *Am. J. Phys.*, **64**, 552–556.

[9237] Pick S. (1997). A tight-binding study of the electronic structure of Pt(111) and chemisorbed CO. *J. Phys. C*, **9**, 141–148.

[9238] Pick S. and Mikušik P. (1993). Tight-binding study of the electronic structure of Pd and Pt overlayers on W (011). The role of *s* electrons. *Chem. Phys. Lett.*, **208**, 97–100.

[9239] Pidun U. and Frenking G. (1996). The bonding of acetylene and ethylene in high-valent and low-valent transition metal compounds. *J. Organomet. Chem.*, **525**, 269–278.

[9240] Pidun U. and Frenking G. (1998). ($HRh(CO)_4$)-catalyzed hydrogenation of Co: A systematic ab initio quantum-chemical investigation of the reaction mechanism. *Chem. Eur. J.*, **4**, 522–540.

[9241] Piekarewicz J. (1993a). Levinson's theorem for Dirac particles. *Phys. Rev. C*, **48**, 2174–2181.

[9242] Piekarewicz J. (1993b). Salpeter's approach to the relativistic two-body problem. *Rev. Mex. Fis.*, **39**, 542–560.

[9243] Pierloot K., De Kerpel J.O.A., Ryde U., Olsson M.H.M. and Roos B.O. (1998). Relation between the structure and spectroscopic properties of blue copper proteins. *J. Am. Chem. Soc.*, **120**, 13156–13166.

[9244] Pierloot K., De Kerpel J.O.A., Ryde U. and Roos B.O. (1997). Theoretical study of the electronic spectrum of plastocyanin. *J. Am. Chem. Soc.*, **119**, 218–226.

[9245] Pierloot K., Persson B.J. and Roos B.O. (1995). Theoretical study of the chemical bonding in $Ni(C_2H_4)$ and ferrocene. *J. Phys. Chem.*, **99**, 3465–3472.

[9246] Pietsch M.A., Couty M. and Hall M.B. (1995). Comparison of Møller-Plesset perturbation methods, complete active space self-consistent field theory, and a new generalized molecular orbital method for oxygen atom transfer from a molybdenum complex to a phosphine. *J. Phys. Chem.*, **99**, 16315–16319.

[9247] Pietsch M.A. and Hall M.B. (1996). Theoretical studies on models for the oxo-transfer reaction of dioxomolybdnenum enzymes. *Inorg. Chem.*, **35**, 1273–1278.

[9248] Pilkuhn H. (1995). A new hyperfine operator in the Dirac equation. *J. Phys. B*, **28**, 4421–4434.

[9249] Pilkuhn H. and Stäudner F. (1993). Double-Dirac equation and decay rates of bound particles. *Phys. Lett. A*, **178**, 156–159.

[9250] Pillin M. (1994). q-deformed relativistic wave equations. *J. Math. Phys.*, **35**, 2804–2817.

[9251] Pindzola M.S. (1993). Parity-violation effects on the Áuger-electron emission from highly charged atomic ions. *Phys. Rev. A*, **47**, 4856–4858.

[9252] Pindzola M.S. and Griffin D.C. (1997). Electron-impact ionization of tungsten ions in the configuration-average distorted-wave approximation. *Phys. Rev. A*, **56**, 1654–1657.

[9253] Pindzola M.S., Robicheaux F.J., Badnell N.R., Chen M.H. and Zimmermann M. (1995). Photorecombination of highly-charged uranium ions. *Phys. Rev. A*, **52**, 420–425.

[9254] Pisani L., Andrė J.M., Andrė M.C. and Clementi E. (1993). Study of relativistic effects in atoms and molecules by the kinetically balanced LCAO approach. Ground state of hydrogen and of hydrogenic atoms in Slater and Gaussian basis functions. *J. Chem. Edu.*, **70**, 894–901.

[9255] Pisani L. and Clementi E. (1994a). Relativistic Dirac-Fock calculations for closed-shell molecules. *J. Comp. Chem.*, **15**, 466–474.

[9256] Pisani L. and Clementi E. (1994b). Relativistic effects on sixth group hydrides. *J. Chem. Phys.*, **101**, 3079–3084.

[9257] Pisani L. and Clementi E. (1995a). Relativistic effects on sixth group monohydrides. *J. Chem. Phys.*, **103**, 9321–9323.

[9258] Pisani L. and Clementi E. (1995b). Dirac-Fock self-consistent-field calculations for closed and open-shell molecules. In *Methods and Techniques in Computational Chemistry: METECC-95*, (Edited by Clementi E. and Gorongiu G.), pp. 219–241, STEF, Cagliari.

[9259] Pissondes J.C. (1999). Quadratic relativistic invariant and metric form in quantum mechanics. *J. Phys. A*, **32**, 2871–2885.

[9260] Pivovarov Y.L. (1998). Coherent excitation of hydrogen-like relativistic heavy ions in a crystal: Structure of electronic levels of an ion and resonance width. *Nucl. Instr. Meth. Phys. Res. B*, **145**, 96–101.

[9261] Pizlo A., Jansen G., Hess B.A. and von Niessen W. (1993). Ionization potential and electron affinity of the Au atom and the AuH molecule by all-electron relativistic configuration interaction and propagator techniques. *J. Chem. Phys.*, **98**, 3945–3951.

[9262] Plante D.R., Johnson W.R. and Sapirstein J. (1994). Relativistic all-order many-body calculations of the $n = 1$ and $n = 2$ states of heliumlike ions. *Phys. Rev. A*, **49**, 3519–3530.

[9263] Plass W., Stoll H., Preuss H. and Savin A. (1995). An ab initio investigation of the molecules X_2, CuX, Cu_2X and CuX_2 (X = Si, Ge, and Sn). *J. Mol. Str.*, **339**, 67–81.

[9264] Plunien G., Beier T., Soff G. and Persson H. (1998). Exact two-loop vacuum polarization correction to the Lamb shift in hydrogenlike ions. *Eur. Phys. J. D*, **1**, 177–185.

[9265] Poblet J.M., Muñoz J., Winkler K., Cancilla M., Hayashi A., Lebrilla C.B. and Balch A.L. (1999). Geometric and electronic structure of metal-cage fullerenes, $C_{59}M$ (M = Pt, Ir) obtained by laser ablation of electrochemically deposited films. *Chem. Comm.*, pp. 493–494.

[9266] Poirier D.M., Knupfer M., Weaver J.H., Andreoni W., Laasonen K., Parrinello M., Kikuchi K. and Achiba Y. (1994). Electronic and geometric structure of La@C_{82} and C_{82}: Theory and experiment. *Phys. Rev. B*, **49**, 17403–17412.

[9267] Polasik M. (1995). Systematic multiconfiguration-Dirac-Fock study of the x-ray spectra accompanying the ionization in collision processes: The structure of the $K\beta_{1,3}L^0M^r$ lines. *Phys. Rev. A*, **52**, 227–235.

[9268] Poliatzky N. (1993a). Levinson's theorem for the Dirac equation. *Phys. Rev. Lett.*, **70**, 2507–2510.

[9269] Poliatzky N. (1993b). Normalization of scattering states scattering phase shifts and Levinson's theorem. *Helv. Phys. Acta*, **66**, 241–263.

[9270] Poliatzky N. (1994). Normalization of scattering states and Levinson's theorem: reply to a comment by R G Newton. *Helv. Phys. Acta*, **67**, 683–689.

[9271] Pollack L., Perdew J.P., He J., Marques M., Nogueira F. and Fiolhais C. (1997). Tests of a density-based local pseudopotential for sixteen simple metals. *Phys. Rev. B*, **55**, 15544–15551.

[9272] Polly R., Dinaev S., Windholz L., Milošević S. and Hess B.A. (1999). Green bands of the CsHg molecule. *J. Chem. Phys.*, **110**, 8992–8999.

[9273] Polly R., Gruber D., Windholz L., Gleichmann M.M. and Hess B.A. (1998). Relativistic all-electron *ab initio* calculations of CsHg potential energy curves including spin-orbit effects. *J. Chem. Phys.*, **109**, 9463–9472.

[9274] Popov V.S., Mur V.D. and Karnakov B.M. (1998). Relativistic version of the imaginary time method. *Phys. Lett. A*, **250**, 20–24.

[9275] Popov Y.V. and Kuz'mina N.M. (1993a). Eikonal calculations for relativistic (e, 2e) experiments. *Vestn. Mosk. Univ. Ser. 3 Fiz. Astron.*, **34(5)**, 26–32.

[9276] Popov Y.V. and Kuz'mina N.M. (1993b). EWBA calculations for relativistic (e, 2e) experiments. *J. Phys. B*, **26**, 1215–1220.

[9277] Porsev S.G. (1997). Calculation of lifetimes of low-lying odd-parity levels of Sm. *Phys. Rev. A*, **56**, 3535–3542.

[9278] Porsev S.G., Rakhlina Y.G. and Kozlov M.G. (1999a). Calculation of hyperfine structure constants for ytterbium. *J. Phys. B*, **32**, 1113–1120.

[9279] Porsev S.G., Rakhlina Y.G. and Kozlov M.G. (1999b). Electric-dipole amplitudes, lifetimes, and polarizabilities of the low-lying levels of atomic ytterbium. *Phys. Rev. A*, **60**, 2781–2785.

[9280] Porteous I.R. (1995). *Clifford Algebras and the Classical Groups*. Cambridge Univ. Press.

[9281] Porter L.E. (1992). Importance of the high-velocity projectile-z^3 term in modified Bethe-Bloch stopping power theory. *Appl. Phys. Lett.*, **61**, 360–362.

[9282] Pöschl W. (1997). Relativistic Hartree-Bogoliubov theory in coordinate space: finite element solution for a nuclear system with spherical symmetry. *Comp. Phys. Comm.*, **103**, 217–250.

[9283] Pöschl W. (1998). B-spline finite elements and their efficiency in solving relativistic mean field equations. *Comp. Phys. Comm.*, **112**, 42–66.

[9284] Pöschl W. and Dietrich K. (1995). The eigenspectrum of the relativistic two-center Hamiltonian in the diatomic basis of bound hydrogenlike Dirac-spinors. *Z. Phys. A*, **351**, 271–279.

[9285] Pöschl W., Vretenar D. and Ring P. (1996). Application of the finite element method in self-consistent relativistic mean field calculations. *Comp. Phys. Comm.*, **99**, 128–148.

[9286] Postnikov A.V., Neumann T., Borstel G. and Methfessel M. (1993). Ferroelectric structure of $KNbO_3$ and $KTaO_3$ from first-principles calculations. *Phys. Rev. B*, **48**, 5910–5918.

[9287] Poteau R., Heully J.L. and Spiegelmann F. (1997). Structure, stability, and vibrational properties of small silver cluster. *Z. Phys. D*, **40**, 479–482.

[9288] Pou-Amerigo R., Merchán M., Nebot-Gil I., Malmqvist P.Å. and Roos B. (1994). The chemical bonds in CuH, Cu_2, NiH, and Ni_2 studied with multiconfigurational second order perturbation theory. *J. Chem. Phys.*, **101**, 4893–4902.

[9289] Poulton J.T., Sigalas M.P., Folting K., Streib W.E., Eisenstein O. and Caulton K.G. (1994). $RuHX(CO)(PR_3)_2$: Can ν_{CO} be a probe for the nature of the Ru-X bond. *Inorg. Chem.*, **33**, 1476–1485.

[9290] Power E.A. and Thirunamachandran T. (1993). Casimir-Polder potential as an interaction between induced dipoles. *Phys. Rev. A*, **48**, 4761–4763.

[9291] Power E.A. and Thirunamachandran T. (1994). Zero-point energy differences and many-body dispersion forces. *Phys. Rev. A*, **50**, 3929–3939.

[9292] Pradhan A.K. (1999). Transition probabilities from the iron project. *Phys. Scripta*, **T83**, 69–75.

[9293] Pratt R.H. and Kim Y.S. (1993). Non-dipolar effects in atomic photoionization. *Romanian J. Phys.*, **38**, 353–370.

[9294] Pratt R.H. and LaJohn L. (1995). The need to include multipole effects beoynd the dipole approximation in the description of photoionization both at nonrelativistic and relativistic energies. *Nucl. Instr. Meth. Phys. Res.*, **99**, 136–139.

[9295] Preisenberger M., Pyykkö P., Schier A. and Schmidbaur H. (1999). Isomerism of aurated phosphine sulfides, thiophosphinates, thiophosphonates, and thiophosphates: Structural and quantum chemical studies. *Inorg. Chem.*, **38**, 5870–5875.

[9296] Prinz H.T., Besch K.H. and Nakel W. (1995). Spin-orbit interaction of the continuum electrons in relativistic (e,2e) measurements. *Phys. Rev. Lett.*, **74**, 243–245.

[9297] Prosser R.T. (1998). On the energy spectrum of the hydrogen atom in a photon field. *J. Math. Phys.*, **39**, 229–277.

[9298] Protopapas M., Keitel C.H. and Knight P.L. (1996). Relativistic mass shift effects in adiabatic intense laser field stabilization of atoms. *J. Phys. B*, **29**, L591–L598.

[9299] Protopapas M., Keitel C.H. and Knight P.L. (1997). Atomic physics with super-high intensity lasers. *Rep. Prog. Phys.*, **60**, 389–486.

[9300] Purton J.A., Bird D.M., Parker S.C. and Bullett D.W. (1999). Comparison of atomistic simulations and pseudopotential calculations of the MgO[100]/Ag[100] and MgO[110]/Ag[110] interfaces. *J. Chem. Phys.*, **110**, 8090–8097.

[9301] Pyper N.C. (1999). Relativistic theory of nuclear shielding in one-electron atoms. 1. Theoretical foundations and first-order terms. *Mol. Phys.*, **97**, 381–390.

[9302] Pyper N.C. and Zhang Z.C. (1999). Relativistic theory of nuclear shielding in one-electron atoms. 2. Analytical and nymerical results. *Mol. Phys.*, **97**, 391–413.

[9303] Pyykkö P. (1995). Predicted chemical bonds between rare gases and Au^+. *J. Am. Chem. Soc.*, **117**, 2067–2070.

[9304] Pyykkö P. (1997). Strong closed-shell interactions in inorganic chemistry. *Chem. Rev.*, **97**, 597–636.

[9305] Pyykkö P., Angermaier K., Assmann B. and Schmidbaur H. (1995). Calculated structures of SAu_3^+ and $S(AuPH_3)_3^+$. *J. Chem. Soc, Chem. Comm.*, pp. 1889–1890.

[9306] Pyykkö P., Li J., Baştuğ T., Fricke B. and Kolb D. (1993). Valence photoelectron spectrum of OsO_4: Evidence for 5p semicore effects?. *Inorg. Chem.*, **32**, 1525–1526.

[9307] Pyykkö P., Li J. and Runeberg N. (1994a). Quasirelativistic pseudopotential study of species isoelectronic to uranyl and the equatorial coordination of uranyl. *J. Phys. Chem.*, **98**, 4809–4813.

[9308] Pyykkö P., Li J. and Runeberg N. (1994b). Predicted ligand dependence of the Au(I)...Au(I) attraction in $(XAuPH_3)_2$. *Chem. Phys. Lett.*, **218**, 133–138.

[9309] Pyykkö P. and Mendizabal F. (1997). Theory of the d^{10}-d^{10} closed-shell attraction: 2. Long-distance behaviour and nonadditive effects in dimers and trimers of type ((X-Au-L)$_n$) (n = 2,3; X = Cl,I,H; L = PH_3, PMe_3, -N≡CH). *Chem. Eur. J.*, **3**, 1458–1465.

[9310] Pyykkö P. and Mendizabal F. (1998). Theory of the d^{10}-d^{10} closed-shell attraction. III. Rings. *Inorg. Chem.*, **37**, 3018–3025.

[9311] Pyykkö P., Runeberg N. and Mendizabal F. (1997a). Theory of the d^{10}-d^{10} closed-shell attraction: 1. Dimers near equilibrium . *Chem. Eur. J.*, **3**, 1451–1457.

[9312] Pyykkö P., Schneider W., Bauer A., Bayler A. and Schmidbaur H. (1997b). An ab initio study of the aggregation of LAuX molecules and $[LAuL]^+[XAuX]^-$ ions. *J. Chem. Soc., Chem. Comm.*, pp. 1111–1112.

[9313] Pyykkö P. and Seth M. (1997). Relativistic effects in nuclear quadrupole coupling. *Theor. Chem. Acc.*, **96**, 92–104.

[9314] Pyykkö P., Straka M. and Tamm T. (1999). Calculations on indium and thallium cyclopentadienyls. Metal-metal interactions and possible new species. *PCCP*, **1**, 3441–3444.

[9315] Pyykkö P. and Tamm T. (1997). Calculated structures of MO_2^{2+}, MN_2, and MP_2 (M = Mo,W). *J. Phys. Chem. A*, **101**, 8107–8114.

[9316] Pyykkö P. and Tamm T. (1998a). Can triple bonds exist between gold and main-group elements? *Theor. Chem. Acc.*, **99**, 113–115.

[9317] Pyykkö P. and Tamm T. (1998b). Theory of the d^{10}-d^{10} closed-shell attraction. 4. $X(AuL)_n^{m+}$ centered systems. *Organometallics*, **17**, 4842–4852.

[9318] Pyykkö P., Tokman M. and Labzowsky L.N. (1998). Estimated valence-level Lamb shifts for group 1 and group 11 metal atoms. *Phys. Rev. A*, **57**, R689–R692.

[9319] Qian T.Z. and Su Z.B. (1994). Spin-orbit interaction and Aharonov-Anandan phase in mesoscopic rings. *Phys. Rev. Lett.*, **72**, 2311–2315.

[9320] Qian T.Z., Yi Y.S. and Su Z.B. (1997). Persistent currents from the competition between Zeeman coupling and spin-orbit interaction. *Phys. Rev. B*, **55**, 4065–4068.

[9321] Qiu Y.H., Li S.C. and Sun Y.S. (1993). Dielectronic spectra for Ne-like titanium from F-like low-lying states. *At. Data Nucl. Data Tables*, **55**, 1–42.

[9322] Qu L.H., Li B.W. and Wang Z.W. (1998a). Energy of $1s2snp\ ^4P$ states for the lithium isoelectronic sequence. *Chin. Phys. Lett.*, **15**, 329–331.

[9323] Qu L.H., Wang Z.W. and Li B.W. (1998b). Accurate calculations for the autoionizing states of the lithium atom. *Phys. Lett. A*, **240**, 65–69.

[9324] Qu Y.Z., Wang J.G., Yuan J.K. and Li J.M. (1998). Radiative dielectronic recombination process: Electron and H-like ions. *Phys. Rev. A*, **57**, 1033–1037.

[9325] Quinet P. (1996). Energy levels and oscillator strengths for transitions in Pd II. *Phys. Scr.*, **54**, 483–495.

[9326] Quinet P. (1997). Transition probabilities for forbidden lines in Cr II. *Phys. Scr.*, **55**, 41–48.

[9327] Quinet P., Palmeri P. and Biémont E. (1999). On the use of the Cowan's code for atomic structure calculations in singly ionized lanthanides. *J. Quant. Spectr. Rad. Transfer*, **62**, 625–646.

[9328] Quiney H.M. and Grant I.P. (1993). Partial-wave mass renormalization in atomic QED calculations. *Phys. Scr.*, **T46**, 132–138.

[9329] Quiney H.M. and Grant I.P. (1994). Atomic self-energy calculations using partial-wave mass renormalization. *J. Phys. B*, **27**, L299–L304.

[9330] Quiney H.M., Laerdahl J.K., Faegri Jr K. and Saue T. (1998a). Ab initio Dirac-Hartree-Fock calculations of chemical properties and PT-odd effects in thallium fluoride. *Phys. Rev. A*, **57**, 920–944.

[9331] Quiney H.M., Skaane H. and Grant I.P. (1997). Relativistic calculation of electromagnetic interactions in molecules. *J. Phys. B*, **30**, L829–L834.

[9332] Quiney H.M., Skaane H. and Grant I.P. (1998b). Hyperfine and PT-odd effects in YbF $^2\Sigma$. *J. Phys. B*, **38**, L85–L95.

[9333] Quiney H.M., Skaane H. and Grant I.P. (1998c). Relativistic, quantum electrodynamic and many-body effects in the water molecule. *Chem. Phys. Lett.*, **290**, 473–480.

[9334] Quiney H.M., Skaane H. and Grant I.P. (1999). Ab initio relativistic quantum chemistry: four-components good, two-components bad! *Adv. Quantum Chem.*, **32**, 1–49.

[9335] Rabilloud F., Spiegelmann F. and Heully J.L. (1999). *Ab initio* calculations of structural and electronic properties of small silver bromide clusters. *J. Chem. Phys.*, **111**, 8925–8933.

[9336] Rabinowitch A.S. (1993). Relativistic quantum physics equation for number of electrons. *Int. J. Theor. Phys.*, **32**, 791–799.

[9337] Radozycki T. and Faisal F.H.M. (1993). Multiphoton ejection of strongly bound relativistic electrons in very intense laser fields. *Phys. Rev. A*, **48**, 2407–2412.

[9338] Raeker A., Blum K. and Bartschat K. (1993). Charge cloud distribution of heavy atoms after excitation by polarized electrons. *J. Phys. B*, **26**, 1491–1508.

[9339] Rahman H.U. (1998). Crystal-field excitations in uranium dioxide. *Phys. Lett. A*, **240**, 306–310.

[9340] Rajagopal A.K. (1994). Time-dependent functional theory of coupled electron and electromagnetic fields in condensed-matter systems. *Phys. Rev. A*, **50**, 3759–3765.

[9341] Rajagopal A.K. and Buot F.A. (1996). Generalized functional theory of interacting coupled Liouvillean quantum fields of condensed matter. *Topics in Current Chem.*, **181**, 173–209.

[9342] Rajagopal A.K. and Mochena M. (1998). Spin-orbit interactions in the many-body theory of magnetic electron systems. *Phys. Rev. B*, **57**, 11582–11591.

[9343] Rak J., Gutowski M., Dokurno P., Thanh H.V. and Blazejowski J. (1994). Theoretical studies on structure, thermochemistry, vibrational spectroscopy, and other features of ZrX_6^{2-} (X=F,Cl,Br,I): Coulombic energy in inorganic and organic hexahalogenozirconates. *J. Chem. Phys.*, **100**, 5810–5820.

[9344] Rakowitz F. (1998). An extrapolation scheme for spin-orbit configuration interaction energies applied to the ground and excited states of thallium hydride. *Chem. Phys.*, **225**, 223–238.

[9345] Rakowitz F. (1999). *Entwicklung, Implementierung und Anwendung effizienter Methoden in der relativistischen Elektronenstrukturtheorie*. Ph.D. thesis, Universität Bonn.

[9346] Rakowitz F., Casarrubios M., Seijo L. and Marian C.M. (1998). *Ab initio* spin-free-state-shifted spin-orbit configuration interaction calculations on singly ionized iridium. *J. Chem. Phys.*, **108**, 7980–7987.

[9347] Rakowitz F. and Marian C.M. (1996). The fine-structure splitting of the thallium atomic ground state: *LS*- versus *jj*-coupling. *Chem. Phys. Lett.*, **257**, 105–110.

[9348] Rakowitz F. and Marian C.M. (1997). An extrapolation scheme for spin-orbit configuration interaction energies applied to the ground and excited electronic states of thallium hydride. *Chem. Phys.*, **225**, 223–238.

[9349] Rakowitz F., Marian C.M. and Seijo L. (1999a). Spin-free relativistic no-pair *ab initio* core model potentials and valence basis sets for the transition metal elements Sc to Hg. II. *J. Chem. Phys.*, **111**, 10436–10443.

[9350] Rakowitz F., Marian C.M., Seijo L. and Wahlgren U. (1999b). Spin-free relativistic no-pair *ab initio* core model potentials and valence basis sets for the transition metal elements Sc to Hg. Part I. *J. Chem. Phys.*, **110**, 3678–3686.

[9351] Ram R.S., Liévin J. and Bernath P.F. (1998). Fourier transform infrared emission spectroscopy and ab initio calculations on RuN. *J. Chem. Phys.*, **109**, 6329–6337.

[9352] Ram R.S., Liévin J. and Bernath P.F. (1999a). Emission spectroscopy and *ab initio* calculations on IrN. *J. Mol. Spectr.*, **197**, 133–146.

[9353] Ram R.S., Liévin J. and Bernath P.F. (1999b). Fourier transform emission spectroscopy and *ab initio* calculations on OsN. *J. Chem. Phys.*, **111**, 3449–3456.

[9354] Ramazashvili Jr R.R. (1993). Electron states in a planar 2D ferromagnet: combined spin-orbit symmetry and half-integer orbital angular momentum of an electron. *JETP Lett.*, **57**, 140–142.

[9355] Ramazashvili Jr R.R. (1993). Electron states in a planar 2D ferromagnet: combined spin-orbit symmetry and half-integer orbital angular momentum of an electron. *Pis. Zhur. Eksp. Teor. Fiz.*, **57**, 130–132.

[9356] Ramer N.J. and Rappe A.M. (1999). Designed nonlocal pseudopotentials for enhanced transferability. *Phys. Rev. B*, **59**, 12471–12478.

[9357] Ramirez-Solis A. (1993). Evolution of the $^1\Sigma^+$ radiative lifetime of copper halides. *Phys. Rev. A*, **47**, 1510–1513.

[9358] Ramirez-Solis A. and Schamps J. (1995). Ab initio study of the three lowest-lying ($X\ ^1\Sigma^+$, $^3\Sigma^+$, and $^1\Sigma^+$) electronic states of AgF. *J. Chem. Phys.*, **102**, 4482–4490.

[9359] Ramirez-Solis A. and Sidonio Castillo A. (1993). C_{3v} versus C_{2v} Cd(^1S, ^3P, ^1P)-CH_4 van der Waals complexes: A variational and perturbational multireference configuration interaction study. *J. Chem. Phys.*, **98**, 8065–8069.

[9360] Ramirez-Solis A. and Sidonio Castillo A. (1994). Accurate spectroscopic constants for the Cd(^1S, ^3P, ^1P)-H_2 van der Waals complexes: a theoretical study. *J. Chem. Phys.*, **100**, 8251–8256.

[9361] Rao D.V., Cesareo R. and Gigante G.E. (1994). L X-ray fluorescence cross sections in the atomic region $46 \leq Z \leq 51$ excited by 6.47, 7.57 and 8.12 keV photons. *Nucl. Instr. Meth. Phys. Res. B*, **86**, 219–224.

[9362] Rao D.V., Cesareo R. and Gigante G.E. (1995). M X-ray fluorescence cross sections and yields in the atomic region $78 \leq Z \leq 82$ excited by 6.47 and 7.57 keV photons. *Radiation Phys. and Chem.*, **46**, 317–320.

[9363] Rao D.V., Cesareo R. and Gigante G.E. (1996). Rayleigh and Compton scattering cross sections for low, medium and high Z elements in the energy region $23.18 \leq E \leq 30.85$ keV. *Phys. Scr.*, **54**, 362–367.

[9364] Raptis S.G., Papadopoulos M.G. and Sadlej A.J. (1999). The correlation, relativistic, and vibrational contributions to the dipole moments, polarizabilities, and first and second hyperpolarizabilities of ZnS, CdS, and HgS. *J. Chem. Phys.*, **111**, 7904–7915.

[9365] Rath B. and Patnaik K. (1993). Effective charge for alkali-like atomic system: doublet separation for lithium like atom $2\ ^2P_{1/2}$ - $2\ ^2P_{3/2}$. *J. Phys. Soc. Japan*, **62**, 4527–4528.

[9366] Rathe U.W., Keitel C.H., Protopapas M. and Knight P.L. (1997). Intense laser-atom dynamics with the two-dimensional Dirac equation. *J. Phys. B*, **30**, L531–L539.

[9367] Raubenheimer H.G., Olivier P.J., Lindeque L., Desmet M., Hrušak J. and Kruger G.J. (1997). Oxidative addition of mono and bis(carbene) complexes derived from imidazolyl and thiazolyl gold(I) compounds. *J. Organomet. Chem.*, **544**, 91–100.

[9368] Ravindran P., Delin A., Ahuja R., Johansson B., Auluck S., Wills J.M. and Eriksson O. (1997). Optical properties of monoclinic SnI_2 from relativistic first-principles theory. *Phys. Rev. B*, **56**, 6851–6861.

[9369] Ravindran P., Nordström L., Ahuja R., Wills J.M., Johansson B. and Eriksson O. (1998). Theoretical investigation of the high-pressure phases of Ce. *Phys. Rev. B*, **57**, 2091–2101.

[9370] Raybaud P., Hafner J., Kresse G. and Toulhoat H. (1997a). Ab initio density functional studies of transition-metal sulphides: II. Electronic structure. *J. Phys. CM*, **9**, 11107–11140.

[9371] Raybaud P., Kresse G., Hafner J. and Toulhoat H. (1997b). Ab initio density functional studies of transition-metal sulphides: I. Crystal structure and cohesive properties. *J. Phys. CM*, **9**, 11085–11106.

[9372] Rayner-Canham G. (1999). *Descriptive Inorganic Chemistry, 2nd Ed.*. W. H. Freeman and Co., New York, 595 p. See pp. 30, 241 for relativistic effects.

[9373] Read J.P. and Buckingham A.D. (1997). Covalency in $ArAu^+$ and related species? *J. Am. Chem. Soc.*, **119**, 9010–9013.

[9374] Reed K.J. and Chen M.H. (1993). Relativistic effects on the polarization of line radiation emitted from He-like and H-like ions following electron-impact excitation. *Phys. Rev. A*, **48**, 3644–3651.

[9375] Reginatto M. (1998). Derivation of the Pauli equation using the principle of minimum Fisher information. *Phys. Lett. A*, **249**, 355–357.

[9376] Reiher M. (1998). *Development and Implementation of Numerical Algorithms for the Solution of Multi-Configuration Self-Consistent Field Equations for Relativistic Atomic Structure Calculations*. Ph.D. thesis, Bielefeld, 190 p.

[9377] Reiher M. and Hinze J. (1999). Self-consistent treatment of the frequency-independent Breit interaction in Dirac-Fock and MCSCF calculations of atomic structures: I. Theoretical considerations. *J. Phys. B*, **32**, 5489–5505.

[9378] Reindl S.L. and Pastor G.M. (1993). Many-body approach to the electronic properties of clusters: Application to doubly charged Pb clusters. *Phys. Rev. B*, **47**, 4680–4690.

[9379] Reinisch H. and Bross H. (1993). Relativistic effects on the Compton profile of polycrystalline gold. *J. Phys: CM.*, **5**, 977–990.

[9380] Reinisch H. and Bross H. (1994). Relativistic density functional calculation of the total energy and Fermi surface of gold. *Z. Phys. D*, **95**, 145–150.

[9381] Reiss H.R. (1998). Relativistic effects in strong electromagnetic fields (Introduction to Focus Issue). *Optics Express*, **2**, 261.

[9382] Rémita S., Archirel P. and Mostafavi M. (1995). Evaluation of the redox potential of $Ag_1^I(CN)_2^-/Ag_1^0(CN)_2^{2-}$ in aqueous solutions. *J. Phys. Chem.*, **99**, 13198–13202.

[9383] Ren J., Whangbo M.H., Dai D. and Li L. (1998). Description of ligand field splitting in terms of density functional theory: split levels of the lowest-lying subterms of the $4f^{n-1}6s^2$ (n=3 - 14) configurations in lanthanide monofluorides LnF (Ln=Pr-Yb). *J. Chem. Phys.*, **108**, 8479–8484.

[9384] Renkema K.B., Bosque R., Streib W.E., Maseras F., Eisenstein O. and Caulton K.G. (1999). Phosphine dissociation mediates C-H cleavage of fluoroarenes by $OsH(C_6H_5)(CO)(P^tBu_2Me)_2$. *J. Am. Chem. Soc.*, **121**, 10895–10907.

[9385] Reynolds G.G. and Carter E.A. (1994). Bimetallic thermochemistry: perturbations in M-H and M-C bonds due to the presence of M'. *J. Phys. Chem.*, **98**, 8144–8153.

[9386] Rez D., Rez P. and Grant I. (1994). Dirac-Fock calculations of X-ray scattering factors and contributions to the mean inner potential for electron cattering. *Acta Cryst.*, **A50**, 481–497.

[9387] Riad Manaa M. (1995). The fragmentation of SH(A $^2\Sigma^+$): ab initio calculations of spin-orbit and Coriolis interactions. *Int. J. Quantum Chem.*, **S29**, 577–584.

[9388] Riad Manaa M. (1999). Photodissociation of NaK: Ab initio spin-orbit interactions of the Na (3 ^2S), K (4 ^2P_J) manifold. *Int. J. Quantum Chem.*, **75**, 693–697.

[9389] Ribbing C. (1992). *Spin-orbit coupling in transition metal systems. A study of octahedral Ni(II)..* Ph.D. thesis, University of Stockholm.

[9390] Ribbing C. and Daniel C. (1994). Spin-orbit coupled excited states in transition metal complexes: A configuration interaction treatment of $HCo(CO)_4$. *J. Chem. Phys.*, **100**, 6591–6596.

[9391] Ribbing C., Gilliams B., Pierloot K., Roos B.O. and Karlström G. (1998). The optical absorption spectrum of the octahedral $RhCl_6^{3-}$ complex: Ab initio calculations of excitation energies and the effect of spin-orbit coupling. *J. Chem. Phys.*, **109**, 3145–3152.

[9392] Ribbing C. and Odelius M. (1993). Normal coordinate analysis of the zero-field splitting in octahedral NiF_6^{4-}. I. Ab initio calculations. *Mol. Phys.*, **78**, 1259–1266.

[9393] Ricca A. and Bauschlicher C.W. (1998). Accurate D_0 values for SiF and SiF^+. *Chem. Phys. Lett.*, **287**, 239–242.

[9394] Ricca A. and Bauschlicher Jr C.W. (1999). Heats of formation for GeH_n (n = 1-4) and Ge_2H_n (n = 1-6). *J. Phys. Chem. A*, **103**, 11121–11125.

[9395] Ricca A., Bauschlicher C.W. and Rosi M. (1994). Second-order Møller-Plesset perturbation theory for systems involving first transition row metals. *J. Phys. Chem.*, **98**, 9498–9502.

[9396] Richardson N.A., Rienstra-Kiracofe J.C. and Schaefer III H.F. (1999a). Examining trends in the tetravalent character of group 14 elements (C, Si, Ge, Sn, Pb) with acids and hydroperoxides. *J. Am. Chem. Soc.*, **121**, 10813–10819.

[9397] Richardson N.A., Rienstra-Kiracofe J.C. and Schaefer III H.F. (1999b). Examination of the stabilities of group 14 (C, Si, Ge, Sn, Pb) congeners of dihydroxycarbene and dioxirane. Comparison to formic acid and hydroperoxycarbene congeners. *Inorg. Chem.*, **38**, 6271–6277.

[9398] Richter M., Oppeneer P.M., Eschrig H. and Johansson B. (1992). Calculated crystal-field parameters of $SmCo_5$. *Phys. Rev. B*, **46**, 13919–13927.

[9399] Richter M., Steinbeck L., Nitzsche U., Oppeneer P.M. and Eschrig H. (1995). On the spatial origin of crystal electric fields in $SmCo_5$. *J. Alloys Compounds*, **225**, 469–473.

[9400] Richter M., Zahn P., Diviš M. and Mertig I. (1996). Giant magnetoresistance in uranium intermetallics: Ab initio calculations for U_2Pd_2In and U_2Pd_2Sn. *Phys. Rev. B*, **54**, 11985–11988.

[9401] Rieger M.M. and Vogl P. (1995). Relativistic self-interaction-free density-functional formalism. *Phys. Rev. A*, **52**, 282–290.

[9402] Riehl J.F., Koga N. and Morokuma K. (1994). Hydride exchange reaction in trimetallic clusters. An ab initio molecular orbital study of $M_3(CO)_9(\mu\text{-H})_3(\mu_3\text{-CH})$ (M=Os, Ru). *J. Am. Chem. Soc.*, **116**, 5414–5424.

[9403] Ring P. (1997). Computer program for the relativistic mean field description of the ground state properties of even-even axially deformed nuclei. *Comp. Phys. Comm.*, **105**, 77–97.

[9404] Ritchie B. and Weatherford C.A. (1999). Quantum classical correspondence in nonrelativistic electrodynamics. *Int. J. Quantum Chem.*, **75**, 655–658.

[9405] Ritze H.H. and Radloff W. (1996). Ab initio study of $AgNH_3$ and its cation. *Chem. Phys. Lett.*, **250**, 415–420.

[9406] Rivas M. (1994). Quantization of generalization spinning particles: New derivation of Dirac's equation. *J. Math. Phys.*, **35**, 3380–3399.

[9407] Rivas-Silva J.F. and Berrondo M. (1996). Emission lifetime calculation of thallium-doped alkali halides. *J. Phys. Chem. Solids*, **57**, 1705–1707.

[9408] Rivelles V.O. (1995). Comment on "Nonlocal symmetry for QED" and "Relativistically covariant symmetry in QED". *Phys. Rev. Lett.*, **75**, 4510.

[9409] Robert V., Petit S. and Borshch S.A. (1999). Electron distribution in $Pt_2(dta)_4I$: From building blocks to the infinite chain. A theoretical investigation of possible Peierls distorsions. *Inorg. Chem.*, **38**, 1573–1578.

[9410] Robinson F.N.H. (1995). *An Introduction to Special Relativity and its Applications*. World Scientific, Singapore, 183 p.

[9411] Robson B.A. and Staudte D.S. (1996). An eight-component relativistic wave equation for spin-1/2 particles. I. *J. Phys. A*, **29**, 157–167.

[9412] Rocchi C. and Sacchetti F. (1995). Radiative corrections to the Compton cross section. *Phys. Rev. B*, **51**, 81–85.

[9413] Rocha W.R. and De Almeida W.B. (1997). Reaction path for the insertion reaction of $SnCl_2$ into the Pt-Cl bond: An ab initio study. *Int. J. Quantum Chem.*, **65**, 643–650.

[9414] Rocha W.R. and De Almeida W.B. (1998). Theoretical study of the olefin insertion reaction in the heterobimetallic $Pt(H)(PH_3)_2(SnCl_3)(C_2H_4)$ compound. *Organometallics*, **17**, 1961–1967.

[9415] Rochefort A. and Fournier R. (1996). Quantum chemical study of CO and NO bonding to Pd_2, Cu_2, and PdCu. *J. Phys. Chem.*, **100**, 13506–13513.

[9416] Rodrigues Jr W.A., Vaz Jr J., Recami E. and Salesi G. (1993). About zitterbewegung and electron structure. *Phys. Lett. B*, **318**, 623–628.

[9417] Rodriguez J.A. and Kuhn M. (1994). Electronic and chemical properties of Ag/Pt(111) and Cu/Pt(111) surfaces: importance of changes in the d electron populations. *J. Phys. Chem.*, **98**, 11251–11255.

[9418] Rodriguez J.A. and Kuhn M. (1995). Electronic properties of Pt in bimetallic systems: photoemission and molecular-orbital studies for Pt-Al surface alloys. *Chem. Phys. Lett.*, **240**, 435–441.

[9419] Rodriguez J.A., Kuhn M. and Hrbek J. (1996). The bonding of sulfur to a Pt(111) surface: photoemission and molecular orbital studies. *Chem. Phys. Lett.*, **251**, 13–19.

[9420] Rodriguez-Fortea A., Alemany P. and Ziegler T. (1999). Density functional calculations of NMR chemical shifts with the inclusion of spin-orbit coupling in tungsten and lead compounds. *J. Phys. Chem. A*, **103**, 8288–8294.

[9421] Rodríguez-Santiago L., Branchadell V. and Sodupe M. (1995). Theoretical study of the bonding of NO_2 to Cu and Ag. *J. Chem. Phys.*, **103**, 9738–9743.

[9422] Rodriquez C.F., Cunje A. and Hopkinson A.C. (1998). Protonation of group 14 oxides: the proton affinity of SnO. *J. Mol. Str. (Theochem)*, **430**, 149–159.

[9423] Rohse R., Klopper W. and Kutzelnigg W. (1993). Configuration interaction calculations with terms linear in the interelectronic coordinate for the ground state of H_3^+. A benchmark study. *J. Chem. Phys.*, **99**, 8830–8839.

[9424] Rolke, J, Zheng Y., Brion C.E., Chakravorty S.J., Davidson E.R. and McCarthy I.E. (1997). Imaging of the HOMO electron density $Cr(CO)_6$, $Mo(CO)_6$ and $W(CO)_6$ by electron momentum spectroscopy: a comparison with Hartree-Fock and DFT calculations. *Chem. Phys.*, **215**, 191–205.

[9425] Romanov D.A. (1993). Origin of linear terms in a quasi-two-dimensional dispersion law. *Fiz. Tverd. Tela.*, **35**, 1421–1426.

[9426] Romanov D.A. (1993). Origin of linear terms in a quasi two-dimensional dispersion law. *Phys. Solid State*, **35**, 717–719.

[9427] Romero R.H. and Aucar G.A. (1998). Relativistic correction of the generalized oscillator strength sum rules. *Phys. Rev. A*, **57**, 2212–2215.

[9428] Romero R.H. and Aucar G.A. (1999). Reply to "Comment on 'Relativistic correction of the generalized oscillator strength sum rules'". *Phys. Rev. A*, **59**, 4849.

[9429] Ron A., Goldberg I.B., Stein J., Manson S.T., Pratt R.H. and Yin R.Y. (1994). Relativistic retardation and multipole effects in photoionization cross sections: Z, n, and l dependence. *Phys. Rev. A*, **50**, 1312–1320.

[9430] Roos B.O. and Andersson K. (1995). Multiconfigurational perturbation theory with level shift - the Cr_2 potential revisited. *Chem. Phys. Lett.*, **245**, 215–223.

[9431] Roos B.O., Andersson K., Fulscher M., Malmqvist P.Å., Serrano-Andrés L., Pierloot K. and Merchán M. (1996). Multiconfigurational perturbation theory: Applications in electron spectroscopy. *Adv. Chem. Phys.*, **93**, 219–331.

[9432] Rosa A., Baerends E.J., van Gisbergen S.J.A., van Lenthe E., Groeneveld J.A. and Snijders J.G. (1999). Electronic spectra of $M(CO)_6$ (M = Cr, Mo, W) revisited by a relativistic TDDFT approach. *J. Am. Chem. Soc.*, **121**, 10356–10365.

[9433] Rosa A., Ricciardi G. and Baerends E.J. (1998). Structural properties of $M(dmit)_2$-based (M=Ni,Pd,Pt; $dmit^{2-}$ = 2-thioxo-1,3-dithiole-4,5-dithiolato) molecular metals. Insights from density functional calculations. *Inorg. Chem.*, **37**, 1368–1379.

[9434] Rosberg M. and Wyart J.F. (1997). The spectrum of singly ionized gold, Au II. *Phys. Scr.*, **55**, 690–706.

[9435] Rösch N., Häberlen O.D. and Dunlap B.I. (1993). Bindungsverhältnisse in endohedralen Metall-Fulleren-Komplexen: f-Orbital-Kovalenz in $Ce@C_{28}$. *Angew. Chem.*, **105**, 78–81.

[9436] Rösch N., Krüger S., Mayer M. and Nasluzov V.A. (1996). The Douglas-Kroll-Hess approach to relativistic density functional theory: methodological aspects and applications to metal complexes and clusters. In *Recent Developments and Applications of Modern Density Functional Theory*, (Edited by Seminario J.M.), pp. 497–566, Elsevier, Amsterdam.

[9437] Rościszewski K., Paulus B., Fulde P. and Stoll H. (1999). *Ab initio* calculation of ground-state properties of rare-gas crystals. *Phys. Rev. B*, **60**, 7905–7910.

[9438] Rosenberg L. (1993). Extremum principles for relativistic atomic structure and scattering: Two-electron ions. *Phys. Rev. A*, **47**, 1771–1777.

[9439] Rosenberg L. (1994a). Infrared radiation by a Dirac electron: First-order correction to the cross-section sum rule. *Phys. Rev. A*, **49**, 4770–4777.

[9440] Rosenberg L. (1994b). Minimum principle for Dirac scattering lengths. *Phys. Rev. A*, **50**, 371–377.

[9441] Rosenberg L. (1995). Effective-potential method for relativistic electron-ion scattering. *Phys. Rev. A*, **51**, 3703–3711.

[9442] Rosenberg L. and Zhou F. (1993). Generalized Volkov wave functions: Application to laser-assisted scattering. *Phys. Rev. A*, **47**, 2146–2155.

[9443] Ross R.B., Gayen S. and Ermler W.C. (1994). Ab initio relativistic effective potentials with spin-orbit operators. V. Ce through Lu. *J. Chem. Phys.*, **100**, 8145–8155.

[9444] Roszak S. and Balasubramanian K. (1993a). Potential energy curves for Pt-CO interactions. *J. Phys. Chem.*, **97**, 11238–11241.

[9445] Roszak S. and Balasubramanian K. (1993b). First-principle interatomic potentials for the carbon monoxide-platinum reaction. *Chem. Phys. Lett.*, **212**, 150–154.

[9446] Roszak S. and Balasubramanian K. (1994a). Organometallic analogues of the Diels-Alder reaction: molybdenum dimer + ethylene; molybdenum dimer + butadiene. *Inorg. Chem.*, **33**, 4169–4172.

[9447] Roszak S. and Balasubramanian K. (1994b). Theoretical study of the interaction of benzene with platinum atom and cation. *Chem. Phys. Lett.*, **234**, 101–106.

[9448] Roszak S. and Balasubramanian K. (1995a). Theoretical study of the Diels-Alder reactions of zirconium dimer with ethylene and butadiene. *Chem. Phys. Lett.*, **99**, 3487–3492.

[9449] Roszak S. and Balasubramanian K. (1995b). A theoretical study of bridged vs atop interactions of Pt_2 with CO. *J. Chem. Phys.*, **103**, 1043–1049.

[9450] Roszak S. and Balasubramanian K. (1995c). Theoretical study of structure and thermodynamic properties of YC_2. *Chem. Phys. Lett.*, **246**, 20–25.

[9451] Roszak S. and Balasubramanian K. (1996a). Theoretical study of structural and thermodynamic properties of yttrium carbides, YC_n (n = 2-6). *J. Phys. Chem.*, **100**, 8254–8259.

[9452] Roszak S. and Balasubramanian K. (1996b). Electronic structure and thermodynamic properties of YIrC and $YIrC_2$. *Chem. Phys. Lett.*, **254**, 274–280.

[9453] Roszak S. and Balasubramanian K. (1996c). Electronic structure and thermodynamic properties of LaC_2. *J. Phys. Chem.*, **100**, 11255–11259.

[9454] Roszak S. and Balasubramanian K. (1997a). Stabilities of isomers of LaC_{12}^+ and LaC_{13}^+. *Chem. Phys. Lett.*, **264**, 80–84.

[9455] Roszak S. and Balasubramanian K. (1997b). Theoretical investigation of structural and thermodynamic properties of lanthanum carbides LaC_n (n=2-6). *J. Chem. Phys.*, **106**, 158–164.

[9456] Roszak S. and Balasubramanian K. (1997c). Theoretical study of the isomerization of TaC_n^+ (n = 7-13) cations. *Chem. Phys. Lett.*, **265**, 553–560.

[9457] Roszak S. and Balasubramanian K. (1997d). Potential energy surfaces for the $Ta^+ + C_2$ reaction. *J. Chem. Phys.*, **106**, 4008–4012.

[9458] Roszak S., Kaufman J.J., Koski W.S., Vijayakumar M. and Balasubramanian K. (1994). Potential energy curves of ground and excited states of tetra halomethanes and the negative ions. *J. Chem. Phys.*, **101**, 2978–2985.

[9459] Roszak S., Koski W.S., Kaufman J.J. and Balasubramanian K. (1997). Structure and energetics of CF_3Cl^-, CF_3Br^-, and CF_3I^- radical anions. *J. Chem. Phys.*, **106**, 7709–7713.

[9460] Roszak S., Majumdara D. and Balasubramanian K. (1999). Theoretical studies of structures and energetics of benzene complexes with Nb^+ and Nb_2^+ cations. *J. Phys. Chem. A*, **103**, 5801–5806.

[9461] Roszak S., Vijayakumar M., Balasubramanian K. and Koski W.S. (1993). A multireference configuration interaction study of photoelectron spectra of carbon tetrahalides. *Chem. Phys. Lett.*, **208**, 225–231.

[9462] Rotenberg E., Chung J.W. and Kevan S.D. (1999). Spin-orbit coupling induced surface band splitting in Li/W(110) and Li/Mo(110). *Phys. Rev. Lett.*, **82**, 4066–4069.

[9463] Rousseau R., Dietrich G., Krückeberg S., Lützenkirchen K., Marx D., Schweikhard L. and Walther C. (1998). Probing cluster structures with sensor molecules: methanol adsorbed onto gold clusters. *Chem. Phys. Lett.*, **295**, 41–46.

[9464] Roy C.L. (1993a). Boundary conditions across a δ-function potential in the one-dimensional Dirac equation. *Phys. Rev. A*, **47**, 3417–3419.

[9465] Roy C.L. (1993b). Some typical features of relativistic tunneling through a double barrier system with delta-function potentials. *Phys. Stat. Sol. B*, **176**, 109–115.

[9466] Roy C.L. (1993c). Some typical features of relativistic boundary conditions across delta-function potential. *Indian J. Pure Appl. Phys.*, **31**, 610–615.

[9467] Roy C.L. (1994). Some special features of relativistic tunneling through multi-barrier systems with δ-function barriers. *Phys. Lett. A*, **189**, 345–350.

[9468] Roy C.L. (1996). Relativistic effects in Landauer resistance of Fibonacci lattice. *J. Phys. Chem. Solids*, **57**, 1825–1830.

[9469] Roy C.L. and Basu C. (1993a). A study of relativistic electrical conduction in disordered systems. *Indian J. Phys. A*, **67**, 99–110.

[9470] Roy C.L. and Basu C. (1993b). Relativistic impacts on electrical resistance of a one-dimensional finite crystal. *Phys. Stat. Sol. B*, **177**, 315–323.

[9471] Roy C.L. and Khan A. (1993a). Relativistic effects on impurity states. *Indian J. Pure Appl. Phys.*, **31**, 303–310.

[9472] Roy C.L. and Khan A. (1993b). Relativistic impacts on tunnelling through multi-barrier systems. *J. Phys: CM*, **5**, 7701–7708.

[9473] Roy C.L., Mendez B. and Dominguez-Adame F. (1994). A relativistic equation for a slowly varying potential. *J. Phys. A*, **27**, 3539–3546.

[9474] Roy D.K. and Singh A. (1994). Relativistic considerations of quantum mechanical tunnel effect. *Indian J. Phys. B*, **68**, 175–184.

[9475] Roychoudhury R. and Panchanan S. (1993). Modified $1/N$ expansion for the Dirac equation for screened Coulomb potential. *Z. Phys. A*, **48**, 1081–1085.

[9476] Rozmej P. and Arvieu R. (1999). The Dirac oscillator. A relativistic version of the Jaynes-Cummings model. *J. Phys. A*, **32**, 5367–5382.

[9477] Rozmej P., Berej W. and Arvieu R. (1997). New mechanism of collapse and revival in wave-packet dynamics due to spin-orbit interaction. *Acta Phys. Pol. B*, **28**, 243–255.

[9478] Rubio J., Zurita S., Barthelat J.C. and Illas F. (1994). Electronic and geometrical structures of Pt_3 and Pt_4. An ab initio one-electron proposal. *Chem. Phys. Lett.*, **217**, 283–287.

[9479] Rudzikas Z. (1997). *Theoretical Atomic Spectroscopy*. Cambridge Univ. Press., 448 p.

[9480] Ruijgrok T.W. (1998). General requirements for a relativistic quantum theory. *Few-Body Syst.*, **25**, 5–27.

[9481] Ruiz E. and Payne M.C. (1998). One-dimensional intercalation compound $2HgS{\cdot}SnBr_2$: Ab initio electronic structure calculations and molecular dynamics simulations. *Chem. Eur. J.*, **4**, 2485–2492.

[9482] Ruiz-Morales Y., Schreckenbach G. and Ziegler T. (1996). Theoretical study of ^{13}C and ^{17}O NMR shielding tensors in transition metal carbonyls based on density functional theory and gauge-including atomic orbitals. *J. Phys. Chem.*, **100**, 3359–3367.

[9483] Ruiz-Morales Y., Schreckenbach G. and Ziegler T. (1997). Calculation of ^{125}Te chemical shifts using gauge-including atomic orbitals and density functional theory. *J. Phys. Chem. A*, **101**, 4121–4127.

[9484] Rumrich K., Momberger K., Soff G., Greiner W., Grun N. and Scheid W. (1991). Nonperturbative character of electron-positron pair production in relativistic heavy-ion collisions. *Phys. Rev. Lett.*, **66**, 2613–2616.

[9485] Rumrich K., Soff G. and Greiner W. (1993). Ionization and pair creation in relativistic heavy-ion collisions. *Phys. Rev. A*, **47**, 215–228.

[9486] Runeberg N. and Pyykkö P. (1998). Relativistic pseudopotential calculations on Xe_2, RnXe, and Rn_2: The van der Waals properties of radon. *Int. J. Quantum Chem.*, **66**, 131–140.

[9487] Runeberg N., Schütz M. and Werner H.J. (1999). The aurophilic attraction as interpreted by local correlation methods. *J. Chem. Phys.*, **110**, 7210–7215.

[9488] Runeberg N., Seth M. and Pyykkö P. (1995). Calculated properties of XeH_2. *Chem. Phys. Lett.*, **246**, 239–244.

[9489] Russo T.V., Martin R.L., Hay P.J. and Rappé A.K. (1995). Vibrational frequencies of transition metal chloride and oxo compounds using effective core potential analytic second derivatives. *J. Chem. Phys.*, **102**, 9315–9321.

[9490] Rutkowski A. (1996). Regular perturbation theory of relativistic corrections: Basic aspects. *Phys. Rev. A*, **53**, 145–151.

[9491] Rutkowski A. (1999). Iterative solution of the one-electron Dirac equation based on the Bloch equation of the 'direct perturbation theory'. *Chem. Phys. Lett.*, **307**, 259–264.

[9492] Rutkowski A. and Kozlowski R. (1997). Double-perturbation approach to the relativistic hydrogenic atom in a static and uniform magnetic field. *J. Phys. B*, **30**, 1437–1448.

[9493] Rutkowski A. and Schwarz W.H.E. (1996). Effective Hamiltonian for near-degenerate states in direct relativistic perturbation theory. I. Formalism. *J. Chem. Phys.*, **104**, 8546–8552.

[9494] Rutkowski A., Schwarz W.H.E. and Koślowski R. (1993). Relativistic virial theorem for diatomic molecules. Application to H_2^+. *Theor. Chim. Acta*, **87**, 75–87.

[9495] Rutkowski A., Schwarz W.H.E., Kozłowski R., Bęczek J. and Franke R. (1998). Effective Hamiltonian for near-degenerate states in relativistic direct perturbation theory. II. H_2^+-like systems. *J. Chem. Phys.*, **109**, 2135–2143.

[9496] Ruud K., Schimmelpfennig B. and Ågren H. (1999). Internal and external heavy-atom effects on phosphorescence radiative lifetimes calculated using a mean-field spin-orbit Hamiltonian. *Chem. Phys. Lett.*, **310**, 215–221.

[9497] Ryder L. (1988). *Quantum Field Theory*. Cambridge U. P., Cambridge, see Ch. 2.

[9498] Ryder L. (1998). Relativistic treatment of inertial spin effects. *J. Phys. A*, **31**, 2465–2469.

[9499] Saalfrank P. (1992). Quantum size effects in thin lead films. *Surf. Sci.*, **274**, 449–456.

[9500] Sadlej A.J. and Snijders J.G. (1994). Spin separation in the regular Hamiltonian approach to solutions of the Dirac equation. *Chem. Phys. Lett.*, **229**, 435–438.

[9501] Sadlej A.J., Snijders J.G., van Lenthe E. and Baerends E.J. (1995). Four component regular relativistic Hamiltonians and the perturbational treatment of Dirac's equation. *J. Chem. Phys.*, **102**, 1758–1766.

[9502] Safonov A.A. and Bagatur'yants A.A. (1991). Use of the silver nonlocal pseudopotential in molecular calculations. *Zh. Fiz. Khim.*, **65**, 712–715.

[9503] Safonov V.L. (1993). Electron in the anisotropic space. *Phys. Stat. Sol. B*, **176**, K55–K57.

[9504] Safronova M.S., Derevianko A. and Johnson W.R. (1998). Relativistic many-body calculations of energy levels, hyperfine constants, and transition rates for sodiumlike ions, Z = 11-16. *Phys. Rev. A*, **58**, 1016–1028.

[9505] Safronova M.S., Johnson W.R. and Derevianko A. (1999). Relativistic many-body calculations of energy levels, hyperfine constants, electric-dipole matrix elements, and static polarizabilities for alkali-metal atoms. *Phys. Rev. A*, **60**, 4476–4487.

[9506] Safronova M.S., Johnson W.R. and Safronova U.I. (1996a). Relativistic many-body calculations of the energies of n=2 states for the berylliumlike isoelectronic sequence. *Phys. Rev. A*, **53**, 4036–4053.

[9507] Safronova M.S., Johnson W.R. and Safronova U.I. (1996b). Relativistic many-body calculations of energies of n=2 states for boronlike ions. *Phys. Rev. A*, **54**, 2850–2862.

[9508] Safronova M.S., Johnson W.R. and Safronova U.I. (1997a). Relativistic many-body calculations of energies of n=3 states of Be-like ions. *J. Phys. B*, **30**, 2375–2393.

[9509] Safronova M.S., Johnson W.R. and Safronova U.I. (1997b). Relativistic many-body calculations of energies of n=3 states of beryllium isoelectronic sequence. *Phys. Scr.*, **T73**, 48–49.

[9510] Safronova M.S., Safronova U.I. and Bruch R. (1994). Z-dependencies of the energy levels of autoionization states for Be-like ions. *Phys. Lett. A*, **194**, 106–112.

[9511] Safronova U.I. (1994). On the calculation of the energies of the $1s^22s^22p^5nl$ and $1s^22s2p^6nl$ (n=3-6, $l = s,p,d,f$) states in Ne-like ions with Z=20-60. Correlative, relativistic and radiative effects. *Opt. Spektr.*, **76**, 183–195.

[9512] Safronova U.I. (1994). On the calculation of the energies of the $1s^22s^22p^5nl$ and $1s^22s2p^6nl$ (n=3-6, $l = s,p,d,f$) states in Ne-like ions with Z=20-60. Correlative, relativistic and radiative effects. *Opt. Spectr.*, **76**, 161–173.

[9513] Safronova U.I. and Bruch R. (1994). Transition and Auger energies of Li-like ions ($1s2lnl'$ configurations). *Phys. Scr.*, **50**, 45–54.

[9514] Safronova U.I., Cornille M. and Dubau J. (1994a). Energy of atomic system for states $1s2s^22p^n$ and $1s^22s^22p^n$ as function of Z. Comparison of two methods: SUPERSTRUCTURE and MZ. *Phys. Scr.*, **49**, 69–79.

[9515] Safronova U.I., Derevianko A., Safronova M.S. and Johnson W.R. (1999a). Relativistic many-body calculations of transition probabilities for the $2l_12l_2[LSJ] - 2l_33l_4[L'S'J']$ lines in Be-like ions. *J. Phys. B*, **32**, 3527–3545.

[9516] Safronova U.I. and Johnson W.R. (1998). Autoionizing rates for doubly excited $2lnl'$ states of He-like ions. *Phys. Scr.*, **58**, 116–125.

[9517] Safronova U.I., Johnson W.R. and Derevianko A. (1999b). Relativistic many-body calculations of magnetic dipole transitions in Be-like ions. *Phys. Scr.*, **60**, 46–53.

[9518] Safronova U.I., Johnson W.R. and Livingston A.E. (1999c). Relativistic many-body calculations of electric-dipole transitions between $n = 2$ states in B-like ions. *Phys. Rev. A*, **60**, 996–1004.

[9519] Safronova U.I., Johnson W.R. and Safronova M.S. (1998a). Relativistic many-body calculations of energies of $n = 3$ states for the boron isoelectronic sequence, Z = 6-30. *At. Data Nucl. Data Tables*, **69**, 183–215.

[9520] Safronova U.I., Johnson W.R., Safronova M.S. and Derevianko A. (1999d). Relativistic many-body calculations of transition probabilities for the $2l_12l_2[LSJ] - 2l_3l_4[L'S'J']$ lines in Be-like ions. *Phys. Scripta*, **59**, 286–295.

[9521] Safronova U.I. and Nilsen J. (1994). Autoionization states of lithium-like ions with large principal quantum number. *J. Quant. Spectr. Radiat. Transfer*, **51**, 853–874.

[9522] Safronova U.I., Safronova M.S. and Bruch R. (1994b). Correlation, relativistic and radiative effects for the energy levels of $1s^22s^22p^5nl$, $1s^22s2p^5nl$ (n=3-6, $l = s,p,d,f$) configurations of Ne-like ions with Z=20-60. *Phys. Scr.*, **49**, 446–462.

[9523] Safronova U.I., Safronova M.S. and Bruch R. (1995). Relative intensity of dielectronic satellite spectra for highly charged He-like ions ($1s2$"nl - $1s^2n'l'$, n, n' = 2, 3) with Z = 6-54. *J. Phys. B*, **28**, 2803–2816.

[9524] Safronova U.I., Safronova M.S., Snyderman N.J. and Pal'chikov V.G. (1994c). Relativistic perturbation theory calculation of two-electron doubly excited states. *Phys. Scr.*, **50**, 29–44.

[9525] Safronova U.I. and Shlyaptseva A.S. (1999). Inner-shell excitation energies and autoionization rates for C-, N-, O-, and F-like ions with Z = 6-54. *Phys. Scr.*, **60**, 36–45.

[9526] Safronova U.I., Shlyaptseva A.S., Cornille M. and Dubau J. (1998b). Autoionization rates for $1s2s^22p^2$, $1s2s2p^3$, $1s2p^4$ states of B-like ions ($6\leq Z\leq 54$) for decay via different channels. *Phys. Scr.*, **57**, 395–409.

[9527] Safronova U.I., Shlyaptseva A.S. and Golovkin I.E. (1997). New atomic data for autoionizing states of Li-like ions with high values of n. *Phys. Scr.*, **T73**, 50–52.

[9528] Safronova U.I., Tolstikhina I.Y., Bruch R., Tanaka T., Hao F. and Schneider D. (1993). Screening theory for transition energies of highly charged ions. *Phys. Scr.*, **47**, 364–382.

[9529] Sághi-Szabo G., Cohen R.E. and Krakauer H. (1998). First-principles study of piezoelectricity in $PbTiO_3$. *Phys. Rev. Lett.*, **80**, 4321–4324.

[9530] Sakai Y., Miyoshi E., Klobukowski M. and Huzinaga S. (1997). Model potentials for main group elements Li through Rn. *J. Chem. Phys.*, **106**, 8084–8092.

[9531] Sakai Y., Miyoshi E. and Tatewaki H. (1998). Model core potentials for the lanthanides. *J. Mol. Str. (Theochem)*, **451**, 143–150.

[9532] Sakai Y., Mogi K. and Miyoshi E. (1999). Theoretical study of low-lying electronic states of TiCl and ZrCl. *J. Chem. Phys.*, **111**, 3989–3994.

[9533] Sakaki S., Kai S. and Sugimoto M. (1999). Theoretical study on σ-bond activation of $(HO)_2B$-XH_3 by $M(PH_3)_2$ (X = C, Si, Ge, or Sn; M = Pd or Pt). Noteworthy contribution of the boryl p_π orbital to M-boryl bonding and activation of the B-X σ-bond. *Organometallics*, **18**, 4825–4837.

[9534] Sakaki S. and Kikuno T. (1997). Reaction of BX_2-BX_2 (X = H or OH) with $M(PH_3)_2$ (M = Pd or Pt). A theoretical study of the characteristic features. *Inorg. Chem.*, **36**, 226–229.

[9535] Sakaki S., Mizoe N. and Sugimoto M. (1998). Theoretical study of platinum(0)-catalyzed hydrosilylation of ethylene. Chalk-Harrod mechanism or modified Chalk-Harrod mechanism. *Organometallics*, **17**, 2510–2523.

[9536] Sakaki S. and Musashi Y. (1996). A theoretical study on CO_2 insertion into an M-H bond (M = Rh and Cu). *Int. J. Quantum Chem.*, **57**, 481–491.

[9537] Sakaki S., Ogawa M. and Kinoshita M. (1995a). A theoretical study on the oxidative addition of an Si-X bond (X = H or Si) to $M(PH_3)_2$ (M = Pd or Pt). A comparison of the reactivity between $Pt(PH_3)_2$ and $Pd(PH_3)_2$. *J. Phys. Chem.*, **99**, 9933–9939.

[9538] Sakaki S., Ogawa M. and Musashi Y. (1995b). A theoretical study on the bond energy and the bonding nature of dinuclear d^{10} metal complexes: $Pt_2(PH_3)_4$, $PtPd(PH_3)_4$, and $Pd_2(PH_3)_4$. *J. Phys. Chem.*, **99**, 17134–17138.

[9539] Sakaki S., Satoh H., Shono H. and Ujino Y. (1996). Ab initio MO study of the geometry, $\eta^3 \rightleftharpoons \eta^1$ conversion and reductive elimination of a palladium(II) η^3-allyl hydride complex and its platinum(II) analogue. *Organometallics*, **15**, 1713–1720.

[9540] Sakaki S., Takeuchi K., Sugimoto M. and Kurosawa H. (1997). Geometries, bonding nature, and relative stabilities of dinuclear palladium(I) π-allyl and mononuclear palladium(II) π-allyl complexes. A theoretical study. *Organometallics*, **16**, 2995–3003.

[9541] Sakamoto J. (1993). Construction of N body bound state solution of the Dirac particles in 1+1 dimensional space-time. *Progr. Theor. Phys.*, **89**, 119–130.

[9542] Salamin Y.I. (1993). On the Dirac equation with anomalous magnetic moment term and a plane electromagnetic field. *J. Phys. A*, **26**, 6067–6071.

[9543] Salamin Y.I. (1994). Evaluation of the diagonal matrix elements $<nl|r^\beta|nl>$ for arbitrary β and fixed l with relativistic hydrogenic functions. *Phys. Scr.*, **51**, 137–140.

[9544] Salcedo L.L. and Ruiz Arriola E. (1996). Wigner transformation for the determination of Dirac operators. *Ann. Phys. (NY)*, **250**, 1–50.

[9545] Salomonson S., Warston H. and Lindgren I. (1996). Many-body calculations of the electron affinity for Ca and Sr. *Phys. Rev. Lett.*, **76**, 3092–3095.

[9546] Salvat F. and Mayol R. (1993). Elastic scattering of electrons and positrons by atoms. Schrödinger and Dirac partial wave analysis. *Comp. Phys. Comm.*, **74**, 358–374.

[9547] Salzner U. and v R Schleyer P. (1993). Generalized anomeric effects and hyperconjugation in $CH_2(OH)_2$, $CH_2(SH)_2$, $CH_2(SeH)_2$ and $CH_2(TeH)_2$. *J. Am. Chem. Soc.*, **115**, 10231–10236.

[9548] Sampson D.H. and Zhang H.L. (1997). Collision strengths for hyperfine-structure transitions: application to $^{57}Fe^{23+}$. *J. Phys. B*, **30**, 1449–1456.

[9549] Samuni U., Kahana S., Fraenkel R., Haas Y., Danovich D. and Shaik S. (1994). The ICN-INC system: experiment and quantum chemical calculations. *Chem. Phys. Lett.*, **225**, 391–397.

[9550] Sanchez A., Abbet S., Heiz U., Schneider W.D., Häkkinen H., Barnett R.N. and Landman U. (1999). When gold is not noble: Nanoscale gold catalysis. *J. Phys. Chem. A*, **103**, 9573–9578.

[9551] Sandars P.G.H. (1993). P and/or T violation. *Phys. Scr.*, **T46**, 16–21.

[9552] Sander W. and Kötting C. (1999). Reactions of difluorovinylidene - a super-electrophilic carbene. *Chem. Eur. J.*, **5**, 24–28.

[9553] Sändig N. and Koch W. (1997). Mechanism of the Ta^+-mediated activation of the C-H bond in methane. *Organometallics*, **16**, 5244–5251.

[9554] Sandratskii L.M. and Kübler J. (1995). Magnetic structures of uranium compounds: effects of relativity and symmetry. *Phys. Rev. Lett.*, **75**, 946–949.

[9555] Sandratskii L.M. and Kübler J. (1999). Relativistic effects in the magnetism of UFe_4Al_8. *Phys. Rev. B*, **60**, R6961–R6964.

[9556] San Miguel M.A., Márquez A. and Fernández Sanz J. (1996). Molecular and electronic structure of zinc carbyne, HZnCH, and zinc stannyne, HZnSnH, from ab initio calculations. *J. Phys. Chem.*, **100**, 1600–1604.

[9557] Sanov A., Faeder J., Parson R. and Lineberger W.C. (1999). Spin-orbit coupling in $I{\cdot}CO_2$ and I·OCS van der Waals complexes: beyond the pseudo-diatomic approximation. *Chem. Phys. Lett.*, **313**, 812–819.

[9558] Sanoyama E., Kobayashi H. and Yabushita S. (1998). Spin-orbit CI study on multiplet terms of trivalent lanthanide cations. *J. Mol. Str. (Theochem)*, **451**, 189–204.

[9559] Santos J.P., Marques J.P., Parente F., Lindroth E., Boucard S. and Indelicato P. (1998a). Multiconfiguration Dirac-Fock calculation of $2s_{1/2}$–$2p_{3/2}$ transition energies in highly ionized bismuth, thorium, and uranium. *Eur. Phys. J. D*, **1**, 149–163.

[9560] Santos J.P., Marques J.P., Parente F., Lindroth E., Indelicato P. and Desclaux J.P. (1999). Relativistic $2s_{1/2}$ (L_1) atomic subshell decay rates and fluorescence yields for Yb and Hg. *J. Phys. B*, **32**, 2089–2097.

[9561] Santos J.P., Parente F. and Indelicato P. (1998b). Application of B-splines finite basis sets to relativistic two-photon decay rates of 2s level in hydrogenic ions. *Eur. Phys. J. D*, **3**, 43–52.

[9562] Sapirstein J. (1993). Theory of many-electron atoms. *Phys. Scr.*, **T46**, 52–60.

[9563] Sapirstein J. (1996a). Quantum electrodynamics. In *Atomic, Molecular and Optical Physics Handbook*, (Edited by Drake G.W.F.), pp. 327–340, AIP, Woodbury, NY.

[9564] Sapirstein J. (1996b). Parity nonconserving effects in atoms. In *Atomic, Molecular and Optical Physics Handbook*, (Edited by Drake G.W.F.), pp. 352–356, AIP, Woodbury, NY.

[9565] Sapirstein J. (1998). Theoretical methods for the relativistic atomic many-body problem. *Rev. Mod. Phys.*, **70**, 55–76.

[9566] Sapirstein J., Cheng K.T. and Chen M.H. (1999). Potential independence of the solution to the relativistic many-body problem and the role of negative-energy states in heliumlike ions. *Phys. Rev. A*, **59**, 259–266.

[9567] Sapirstein J. and Johnson W.R. (1996). The use of basis splines in theoretical atomic physics. *J. Phys. B*, **29**, 5213–5225.

[9568] Sapirstein J. and Yennie D. (1990). Theory of hydrogenic bound states. In *Quantum Electrodynamics*, (Edited by Kinoshita T.), World Scientific, Singapore.

[9569] Sargent A.L. and Titus E.P. (1998). C-S and C-H bond activation of thiophene by Cp*Rh(PMe_3): A DFT theoretical investigation. *Organometallics*, **17**, 65–77.

[9570] Sasaki S., Katsuki A., Akiyama K. and Tero-Kubota S. (1997). Spin-orbit coupling induced electron polarization: Influence of heavy atom position. *J. Am. Chem. Soc.*, **119**, 1323–1327.

[9571] Saue T. (1995). *Principles and applications of relativistic molecular calculations.*. Ph.D. thesis, Univ. Oslo.

[9572] Saue T., Faegri K. and Gropen O. (1996). Relativistic effects on the bonding of heavy and superheavy hydrogen halides. *Chem. Phys. Lett.*, **263**, 360–366.

[9573] Saue T., Faegri K., Helgaker T. and Gropen O. (1997). Principles of direct 4-component relativistic SCF: application to caesium auride. *Mol. Phys.*, **91**, 937–950.

[9574] Saue T. and Jensen H.J.A. (1999). Quaternion symmetry in relativistic molecular calculations: The Dirac-Hartree-Fock method. *J. Chem. Phys.*, **111**, 6211–6222.

[9575] Sauter M., Ott H. and Nakel W. (1998). Spin asymmetry in relativistic (e,2e) processes: atomic-number dependence. *J. Phys. B*, **31**, L967–L970.

[9576] Savchenko O.Y. (1996). Dirac particle with an anomalous magnetic moment in a circularly polarized wave. *Zh. Eksp. Teor. Fiz.*, **109**, 1234–1239.

[9577] Schädel M. (1995). Chemistry of the transactinide elements. *Radiochim. Acta*, **70/71**, 207–223.

[9578] Schädel M., Brüchle W., Dressler R., Eichler B., Gäggeler H.W., Günther R., Gregorich K.E., Hoffman D.C., Hübener S., Jost D.T., Kratz J.V., Paulus W., Schumann D., Timokhin S., Trautmann N., Türler A., Wirth G. and Yakushev A. (1997a). Chemical properties of element 106 (seaborgium). *Nature*, **388**, 55–57.

[9579] Schädel M., Brüchle W., Schausten B., Schimpf E., Jäger E., Wirth G., Günther R., Kratz J.V., Paulus W., Seibert A., Thörle P., Trautmann N., Zauner S., Schumann D., Andrassy M., Misiak R., Gregorich K.E., Hoffman D.C., Lee D.M., Sylwester E.R., Nagame Y. and Oura Y. (1997b). First aqueous chemistry with seaborgium (element 106). *Radiochim. Acta*, **77**, 149–159.

[9580] Schaden M., Spruch L. and Zhou F. (1998). Unified treatment of some Casimir energies and Lamb shifts: A dielectric between two ideal conductors. *Phys. Rev. A*, **57**, 1108–1120.

[9581] Schäfer A. and Ahlrichs R. (1994). Ab initio study of structures and energetics of small copper-selenium clusters. *J. Am. Chem. Soc.*, **116**, 10686–10692.

[9582] Schäfer A., Huber C., Gauss J. and Ahlrichs R. (1993). An *ab initio* investigation of Cu_2Se and Cu_4Se_2. *Theor. Chim. Acta*, **87**, 29–40.

[9583] Schäfer A., Kollwitz M. and Ahlrichs R. (1996). Electronic excitation energies in copper selenide clusters. *J. Chem. Phys.*, **104**, 7113–7121.

[9584] Schäfer A. and Reinhardt J. (1995). Vacuum polarization as a test of C and CPT invariance. *Phys. Rev. A*, **51**, 838–840.

[9585] Schamps J., Bencheikh M., Barthelat J.C. and Field R.W. (1995). The electronic structure of LaO: ligand field versus ab initio calculations. *J. Chem. Phys.*, **103**, 8004–8013.

[9586] Schaphorst S.J., Kodre A.F., Ruscheinski J., Crasemann B., Åberg T., Tulkki J., Chen M.H., Azuma Y. and Brown G.S. (1993). Multielectron inner-shell photo excitation in absorption spectra of Kr: theory and experiment. *Phys. Rev. A*, **47**, 1953–1966.

[9587] Schautz A., Flad H.J. and Dolg M. (1998). Quantum Monte Carlo study of Be_2 and group 12 dimers M_2 (M= Zn, Cd, Hg). *Theor. Chem. Acc.*, **99**, 231–240.

[9588] Scheer M., Müller J. and Häser M. (1996). Complexes containing phosphorus and arsenic as terminal ligands. *Angew. Chem. Int. Ed. Engl.*, **35**, 2492–2496.

[9589] Scherer S., Poulis G.I. and Fearing H.W. (1994). Low-energy Compton scattering by a proton: Comparison of effective hamiltonians with relativistic corrections. *Nucl. Phys. A*, **570**, 686–700.

[9590] Schimmelpfennig B., Maron L., Wahlgren U., Teichteil C., Fagerli H. and Gropen O. (1998a). On the efficiency of an effective Hamiltonian in spin-orbit CI calculations. *Chem. Phys. Lett.*, **286**, 261–266.

[9591] Schimmelpfennig B., Maron L., Wahlgren U., Teichteil C., Fagerli H. and Gropen O. (1998b). On the combination of ECP-based CI calculations with all-electron spin-orbit mean-field integrals. *Chem. Phys. Lett.*, **286**, 267–271.

[9592] Schiødt N.C., Sommer-Larsen P., Folmer Nielsen M., Larsen J. and Bechgaard K. (1995). Preparation and electronic structure of substituted aromatic dithiolene complexes of gold(III). *Inorg. Chem.*, **34**, 3688–3694.

[9593] Schmid R.N., Engel E., Dreizler R.M., Blaha P. and Schwarz K. (1999). Full potential linearized-augmented-plane-wave calculations for 5d transition metals using the relativistic generalized gradient approximation. *Adv. Quantum Chem.*, **33**, 209–223.

[9594] Schmidt K. and Springborg M. (1997). One-dimensional chains of metal atoms. *Solid State Comm.*, **104**, 413–417.

[9595] Schmidt K.M. (1993). On the genericity of nonvanishing instability intervals in periodic Dirac systems. *Ann. L'Inst. Henri Poincaré, Phys. Theor.*, **59**, 315–326.

[9596] Schmiedeskamp B., Vogt B. and Heinzmann U. (1993). Spin polarization in the photo emission from the 5d-core levels of a thallium film. *Z. Phys. B*, **90**, 197–200.

[9597] Schmitz H., Boucke K. and Kull H.J. (1998). Three-dimensional relativistic calculation of strong-field photoionization by the phase-space-averaging method. *Phys. Rev. A*, **57**, 467–475.

[9598] Schneider S.M. (1995). *Die Hyperfeinstrukturaufspaltung von Einelektronenatomen*. Ph.D. thesis, Johann Wolfgang Goethe-Universität, Frankfurt am Main.

[9599] Schneider S.M., Greiner W. and Soff G. (1993a). Källen-Sabry energy shift for hydrogen-like atoms with finite size nuclei. *J. Phys. B*, **26**, L529–L534.

[9600] Schneider S.M., Greiner W. and Soff G. (1994a). Vacuum-polarization contribution to the hyperfine-structure splitting of hydrogenlike atoms. *Phys. Rev. A*, **50**, 118–122.

[9601] Schneider S.M., Greiner W. and Soff G. (1994b). The transition time for the ground state hyperfine splitting of $^{209}Bi^{82+}$. *Z. Phys. D*, **31**, 143–144.

[9602] Schneider S.M., Schaffner J., Greiner W. and Soff G. (1993b). The hyperfine structure of $^{209}_{83}Bi^{82+}$. *J. Phys. B*, **26**, L581–L584.

[9603] Schneider W.F., Strittmatter R.J., Bursten B.E. and Ellis D.E. (1991). Relativistic DV-Xα studies of three-coordinate actinide complexes. In *Density Functional Methods in Chemistry*, (Edited by Labanowski J.K. and Andzelm J.W.), Springer-Verlag.

[9604] Schoeller W.W. and Sundermann A. (1998). Ring structure formation in transition-metal nitrido chlorides by donor–acceptor formation. *Inorg. Chem.*, **37**, 3034–3039.

[9605] Schoeller W.W., Sundermann A. and Reiher M. (1999a). Bonding properties of aminidate complexes of the Group 14 elements silicon, germanium, tin, and lead in their divalent and tetravalent oxidation states. *Inorg. Chem.*, **38**, 29–37.

[9606] Schoeller W.W., Sundermann A., Reiher M. and Rozhenko A. (1999b). On the bonding properties of diphosphanylmethanide complexes with the Group-14 elements silicon, germanium, tin, and lead in their divalent oxidation states. *Eur. J. Inorg. Chem.*, **1**, 1155–1159.

[9607] Schön J. and Köppel H. (1998). Geometric phases and quantum dynamics in spin-orbit coupled systems. *J. Chem. Phys.*, **108**, 1503–1513.

[9608] Schreckenbach G., Hay P.J. and Martin R.L. (1998). Theoretical study of stable trans and cis isomers in $[UO_2(OH)_4]^{2-}$ using relativistic density functional theory. *Inorg. Chem.*, **37**, 4442–4451.

[9609] Schreckenbach G., Hay P.J. and Martin R.L. (1999). Density functional calculations on actinide compounds: Survey of recent progress and application to $[UO_2X_4]^{2-}$ (X=F, Cl, OH) and AnF_6 (An=U, Np, Pu). *J. Comp. Chem.*, **20**, 70–90.

[9610] Schreckenbach G. and Ziegler T. (1996). The calculation of NMR shielding tensors based on density functional theory and the frozen-core approximation. *Int. J. Quantum Chem.*, **60**, 753–766.

[9611] Schreckenbach G. and Ziegler T. (1997). Calculation of NMR shielding tensors based on density functional theory and a scalar relativistic Pauli-type Hamiltonian. The application to transition metal complexes. *Int. J. Quantum Chem.*, **61**, 899–918.

[9612] Schreckenbach G. and Ziegler T. (1998). Density functional calculations of NMR chemical shifts and ESR g-tensors. *Theor. Chem. Acc.*, **99**, 71–82.

[9613] Schreckenbach G., Ziegler T. and Li J. (1995). The implementation of analytical energy gradients based on quasi-relativistic density functional method: the application to metal carbonyls. *Int. J. Quantum Chem.*, **56**, 477–488.

[9614] Schreiter E.R., Stevens J.E., Ortwerth M.F. and Freeman R.G. (1999). A room-temperature molten salt prepared from $AuCl_3$ and 1-ethyl-3-methylimidazolium chloride. *Inorg. Chem.*, **38**, 3935–3937.

[9615] Schröder D., Diefenbach M., Klapötke T.M. and Schwarz H. (1999). UF^{3+} – A thermochemically stable diatomic trication with a covalent bond. *Angew. Chem. Int. Ed. Engl.*, **38**, 137–140.

[9616] Schröder D., Diefenbach M., Klapötke T.M. and Schwarz H. (1999). UF^{3+} – ein thermochemisch stabiles binäres Trikation mit kovalenter Bindung. *Angew. Chem.*, **111**, 206–209.

[9617] Schröder D., Harvey J.N., Aschi M. and Schwarz H. (1998a). Experimental and computational study of neutral xenon halides (XeX) in the gas phase for X=F, Cl, Br, and I. *J. Chem. Phys.*, **108**, 8446–8455.

[9618] Schröder D., Harvey J.N. and Schwarz H. (1998b). Long-lived, multiply charged diatomic TiF^{n+} ions (n = 1-3). *J. Phys. Chem.*, **102**, 3639–3642.

[9619] Schröder D., Hrušák J., Hertwig R.H., Koch W., Schwerdtfeger P. and Schwarz H. (1995). Experimental and theoretical studies of gold(I) complexes $Au(L)^+$ (L = H_2O, CO, NH_3, C_2H_4, C_3H_6, C_4H_6, C_6H_6, C_6F_6). *Organometallics*, **14**, 312–316.

[9620] Schröder D. and Schwarz H. (1995). C-H and C-C bond activation by bare transition-metal oxide cations in the gas phase. *Angew. Chem. Int. Ed. Engl.*, **34**, 1973–1995.

[9621] Schröder D., Schwarz H., Hrušák J. and Pyykkö P. (1998c). Cationic gold(I) complexes of xenon and of ligands containing the donor atoms oxygen, nitrogen, phosphorus, and sulfur. *Inorg. Chem.*, **37**, 624–632.

[9622] Schröder D., Schwarz H., Löbrecht B., Koch W. and Ogawa S. (1998d). Generation of 1,3-dithia-, 1-selena-3-thia-, and 1,3-diselena-2-tellurole by neutralization-reionization mass spectrometry. *Eur. J. Inorg. Chem.*, pp. 983–987.

[9623] Schulz A. and Klapötke T.M. (1997). Theoretical evidence for two new intermediate xenon species: Xenon azide fluoride, $FXe(N_3)$, and xenon isocyanate fluoride, FXe(NCO). *Inorg. Chem.*, **36**, 1929–1933.

[9624] Schulz A., Tornieporth-Oetting I.C. and Klapötke T.M. (1995). Experimental and theoretical vibrational studies of covalent X-N_3 azides (X = H, F , Cl, Br, I). Application of the density functional theory and comparison with ab initio results. *Inorg. Chem.*, **34**, 4343–4346.

[9625] Schulze K., Anton J., Sepp W.D. and Fricke B. (1999). An analysis of the MO x-ray spectra in U^{92+}-Pb collisions. *Phys. Scripta*, **T80**, 430–431.

[9626] Schwabl F. (1997). *Quantenmechanik für Fortgeschrittene*. Springer, Berlin, 431 pp.

[9627] Schwarz J., Heinemann C., Schröder D. and Schwarz H. (1996a). Unparalleled, enormous metal-carbon bond strength in $PdCH_2I^+$. *Helv. Chim. Acta*, **79**, 1–5.

[9628] Schwarz J., Schröder D., Schwarz H., Heinemann C. and Hrušák J. (1996b). Theory-enforced re-investigation of the origin of the large metal-carbon bond strength in $PdCH_2I^+$ and its reactions with unsaturated hydrocarbons. *Helv. Chim. Acta*, **79**, 1110–1120.

[9629] Schwarz K., Ripplinger H. and Blaha P. (1996). Electric field gradient calculations of various borides. *Z. Naturf.*, **51a**, 527–533.

[9630] Schwarz W.H.E., Rutkowski A. and Wang S.G. (1996). Understanding relativistic effects of chemical bonding. *Int. J. Quantum Chem.*, **57**, 641–653.

[9631] Schweber S.S. (1994). *QED and the Men Who Made It: Dyson, Feynman, Schwinger, and Tomonaga*. Princeton Univ. Press, Princeton, NJ, 732 pp.

[9632] Schweizer W., Fassbinder P. and González-Férez R. (1999). Model potentials for alkali metal atoms and Li-like ions. *At. Data Nucl. Data Tables*, **72**, 33–55.

[9633] Schwerdtfeger P. (1995). Spectroscopic properties for the $^1\Sigma^+$ ground state of AuCl. A scalar relativistic coupled cluster study. *Mol. Phys.*, **86**, 359–368.

[9634] Schwerdtfeger P. (1996). Second-order Jahn-Teller distortions in group 17 fluorides EF_3 (E = Cl, Br, I, and At). Large relativistic bond angle changes in AtF_3. *J. Phys. Chem.*, **100**, 2968–2973.

[9635] Schwerdtfeger P. and Bowmaker G.A. (1994). Relativistic effects in gold chemistry. V. Group 11 dipole polarizabilities and weak bonding in monocarbonyl compounds. *J. Chem. Phys.*, **100**, 4487–4497.

[9636] Schwerdtfeger P., Boyd P.D.W., Brienne S., McFeaters J.S., Dolg M., Liao M.S. and Schwarz W.H.E. (1993). The mercury-mercury bond in inorganic and organometallic compounds. A theoretical study. *Inorg. Chim. Acta*, **213**, 233–246.

[9637] Schwerdtfeger P., Boyd P.D.W., Fischer T., Hunt P. and Liddell M. (1994a). Trends in inversion barriers of group 15 compounds. 2. Ab-initio and density functional calculations on group 15 fluorides. *J. Am. Chem. Soc.*, **116**, 9620–9633.

[9638] Schwerdtfeger P., Bruce A.E. and Bruce M.R.M. (1998). Theoretical studies on the photochemistry of the cis-to-trans conversion in dinuclear gold halide bis(diphenylphosphino)ethylene complexes. *J. Am. Chem. Soc.*, **120**, 6587–6597.

[9639] Schwerdtfeger P., Fischer T., Dolg M., Igel-Mann G., Nicklass A., Stoll H. and Haaland A. (1995a). The accuracy of the pseudopotential approximation. I. Analysis of the spectroscopic constants for the electronic ground states of InCl and $InCl_3$ using various three valence electron pseudopotentials for indium. *J. Chem. Phys.*, **102**, 2050–2062.

[9640] Schwerdtfeger P. and Hunt P. (1999). Symmetry-broken inversion structures for Group 15 EX_3 halides. In *Advances in Molecular Structure Research*, (Edited by Hargittai M. and Hargittai I.), vol. 5, JAI Press, Inc., Stamford, CT.

[9641] Schwerdtfeger P. and Ischtwan J. (1993). Theoretical investigations on thallium halides: relativistic and electron correlation effects in TlX and TlX_3 compounds (X=F, Cl, Br, and I). *J. Comp. Chem.*, **14**, 913–921.

[9642] Schwerdtfeger P., Li J. and Pyykkö P. (1994b). The polarisability of Hg and ground-state interaction potential of Hg_2. *Theor. Chim. Acta*, **87**, 313–320.

[9643] Schwerdtfeger P., McFeaters J.S. and Liddell M.J. (1995b). Spectroscopic properties for the ground states of AuF, AuF^+, AuF_2, and Au_2F_2: A pseudopotential scalar relativistic Møller-Plesset and coupled cluster study. *J. Chem. Phys.*, **103**, 245–252.

[9644] Schwerdtfeger P., McFeaters J.S., Stephens R.L., Liddell M.J., Dolg M. and Hess B.A. (1994c). Can AuF be synthesized? A theoretical study using relativistic configuration interaction and plasma modeling techniques. *Chem. Phys. Lett.*, **218**, 362–366.

[9645] Schwerdtfeger P., Pernpointner M. and Laerdahl J.K. (1999). The accuracy of current density functionals for the calculation of electric field gradients: A comparison with *ab initio* methods for HCl and CuCl. *J. Chem. Phys.*, **111**, 3357–3364.

[9646] Schwerdtfeger P. and Seth M. (1998). Relativistic effects of the superheavy elements. In *Encyclopedia of Computational Chemistry*, (Edited by von Ragué Schleyer et al. P.), vol. 4, pp. 2480–2499, Wiley, New York.

[9647] Scofield J.H. and Nilsen J. (1994). Hyperfine splittings of neonlike lasing lines. *Phys. Rev. A*, **49**, 2381–2388.

[9648] Seaborg G.T. (1996). Evolution of the modern periodic table. *J. Chem. Soc, Dalton Trans.*, pp. 3899–3907.

[9649] Segev B. and Wells J.C. (1998). Light-fronts approach to electron-positron pair production in ultrarelativistic heavy-ion collisions. *Phys. Rev. A*, **57**, 1849–1861.

[9650] Seifert G., Kaschner R., Schöne M. and Pastore G. (1998). Density functional calculations for Zintl systems: structure, electronic structure and electrical conductivity of liquid NaSn alloys. *J. Phys. CM*, **10**, 1175–1198.

[9651] Seijo L. (1995). Relativistic ab initio model potential calculations including spin-orbit effects through the Wood-Boring Hamiltonian. *J. Chem. Phys.*, **102**, 8078–8088.

[9652] Seijo L. and Barandiarán Z. (1996). Applications of the group-function theory to the field of materials science. *Int. J. Quantum Chem.*, **60**, 617–634.

[9653] Seke J. (1993). Gauge independence of the non-relativistic Lamb shift including retardation effects. *Nuovo Cim. D*, **15**, 691–694.

[9654] Seke J. (1994). Spontaneous decay of an unstable atomic state in non-relativistic QED: a complete treatment including gauge invariance renormalization and non-Markovian time behaviour. *J. Phys. A*, **27**, 263–274.

[9655] Seke J. (1996). Complete Lamb-shift calculation to order α^5 by applying the methods of non-relativistic quantum electrodynamics. *Nuovo Cim. D*, **18**, 533–545.

[9656] Seke J. (1998). Spontaneous decay of an unstable atomic state: New method for a unified nonrelativistic-relativistic complete treatment. *Phys. Lett. A*, **244**, 111–119.

[9657] Sekine R., Nakamatsu H., Mukoyama T., Onoe J., Hirata M., Kurihara M. and Adachi H. (1997). Electronic structures of metal carbides TiC and UC: similarity and dissimilarity. *Adv. Quantum Chem.*, **29**, 123–136.

[9658] Sekiya M., Sasaki F. and Tatewaki H. (1997). 6s and 4f ionized states of lanthanide calculated by the configuration-interaction method. *Phys. Rev. A*, **56**, 2731–2740.

[9659] Sekkat Z., Komiha N. and Chraibi M. (1995). A representation of effective potentials of molecular fragments (NH_3, PH_3, AsH_3): applications to AH_3-EH_3 systems, where A is B, Al, Ga, In or Tl, and E is N, P or As. *J. Mol. Str. (Theochem)*, **358**, 219–228.

[9660] Sellers H. (1992). Differences in bonding, structures and vibrational force constants of mercapto and methylthio on platinum(111) surfaces. *Surf. Sci.*, **264**, 177–184.

[9661] Sellers H. (1993a). On analytic potential functions for reactions on metal surfaces: the case of H_2 to 2H on the liquid mercury surface. *J. Chem. Phys.*, **98**, 627–633.

[9662] Sellers H. (1993b). On the chemisorption and dissociation of $HSCH_3$ on the Au(111) surface. *Surface Sci.*, **294**, 99–107.

[9663] Sellers H. (1993c). A bond order conservation-Morse potential model of adsorbate-surface interactions: dissociation of the hydrogen, oxygen, and fluorine molecules on the liquid mercury surface. *J. Chem. Phys.*, **99**, 650–655.

[9664] Sellers H. (1994). On analytical potential functions and molecular dynamics for reactions on metal surfaces. *Surf. Sci.*, **310**, 281–291.

[9665] Sellers H., Ulman A., Shnidman Y. and Eilers J.E. (1993). Structure and binding alkanethiolates on gold and silver surfaces: Implications for self-assembled monolayers. *J. Am. Chem. Soc.*, **115**, 9389–9401.

[9666] Semay C. (1993). Virial theorem for two-body Dirac equation. *J. Math. Phys.*, **34**, 1791–1793.

[9667] Semay C. and Ceuleneer R. (1993). Two-body Dirac equation and Regge trajectories. *Phys. Rev. D*, **48**, 4361–4369.

[9668] Semay C., Ceuleneer R. and Silvestre-Brac B. (1993). Two-body Dirac equation with diagonal central potentials. *J. Math. Phys.*, **34**, 2215–2225.

[9669] Seminario J.M. and Tour J.M. (1997). Systematic study of the lowest energy states of Au_n (n = 1-4) using DFT. *Int. J. Quantum Chem.*, **65**, 749–758.

[9670] Senda Y., Shimojo F. and Hoshino K. (1999). Composition dependence of the structure and the electronic states of liquid K-Pb alloys: *ab initio* molecular-dynamics simulations. *J. Phys. CM*, **11**, 5387–5398.

[9671] Seo D.K. and Hoffmann R. (1999). What determines the structures of the Group 15 elements? *J. Solid State Chem.*, **147**, 26–37.

[9672] Seong S.Y. and Anderson A.B. (1996). Water dissociation on Pt(111) and (100) anodes: Molecular orbital theory. *J. Phys. Chem.*, **100**, 11744–11747.

[9673] Serebrennikov Y.A. and Steiner U.E. (1994). Adiabatic rotation of effective spin. II. Spin-rotational relaxation. *J. Chem. Phys.*, **100**, 7508–7514.

[9674] Sergheyev A.G. (1997). A relativistic Coulomb problem for the modified Stueckelberg equation. *Ukr. Fiz. Zh.*, **42**, 1171–1174.

[9675] Seshadri R., Felser C., Thieme K. and Tremel W. (1998). Metal-metal bonding and metallic behavior in some ABO_2 delafossites. *Chem. Mater.*, **10**, 2189–2196.

[9676] Seth M. (1998). *The chemistry of superheavy elements*. Ph.D. thesis, University of Auckland, 181 p.

[9677] Seth M., Cooke F., Schwerdtfeger P., Heully J.L. and Pelissier M. (1998a). The chemistry of the superheavy elements. II. The stability of high oxidation states in group 11 elements: Relativistic coupled cluster calculations for the di-, tetra- and hexafluoro metallates of Cu, Ag, Au, and element 111. *J. Chem. Phys.*, **109**, 3935–3943.

[9678] Seth M., Dolg M., Fulde P. and Schwerdtfeger P. (1995). Lanthanide and actinide contractions: relativistic and shell structure effects. *J. Am. Chem. Soc.*, **117**, 6597–6598.

[9679] Seth M., Faegri K. and Schwerdtfeger P. (1998b). The stability of the oxidation state +4 in Group 14 compounds from carbon to element 114. *Angew. Chem. Int. Ed. Engl.*, **37**, 2493–2496.

[9680] Seth M., Faegri K. and Schwerdtfeger P. (1998b). The stability of the oxidation state +4 in Group 14 compounds from carbon to element 114 (in German). *Angew. Chem.*, **110**, 2669–2672.

[9681] Seth M., Fischer T.H. and Schwerdtfeger P. (1996a). Relativistic pseudopotential calculations of the ground-state spectroscopic properties of HBr. *J. Chem. Soc., Faraday Trans.*, **92**, 167–174.

[9682] Seth M., Pernpointner M., Bowmaker G.A. and Schwerdtfeger P. (1999a). Vibrational-rotational dependence of molecular properties. Electric field gradients for HCl, LiCl, NaCl and KCl. *Mol. Phys.*, **96**, 1767–1780.

[9683] Seth M., Schwerdtfeger P. and Dolg M. (1997). The chemistry of the superheavy elements. I. Pseudopotentials for 111 and 112 and relativistic coupled cluster calculations for $(112)H^+$, $(112)F_2$, and $(112)F_4$. *J. Chem. Phys.*, **106**, 3623–3632.

[9684] Seth M., Schwerdtfeger P., Dolg M., Faegri K., Hess B.A. and Kaldor U. (1996b). Large relativistic effects in molecular properties of the hydride of superheavy element 111. *Chem. Phys. Lett.*, **250**, 461–465.

[9685] Seth M., Schwerdtfeger P. and Faegri K. (1999b). The chemistry of superheavy elements. III. Theoretical studies on element 113 compounds. *J. Chem. Phys.*, **111**, 6422–6433.

[9686] Severin L., Richter M. and Steinbeck L. (1997). Self-interaction correction and relativistic exchange on the core states and core hyperfine fields in Fe, Co, and Ni. *Phys. Rev. B*, **55**, 9211–9214.

[9687] Shabad A.Y. (1991). *Polarization of the Vacuum and a Quantum Relativistic Gas in an External Field*. Nova Science Publ.

[9688] Shabaev V.M. (1993a). Finite nuclear size corrections to the energy levels of the multicharged ions. *J. Phys. B*, **26**, 1103–1108.

[9689] Shabaev V.M. (1993b). Schrodinger-like equation for the relativistic few-electron atom. *J. Phys. B*, **26**, 4703–4718.

[9690] Shabaev V.M. (1994a). Hyperfine structure of hydrogen-like ions. *J. Phys. B*, **27**, 5825–5832.

[9691] Shabaev V.M. (1994b). Quantum electrodynamic theory of recombination of an electron with a highly charged ion. *Phys. Rev. A*, **50**, 4521–4534.

[9692] Shabaev V.M. (1998a). QED theory of the nuclear recoil effect in atoms. *Phys. Rev. A*, **57**, 59–67.

[9693] Shabaev V.M. (1998b). Transition probability between the hyperfine structure components of hydrogenlike ions and bound-electron g-factor. *Can. J. Phys.*, **76**, 907–910.

[9694] Shabaev V.M. (1999). Hyperfine structure of highly charged ions. In *Atomic Physics with Heavy Ions*, (Edited by Beyer H.F. and Shevelko V.P.), vol. 26, pp. 139–159, Springer, Berlin.

[9695] Shabaev V.M. and Artemyev A.N. (1994). Relativistic nuclear recoil corrections to the energy levels of multicharged ions. *J. Phys. B*, **27**, 1307–1314.

[9696] Shabaev V.M., Artemyev A.N., Beier T., Plunien G., Yerokhin V.A. and Soff G. (1998a). Recoil correction to the ground-state energy of hydrogenlike atoms. *Phys. Rev. A*, **57**, 4235–4239.

[9697] Shabaev V.M., Artemyev A.N., Beier T., Plunien G., Yerokhin V.A. and Soff G. (1999). Relativistic nuclear recoil corrections to the energy levels of hydrogenlike ions. *Phys. Scr.*, **T80**, 493–494.

[9698] Shabaev V.M., Artemyev A.N., Beier T. and Soff G. (1998b). Relativistic recoil correction to hydrogen energy levels. *J. Phys. B*, **31**, L337–L339.

[9699] Shabaev V.M., Artemyev A.N. and Yerokhin V.A. (1998c). QED and nuclear effects in highly charged ions. In *Trapped Charged Particles and Fundamental Physics, AIP Conf. Proc. 457*, (Edited by Dubin D.H.E. and Schneider D.), pp. 22–31.

[9700] Shabaev V.M. and Fokeeva I.G. (1994). Calculation formulas for the reductible part of the two-photon-exchange diagrams in the QED of multicharged ions. *Phys. Rev. A*, **49**, 4489–4501.

[9701] Shabaev V.M., Shabaeva M.B. and Tupitsyn I.I. (1995). Hyperfine structure of hydrogenlike and lithiumlike atoms. *Phys. Rev. A*, **52**, 3686–3690.

[9702] Shabaev V.M., Shabaeva M.B. and Tupitsyn I.I. (1997a). Hyperfine structure of lithium-like ions. *Astron. Astrophys. Transactions*, **12**, 243–246.

[9703] Shabaev V.M., Shabaeva M.B., Tupitsyn I.I. and Yerokhin V.A. (1998d). Hyperfine structure of highly charged ions. *Hyperfine Int.*, **114**, 129–133.

[9704] Shabaev V.M., Shabaeva M.B., Tupitsyn I.I., Yerokhin V.A., Artemyev A.N., Kühl T., Tomaselli M. and Zherebtsov O.M. (1998e). Transition energy and lifetime for the ground-state hyperfine splitting of high-Z lithiumlike ions. *Phys. Rev. A*, **57**, 149–156.

[9705] Shabaev V.M. and Yerokhin V.A. (1996). Self-energy contribution to the ground state hyperfine splitting of Bi^{82+}. *JETP Letters*, **63**, 316–318.

[9706] Shabaev V.M., Tomaselli M., Kühl T., Artemyev A.N. and Yerokhin V.A. (1997b). Ground-state hyperfine splitting of high-Z hydrogenlike ions. *Phys. Rev. A*, **56**, 252–255.

[9707] Shabayeva M.B. (1993). Hyperfine splitting of the ground state of Bi^{80+} (in Russian). *Opt. Spektr.*, **74**, 1042–1045.

[9708] Shabayeva M.B. (1993). Hyperfine splitting of the ground state of Bi^{80+}. *Opt. Spectr.*, **74**, 619–620.

[9709] Shabaeva M.B. and Shabaev V.M. (1995). Interelectronic interaction contribution to the hyperfine structure of highly charged lithiumlike ions. *Phys. Rev. A*, **52**, 2811–2819.

[9710] Shaffer C.D. and Pratt R.H. (1997). Comparison of relativistic partial-wave calculations of triply differential electron-atom bremsstrahlung with simpler theories. *Phys. Rev. A*, **56**, 3653–3658.

[9711] Shankar R. and Mathur H. (1994). Thomas precession Berry potential and the meron. *Phys. Rev. Lett.*, **73**, 1565–1569.

[9712] Sharma A., Moshinsky M. and Smirnov Y.F. (1998). Supermultiplets and relativistic problems: III. The non-relativistic limit for a particle of arbitrary spin in an external field. *J. Phys. A*, **31**, 10017–10028.

[9713] Sharp S.B. and Gellene G.I. (1997). Ab initio calculations of the ground states of polyiodide anions. *J. Phys. Chem. A*, **101**, 2192–2197.

[9714] Sharpe A.G. (1992). *Inorganic Chemistry, 3rd Ed.*. Longman, Singapore, see pp. 81-83.

[9715] Shelimov K.B., Safonov A.A. and Bagatur'yants A.A. (1993). Ab initio calculations of an M center on the AgBr (100) surface. Formation, electronic structure and spectrum of a primary Ag_2 cluster. *Chem. Phys. Lett.*, **201**, 84–88.

[9716] Shelyuto V.A. (1996). One-loop radiative corrections in a gauge with improved infrared and ultraviolet properties. *JETP*, **83**, 635–641.

[9717] Shelyuto V.A. (1996). One-loop radiative corrections in a gauge with improved infrared and ultraviolet properties (in Russian). *Zh. Eksp. Teor. Fiz.*, **110**, 1153–1167.

[9718] Shen B.R., Fan K.N., Wang W.N. and Deng J.F. (1999). Ab initio study on the adsorption and oxidation of HCHO with Ag_2 cluster. *J. Mol. Str. (Theochem)*, **469**, 157–161.

[9719] Shen M.Z., Schaefer III H.F. and Partridge H. (1993). Tungsten hexahydride (WH_6). An equilibrium geometry far from octahedral. *J. Chem. Phys.*, **98**, 508–521.

[9720] Shi Q.C., Xu K.Z., Chen Z.J., Cho H. and Li J.M. (1998a). Relativistic effects on the $4p-5s$ excitations of krypton. *Phys. Rev. A*, **57**, 4980–4982.

[9721] Shi Q.C., Zhang S.M., Cho H., Xu K.Z., Li J.M. and Kais S. (1998b). Relativistic structure description and relaxation effect on krypton $4p^5(^2P_{3/2,1/2})5s$ excitation at small squared momentum transfer. *J. Phys. B*, **31**, 4123–4135.

[9722] Shick A.B., Drchal V., Kudrnovský J. and Weinberger P. (1996). Electronic structure and magnetic properties of random alloys: Fully relativistic spin-polarized linear muffin-tin-orbital method. *Phys. Rev. B*, **54**, 1610–1621.

[9723] Shick A.B. and Gubanov V.A. (1993). Electronic structure of U, Np, Pu impurities in thorium: Spin-polarized relativistic calculations. *Europhys. Lett.*, **21**, 599–604.

[9724] Shick A.B. and Gubanov V.A. (1994). Electronic structure and hyperfine interactions for light actinide impurities in bcc Fe: spin-polarized relativistic calculations. *Phys. Rev. B*, **49**, 12860–12863.

[9725] Shick A.B., Ketterson J.B., Novikov D.L. and Freeman A.J. (1999). Electronic structure, phase stability, and semimetal-semiconductor transitions in Bi. *Phys. Rev. B*, **60**, 15484–15487.

[9726] Shik A.B., Solov'ev I.V., Antropov V.P., Likhtenshtein A.I. and Gubanov V.A. (1992). Self-consistent spin-polarization relativistic approach to an analysis of the electron structure and magnetic properties of substitutional impurities in regular crystals. *Phys. Metals Metallography*.

[9727] Shik A.B., Solov'ev I.V., Antropov V.P., Likhtenshtein A.I. and Gubanov V.A. (1992). Self-consistent spin-polarization relativistic approach to an analysis of the electron structure and magnetic properties of substitutional impurities in regular crystals (in Russian). *Fiz. Metallov Metalloved*, **(1)**, 61–70.

[9728] Shim I. and Gingerich K.A. (1997). Electronic states and nature of bonding in the molecule MoC by all electron ab initio calculations. *J. Chem. Phys.*, **106**, 8093–8100.

[9729] Shim I. and Gingerich K.A. (1999). All-electron ab initio investigations of the electronic states of the NiC molecule. *Chem. Phys. Lett.*, **303**, 87–95.

[9730] Shim I., Mandix K. and Gingerich K.A. (1997). Electronic states and nature of bonding in the molecule RhN by all-electron ab initio calculations. *J. Mol. Str. (Theochem)*, **393**, 127–139.

[9731] Shin G.R. and Rafelski J. (1993). Relativistic classical limit of quantum theory. *Phys. Rev. A*, **48**, 1869–1874.

[9732] Shishidou T., Oguchi T. and Jo T. (1999). Hartree-Fock study on the $5f$ orbital magnetic moment of US. *Phys. Rev. B*, **59**, 6813–6823.

[9733] Shishkin G.V. and Villalba V.M. (1993). Electrically neutral Dirac particles in the presence of external fields: Exact solutions. *J. Math. Phys.*, **34**, 5037–5049.

[9734] Shivamoggi B.K. and Mulser P. (1993). Relativistic Thomas-Fermi model for atoms in a very strong magnetic field. *Europhys. Lett.*, **22**, 657–662.

[9735] Shivamoggi B.K. (1995). Relativistic Thomas-Fermi formulations with thermal effects for an atom within and without a very strong magnetic field. *Phys. Rev. A*, **51**, 185–190.

[9736] Shukla A. and Banerjee A. (1994). The relativistic valence shell effective Hamiltonian method for atomic and molecular systems. *J. Chem. Phys.*, **100**, 3695–3705.

[9737] Shukla A., Das B.P. and Andriessen J. (1994a). Relativistic many-body calculation of the electric dipole moment of atomic rubidium due to parity and time-reversal violation. *Phys. Rev. A*, **50**, 1155–1171.

[9738] Shukla A., Das B.P. and Mukherjee D. (1994b). Application of the coupled-cluster approach to the electric dipole moment of atoms and molecules due to parity and time-reversal violation. *Phys. Rev. A*, **50**, 2096–2107.

[9739] Shukla A., Dolg M., Flad H.J., Banerjee A. and Mohanty A.K. (1997). Relativistic configuration-interaction study of valence-electron correlation effects on the fine-structure splitting in the Pb isoelectronic series. *Phys. Rev. A*, **55**, 3433–3439.

[9740] Shurki A., Hiberty P.C. and Shaik S. (1999). Charge-shift bonding in group IVB halides: A valence bond study of MH_3–Cl (M=C, Si, Ge, Sn, Pb) molecules [Errata: *ibid.* 121 (1999) 9768. *J. Am. Chem. Soc.*, **121**, 822–834.

[9741] Sicilia E., Toscano M., Mineva T. and Russo N. (1997). Density functional investigation of the molecular geometries, harmonic vibrational frequencies,singlet-triplet energy separations, adiabatic ionization potentials, and electron affinities of XY_2 (X = Si, Ge, Sn; Y = F, Cl) systems. *Int. J. Quantum Chem.*, **61**, 571–577.

[9742] Siegbahn P.E.M. (1993a). A comparison of the bonding in the second-row transition-metal oxides and carbenes. *Chem. Phys. Lett.*, **201**, 15–23.

[9743] Siegbahn P.E.M. (1993b). Binding in second-row transition metal dioxides, trioxides, tetraoxides, peroxides, and superoxides. *J. Phys. Chem.*, **97**, 9096–9102.

[9744] Siegbahn P.E.M. (1994a). The activation of the C-H bond in acetylene by second row transition metal atoms. *Theor. Chim. Acta*, **87**, 277–292.

[9745] Siegbahn P.E.M. (1994b). The bonding in second row transition metal dihydrides, difluorides, and dichlorides. *Theor. Chim. Acta*, **87**, 441–452.

[9746] Siegbahn P.E.M. (1994c). Second row transition metal mixed hydride-halide triatomic molecules. *Theor. Chim. Acta*, **88**, 413–424.

[9747] Siegbahn P.E.M. (1994d). Halide ligand effects on olefin insertion into metal-hydrogen bonds for second row transition metal complexes. *J. Organomet. Chem.*, **478**, 83–93.

[9748] Siegbahn P.E.M. (1995). Trends of metal-carbon bond strengths in transition metal complexes. *J. Phys. Chem.*, **99**, 12723–12729.

[9749] Siegbahn P.E.M. (1996). Electronic structure calculations for molecules containing transition metals. *Adv. Chem. Phys.*, **93**, 333–387.

[9750] Siegbahn P.E.M., Blomberg M.R.A. and Svensson M. (1993). The effects of covalent ligands on the oxidative addition reaction between second-row transition-metal atoms and methane. *J. Am. Chem. Soc.*, **115**, 4191–4200.

[9751] Siegbahn P.E.M. and Crabtree R.H. (1996). Solvent effects on the relative stability of the $PdCl_2(H_2O)_n$ and $PdHCl(H_2O)_n$ cis and trans isomers. *Mol. Phys.*, **89**, 279–296.

[9752] Siegbahn P.E.M., Svensson M. and Crabtree R.H. (1995). A theoretical study of mercury photosensitized reactions. *J. Am. Chem. Soc.*, **117**, 6758–6765.

[9753] Siegel W., Migdalek J. and Kim Y.K. (1998). Dirac-Fock oscillator strengths for E1 transitions in the sodium isoelectronic sequence (Na I - Ca X). *At. Data Nucl. Data Tables*, **68**, 303–322.

[9754] Sienkiewicz J.E. (1997). Differential cross sections for elastic scattering of electrons by mercury. *J. Phys. B*, **30**, 1261–1267.

[9755] Sienkiewicz J.E. and Baylis W.E. (1997). Relativistic multiconfiguration approach to the spin polarization of slow electrons elastically scattered from krypton. *Phys. Rev. A*, **55**, 1108–1112.

[9756] Sienkiewicz J.E., Fritzsche S. and Grant I.P. (1995). Relativistic configuration-interaction approach to the elastic low-energy scattering of electrons from atoms. *J. Phys. B*, **28**, L633–L636.

[9757] Sierraalta A. (1994). Electronic density topology of metal-metal quadruple bond in some Mo complexes. *Chem. Phys. Lett.*, **227**, 557–560.

[9758] Sierraalta A. and Ruette F. (1994). A comparative study of effective core potential and full-electron calculations in Mo compounds. I. An analysis of topological properties of bond charge distribution. *J. Comp. Chem.*, **15**, 313–321.

[9759] Sierraalta A. and Ruette F. (1996). The Laplacian of the electron density at the valence-shell charge concentration (VSCC): A comparative study of effective core potential and full-electron calculations in Mo compounds. II. *Int. J. Quantum Chem.*, **60**, 1015–1026.

[9760] Sigalas M.M., Rose J.H., Papaconstantopoulos D.A. and Shore H.B. (1998). Scaling lengths of elemental metals. *Phys. Rev. B*, **58**, 13438–13441.

[9761] Sigg T. and Sorg M. (1998). Exchange effects in relativistic Schrödinger theory. *Nuovo Cim. B*, **113**, 1261–1272.

[9762] Simionovici A., Dietrich D.D., Keville R., Cowan T., Beiersdorfer P., Chen M.H. and Blundell S.A. (1993). Soft-x-ray spectroscopy of $\Delta n = 0$, n=3 transitions in highly stripped lead. *Phys. Rev. A*, **48**, 3056–3061.

[9763] Şimşek M. (1999). Negative-energy levels of the Dirac equation in N dimensions. *Phys. Lett. A*, **259**, 215–219.

[9764] Simulik V.M. (1997). The hydrogen spectrum in classical electrodynamics (in Ukrainian). *Ukr. Fiz. Zh.*, **42**, 406–407.

[9765] Singh P.P. (1994a). Relativistic effects in mercury: Atom, clusters and bulk. *Phys. Rev. B*, **49**, 4954–4958.

[9766] Singh P.P. (1994b). From hexagonal close packed to rhombohedral sructure: Relativistic effects in Zn, Cd, and Hg. *Phys. Rev. Lett.*, **72**, 2446–2449.

[9767] Sjøvoll M., Fagerli H., Gropen O., Almlöf J., Olsen J. and Helgaker T. (1998a). Spin-orbit and correlation effects in platinum hydride (PtH). *Int. J. Quantum Chem.*, **68**, 53–64.

[9768] Sjøvoll M., Fagerli H., Gropen O., Almlöf J., Saue T., Olsen J. and Helgaker T. (1997a). Extensive relativistic calculations on the palladium hydride molecule. *J. Chem. Phys.*, **107**, 5496–5501.

[9769] Sjøvoll M., Fagerli H., Gropen O., Almlöf J., Schimmelpfennig B. and Wahlgren U. (1998b). An efficient treatment of kinematic factors in pseudo-relativistic calculations of electronic structure. *Theor. Chem. Accounts*, **99**, 1–7.

[9770] Sjøvoll M., Gropen O. and Olsen J. (1997b). A determinantal approach to spin-orbit configuration interaction. *Theor. Chem. Acc.*, **97**, 301–312.

[9771] Skaane H. (1998). *Relativistic quantum theory and its applications to atoms and molecules.* Ph.D. thesis, Oxford Univ., 218 p.

[9772] Skarzhinsky V.D. and Audretsch J. (1997). Scattering of scalar and Dirac particles by a magnetic tube of finite radius. *J. Phys. A*, **30**, 7603–7620.

[9773] Skripnikova O. and Zapriagaev S. (1997). The correlation and relativistic corrections in transition probabilities in the He-like ions. *Phys. Scr.*, **T73**, 38–40.

[9774] Skylaris C.K., Gagliardi L., Handy N.C., Ioannou A.G., Spencer S., Willetts A. and Simper A.M. (1998). An efficient method for calculating effective core potential integrals which involve projection operators. *Chem. Phys. Lett.*, **296**, 445–451.

[9775] Skyrme T.H.R. (1971). Kinks and the Dirac equation. *J. Math. Phys.*, **12**, 222–230.

[9776] Smart B.A. and Schiesser C.H. (1994). On the stability of trivalent chalcogen radicals - a pseudopotential study of homolytic substitution by a methyl radical at methanethiol methaneselenol and methanetellurol. *J. Chem. Soc., Perkin Trans. 2*, pp. 2269–2270.

[9777] Smelyansky V.I., Lee M.J.G. and Perz J.M. (1993). Relativistic calculation of conduction-electron g-factors in noble metals. *J. Phys: CM*, **5**, 6061–6066.

[9778] Smentek L. (1998). Theoretical description of the spectroscopic properties of rare earth ions in crystals. *Phys. Rep.*, **297**, 155–237.

[9779] Smentek L. (1999). Relativistic contributions to the amplitudes of electric dipole transitions in rare earth ions in crystals: I. Spin-orbit interaction. *J. Phys. B*, **32**, 593–606.

[9780] Smentek L. and Hess Jr. B.A. (1998). Are the odd-rank crystal field parameters independent quantities? *J. Alloys Compounds*, **275-277**, 170–173.

[9781] Šmit Ž. and Orlić I. (1994). First-order theories for adiabatic L-shell ionization by protons. *Phys. Rev. A*, **50**, 1301–1308.

[9782] Snijders J.G. and Sadlej A.J. (1996). Perturbation versus variation treatment of regular relativistic Hamiltonians. *Chem. Phys. Lett.*, **252**, 51–61.

[9783] Snygg J. (1997). *Clifford Algebra. A Computational Tool for Physicists.* Oxford Univ. Press, Oxford, 352 pp.

[9784] Söderlind P. (1998). Theory of the crystal structures of cerium and the light actinides. *Adv. Phys.*, **47**, 959–998.

[9785] Söderlind P. and Eriksson O. (1997). Pressure-induced phase transitions in Pa metal from first-principles theory. *Phys. Rev. B*, **56**, 10719–10721.

[9786] Söderlind P., Eriksson O., Johansson B. and Wills J.M. (1994). Electronic properties of f-electron metals using the generalized gradient approximation. *Phys. Rev. B*, **50**, 7291–7294.

[9787] Söderlind P., Eriksson O., Johansson B. and Wills J.M. (1995). Theoretical investigation of the high-pressure crystal structures of Ce and Th. *Phys. Rev. B*, **52**, 13169–13176.

[9788] Söderlind P. and Moriarty J.A. (1998). First-principles theory of Ta up to 10 Mbar pressure: structural and mechanical properties. *Phys. Rev. B*, **57**, 10340–10350.

[9789] Söderlind P., Wills J.M., Johansson B. and Eriksson O. (1997). Structural properties of plutonium from first-principles theory. *Phys. Rev. B*, **78**, 1997–2004.

[9790] Soff G. (1993). Radiative corrections in strong Coulomb fields. *Phys. Scr.*, **T46**, 266–269.

[9791] Soff G., Beier T., Greiner M., Persson H. and Plunien G. (1998). Quantum electrodynamics of strong field: status and perspectives. *Adv. Quantum Chem.*, **30**, 125–161.

[9792] Sohlberg K. and Yarkony D.R. (1997a). New and unusual bonding in open shell van der Waals molecules revealed by the heavy atom effect: The case of BAr. *J. Phys. Chem. A*, **101**, 3166–3173.

[9793] Sohlberg K. and Yarkony D.R. (1997b). On the relation between bonding and the spin-orbit interaction in BNe: the C $^2\Delta$ and 1 $^4\Pi$ states. *J. Phys. Chem. A*, **101**, 9520–9524.

[9794] Sökeland F., Westphal C., Dreiner S. and Zacharias H. (1999). On the separability of relativistic electron propagators. *Eur. Phys. J. B*, **9**, 577–582.

[9795] Solanki A.K., Kashyap A., Nautiyal T., Auluck S. and Khan M.A. (1997). Band structure and optical properties of HgI_2. *Phys. Rev. B*, **55**, 9215–9218.

[9796] Soldner T., Tröger W., Butz T., Blaha P. and Schwarz K. (1998a). Calculation of electric field gradients in isolated molecules using the FPLAPW-code WIEN95. *Z. Naturf.*, **53a**, 411–418.

[9797] Soldner T., Tröger W., Butz T., Blaha P., Schwarz K. and the ISOLDE-Collaboration (1998b). Measurement and calculation of electric field gradients in Hg-mercaptides. *Z. Naturf.*, **53a**, 404–410.

[9798] Soliman S.S.M. and Abelraheem S.K. (1999). Modification of band structure in intense laser fields. *Arab J. Nucl. Sci. and Appl.*, **32**, 171–184.

[9799] Solov'ev I.V. (1992). Electron structure and magnetic properties of U, Np and Pu monocarbides and mononitrides. *Fiz. Metallov Metallovedenie*, **(No. 11)**, 17–27.

[9800] Sommerer A.J. (1993). *Relativistic Two-Body Wave Equations*. Ph.D. thesis, Iowa State Univ., 149 p.

[9801] Sommerfield C.M. (1958). The magnetic moment of the electron. *Ann. Phys. (NY)*, **5**, 26–57.

[9802] Song J. and Hall M.B. (1993). Theoretical studies of inorganic and organometallic reaction mechanisms. 6. Methane activation on transient cyclopentadienylcarbonylrhodium. *Organometallics*, **12**, 3118–3126.

[9803] Sørensen A.H. and Belkacem A. (1994). Relativistic Coulomb wave functions in momentum space. *Phys. Rev. A*, **49**, 81–88.

[9804] Sousa C., de Jong W.A., Broer R. and Nieuwpoort W.C. (1997a). Theoretical characterization of the low-lying excited states of the CuCl molecule. *J. Chem. Phys.*, **106**, 7162–7169.

[9805] Sousa C., de Jong W.A., Broer R. and Nieuwpoort W.C. (1997b). Charge transfer and relativistic effects in the low-lying electronic states of CuCl, CuBr and CuI. *Mol. Phys.*, **92**, 677–686.

[9806] Souter P.F., Kushto G.P., Andrews L. and Neurock M. (1997a). Experimental and theoretical evidence for the isolation of thorium hydride molecules in argon matrices. *J. Phys. Chem.*, **101**, 1287–1291.

[9807] Souter P.F., Kushto G.P., Andrews L. and Neurock M. (1997b). Experimental and theoretical evidence for the formation of several uranium hydride molecules. *J. Am. Chem. Soc.*, **119**, 1682–1687.

[9808] Spencer S., Gagliardi L., Handy N.C., Ioannou A.G., Skylaris C.K. and Willetts A. (1999). Hydration of UO_2^{2+} and PuO_2^{2+}. *J. Phys. Chem. A*, **103**, 1831–1837.

[9809] Šponer J., Burda J.V., Sabat M., Leszczynski J. and Hobza P. (1998). Interaction between the guanine-cytosine Watson-Crick DNA base pair and hydrated group IIa (Mg^{2+}, Ca^{2+}, Sr^{2+}, Ba^{2+}) and group IIb (Zn^{2+}, Cd^{2+}, Hg^{2+}) metal cations. *J. Phys. Chem. A*, **102**, 5951–5957.

[9810] Šponer J., Sabat M., Burda J.V., Leszczynski J. and Hobza P. (1999). Interaction of the adenine-thymine Watson-Crick and adenine-adenine reverse-Hoogsteen DNA base pairs with hydrated group IIa (Mg^{2+}, Ca^{2+}, Sr^{2+}, Ba^{2+}) and IIb (Zn^{2+}, Cd^{2+}, Hg^{2+}) metal cations: Absence of the base pair stabilization by metal-induced polarization effects. *J. Phys. Chem. B*, **103**, 2528–2534.

[9811] Springborg M. (1999). Dimensionality effects on the electronic and structural properties of PtS_2 chains. *Chem. Phys.*, **246**, 347–361.

[9812] Springborg M. and Albers R.C. (1996). Electronic structure of Pt in polyyne. *Phys. Rev. B*, **53**, 10626–10631.

[9813] Spruch L., Babb J.F. and Fei Z. (1994). Simple derivation of the asymptotic Casimir interaction of a pair of finite systems. *Phys. Rev. A*, **49**, 2476–2482.

[9814] Srinivas G.N. and Jemmis E.D. (1997). H-bridged structures for tetrahedranes A_4H_4 (A = C,Si,Ge,Sn, and Pb). *J. Am. Chem. Soc.*, **119**, 12968–12973.

[9815] Srivastava R., Blum K., McEachran R.P. and Stauffer A.D. (1996a). Excitation of the 6 $^{1,3}P_1$ states of mercury by polarized electrons. *J. Phys. B*, **29**, 3513–3527.

[9816] Srivastava R., Blum K., McEachran R.P. and Stauffer A.D. (1996b). State multipoles and Stokes parameters for the 5 $^{1,3}D_2$ excitation of cadmium. *Z. Phys. D*, **37**, 141–147.

[9817] Srivastava R., McEachran R.P. and Stauffer A.D. (1995a). Relativistic distorted-wave calculation of the excitation of the $^3D_3$ state of heavy noble gases. *J. Phys. B*, **28**, 869–877.

[9818] Srivastava R., McEachran R.P. and Stauffer A.D. (1995b). Relativistic distorted-wave calculation of electron excitation of ytterbium. *J. Phys. B*, **28**, 885–891.

[9819] Srivastava R., Zeeman V., McEachran R.P. and Stauffer A.D. (1995c). Excitation of copper by electron impact in the relativistic distorted-wave approximation. *J. Phys. B*, **28**, 1059–1066.

[9820] Srivastava R., Zuo T., McEachran R.P. and Stauffer A.D. (1993). Differential cross sections for the excitation of mercury. *J. Phys. B*, **26**, 1025–1030.

[9821] Stachiotti G., Corà F., Catlow C.R.A. and Rodriguez C.O. (1997). First-principles investigation of of ReO_3 and related oxides. *Phys. Rev. B*, **55**, 7508–7514.

[9822] Stahler S., Schutz G. and Ebert H. (1993). Magnetic K-edge absorption in $3d$ elements and its relation to local magnetic structure. *Phys. Rev. B*, **47**, 818–826.

[9823] Stahlhofen A.A. (1997). Algebraic solutions of relativistic Coulomb problems. *Helv. Phys. Acta*, **70**, 372–386.

[9824] Staruszkiewicz A. (1998). Quantum mechanics of the electric charge. *Acta Phys. Pol.*, **29**, 929–936.

[9825] Staudte D.S. (1996). An eight-component relativistic wave equation for spin-1/2 particles II. *J. Phys. A*, **29**, 169–192.

[9826] Stegmann R. and Frenking G. (1998). Mechanism of the acetylene-vinylidene rearrangement in the coordination sphere of a transition metal. *Organometallics*, **17**, 2089–2095.

[9827] Stegmann R., Neuhaus A. and Frenking G. (1993). Theoretical studies of organometallic compounds. 5. Alkyne and vinylidene complexes of molybdenum and tungsten in high-oxidation states. *J. Am. Chem. Soc.*, **115**, 11930–11938.

[9828] Stein M. (1993). Pseudo-potential approach to the relativistic treatment of alkali atoms. *J. Phys. B*, **26**, 2087–2097.

[9829] Steinbeck L., Richter M., Eschrig H. and Nitzsche U. (1994). Calculated crystal-field parameters for rare-earth impurities in noble metals. *Phys. Rev. B*, **49**, 16289–16292.

[9830] Steinbeck L., Richter M., Nitsche U. and Eschrig H. (1996). Ab initio calculation of electronic structure, crystal field, and intrinsic magnetic properties of Sm_2Fe_{17}, $Sm_2Fe_{17}N_3$, $Sm_2Fe_{17}C_3$, and Sm_2Co_{17}. *Phys. Rev. B*, **53**, 7111–7127.

[9831] Steinbrenner U., Bergner A., Dolg M. and Stoll H. (1994). On the transferability of energy adjusted pseudopotentials:a calibration study for XH_4 (X=C, Si, Ge, Sn, Pb). *Mol. Phys.*, **82**, 3–11.

[9832] Steiner U.E. and Serebrennikov Y.A. (1994). Adiabatic rotation of effective spin. I. New insight into spin-rotational interaction. *J. Chem. Phys.*, **100**, 7503–7507.

[9833] Stener M., Albert K. and Rösch N. (1999). Relativistic density functional study on the bimetallic cluster $[Pt_3Fe_3(CO)_{15}]^{n-}$ ($n = 0, 1, 2$). *Inorg. Chim. Acta*, **286**, 30–36.

[9834] Stevens W.J., Basch H. and Krauss M. (1984). Compact effective potentials and efficient shared-exponent basis sets for the first- and second-row atoms. *J. Chem. Phys.*, **81**, 6026–6033.

[9835] Stewart G.M., Tiekink E.R.T. and Buntine M.A. (1997). Structural aspects of the coordination of triethylphosphinegold(I) to 2-thiouracil: A comparison between theory and experiment. *J. Phys. Chem. A*, **101**, 5368–5373.

[9836] Stöcker H., Gallmann A. and Hamilton J.H. (1998). Structure of Vacuum and Elementary Matter (Ed.). World Scientific, Singapore.

[9837] Stoitsov M., Ring P., Vretenar D. and Lalazissis G.A. (1998). Solution of relativistic Hartree-Bogoliubov equations in configurational representation: Spherical neutron halo nuclei. *Phys. Rev. C*, **58**, 2086–2091.

[9838] Stoll H. and Werner H.J. (1996). The Cr_2 potential curve: a multireference pair functional treatment. *Mol. Phys.*, **88**, 793–802.

[9839] Stowasser R. and Hoffmann R. (1999). What do the Kohn-Sham orbitals and eigenvalues mean? *J. Am. Chem. Soc.*, **121**, 3414–3420.

[9840] Strange P. (1998). *Relativistic Quantum Mechanics with Applications in Condensed Matter and Atomic Physics*. Cambridge Univ. Press, 594 pp.

[9841] Stranger R., Macgregor S.A., Lovell T., McGrady J.E. and Heath G.A. (1996a). Density-functional study of the ground- and excited-spin states of $[M_2Cl_9]^{3-}$ (M = Mo or W) face-shared dimers: consequences for structural variation in $A_3M_2Cl_9$ complexes. *J. Chem. Soc., Dalton Trans.*, pp. 4485–4491.

[9842] Stranger R., Medley G.A., McGrady J.E., Garrett J.M. and Appleton T.G. (1996b). Electronic structure of $[Pt_2(\mu\text{-}O_2CCH_3)_4(H_2O)_2]^{2+}$ using the quasi-relativistic Xα-SW method: Analysis of metal-metal bonding, assignment of electronic spectra, and comparison with $Rh_2(\mu\text{-}O_2CCH_3)_4(H_2O)_2$. *Inorg. Chem.*, **35**, 2268–2275.

[9843] Strömberg D., Strömberg A. and Wahlgren U. (1991). Relativistic quantum calculations on some mercury sulfide molecules. *Water, Air, & Soil Pollution*, **56**, 681–.

[9844] Strömberg S., Svensson M. and Zetterberg K. (1997a). Binding of ethylene to anionic, neutral, and cationic nickel(II), palladium(II), and platinum(II) cis/trans chloride ammonia complexes. A theoretical study. *Organometallics*, **16**, 3165–3168.

[9845] Strömberg S., Zetterberg K. and Siegbahn P.E.M. (1997b). Trends within a triad: comparison between σ-alkyl complexes of nickel, palladium and platinum with respect to association of ethylene, migratory insertion and β-hydride elimination. A theoretical study. *J. Chem. Soc., Dalton Trans.*, pp. 4147–4152.

[9846] Strout D.L. and Hall M.B. (1996). Small yttrium-carbon and lanthanum-carbon clusters: Rings are most stable. *J. Phys. Chem.*, **100**, 18007–18009.

[9847] Stückl A.C., Daul C.A. and Güdel H.U. (1997). Excited-state energies and distortions of d^0 transition metal tetraoxo complexes: A density functional study. *J. Chem. Phys.*, **107**, 4606–4617.

[9848] Styszyński J., Cao X.P., Malli G.L. and Visscher L. (1997). Relativistic all-electron Dirac-Fock-Breit calculations on xenon fluorides (XeF_n, n = 1,2,4,6). *J. Comp. Chem.*, **18**, 601–608.

[9849] Su M.D. (1995a). Mechanism of photochemical rearrangements of 3-substituted cyclopropenes to cyclopentadienes. *Tetrahedron*, **51**, 12109–12118.

[9850] Su M.D. (1995b). Mechanism for the photorearrangements of cyclohexadienes. *J. Org. Chem.*, **60**, 6621–6623.

[9851] Su M.D. (1995c). The mechanism of photochemical rearrangement of vinylcyclopropanes to cyclopentenes. *Tetrahedron*, **51**, 5871–5876.

[9852] Su M.D. (1996a). Role of spin-orbit coupling and symmetry in triplet carbenic addition chemistry. *J. Phys. Chem.*, **100**, 4339–4349.

[9853] Su M.D. (1996b). The role of spin-orbit coupling and symmetry in photochemical rearrangements of α, β-unsaturated cyclic ketones. *Chem. Phys.*, **205**, 277–308.

[9854] Su M.D. (1996c). The role of spin-orbit coupling and symmetry in oxadi-π-methane rearrangements and some related photochemical reactions. *J. Org. Chem.*, **61**, 3080–3087.

[9855] Su M.D. and Chu S.Y. (1997a). An energetically feasible mechanism for the activation of the C-H bond by the 16-electron $CpM(PH_3)(CH_3)^+$ (M=Rh,Ir) complex. A theoretical study. *J. Am. Chem. Soc.*, **119**, 5373–5383.

[9856] Su M.D. and Chu S.Y. (1997b). C-F bond activation by the 14-electron $M(X)(PH_3)_2$ (M=Rh,Ir; X = CH_3, H, Cl) complex. A density functional study. *J. Am. Chem. Soc.*, **119**, 10178–10185.

[9857] Su M.D. and Chu S.Y. (1997c). A new aspect for the insertion of the 16-electron species (η^5-C_5H_5ML into saturated hydrocarbons. A (η^5-C_5H_5)ML +CH_4 (M = Rh, Ir; L = CO, SH_2, PH_3) case study. *J. Phys. Chem. A*, **101**, 6798–6806.

[9858] Su M.D. and Chu S.Y. (1997d). Theoretical model for insertion of the 16-electron species (η^5-C_5H_5)M(L) into saturated hydrocarbons. A (η^5-C_5H_5)M(CO) + CH_4 (M = Ru^-,Os^-,Rh,Ir,Pd^+,Pt^+) case study. *Organometallics*, **16**, 1621–1627.

[9859] Su M.D. and Chu S.Y. (1998a). Theoretical model for oxidative addition of the O-H bond to platinum(0) complexes. *Chem. Phys. Lett.*, **282**, 25–28.

[9860] Su M.D. and Chu S.Y. (1998b). Singlet-triplet splitting and the activation of C–H bond for (η^5-C_5H_5)M(CO) isoelectronic fragments: A theoretical study. *Int. J. Quantum Chem.*, **70**, 961–971.

[9861] Su M.D. and Chu S.Y. (1998c). Substituent effects on oxidative addition for coordinative unsaturated d^8 ML_3. Mechanistic and thermodynamic considerations. *J. Phys. Chem. A*, **102**, 10159–10166.

[9862] Su M.D. and Chu S.Y. (1998d). Theoretical study of oxidative addition and reductive elimination of 14-electron d^{10} ML_2 complexes: A ML_2 +CH_4 (M = Pd, Pt; L = CO, PH_3, L_2 = $PH_2CH_2CH_2PH_2$) case study. *Inorg. Chem.*, **37**, 3400–3406.

[9863] Su M.D. and Chu S.Y. (1999a). A theoretical model for the orientation of 16-electron [CpML] insertion into the C-H bond of propane and cyclopropane and its regio- and stereoselectivity. *Chem. Eur. J.*, **5**, 198–207.

[9864] Su M.D. and Chu S.Y. (1999b). Resistivity of the C–X (X=F, Cl, Br, and I) bond activation in CX_4 by an iridium(I) complex from a theoretical viewpoint. *J. Am. Chem. Soc.*, **121**, 1045–1058.

[9865] Su M.D. and Chu S.Y. (1999c). Density functional theory of C–H bond activation by transition-metal complex: A (η^5-C_5H_5)ML (M=Rh, Ir; L=CH_2, CO, SH_2, PH_3) + CH_4 case study. *Int. J. Quantum Chem.*, **72**, 405–410.

[9866] Su M.D. and Chu S.Y. (1999d). A correlation between C-H bond activation barrier and singlet-triplet energy gap of transition metal complex - density functional study on CpML insertion into CH_4. *J. Chin. Chem. Soc. (Taipei)*, **46**, 403–407.

[9867] Su Q., Smetanko B.A. and Grobe R. (1998). Relativistic suppression of wave packet spreading. *Optics Express*, **2**, 277–281.

[9868] Su R.K., Siu G.G. and Chou X. (1993). Barrier penetration and Klein paradox. *J. Phys. A*, **26**, 1001–1005.

[9869] Su Z.M., Zhang H.X. and Che C.M. (1997). An ab initio study on the conformation and gold(I)-gold(I) interaction of the isomeric $H_2C[P(Ph)_2AuX]_2$ and $HC[P(Ph)_2AuX]_3$ (X=I,Cl). *Chem. J. Chinese Univ.*, **18**, 1171–1179.

[9870] Su Z.W. and Coppens P. (1997). Relativistic x-ray elastic scattering factors for neutral atoms Z=1-54 from multiconfiguration Dirac-Fock wavefunctions in the 0-12 $A^{-1}\sin\theta/\lambda$ range, and six Gaussian analytical expressions in the 0-6 A^{-1} range. *Acta Cryst. A*, **53**, 749–762.

[9871] Subramanian V. and Ramasami T. (1995). Calculation of the ionization energy of the tetrabromides and tetraiodides of Zr(IV) and Hf(IV) using the MS-Xα method. *J. Mol. Str.*, **332**, 177–182.

[9872] Subramanian V., Vijayakumar M. and Ramasami T. (1993). Calculation of charge transfer energies and ionization potentials of MoS_4^{2-} and WS_4^{2-} complexes using the multiple scattering Xα method. *J. Mol. Str. (Theochem)*, **284**, 157–162.

[9873] Sucher J. (1995). Confinement in relativistic potential models. *Phys. Rev. D*, **51**, 5965–5966.

[9874] Sucher J. (1998). What is the force between electrons? *Adv. Quantum Chem.*, **30**, 433–443.

[9875] Sugawara-Tanabe K. and Arima A. (1998). Hidden pseudospin symmetry in the Dirac equation. *Phys. Rev. C*, **58**, R3065–R3068.

[9876] Sugawara-Tanabe K., Meng J., Yamaji S. and Arima A. (1999). The pseudo-spin symmetry in a Dirac equation. *J. Phys. G*, **25**, 811–813.

[9877] Sugimoto M., Horiuchi F. and Sakaki S. (1997). Pt^I-Pt^I bond energy in dinuclear Pt^I complexes. A theoretical study. *Chem. Phys. Lett.*, **274**, 543–548.

[9878] Sugimoto M., Yamasaki I., Mizoe N., Anzai M. and Sakaki S. (1999). Acetylene insertion reactions into Pt(II)-H and Pt(II)-SiH_3 bonds. An ab initio MO study and analysis based on the vibronic coupling model. *Theor. Chem. Acc.*, **102**, 377–384.

[9879] Sumi T., Miyoshi E. and Sakai Y. (1998). Molecular-orbital and molecular-dynamics study of mercury. *Phys. Rev. B*, **57**, 914–918.

[9880] Sumi T., Miyoshi E. and Tanaka K. (1999). Molecular-dynamics study of liquid mercury in the density region between the metal and nonmetal. *Phys. Rev. B*, **59**, 6153–6158.

[9881] Sundermann A. and Schoeller W.W. (1999). Electronic structure metallacyclophosphazene and metallacyclothiazene complexes. *Inorg. Chem.*, **38**, 6261–6270.

[9882] Sundholm D. (1994). Fully numerical solutions of molecular Dirac equations for highly charged one-electron homonuclear diatomic molecules. *Chem. Phys. Lett.*, **223**, 469–473.

[9883] Sundholm D. (1995). Core-valence correlation effects on the ground-state electron affinities of strontium and barium. *J. Phys. B*, **28**, L399–L404.

[9884] Sundholm D. and Olsen J. (1994). Core-valence correlation effects on the ground state electron affinity of calcium. *Chem. Phys. Lett.*, **217**, 451–455.

[9885] Sundholm D. and Ottschofski E. (1997). Relativistic multiconfiguration Hartree-Fock by means of Direct Perturbation Theory. *Int. J. Quantum Chem.*, **65**, 151–158.

[9886] Sundholm D., Tokman M., Pyykkö P., Eliav E. and Kaldor U. (1999). *Ab initio* calculations of the ground-state electron affinities of gallium and indium. *J. Phys. B*, **32**, 5853–5859.

[9887] Sunnergren P. (1998). *Complete One-Loop QED Calculations for Few-Electron Ions. Applications to Electron-Electron Interaction, the Zeeman Effect and Hyperfine Structure*. Ph.D. thesis, Göteborg, 141 p.

[9888] Sunnergren P., Persson H., Salomonson S., Schneider S.M., Lindgren I. and Soff G. (1998). Radiative corrections to the hyperfine structure splitting of hydrogenlike systems. *Phys. Rev. A*, **58**, 1055–1069.

[9889] Sushkov O.P. (1993). The possibility to observe P and PT violation using the NMR frequency shift in a laser beam. *Phys. Scr.*, **T46**, 193–197.

[9890] Suzuki A. and Nogami Y. (1993). Variable-phase approach to the Dirac equation-bound states. *Nuovo Cim. B*, **108**, 303–312.

[9891] Suzuki S. and Nakao K. (1999). A fully relativistic full-potential LCAO method for solids. *J. Phys. Soc. Japan*, **68**, 1982–1987.

[9892] Suzumura T., Nakajima T. and Hirao K. (1999). Ground-state properties of MH, MCl, and M_2 (M = Cu, Ag, and Au) calculated by a scalar relativistic density functional theory. *Int. J. Quantum Chem.*, **75**, 757–766.

[9893] Svane A. (1994). Electronic structure of cerium in the self-interaction corrected local spin density approximation. *Phys. Rev. Lett.*, **72**, 1248–1251.

[9894] Svane A. (1996). Electronic structure of cerium in the self-interaction-corrected local-spin-density approximation. *Phys. Rev. B*, **53**, 4275–5286.

[9895] Svane A., Christensen N.E., Rodriguez C.O. and Methfessel M. (1997). Calculations of hyperfine parameters in tin compounds. *Phys. Rev. B*, **55**, 12572–12577.

[9896] Svensson M., Humbel S., Froese R.D.J., Matsubara T., Sieber S. and Morokuma K. (1996). ONIOM: A multilayered integrated MO + MM method for geometry optimizations and single point energy predictions. A test for Diels-Alder reactions and $Pt(P(t\text{-}Bu)_3)_2 + H_2$ oxidative addition. *J. Phys. Chem.*, **100**, 19357–19363.

[9897] Svensson P.H., Rosdahl J. and Kloo L. (1999). Metal iodides in polyiodide networks – The structural chemistry of complex gold iodides with excess iodine. *Chem. Eur. J.*, **5**, 305–311.

[9898] Swang O., Baerends E.J., Faegri Jr. K. and Gropen O. (1996). Theoretical cluster model studies of bimetallic heterogeneous catalysis: dissociation of hydrogen on pure and rhenium-doped Pt(1,0,0). *J. Mol. Struct. (Theochem)*, **388**, 321–329.

[9899] Swang O., Faegri Jr. K. and Gropen O. (1993). Theoretical cluster model studies of bimetallic heterogeneous catalysis: dissociation of hydrogen on a rhenium-doped nickel cluster. *Chem. Phys. Lett.*, **207**, 397–402.

[9900] Swang O., Faegri Jr. K. and Gropen O. (1994). Theoretical study of methane activation by Re, Os, Ir, and Pt. *J. Phys. Chem.*, **98**, 3006–3009.

[9901] Szilagyi R.K. and Frenking G. (1997). Structure and bonding of the isoelectronic hexacarbonyls $[Hf(CO)_6]^{2-}$, $[Ta(CO)_6]^-$, $W(CO)_6$, $[Re(CO)_6]^+$, $[Os(CO)_6]^{2+}$, and $[Ir(CO)_6]^{3+}$: A theoretical study. *Organometallics*, **16**, 4807–4815.

[9902] Szmytkowski R. (1993a). Theoretical study of low-energy positron scattering on alkaline-earth atoms in the relativistic polarized orbital approximation. *J. Physique II*, **3**, 183–189.

[9903] Szmytkowski R. (1993b). The elastic positron scattering from mercury in the relativistic polarized orbital method. *J. Phys. B*, **26**, 535–545.

[9904] Szmytkowski R. (1993c). Elastic positron scattering from zinc and cadmium in the relativistic polarized orbital approximation. *Acta Phys. Pol. A*, **84**, 1035–1040.

[9905] Szmytkowski R. (1994). Elastic positron scattering from krypton and xenon in the relativistic polarized orbital approximation. *Acta Phys. Pol. A*, **86**, 309–314.

[9906] Szmytkowski R. (1995). The relativistic multi-channel variable phase method for solving asymptotic equations in electron-atom and electron-ion scattering. *Comp. Phys. Comm.*, **90**, 244–250.

[9907] Szmytkowski R. (1997). The Dirac-Coulomb Sturmians and the series expansion of the Dirac-Coulomb Green function: application to the relativistic polarizability of the hydrogen-like atom (Erratum: Ibid. p. 2747). *J. Phys. B*, **30**, 825–861.

[9908] Szmytkowski R. (1998a). The continuum Schrödinger–Coulomb and Dirac–Coulomb Sturmian functions [Errata: 31, 7415-7416]. *J. Phys. A*, **31**, 4963–4990.

[9909] Szmytkowski R. (1998b). Unified construction of variational R-matrix methods for the Dirac equation. *Phys. Rev. A*, **57**, 4351–4364.

[9910] Szmytkowski R. and Alhasan A.M. (1995). Relativistic calculations of static electric polarizabilities for alkaline-earth-metals and their isoelectronic sequences. *Phys. Scr.*, **52**, 309–312.

[9911] Szmytkowski R. and Hinze J. (1996). Convergence of the non-relativistic and relativistic R-matrix expansions at the reaction volume boundary. *J. Phys. B*, **29**, 761–777.

[9912] Szmytkowski R. and Sienkiewicz J.E. (1994a). Spin polarization of slow electrons elastically scattered from mercury cadmium and zinc atoms. *J. Phys. B*, **27**, 555–563.

[9913] Szmytkowski R. and Sienkiewicz J.E. (1994b). Spin polarization of slow electrons elastically scattered from xenon atoms. *J. Phys. B*, **27**, 2277–2282.

[9914] Szmytkowski R. and Sienkiewicz J.E. (1994c). Elastic scattering of electrons by strontium and barium atoms. *Phys. Rev. A*, **50**, 4007–4012.

[9915] Szymanowski S. and Maquet A. (1998). Relativistic signatures in laser-assisted scattering at high field intensities. *Optics Express*, **2**, 262–267.

[9916] Szymanowski C., Véniard V., Taïeb R. and Maquet A. (1997). Relativistic calculation of two-photon bound-bound transition amplitudes in hydrogenic atoms. *Phys. Rev. A*, **56**, 700–711.

[9917] Szyperski T. and Schwerdtfeger P. (1989). Theoretische Betrachtungen über die Stabilität der Trioxo(η^5-cyclopentadienyl)-Verbindungen der Übergangsmetalle Mangan, Technetium und Rhenium. *Angew. Chem.*, **101**, 1271–1274.

[9918] Taatjes C.A., Mastenbroek J.W.G., van den Hoek G., Snijders J.G. and Stolte S. (1993). Polarization-resolved (2+1) resonance-enhanced multiphotonionization spectroscopy of CF_3I (6s) Rydberg states. *J. Chem. Phys.*, **98**, 4355–4371.

[9919] Tachiev G. and Froese Fischer C. (1999). Breit-Pauli energy levels, lifetimes, and transition data: beryllium-like spectra. *J. Phys. B*, **32**, 5805–5823.

[9920] Tadjeddine M., Flament J.P. and Teichteil C. (1987). Non-empirical spin-orbit calculation of the CH_3I ground state. *Chem. Phys.*, **118**, 45–55.

[9921] Taïeb R., Véniard V. and Maquet A. (1998). Signature of relativistic effects in atom-laser interactions at ultrahigh intensities. *Phys. Rev. Lett.*, **81**, 2882–2885.

[9922] Takabayasi T. (1955). On the structure of Dirac wave function. *Progr. Theor. Phys.*, **13**, 106–108.

[9923] Takabayasi T. and Vigier J.P. (1957). Relativistic hydrodynamics of the Dirac matter. *Progr. Theor. Phys., Suppl.*, **4**, 1–.

[9924] Takahashi O., Saito K. and Yabushita S. (1999). Simple SCF method with spin-orbit interaction: SOSCF method. *Int. J. Quantum Chem.*, **74**, 515–530.

[9925] Takahashi O., Saito K., Yamamoto S. and Nishimura N. (1993). A theoretical study of cadmium-ethylenediamine and -ammonia exciplexes. Comparison with experiments. *Chem. Phys. Lett.*, **207**, 379–383.

[9926] Takashima H., Hada M. and Nakatsuji H. (1995). Spin-orbit effect on the magnetic shielding constant using the ab initio UHF method: gallium and indium tetrahalides. *Chem. Phys. Lett.*, **235**, 13–16.

[9927] Talman J.D. (1994). Spurious solutions arising in matrix approximations to the Dirac-Coulomb problem. *Phys. Rev. A*, **50**, 3525–3526.

[9928] Tamura E., van Ek J., Froeba M. and Wong J. (1995). X-ray absorption near edge structure in metals: relativistic effects and core-hole screening. *Phys. Rev. Lett.*, **74**, 4899–4902.

[9929] Tamura E., Waddill G.D., Tobin J.G. and Sterne P.A. (1994). Linear and circular dichroism in angle resolved Fe 3p photoemission. *Phys. Rev. Lett.*, **73**, 1533–1536.

[9930] Tan A.L., Chiew M.L. and Hor T.S.A. (1997). ($Pt_2(PPh_3)_4(\mu\text{-}S)_2$) as a metalloligand toward main-group Lewis acids: theoretical study of intermetallic complexes with Tl(I), Pb(II), In(III) and Ga(III). *J. Mol. Str. (Theochem)*, **393**, 189–196.

[9931] Tan H., Dai D.G. and Balasubramanian K. (1998a). Spectroscopic properties and potential energy curves for 15 electronic states of palladium carbide (PdC). *Chem. Phys. Lett.*, **286**, 375–381.

[9932] Tan H., Liao M.Z., Dai D.G. and Balasubramanian K. (1998b). Potential energy surfaces of NbCO. *Chem. Phys. Lett.*, **297**, 173–180.

[9933] Tan H., Liao M.Z. and Balasubramanian K. (1997a). Electronic states and potential energy curves of iridium carbide (IrC). *Chem. Phys. Lett.*, **280**, 219–226.

[9934] Tan H., Liao M.Z. and Balasubramanian K. (1997b). Electronic states and potential energy curves of rhodium carbide (RhC). *Chem. Phys. Lett.*, **280**, 423–429.

[9935] Tan H., Liao M.Z. and Balasubramanian K. (1998c). Potential energy surfaces of RuCO. *Chem. Phys. Lett.*, **284**, 1–5.

[9936] Tan H., Liao M.Z. and Balasubramanian K. (1998d). Potential energy surfaces of OsCO. *Chem. Phys. Lett.*, **290**, 458–464.

[9937] Tan H., Liao M.Z., Dai D.G. and Balasubramanian K. (1998e). Potential energy surfaces for Mo + CO and W + CO. *J. Phys. Chem. A*, **102**, 6801–6806.

[9938] Tan H., Liao M.Z., Dai D.G. and Balasubramanian K. (1999). Potential energy surfaces for Tc + CO, Re + CO, and Ta + CO and periodic trends of the second- and third-row transition metals interaction with CO. *J. Phys. Chem. A*, **103**, 3495–3504.

[9939] Tan M.L., Zhu Z.H., Zhao Y.K. and Chen X.F. (1996a). Relativistic multiconfiguration calculation of fine-structure energy levels. *Wuli Xuebao*, **45**, 1609–1614.

[9940] Tan M.L., Zhu Z.H., Zhao Y.K., Peng H.S. and Cheng X.F. (1996b). Calculation of energy levels and transitions for the highly stripped ion Bi^{54+} (in Chinese). *Yuanzi Yu Fenzi Wuli Xuebao*, **13**, 189–194.

[9941] Tang D. and Dorignac D. (1994). The calculation of scattering factors in HERM image simulation. *Acta Cryst.*, **A50**, 45–52.

[9942] Tang Z. and Finkelstein D. (1994). Relativistically covariant symmetry in QED. *Phys. Rev. Lett.*, **73**, 3055–3057.

[9943] Tangerman R.D. and Tjon J.A. (1993). Exact supersymmetry in the nonrelativistic hydrogen atom. *Phys. Rev. A*, **48**, 1089–1092.

[9944] Tanpipat N. and Baker J. (1996). Density functional calculations on WH_6 and WF_6. *J. Phys. Chem.*, **100**, 19818–19823.

[9945] Tarczay G., Császár A.G., Klopper W., Szalay V., Allen W.D. and Schaefer III H.F. (1999). The barrier to linearity of water. *J. Chem. Phys.*, **110**, 11971–11981.

[9946] Tashkova M. and Donev S. (1993). Relativistic electron tunneling through a one-dimensional structured barrier. *Phys. Stat. Sol. B*, **178**, K83–K86.

[9947] Tatchen J. and Marian C.M. (1999). On the performance of approximate spin-orbit Hamiltonians in light conjugated molecules: the fine-structure splitting of HC_6H^+, NC_5H^+, and NC_4N^+. *Chem. Phys. Lett.*, **313**, 351–357.

[9948] Tatewaki H. and Matsuoka O. (1997). All-electron Dirac-Fock-Roothaan calculations on the electronic structure of the GdF molecule. *J. Chem. Phys.*, **106**, 4558–4565.

[9949] Tatewaki H. and Matsuoka O. (1998). All-electron Dirac-Fock-Roothaan calculations for the electronic structures of the GdF_2 molecule. *Chem. Phys. Lett.*, **283**, 161–166.

[9950] Tatewaki H., Sekiya M., Sasaki F., Matsuoka O. and Koga T. (1995). $6s$ and $4f$ ionized states of the lanthanides calculated by numerical and analytical Hartree-Fock methods. *Phys. Rev. A*, **51**, 197–203.

[9951] Taylor M.B. and Gyorffy B.L. (1993). A ferromagnetic monolayer with model spin-orbit and dipole-dipole interactions. *J. Phys: CM*, **5**, 4527–4540.

[9952] Teichteil C. and Pélissier M. (1994). Relativistic calculations of excited states of molecular iodine. *Chem. Phys.*, **180**, 1–18.

[9953] Teles J.H., Brode S. and Chabanas M. (1998). Cationic gold(I) complexes: highly efficient catalysts for the addition of alcohols to alkynes. *Angew. Chem. Int. Ed. Engl.*, **37**, 1415–1418.

[9954] Temmerman W.M., Szotek Z., Svane A., Strange P., Winter H., Delin A., Johansson B., Eriksson O., Fast L. and Wills J.M. (1999). Electronic configuration of Yb compounds. *Phys. Rev. Lett.*, **83**, 3900–3903.

[9955] Teng H.G., Xu Z.Z., Shen B.F. and Zhang W.Q. (1994a). Concise rate coefficient formulas for dielectronic recombination of the He-like isoelectronic sequence. *Phys. Scr.*, **49**, 696–698.

[9956] Teng H.G., Xu Z.Z., Zhang W.Q. and Shen B.F. (1994b). Analytic formulas of dielectronic recombination rate coefficients for ions of the F-like isoelectronic sequence. *Phys. Scr.*, **50**, 55–60.

[9957] Terpstra H.J., de Groot R.A. and Haas C. (1995). Electronic structure of the lead monoxides: band-structure calculations and photoelectron spectra. *Phys. Rev. B*, **52**, 11690–11696.

[9958] Terpstra H.J., de Groot R.A. and Haas C. (1997). The electronic structure of the mixed valence compound Pb_3O_4. *J. Phys. Chem. Solids*, **58**, 561–566.

[9959] Teychenné D., Bonnaud G. and Bobin J.L. (1994). Oscillatory relativistic motion of a particle in a power-law or sinusoidal-shaped potential well. *Phys. Rev. E*, **49**, 3253–3263.

[9960] Tezuka H. (1995). Confinement by polynomial potentials. *Z. Phys. C*, **65**, 101–104.

[9961] Thiel J., Hoffstadt J., Grün N. and Scheid W. (1994). Electron-positron pair creation in relativistic atomic heavy collisions. In *Frontier Topics in Nuclear Physics*, (Edited by Scheid W. and Sandulescu A.), pp. 453–464, Plenum Press, New York.

[9962] Thiel J., Hoffstadt J., Grün N. and Scheid W. (1995). Fermionic and bosonic coupled channel calculations in momentum space for electron-positron pair creation in relativistic heavy-ion collisions. *Z. Phys. D*, **34**, 21–28.

[9963] Thiel W. and Voityuk A.A. (1996). Extension of MNDO to d orbitals: parameters and results for the second-row elements and for the zinc group. *J. Phys. Chem.*, **100**, 616–626.

[9964] Thøgersen J., Scheer M., Steele L.D., Haugen H.K. and Wijesundera W.P. (1996). Two-photon detachment of negative ions via magnetic dipole transitions. *Phys. Rev. Lett.*, **76**, 2870–2873.

[9965] Thole B.T., Wang X.D., Harmon B.N., Li D.Q. and Dowben P.A. (1993). Multiplet fine structure in the photoemission of the gadolinium and terbium 5p levels. *Phys. Rev. B*, **47**, 9098–9101.

[9966] Thumm U. (1993). The Dirac R-matrix method for scattering of slow electrons from alkali-metal-like targets. *AIP Conf. Proc.*, **295**, 263–275.

[9967] Thumm U., Bartschat K. and Norcross D.W. (1993). Relativistic effects in spin-polarization parameters for low-energy electron-Cs scattering. *J. Phys. B*, **26**, 1587–1598.

[9968] Thumm U., Baştuğ T. and Fricke B. (1995). Target-electronic-structure dependence in highly-charged-ion-C_{60} collisions. *Phys. Rev. A*, **52**, 2955–2964.

[9969] Thumm U. and Norcross D.W. (1993). Angle-differential and momentum-transfer cross sections for low-energy electron-Cs scattering. *Phys. Rev. A*, **47**, 305–316.

[9970] Timberlake J.M., Green J., Christou V. and Arnold J. (1998). Electronic structure of chalcogenols: photoelectron spectroscopic and theoretical studies of tris(trimethylsilyl) chalcogenols. *J. Chem. Soc., Dalton Trans.*, pp. 4029–4033.

[9971] Timms D.N. and Cooper M.J. (1993). The electron momentum distribution in lead. *Z. Natur. A*, **48**, 343–347.

[9972] Timoshkin A.Y., Suvorov A.V., Bettinger H.F. and Schaefer III H.F. (1999). Role of the terminal atoms in the donor-acceptor complexes MX_3-D (M = Al, Ga, In; X = F, Cl, Br, I; D = YH_3, YX_3, X^-; Y = N, P, As). *J. Am. Chem. Soc.*, **121**, 5687–5699.

[9973] Titov A.V. (1992). Matrix elements of the $U(2n)$ generators in the spin-orbit basis. *Int. J. Quantum Chem.*, **42**, 1711–1716.

[9974] Titov A.V. (1996). A two-step method of calculation of the electronic structure of molecules with heavy atoms: theoretical aspect. *Int. J. Quantum Chem.*, **57**, 453–463.

[9975] Titov A.V. and Mosyagin N.S. (1995). Self-consistent relativistic effective core potentials for transition metal atoms: Cu, Ag, and Au. *Structural Chem.*, **6**, 317–321.

[9976] Titov A.V. and Mosyagin N.S. (1999). Generalized relativistic effective core potential: Theoretical grounds. *Int. J. Quantum Chem.*, **71**, 359–401.

[9977] Titov A.V., Mosyagin N.S. and Ezhov V.F. (1996). P,T-odd spin-rotational Hamiltonian for the YbF molecule. *Phys. Rev. Lett.*, **77**, 5346–5349.

[9978] Tiwary S.N. (1993). Relativistic and nonrelativistic oscillator strengths in the Na isoelectronic sequence. *Int. J. Theor. Phys.*, **32**, 2047–2051.

[9979] Tiwary S.N. and Kandpal P. (1994). Effect of configuration interaction and relativity on oscillator strengths for sodium-like systems. *Nuovo Cim. D*, **16**, 339–358.

[9980] Tkachuk V.M. and Roy P. (1999). Supersymmetry of a spin $\frac{1}{2}$ particle on the real line. *Phys. Lett. A*, **263**, 245–249.

[9981] Tkachuk V.M. and Vakarchuk S.I. (1996). The N=4 supersymmetry of electron in the magnetic field. *Zh. Fyiz. Dosl.*, **1(1)**, 3941.

[9982] Todorov I.T. (1971). Quasipotential equation corresponding to the relativistic eikonal approximation. *Phys. Rev. D*, **3**, 2351–2356.

[9983] Tokitoh N., Arai Y., Okazaki R. and Nagase S. (1997). Synthesis and characterization of a stable dibismuthene: Evidence for a Bi-Bi double bond. *Science*, **277**, 78–80.

[9984] Tolstikhina I.Y., Tawara H. and Safronova U.I. (1994). K-X-ray satellite energies and radiative probabilities of Ar ions formed under slow Ar^{17+} ions upon metallic surface. *Nucl. Instr. Meth. Phys. Res. B*, **86**, 241–244.

[9985] Tomàs J., Lledós A. and Jean Y. (1998a). The Kubas complex revisited. A theoretical study of dihydrogen addition and structure of the dihydride form. *Organometallics*, **17**, 190–195.

[9986] Tomàs J., Lledós A. and Jean Y. (1998b). A theoretical insight into the ability of group 6 ML_5 metal fragments to break the H-H bond. *Organometallics*, **17**, 4932–4939.

[9987] Tomaselli M., Kühl T., Seelig P., Holbrow C. and Kankeleit E. (1998). Hyperfine splittings of hydrogenlike ions and the dynamic-correlation model for one-hole nuclei. *Phys. Rev. C*, **58**, 1524–1534.

[9988] Tomasulo A. and Ramakrishna M.V. (1996a). Spin-orbital angular momentum coupling effects on the electronic structure of nanocrystalline semiconductor clusters. *Chem. Phys.*, **210**, 55–70.

[9989] Tomonaga S.I. (1997). *The Story of Spin.* The University of Chicago Press, 258 p.

[9990] Tomasulo A. and Ramakrishna M.V. (1996b). Quantum confinement effects in semiconductor clusters. II. *J. Chem. Phys.*, **105**, 3612–3626.

[9991] Tong X.M. and Chu S.I. (1998). Relativistic density-functional theory with the optimized effective potential and self-interaction correction: Application to atomic structure calculations (Z=2-106). *Phys. Rev. A*, **57**, 855–863.

[9992] Tong X.M., Liu L. and Li J.M. (1994a). Relativistic effect of atomic radiative processes. *Phys. Rev. A*, **49**, 4641–4644.

[9993] Tong X.M., Zou Y., Li J.M. and Liu X.W. (1994b). Relativistic theoretical calculation of O III: radiative transition rates and the Bowen mechanism in planetary nebulae. *Chinese Phys. Lett.*, **11**, 69–72.

[9994] Tornaghi E., Andreoni W., Carloni P., Hutter J. and Parrinello M. (1995). Carboplatin versus cisplatin: density functional approach to their molecular properties. *Chem. Phys. Lett.*, **246**, 469–474.

[9995] Toronto D.V., Balch A.L. and Tinti D.S. (1994). Photoluminesence from electron-deficient mixed-metal clusters with $[Pt_2Au]^{3+}$ and $[Pt_2Hg]^{4+}$ cores. *Inorg. Chem.*, **33**, 2507–2508.

[9996] Tornieporth-Oetting I.C., Klapötke T.M., Schulz A., Buzek P. and v R Schleyer P. (1993). The $I(N_3)_2^+$ cation: preparation identification by Raman spectroscopy and ab initio quantum mechanical studies. *Inorg. Chem.*, **32**, 5640–5642.

[9997] Torrent M., Deng L.Q. and Ziegler T. (1998). A density functional study of [2+3] versus [2+2] addition of ethylene to chromium-oxygen bonds in chromyl chloride. *Inorg. Chem.*, **37**, 1307–1314.

[9998] Torrent M., Gili P., Duran M. and Solá M. (1997). Molybdenum(VI) dioxodihalides: Agreement with experiment and prediction of unknown properties through density functional theory. *Int. J. Quantum Chem.*, **61**, 405–414.

[9999] Torres del Castillo G.F. and Cortés-Cuautli L.C. (1998). Solution of the Dirac equation in the field of a magnetic monopole. *J. Math. Phys.*, **38**, 2996–3006.

[10000] Tossell J.A. (1994). The speciation of antimony in sulfidic solutions: A theoretical study. *Geochim. Cosmochim. Acta*, **58**, 5093–5104.

[10001] Tossell J.A. (1996). The speciation of gold in aqueous solution: A theoretical study. *Geochim. Cosmochim. Acta*, **60**, 17–29.

[10002] Tossell J.A. (1997). Theoretical studies on arsenic oxide and hydroxide species in minerals and in aqueous solution. *Geochim. Cosmochim. Acta*, **61**, 1613–1623.

[10003] Tossell J.A. (1998a). A theoretical study of the decomposition of gold(I) complexes. *Chem. Phys. Lett.*, **286**, 73–78.

[10004] Tossell J.A. (1998b). Theoretical study of the photodecomposition of methyl Hg complexes. *J. Phys. Chem. A*, **102**, 3587–3591.

[10005] Tossell J.A. (1999). Theoretical studies on the formation of mercury complexes in solution and the dissolution and reactions of cinnabar. *American Mineralogist*, **84**, 877–883.

[10006] Toyama F.M. and Nogami Y. (1999). Harmonic oscillators in relativistic quantum mechanics. *Phys. Rev. A*, **59**, 1056–1062.

[10007] Toyama F.M., Nogami Y. and Coutinho F.A.B. (1997). Behaviour of wavepackets of the 'Dirac oscillator': Dirac representation versus Foldy-Wouthuysen representation. *J. Phys. A*, **30**, 2585–2595.

[10008] Toyama F.M., Nogami Y. and Zhao Z. (1993). Relativistic extension of the Kay-Moses method for constructing transparent potentials in quantum mechanics. *Phys. Rev. A*, **47**, 897–902.

[10009] Toyota S. and Ōki M. (1995). Mechanism of sulfur inversion in sulfide complexes of gold(I) and gold(III). *Bull. Chem. Soc. Jpn.*, **68**, 1345–1351.

[10010] Trail J.R. and Bird D.M. (1999a). Core reconstruction in pseudopotential calculations. *Phys. Rev. B*, **60**, 7863–7874.

[10011] Trail J.R. and Bird D.M. (1999b). Accurate structure factors from pseudopotential methods. *Phys. Rev. B*, **60**, 7875–7880.

[10012] Treboux G. and Barthelat J.C. (1993). X-X direct bonds versus bridged structures in group 13 X_2H_2 potential energy surfaces. *J. Am. Chem. Soc.*, **115**, 4870–4878.

[10013] Triebel H. (1980). *Höhere Analysis, 2. Auflage.*. Verlag Harri Deutsch, Thun und Frankfurt am Main, see pp. 554-569.

[10014] Trigueiros A.G., Mania A.J., Pettersson S.G. and Reyna-Almados J.G. (1993). Levels of the $4p^4$ configuration of Ge-like Kr V. *Phys. Rev. A*, **48**, 3595–3597.

[10015] Troullier N. and Martins J.L. (1991). Efficient pseudopotentials for plane-wave calculations. *Phys. Rev. B*, **43**, 1993–2006.

[10016] Tse J.S., Frapper G., Ker A., Rousseau R. and Klug D.D. (1999). Phase stability and electronic structure of K-Ag intermetallics at high pressure. *Phys. Rev. Lett.*, **82**, 4472–4475.

[10017] Tseng H.K. (1994). Calculation of pair production by photons of energies near threshold on atoms. *Phys. Rev. A*, **50**, 343–348.

[10018] Tseng H.K. (1995). Relativistic calculation of pair-production positron energy-angle distributions for low-energy photons on atoms. *Phys. Rev. A*, **52**, 369–374.

[10019] Tseng H.K. (1997a). Relativistic calculation of an elementary process of electron bremsstrahlung from atoms. *J. Phys. B*, **30**, L317–L321.

[10020] Tseng H.K. (1997b). Pair-production polarization correlations of intermediate energy photons on atoms. *J. Phys. B*, **30**, 557–564.

[10021] Tsuchiya T., Taketsugu T., Nakano H. and Hirao K. (1999). Theoretical study of electronic and geometric structures of a series of lanthanide trihalides LnX_3 (Ln=La-Lu; X=Cl, F). *J. Mol. Str. (Theochem)*, **461-462**, 203–222.

[10022] Tsushima S. and Suzuki A. (1999). Ab initio effective core potential study of equatorially coordinated uranyl species: effect of hydration to the calculated properties. *J. Mol. Str. (Theochem)*, **487**, 33–38.

[10023] Tu X.Y. and Dai S.S. (1994). Relativistic pseudopotential CI ab initio calculations on the low-lying states of XeH. *Chinese Sci. Bull.*, **39**, 284–287.

[10024] Tuan D.F.T. and Pitzer R.M. (1995a). Electronic structure of $Hf@C_{28}$ and its ions. 1. SCF calculations. *J. Phys. Chem.*, **99**, 9762–9767.

[10025] Tuan D.F.T. and Pitzer R.M. (1995b). Electronic structure of $Hf@C_{28}$ and its ions. 2. CI calculations. *J. Phys. Chem.*, **99**, 15069–15073.

[10026] Tuan D.F.T. and Pitzer R.M. (1996). Electronic structures of $C_{28}H_4$ and $Hf@C_{28}H_4$ and their ions. SCF calculations. *J. Phys. Chem.*, **100**, 6277–6283.

[10027] Tulkki J. (1993). Combined effect of relaxation and channel interaction on outer-shell photoionization in Ar K^+ and Ca^{2+}. *Phys. Rev. A*, **48**, 2048–2053.

[10028] Tulkki J., Aksela H. and Kabachnik N.M. (1993a). Evolution of the anisotropy of the Auger decay of $2p_{3/2}^{-1}ns$ resonances into anisotropy of L_3MM Auger decay in Ar. *Phys. Rev. A*, **48**, 2957–2961.

[10029] Tulkki J., Aksela H. and Kabachnik N.M. (1994). Influence of the initial- and final-state configuration interaction on the anisotropy of the resonant Auger decay of Kr $3d^{-1}5p$ and Xe $4d^{-1}6p$ states. *Phys. Rev. A*, **50**, 2366–2375.

[10030] Tulkki J., Kabachnik N.M. and Aksela H. (1993b). Effects of channel interaction, exchange and relaxation on the angular distribution and spin polarization of Auger electrons from noble-gas atoms. *Phys. Rev. A*, **48**, 1277–1291.

[10031] Tulkki J. and Mäntykenttä A. (1993). Subshell-dependent relaxation in the Auger effect. *Phys. Rev. A*, **47**, 2995–2999.

[10032] Tulub A.V., Pak M.V. and Brattsev V.F. (1997). Spectrum of hyperfine interaction with allowance for relativistic effects and quantum electrodynamic corrections (in Russian). *Opt. Spektr.*, **82**, 533–535.

[10033] Tulub A.V., Pak M.V. and Brattsev V.F. (1997). Spectrum of hyperfine interaction with allowance for relativistic effects and quantum electrodynamic corrections. *Opt. Spectr.*, **82**, 491–493.

[10034] Tupitsyn I.I., Mosyagin N.S. and Titov A.V. (1995). Generalized relativistic effect core potential. I. Numerical calculations for atoms Hg through Bi. *J. Chem. Phys.*, **103**, 6548–6555.

[10035] Turi Nagy L., Liška M. and Tunega D. (1994). Parametrization of the scaling factor in the quasi-relativistic INDO method based on the equilibrium geometries and vibrational frequencies of electroneutral closed-shell XY_4 molecules possessing the T_d stmmetry. *Collect. Czech. Chem. Commun.*, **59**, 1901–1910.

[10036] Türler A. (1996). Gas phase chemistry experiments with transactinide elements. *Radiochim. Acta*, **72**, 7–17.

[10037] Turski P. (1999). On the ground states of copper, silver and gold silicides. *Chem. Phys. Lett.*, **315**, 115–118.

[10038] Turski P. and Barysz M. (1999). Electronic states of the copper silicide and its ions. *J. Chem. Phys.*, **111**, 2973–2977.

[10039] Tyson T.A. (1994). Relativistic effects in the x-ray-absorption fine structure. *Phys. Rev. B*, **49**, 12578–12589.

[10040] Tzara C. (1985). A study of the relativistic Coulomb problem in momentum space. *Phys. Lett.*, **111A**, 343–348.

[10041] Uddin J., Dapprich S., Frenking G. and Yates B.F. (1999). Nature of the metal-alkene bond in platinum complexes of strained olefins. *Organometallics*, **18**, 457–465.

[10042] Uhl W., Jantschak A., Saak W., Kaupp M. and Wartchow R. (1998). Systematic experimental and quantum chemical investigation into the structures, the stability, and the spectroscopic properties of alkylindium(I) compounds: Tetrameric $In_4[C(SiMeRR')_3]_4$ versus monomeric $InC(SiMeRR')_3$ derivatives. *Organometallics*, **17**, 5009–5017.

[10043] Ujaque G., Maseras F. and Ledós A. (1996). A theoretical evaluation of steric and electronic effects on the structure of $OsO_4(NR_3)$, (NR_3 = bulky chiral alkaloid derivative) complexes. *Theor. Chim. Acta*, **94**, 67–73.

[10044] Ujaque G., Maseras F. and Ledós A. (1999). Theoretical study on the origin of enantioselectivity in the bis(dihydroquinidine)-3,6-pyridazine·osmium tetroxide-catalyzed dihydroxylation of styrene. *J. Am. Chem. Soc.*, **121**, 1317–1323.

[10045] Újfalussy B., Szunyogh L. and Weinberger P. (1995). Magnetism of $4d$ and $5d$ adlayers on Ag(001) and Au(001): Comparison between a nonrelativistic and a fully relativistic approach. *Phys. Rev. B*, **51**, 12836–12839.

[10046] Újsághy O. and Zawadowski A. (1998). Spin-orbit-induced magnetic anisotropy for impurities in metallic samples. II. Finite-size dependence in the Kondo resistivity. *Phys. Rev. B*, **57**, 11609–11622.

[10047] Ulvenlund S., Rosdahl J., Fischer A., Schwerdtfeger P. and Kloo L. (1999). Hard acid and soft base stabilisation of di- and trimercury cations in benzene solution - A spectroscopic, x-ray scattering, and quantum chemical study. *Eur. J. Inorg. Chem.*, pp. 633–642.

[10048] Ulvenlund S., Ståhl K. and Bengtsson-Kloo L. (1996). Structural and quantum chemical study of Bi_5^{3+} and isoelectronic main-group metal clusters. The crystal structure of pentabismuth(3+) tetrachlorogallate(III) refined from x-ray powder diffraction data and synthetic attempts on its antimony analogue. *Inorg. Chem.*, **35**, 223–230.

[10049] Umemoto K. and Saito S. (1996). Electronic configurations of superheavy elements. *J. Phys. Soc. Japan*, **65**, 3175–3179.

[10050] Umland J.B. (1993). *General Chemistry.*. West.

[10051] Umland J.B. and Bellama J.M. (1996). *General Chemistry, 2nd Ed.*. Brooks/Cole, Pacific Grove, see p. 245, The Dirac equation.

[10052] Unterberger A. (1998). A calculus of observables on a Dirac particle. *Ann. Inst. Henri Poincaré, Phys. Théor.*, **69**, 189–239.

[10053] Urban M. and Sadlej A.J. (1995). Electronic structure and electric properties of the alkali metal dimers. *J. Chem. Phys.*, **103**, 9692–9704.

[10054] Urban M. and Sadlej A.J. (1997). Binding of aluminium to coinage metals: electron correlation and relativistic effects. *Mol. Phys.*, **92**, 587–600.

[10055] Usón R., Forniés J., Tomás M., Casas J.M., Cotton F.A., Falvello L.R. and Feng X. (1993). Synthesis and structural characterization of the Pt_2(II,III) complex $(NBu_4)[(C_6F_5)_2Pt(\mu\text{-}C_6F_5)_2Pt(C_6F_5)_2]$ and the Pt_2(III,III) complex $(NBu_4)[(C_6F_5)_2Pt(\mu\text{-}C_6F_5Cl)(\mu\text{-}C_6F_5)Pt(C_6F_5)_2]$. Novel ligand reactivity of a bridging C_6F_5 group. *J. Am. Chem. Soc.*, **115**, 4145–4154.

[10056] Uylings P.H.M. and Raassen A.J.J. (1995). Accurate calculation of transition probabilities using orthogonal operators. *J. Phys. B*, **28**, L209–L212.

[10057] Uzan O. and Martin J.M.L. (1998). Can Si=O bonds be stabilized by Rh/Ir complexes? A density functional theory study. *Chem. Phys. Lett.*, **290**, 535–542.

[10058] Vaara J. and Hiltunen Y. (1997). Deuterium quadrupole coupling tensors in methyl halides: Ab initio effective core potential and liquid crystal nuclear magnetic resonance study. *J. Chem. Phys.*, **107**, 1744–1752.

[10059] Vaara J., Ruud K. and Vahtras O. (1999a). Second- and third-order spin-orbit contributions to nuclear shielding tensors. *J. Chem. Phys.*, **111**, 2900–2909.

[10060] Vaara J., Ruud K. and Vahtras O. (1999b). Correlated response calculations of the spin-orbit interaction contribution to nuclear spin-spin couplings. *J. Comp. Chem.*, **20**, 1314–1327.

[10061] Vaara J., Ruud K., Vahtras O., Ågren H. and Jokisaari J. (1998). Quadratic response calculations of the electronic spin-orbit contribution to nuclear shielding tensors. *J. Chem. Phys.*, **109**, 1212–1222.

[10062] Valerio G. and Toulhoat H. (1996). Local, gradient-corrected, and hybrid density functional calculations on Pd_n clusters for n = 1-6. *J. Phys. Chem.*, **100**, 10827–10830.

[10063] Valerio G. and Toulhoat H. (1997). Atomic sulfur and chlorine interaction with Pd_n clusters (n=1-6): A density functional study. *J. Phys. Chem.*, **101**, 1969–1974.

[10064] Valiev K.V. and Pazdzerskij V.A. (1996). Dirac equation for singular point potential. *Uzbek. Fiz. Zh.*, **1**, 18–21.

[10065] Vallet V., Maron L., Schimmelpfennig B., Leininger T., Teichteil C., Gropen O., Grenthe I. and Wahlgren U. (1999a). Reduction behavior of the early actinyl ions in aqueous solution. *J. Phys. Chem. A*, **103**, 9285–9289.

[10066] Vallet V., Schimmelpfennig B., Maron L., Teichteil C., Leininger T., Gropen O., Grenthe I. and Wahlgren U. (1999b). Reduction of uranyl by hydrogen: an ab initio study. *Chem. Phys.*, **244**, 185–193.

[10067] Vanderbilt D. (1990). Soft self-consistent pseudopotentials in a generalized eigenvalue formalism. *Phys. Rev. B*, **41**, 7892–7895.

[10068] van der Lugt W. (1996). Polyanions in liquid ionic alloys: a decade of research . *J. Phys. C*, **8**, 6115–6138.

[10069] van de Walle C.G. and Blöchl P.E. (1993). First-principles calculations of hyperfine parameters. *Phys. Rev. B*, **47**, 4244–4255.

[10070] van Leeuwen R., van Lenthe E., Baerends E.J. and Snijders J.G. (1994). Exact solutions of regular approximate relativistic wave equations for hydrogen-like atoms. *J. Chem. Phys.*, **101**, 1272–1281.

[10071] van Lenthe E. (1996). *The ZORA Equation.*. Ph.D. thesis, Vrije Universiteit, Amsterdam, 101 p.

[10072] van Lenthe E., Baerends E.J. and Snijders J.G. (1993). Relativistic regular two-component Hamiltonians. *J. Chem. Phys.*, **99**, 4597–4610.

[10073] van Lenthe E., Baerends E.J. and Snijders J.G. (1994a). Relativistic total energy using regular approximations. *J. Chem. Phys.*, **101**, 9783–9792.

[10074] van Lenthe E., Baerends E.J. and Snijders J.G. (1995). Solving the Dirac equation, using the large component only, in a Dirac-type Slater orbital basis set. *Chem. Phys. Lett.*, **236**, 235–241.

[10075] van Lenthe E., Baerends E.J. and Snijders J.G. (1996a). Construction of the Foldy-Wouthuysen transformation and solution of the Dirac equation using large components only. *J. Chem. Phys.*, **105**, 2373–2377.

[10076] van Lenthe E., Ehlers A. and Baerends E.J. (1999). Geometry optimizations in the zero order regular approximation for relativistic effects. *J. Chem. Phys.*, **110**, 8943–8953.

[10077] van Lenthe E., Snijders J.G. and Baerends E.J. (1996b). The zero-order regular approximation for relativistic effects: The effects of spin-orbit coupling in closed shell molecules. *J. Chem. Phys.*, **105**, 6505–6516.

[10078] van Lenthe E., van der Avoird A. and Wormer P.E.S. (1998). Density functional calculations of molecular hyperfine interactions in the zero order regular approximation for relativistic effects. *J. Chem. Phys.*, **108**, 4783–4796.

[10079] van Lenthe E., van Leeuwen R., Baerends E.J. and Snijders J.G. (1994b). Relativistic regular two-component Hamiltonians. In *New Challenges in Computational Quantum Chemistry*, (Edited by Broer R., Aerts P.J.C. and Bagus P.S.), pp. 93–111, Univ. Groningen.

[10080] van Lenthe E., van Leeuwen R., Baerends E.J. and Snijders J.G. (1996c). Relativistic regular two-component Hamiltonians. *Int. J. Quantum Chem.*, **57**, 281–293.

[10081] van Lenthe E., Wormer P.E.S. and van der Avoird A. (1997). Density functional calculations of molecular g-tensors in the zero-order regular approximation for relativistic effects. *J. Chem. Phys.*, **107**, 2488–2498.

[10082] van Wijngaarden W.A. and Li J. (1994). Polarizabilities of cesium s, p, d, and f states. *J. Quant. Spectr. Rad. Transfer*, **52**, 555–562.

[10083] van Wüllen C. (1995). A relativistic Kohn-Sham density functional procedure by means of direct perturbation theory. *J. Chem. Phys.*, **103**, 3589–3599.

[10084] van Wüllen C. (1996a). On the use of common effective core potentials in density functional calculations. I. Test calculations on transition-metal carbonyls. *Int. J. Quantum Chem.*, **58**, 147–152.

[10085] van Wüllen C. (1996b). A relativistic Kohn-Sham density functional procedure by means of direct perturbation theory. II. Application to the molecular structure and bond dissociation energies of transition metal carbonyls and related complexes. *J. Chem. Phys.*, **105**, 5485–5493.

[10086] van Wüllen C. (1997). Molecular structure and binding energies of monosubstituted hexacarbonyls of chromium, molybdenum, and tungsten: Relativistic density functional study. *J. Comp. Chem.*, **18**, 1985–1992.

[10087] van Wüllen C. (1998). Molecular density functional calculations in the regular relativistic approximation: method, application to coinage metal diatomics, hydrides, fluorides and chlorides, and comparison with first-order relativistic calculations. *J. Chem. Phys.*, **109**, 392–399.

[10088] van Wüllen C. (1999). Relativistic all-electron density functional calculations. *J. Comp. Chem.*, **20**, 51–62.

[10089] Varella do N. M.T., Bettega M.H.F. and Lima M.A.P. (19997). Cross sections for rotational excitations of CH_4, SiH_4, GeH_4, SnH_4 and PbH_4 by electron impact. *Z. Phys. D*, **39**, 59–67.

[10090] Varga S., Engel E., Sepp W.D. and Fricke B. (1999). Systematic study of the Ib diatomic molecules Cu_2, Ag_2, and Au_2 using advanced relativistic density functionals. *Phys. Rev. A*, **59**, 4288–4294.

[10091] Vargas A., Waldés J.E. and Arista N.R. (1996). Energy loss of slow protons channeled in Au. *Phys. Rev. A*, **53**, 1638–1643.

[10092] Vaz Jr J. and Rodriguez Jr W.A. (1993). Zitterbewegung and the electromagnetic field of the electron. *Phys. Lett. B*, **319**, 203–208.

[10093] Velasco A.M., Lavin C. and Martin I. (1997). Relativistic calculations of oscillator strengths in neutral fluorine. *J. Quant. Spectr. Rad. Transfer*, **57**, 509–517.

[10094] Veldkamp A. and Frenking G. (1993a). Quantum-mechanical ab initio investigation of the transition-metal compounds OsO_4, OsO_3F_2, OsO_2F_4, $OsOF_6$, and OsF_8. *Chem. Ber.*, **126**, 1325–1330.

[10095] Veldkamp A. and Frenking G. (1993b). Theoretical studies of organometallic compounds. 6. Structures and bond energies of $M(CO)_n^+$, MCN, and $M(CN)_2^-$ (M=Ag, Au; n=1-3). *Organometallics*, **12**, 4613–4622.

[10096] Veldkamp A. and Frenking G. (1994). Structures and bond energies of the noble gas complexes NgBeO (Ng = Ar, Kr, Xe). *Chem. Phys. Lett.*, **226**, 11–16.

[10097] Verner D.A., Yakovlev D.G., Band I.M. and Trzhaskovskaya M.B. (1993). Subshell photoionization cross sections and ionization energies of atoms and ions from He to Zn. *At. Data Nucl. Data Tables*, **55**, 233–280.

[10098] Victora R.H. (1997). Calculated electronic structure of silver halide crystals. *Phys. Rev. B*, **56**, 4417–4421.

[10099] Vidolova-Angelova E.P. (1993). Energies of Sr states with excited valence shell. *Z. Phys. D*, **28**, 91–95.

[10100] Vijayakumar M. and Gopinathan M.S. (1993). Multiple-scattered, relativistic and correlated method (MS-RCΞ) for molecules. *J. Chem. Phys.*, **98**, 4009–4014.

[10101] Vijayakumar M. and Gopinathan M.S. (1995). Study of the importance of relativistic, correlation, and relaxation effects on ionization energy of atoms by a relativistic and correlated local density method. *J. Chem. Phys.*, **103**, 6576–6580.

[10102] Vijayakumar M. and Gopinathan M.S. (1996). Spin-orbit coupling constants of transition metal atoms and ions in density functional theory. *J. Mol. Str. (Theochem)*, **361**, 15–19.

[10103] Vijayakumar M., Roszak S. and Balasubramanian K. (1993). An efficient relativistic multireference configuration interaction method. *Chem. Phys. Lett.*, **215**, 87–92.

[10104] Vilkas M.J., Ishikawa Y. and Koc K. (1998a). Second-order multiconfigurational Dirac-Fock calculations on boronlike ions. *Int. J. Quantum Chem.*, **70**, 813–823.

[10105] Vilkas M.J., Ishikawa Y. and Koc K. (1999). Relativistic multireference many-body perturbation theory for quasidegenerate systems: Energy levels of ions of the oxygen isoelectronic sequence. *Phys. Rev. A*, **60**, 2808–2821.

[10106] Vilkas M.J., Koc K. and Ishikawa Y. (1997). Second-order multiconfigurational Dirac-Fock self-consistent field and multireference configuration interaction calculations on beryllium and beryllium-like Ne^{6+}. *Chem. Phys. Lett.*, **280**, 167–176.

[10107] Vilkas M.J., Koc K. and Ishikawa Y. (1998b). Relativistic multireference Møller-Plesset perturbation theory based on multiconfigurational Dirac-Fock reference functions. *Chem. Phys. Lett.*, **296**, 68–76.

[10108] Vilkas M.J., Merkelis G., Kisielius R., Gaigalas G., Bernotas A. and Rudzikas Z. (1994). Ab initio calculation of El transitions in the oxygen isoelectronic sequence. *Phys. Scr.*, **49**, 592–600.

[10109] Villalba V.M. (1994a). Exact solution of the two-dimensional Dirac oscillator. *Phys. Rev. A*, **49**, 586–587.

[10110] Villalba V.M. (1994b). The angular momentum operator in the Dirac equation. *Eur. J. Phys.*, **15**, 191–196.

[10111] Villalba V.M. (1994c). Bound states of the hydrogen atom in the presence of a magnetic monopole field and an Aharonov-Bohm potential. *Phys. Lett. A*, **193**, 218–222.

[10112] Villalba V.M. (1995). Exact solution of the Dirac equation for a Coulomb and scalar potential in the presence of an Aharonov-Bohm and a magnetic monopole fields. *J. Math. Phys.*, **36**, 3332–3344.

[10113] Villalba V.M. (1996). On the relativistic hydrogen atom . *Rev. Mex. Fis.*, **42(1)**, 1–11.

[10114] Villalba V.M. (1997). Exact solution of the Dirac equation in the presence of pseudoscalar potentials. *Nuovo Cim. B*, **112**, 109–116.

[10115] Villalba V.M. and Pino R. (1998). Analytic solution of a relativistic two-dimensional hydrogen-like atom in a constant magnetic field. *Phys. Lett. A*, **238**, 49–53.

[10116] Visscher L. (1993). *Relativity and electron correlation in chemistry.*. Ph.D. thesis, Groningen.

[10117] Visscher L. (1994). Relativity and electron correlation in chemistry. In *New Challenges in Computational Quantum Chemistry*, (Edited by Broer R., Aerts P.J.C. and Bagus P.S.), pp. 134–144, Univ. Groningen.

[10118] Visscher L. (1996). On the construction of double group molecular symmetry functions. *Chem. Phys. Lett.*, **253**, 20–26.

[10119] Visscher L. (1997). Approximate molecular relativistic Dirac-Coulomb calculations using a simple Coulombic correction. *Theor. Chem. Acc.*, **98**, 68–70.

[10120] Visscher L., de Jong W.A., Visser O., Aerts P.J.C., Merenga H. and Nieuwpoort W.C. (1995). Relativistic quantum chemistry: the MOLDFIR program package. In *METECC-95*, (Edited by Clementi E. and Corongiu G.), pp. 169–218, STEF, Cagliari.

[10121] Visscher L. and Dyall K.G. (1995). MP2 studies of relativistic effects on the linear stationary points of the H + Cl_2 → HCl + Cl and Cl + HCl → ClH + Cl reactions. *Chem. Phys. Lett.*, **239**, 181–185.

[10122] Visscher L. and Dyall K.G. (1996). Relativistic and correlation effects on molecular properties. I. The dihalogens F_2, Cl_2, Br_2, I_2, and At_2. *J. Chem. Phys.*, **104**, 9040–9046.

[10123] Visscher L. and Dyall K.G. (1997). Dirac-Fock atomic electronic structure calculations using different nuclear charge distributions. *At. Data Nucl. Data Tables*, **67**, 207–224.

[10124] Visscher L., Enevoldsen T., Saue T., Aagaard Jensen H.J. and Oddershede J. (1999). Full four-component relativistic calculations of NMR shielding and indirect spin-spin coupling tensors in hydrogen halides. *J. Comp. Chem.*, **20**, 1262–1273.

[10125] Visscher L., Enevoldsen T., Saue T. and Oddershede J. (1998). Molecular relativistic calculations of the electric field gradients at the nuclei in the hydrogen halides. *J. Chem. Phys.*, **109**, 9677–9684.

[10126] Visscher L., Lee T.J. and Dyall K.G. (1996a). Formulation and implementation of a relativistic unrestricted coupled-cluster method including noniterative connected triples. *J. Chem. Phys.*, **105**, 8769–8776.

[10127] Visscher L. and Nieuwpoort W.C. (1994). Relativistic and electron correlation effects on the d-d spectrum of transition metal fluorides. *Theor. Chim. Acta*, **88**, 447–472.

[10128] Visscher L., Saue T., Nieuwpoort W.C., Faegri K. and Gropen O. (1993). The electronic structure of the PtH molecule: fully relativistic configuration interaction calculations of the ground and excited states. *J. Chem. Phys.*, **99**, 6704–6715.

[10129] Visscher L., Saue T. and Oddershede J. (1997). The 4-component random phase approximation method applied to the calculation of frequency-dependent dipole polarizabilities. *Chem. Phys. Lett.*, **274**, 181–188.

[10130] Visscher L., Styszynski J. and Nieuwpoort W.C. (1996b). Relativistic and correlation effects on molecular properties. II. The hydrogen halides HF, HCl, HBr, HI, and HAt. *J. Chem. Phys.*, **105**, 1987–1994.

[10131] Visscher L. and van Lenthe E. (1999). On the distinction between scalar and spin-orbit relativistic effects. *Chem. Phys. Lett.*, **306**, 357–365.

[10132] Visscher L., Visser O., Aerts P.J.C., Merenga H. and Nieuwpoort W.C. (1994). Relativistic quantum chemistry: the MOLFDIR program package. *Comp. Phys. Comm.*, **81**, 120–144.

[10133] Visser R., Andriessen J., Dorenbos P. and Van Eijk C.W.E. (1993). Ce^{n+} energy levels in alkaline-earth fluorides and cerium-electron cerium-hole interactions. *J. Phys: CM*, **5**, 5887–5910.

[10134] Vitos L., Kollár J. and Skriver H.L. (1997). *Ab initio* full charge-density study of the atomic volume of α-phase Fr, Ra, Ac, Th, Pa, U, Np, and Pu. *Phys. Rev. B*, **55**, 4947–4952.

[10135] Vogel D., Krüger P. and Pollmann J. (1998). Ab initio electronic structure of silver halides calculated with self-interaction and relaxation-corrected pseudopotentials. *Phys. Rev. B*, **58**, 3865–3869.

[10136] Volkova L.M. and Magarill S.A. (1999). On the formation of polyatomic cations of mercury (in Russian). *Zh. Str. Khim.*, **40**, 314–323.

[10137] Von Grünberg H.H., Gersonde I.H. and Gabriel H. (1993). Using diatomics-in-molecules in calculating semiempirical band structures. Application to Ar, Kr, Xe crystals. *Z. Phys. D*, **28**, 145–151.

[10138] von R Schleyer P. and Kapp J. (1996). Hypermetallation is ubiquitous: MX_6 molecules (M = C-Pb, X = Li-K). *Chem. Phys. Lett.*, **255**, 363–366.

[10139] Vonsovskii S.V. and Svirskii M.S. (1993). The Klein paradox and the Zitterbewegung of an electron in a field with a constant scalar potential. *Physics-Uspekhi.*, **36**, 436–439.

[10140] Vonsovskii S.V. and Svirskii M.S. (1993). The Klein paradox and the Zitterbewegung of an electron in a field with a constant scalar potential (in Russian). *Usp. Fiz. Nauk.*, **163**, 115–118.

[10141] Vonsovskii S.V. and Svirskii M.S. (1997). Zitterbewegung and uncertainties of velocity and acceleration in Dirac theory (in Russian). *Fiz. El. Chast. i At. Yadr.*, **28**, 162–190.

[10142] Vonsovskii S.V., Svirskii M.S. and Svirskaya L.M. (1993). Electron Zitterbewegung in a linear crystal with alternating parity (in Russian). *Teor. Mat. Fiz.*, **94**, 343–352.

[10143] Vonsovskii S.V., Svirskii M.S. and Svirskaya L.M. (1993). Electron Zitterbewegung in a linear crystal with alternating parity. *Theor. Math. Phys.*, **94**, 243–249.

[10144] Vosko S.H. and Chevary J.A. (1993). Prediction of a further irregularity in the electron filling of subshell: Lu^- $(Xe)4f^{14}5d^16s^26p^1$ and its relation to the group IIIB anions. *J. Phys. B*, **26**, 873–887.

[10145] Voss D., Krüger P., Mazur A. and Pollmann J. (1999). Atomic and electronic structure of WSe_2 from *ab initio* theory: Bulk crystal and thin film systems. *Phys. Rev. B*, **60**, 14311–14317.

[10146] Vrejoiu C., Zota S., Mezincescu N. and Boca M. (1999). Retardation corrections to the angular distributions for the free-bound transitions in H-like atoms. *Eur. Phys. J. D*, **7**, 503–513.

[10147] Vrbik J. (1994). Dirac equation and Clifford algebra. *J. Math. Phys.*, **35**, 2309–2314.

[10148] Vyboishchikov S.F. and Frenking G. (1998a). Structure and bonding of low-valent (Fischer-type) and high-valent (Schrock-type) transition metal carbene complexes. *Chem. Eur. J.*, **4**, 1428–1438.

[10149] Vyboishchikov S.F. and Frenking G. (1998b). Structure and bonding of low-valent (Fischer-type) and high-valent (Schrock-type) transition metal carbyne complexes. *Chem. Eur. J.*, **4**, 1439–1448.

[10150] Vyboishchikov S.F. and Frenking G. (1999). The Lewis basicity of nitrido complexes. Theoretical investigation of the structure and bonding of $Cl_2(PH_3)_3ReN$-X (X = BH_3, BCl_3, BBr_3, AlH_3, $AlCl_3$, $AlBr_3$, GaH_3, $GaCl_3$, $GaBr_3$, O, S, Se, Te). *Theor. Chem. Acc.*, **102**, 300–308.

[10151] Wachter A., Bali G. and Schilling K. (1997). Relativistic corrections to the central $q\bar{q}$ potential from pure $SU(3)$ lattice gauge theory. *Nucl. Phys. B (Proc. Suppl.)*, **53**, 286–288.

[10152] Wahlgren U., Moll H., Grenthe I., Schimmelpfennig B., Maron L., Vallet V. and Gropen O. (1999). Structure of uranium(VI) in strong alkaline solutions: A combined theoretical and experimental investigation. *J. Phys. Chem. A*, **103**, 8257–8264.

[10153] Wahlgren U., Schimmelpfennig B., Jusuf S., Strömsnes H., Gropen O. and Maron L. (1998). A local approximation for relativistic scalar operators applied to the uranyl ion and to Au_2. *Chem. Phys. Lett.*, **287**, 525–530.

[10154] Wahlgren U. and Siegbahn P.E.M. (1994). On the use of small d-spaces in SCF and CI calculations on transition metals. *Theor. Chim. Acta*, **87**, 267–275.

[10155] Wahlgren U., Sjøvoll M., Fagerli H., Gropen O. and Schimmelpfennig B. (1997). Ab initio calculations of the ${}^2P_{1/2}$-${}^2P_{3/2}$ splitting in the thallium atom. *Theor. Chem. Acc.*, **97**, 324–330.

[10156] Wang B.X. and Lange H. (1999). Attractors for the Klein-Gordon-Schrödinger equation. *J. Math. Phys.*, **40**, 2445–2457.

[10157] Wang C.R., Lo K.K.W. and Yam V.W.W. (1996). Ab initio study of luminescent chalcogenido silver(I) clusters $(Ag_4(\mu\text{-}H_2PCH_2PH_2)_4(\mu_4\text{-}E)\ )^{2+}$. *Chem. Phys. Lett.*, **262**, 91–96.

[10158] Wang H. and Carter E.A. (1993). Metal-metal bonding in Engel-Brewer intermetallics: "Anomalous" charge transfer in $ZrPt_3$. *J. Am. Chem. Soc.*, **115**, 2357–2362.

[10159] Wang P., MacFarlane J.J. and Moses G.A. (1993). Relativistic-configuration-interaction calculations of Kα satellite properties for aluminum plasmas created by intense proton beams. *Phys. Rev. E*, **48**, 3934–3942.

[10160] Wang Q.S. and Stedman G.E. (1993). Spin-assisted matter-field coupling and lanthanide transition intensities. *J. Phys. B*, **26**, 1415–1423.

[10161] Wang L.R., Chi H.C. and Huang K.N. (1999). Spin-orbit and core-shielding effects on double-excitation resonances of Zn. *Phys. Rev. Lett.*, **83**, 702–705.

[10162] Wang S., Trajmar S. and Zetner P.W. (1994). Cross sections for electron scattering by ground state Ba: elastic scattering and excitation of the $6s6p\ {}^1P_1$ level. *J. Phys. B*, **27**, 1613–1621.

[10163] Wang S.G. (1994). *Relativistische Effekte und Chemische Bindung (in English)*. Ph.D. thesis, Universität Siegen, 262 p.

[10164] Wang S.G., Pan D.K. and Schwarz W.H.E. (1995). Density functional calculations of lanthanide oxides. *J. Chem. Phys.*, **102**, 9296–9308.

[10165] Wang S.G. and Schwarz W.H.E. (1995a). Lanthanide diatomics and lanthanide contractions. *J. Phys. Chem.*, **99**, 11687–11695.

[10166] Wang S.G. and Schwarz W.H.E. (1995b). Relativistic effects of p-block molecules. *J. Mol. Str. (Theochem)*, **338**, 347–362.

[10167] Wang S.G. and Schwarz W.H.E. (1997). On oxides of monovalent transition metals. *J. Alloys Compounds*, **246**, 131–138.

[10168] Wang S.G. and Schwarz W.H.E. (1998). Density functional study of first row transition metal dihalides. *J. Chem. Phys.*, **109**, 7252–7262.

[10169] Wang W.J. (1993). Multiconfiguration Dirac-Fock calculation of fine-structure energy levels and transition wavelengths in N-like CoXXI, NiXXII, CuXXIII and ZnXXIV ions. *Nucl. Instr. Meth. Phys. Res. B*, **73**, 159–162.

[10170] Wang W.J., Jiang R.B. and Wang X.D. (1996). GRASP calculations of some radiation lifetimes and wavelengths for O-like ions with $52 \leq Z \leq 79$. *Nucl. Sci. Techn.*, **7**, 154–156.

[10171] Wang X.D., Leung T.C., Harmon B.N. and Carra P. (1993). Circular magnetic X-ray dichroism in the heavy rare-earth metals. *Phys. Rev. B*, **47**, 9087–9090.

[10172] Wang Y.X. and Dolg M. (1998). Pseudopotential study of the ground and excited states of Yb_2. *Theor. Chem. Acc.*, **100**, 124–133.

[10173] Wang Y.X., Schautz F., Flad H.J. and Dolg M. (1999). On the importance of 5d orbitals for covalent bonding in ytterbium clusters. *J. Phys. Chem. A*, **103**, 5091–5098.

[10174] Wang Z.C. and Li B.Z. (1999). Geometric phase in relativistic quantum theory. *Phys. Rev. A*, **60**, 4313–4317.

[10175] Wang Z.W. and Chung K.T. (1994). Dipole polarizabilities for the ground states of lithium-like systems from Z = 3 to 50. *J. Phys. B*, **27**, 855–864.

[10176] Wang Z.W. and Ge Z.M. (1997). Ionization potential of the lithium-like ground states from scandium to zinc. *Phys. Scr.*, **T73**, 53–55.

[10177] Wang Z.W., Zhu X.W. and Chung K.T. (1993). Energy and fine structure of $1s^2np$ states (n=2, 3, 4 and 5) for the lithium isoelectronic sequence. *Phys. Scr.*, **47**, 65–74.

[10178] Warrier L.S. and Gambhir Y.K. (1994). Single particle spectrum and spin-orbit splittings in relativistic mean field theory. *Phys. Rev. C*, **49**, 871–877.

[10179] Warttmann I. and Häfelinger G. (1998). Ab initio MO optimizations of osmiumtetracarbonyldihydride and metallacyclophanes with two osmium atoms and their molecular complexes with different guests. *Z. Naturf.*, **53 b**, 1223–1235.

[10180] Watanabe R., Muto K., Oda T., Niwa T., Ohtsubo H., Morita R. and Morita M. (1993). Asymmetry and energy spectrum of electrons in bound-muon decay. *At. Data Nucl. Data Tables*, **54**, 165–178.

[10181] Watanabe Y. and Matsuoka O. (1997). All-electron Dirac-Fock-Roothaan calculations for the ThO molecule. *J. Chem. Phys.*, **107**, 3738–3739.

[10182] Watanabe Y. and Matsuoka O. (1998). Dirac-Fock-Roothaan calculations using a relativistic reduced frozen-core approximation. *J. Chem. Phys.*, **109**, 8182–8187.

[10183] Watari N. and Ohnishi S. (1997). Electronic structure of H adsorbed on Pt_{13} clusters. *J. Chem. Phys.*, **106**, 7531–7540.

[10184] Watari N. and Ohnishi S. (1998). Atomic and electronic structures of Pd_{13} and Pt_{13}. *Phys. Rev. B*, **58**, 1665–1677.

[10185] Watkins G.D. and Williams P.M. (1995). Vacancy model for substitutional Ni^-, Pd^-, Pt^-, and Au^0 in silicon. *Phys. Rev. B*, **52**, 16575–16580.

[10186] Watson G.W., Parker S.C. and Kresse G. (1999). *Ab initio* calculation of the origin of the distortion of α-PbO. *Phys. Rev. B*, **59**, 8481–8486.

[10187] Watson R.E. and Weinert M. (1994). Charge transfer in gold–alkali-metal systems. *Phys. Rev. B*, **49**, 7148–7154.

[10188] Waxman D. (1994). The Fredholm determinant for a Dirac operator. *Ann. Phys. (NY)*, **231**, 256–269.

[10189] Webb J.K., Flambaum V.V., Churchill C.W., Drinkwater M.J. and Barrow J.D. (1999). Search for time variation of the fine structure constant. *Phys. Rev. Lett.*, **82**, 884–887.

[10190] Webb S.P. and Gordon M.S. (1998). The effect of spin-orbit coupling on the magnetic properties of $H_2Ti(\mu\text{-}H)_2TiH_2$. *J. Chem. Phys.*, **109**, 919–927.

[10191] Wei S.H. and Zunger A. (1997). Electronic and structural anomalies in lead chalcogenides. *Phys. Rev. B*, **55**, 13605–13610.

[10192] Weigend F., Häser M., Patzelt H. and Ahlrichs R. (1998). RI-MP2: optimized auxiliary basis sets and demonstration of efficiency. *Chem. Phys. Lett.*, **294**, 143–152.

[10193] Weinberger P., Blaas C., Bennett B.I. and Boring A.M. (1993a). Fully relativistic effective pair interactions for the Au-Pd system. *Phys. Rev. B*, **47**, 10158–10163.

[10194] Weinberger P., Drchal V., Szunyogh L., Fritscher J. and Bennett B.I. (1994). Electronic and structural properties of Cu-Au alloys. *Phys. Rev. B*, **49**, 13366–13372.

[10195] Weinberger P., Kudrnovsky J., Redinger J., Bennett B.I. and Boring A.M. (1993b). Calculation of equilibrium lattice parameters and heat of mixing for the system Au/Pd by the relativistic Korringa-Kohn-Rostoker coherent-potential-approximation method. *Phys. Rev. B*, **48**, 7866–7871.

[10196] Weinberger P., Levy P.M., Banhart J., Szunyogh L. and Újfalussy B. (1996). 'Band structure' and electrical conductivity of disordered layered systems. *J. Phys. C*, **8**, 7677–7688.

[10197] Weiss A.W. and Kim Y.K. (1995). Relativistic modifications of charge expansion theory. *Phys. Rev. A*, **51**, 4487–4493.

[10198] Wen J.H., Li K.H., and Chen J. (1998). Monte Carlo simulations for high harmonics of hydrogen atom in the ultrastrong laser field. *Chin. J. At. Mol. Phys.*, **15**, 491–496.

[10199] Wesendrup R., Laerdahl J.K. and Schwerdtfeger P. (1999a). Relativistic effects in gold chemistry. VI. Coupled cluster calculations for the isoelectronic series $AuPt^-$, Au_2, and $AuHg^+$. *J. Chem. Phys.*, **110**, 9457–9462.

[10200] Wesendrup R., Pernpointner M. and Schwerdtfeger P. (1999b). Coulomb-stable triply charged diatomic: HeY^{3+}. *Phys. Rev. A*, **60**, R3347–R3349.

[10201] Wessels P.P.F., Caspers W.J. and Wiegel F.W. (1999). Discretizing the one-dimensional Dirac equation. *Europhys. Lett.*, **46**, 123–126.

[10202] Westerberg J. and Blomberg M.R.A. (1998). Methane activation by naked Rh^+ atoms. A theoretical study. *J. Phys. Chem. A*, **102**, 7303–7307.

[10203] Wetzel T.L., Moran T.F. and Borkman R.F. (1994). Structures and energies of sodium halide ions and neutral clusters computed with ab initio effective core potentials. *J. Phys. Chem.*, **98**, 10042–10047.

[10204] Wezenbeek E.M., Baerends E.J. and Ziegler T. (1995). Theoretical study of the relativistic effects on the bonds between $HfCl_3$ and H between $ThCl_3$ and H. *Inorg. Chem.*, **34**, 238–246.

[10205] Whelan C.T., Ast H., Keller S., Walters H.R.J. and Dreizler R.M. (1995). Triple differential cross sections in energy-sharing symmetric geometry for gold and uranium at relativistic impact energies. *J. Phys. B*, **28**, L33–L39.

[10206] Whelan C.T., Ast H., Walters H.R.J., Keller S. and Dreizler R.M. (1996). Relativistic-energy-sharing (e,2e) collisions in coplanar constant $\Theta_{1,2}$ geometry. *Phys. Rev. A*, **53**, 3262–3270.

[10207] Whitten J.L. (1997). Photodissociation of methane on platinum. *Chem. Phys.*, **225**, 189–195.

[10208] Whitten J.L. (1999). Theoretical studies of surface reactions on metals: I. Ethyl to ethylene conversion on platinum II. Photodissociation of methane on platinum. *J. Vac. Sci. Techn. A*, **17**, 1710–1716.

[10209] Wijesundera W.P. (1997). Theoretical study of the negative ions of boron, aluminum, gallium, indium, and thallium. *Phys. Rev. A*, **55**, 1785–1791.

[10210] Wijesundera W.P. and Parpia F.A. (1998). Negative ions of carbon, nitrogen, and phosphorus. *Phys. Rev. A*, **57**, 3462–3468.

[10211] Wijesundera W.P., Vosko S.H. and Parpia F.A. (1995). Relativistic and correlation effects in the ground state of atomic lawrencium. *Phys. Rev. A*, **51**, 278–282.

[10212] Wijesundera W.P., Vosko S.H. and Parpia F.A. (1996). Relativistic calculations of the fine-structure intervals of the $ns^2np\ ^2P$ terms and the ionization energies of the Ca^-, Sr^- and Ba^- ions. *J. Phys. B*, **29**, 379–387.

[10213] Wildman S.A., DiLabio G.A. and Christiansen P.A. (1997). Accurate relativistic effective potentials for the sixth-row main group elements. *J. Chem. Phys.*, **107**, 9975–9979.

[10214] Willatzen M., Cardona M. and Christensen N.E. (1995). Spin-orbit coupling parameters and electron g factor of II-VI zinc-blende materials. *Phys. Rev. B*, **51**, 17992–17994.

[10215] Williams D.S., Schofield M.H. and Schrock R.R. (1993). Synthesis of d^2 complexes that contain $[W(NAr)_2]$ and $[Re(NAr)_2]$ cores, SCF-Xα-SW calculations, and a discussion of the $MCp_2/M'(NR)_2$ isolobal relationship. *Organometallics*, **12**, 4560–4571.

[10216] Willis B.L. (1993). Classical description of the absence of bound states for strong Coulomb fields. *Nuovo Cim. B*, **108**, 217–218.

[10217] Wills A.A., Gorczyca T.W., Berrah N., Langer B., Felfli Z., Kukk E., Bozek J.D., Nayandin O. and Alshehri M. (1998). Importance of spin-orbit effects in parity-unfavored photoionization of neon, observed using a two-dimensional photoelectron imaging technique. *Phys. Rev. Lett.*, **80**, 5085–5088.

[10218] Willson S.P. and Andrews L. (1999). Characterization of the reaction products of laser-ablated late lanthanide metal atoms with molecular oxygen: Infrared spectra of LnO, LnO^+, LnO^-, LnO_2, LnO_2^-, LnO_3^-, and $(LnO)_2$ in solid argon. *J. Phys. Chem. A*, **103**, 6972–6983.

[10219] Wittborn C. and Wahlgren U. (1995). New relativistic effective core potentials for heavy elements. *Chem. Phys.*, **201**, 357–362.

[10220] Wolff S.K., Jayatilaka D. and Chandler G.S. (1995). An ab initio calculation of magnetic structure factors for Cs_3CoCl_5 including spin-orbit and finite magnetic field effects. *J. Chem. Phys.*, **103**, 4562–4571.

[10221] Wolff S.K. and Ziegler T. (1998). Calculation of DFT-GIAO NMR shifts with the inclusion of spin-orbit coupling. *J. Chem. Phys.*, **109**, 895–905.

[10222] Wolff S.K., Ziegler T., van Lenthe E. and Baerends E.J. (1999). Density functional calculations of nuclear magnetic shieldings using the zeroth-order regular approximation (ZORA) for relativistic effects: ZORA nuclear magnetic resonance. *J. Chem. Phys.*, **110**, 7689–7698.

[10223] Wolniewicz L. (1993). Relativistic energies of the ground state of the hydrogen molecule. *J. Chem. Phys.*, **99**, 1851–1868.

[10224] Wolniewicz L. (1995). Lowest order relativistic corrections to the energies of the B $^1\Sigma_u$ state of H_2. *Chem. Phys. Lett.*, **233**, 647–650.

[10225] Wolniewicz L. (1998). Relativistic corrections to the energies of the EF, GK, and $H\overline{H}$ $^1\Sigma_g$ states of the hydrogen molecule. *J. Chem. Phys.*, **109**, 2254–2256.

[10226] Woon D.E. and Dunning Jr T.H. (1993). Calculation of the electron affinities of the second row atoms: Al-Cl. *J. Chem. Phys.*, **99**, 3730–3737.

[10227] Wright J.S., Carpenter D.J., Alekseyev A.B., Liebermann H.P., Lingott R. and Buenker R.J. (1997). Thermodynamically stable diatomic dications: potential curves and radiative lifetimes for $CaCl^{2+}$ including relativistic effects. *Chem. Phys. Lett.*, **266**, 391–396.

[10228] Wright K., Hillier I.H., Vincent M.A. and Kresse G. (1999). Dissociation of water on the surface of galena (PbS): A comparison of periodic and cluster models. *J. Chem. Phys.*, **111**, 6942–6946.

[10229] Wu C.J., Yang L.H., Klepeis J.E. and Mailhiot C. (1995). Ab initio pseudopotential calculations of the atomic and electronic structure of the Ta (100) and (110) surfaces. *Phys. Rev. B*, **52**, 11784–11792.

[10230] Wu C.M., Chi H.C. and Huang K.N. (1994). Photoionization of strontium above the $5p_{3/2}$ threshold using the multiconfiguration relativistic random-phase approximation. *J. Phys. B*, **27**, 3927–3937.

[10231] Wu L.J. (1996). Pseudoconfigurational aprroach in relativistic many-body perturbation-theory calculations. *Phys. Rev. A*, **53**, 139–144.

[10232] Wu R.Q. (1995). The bonding mechanism at bimetallic interface: Pd/Ta(110). *Chem. Phys. Lett.*, **238**, 99–103.

[10233] Wu Y.D. and Sun J. (1998). Transition structures of epoxidation by $CH_3Re(O)_2(O_2)$ and $CH_3Re(O)(O_2)_2$ and their water adducts. *J. Org. Chem.*, **63**, 1752–1753.

[10234] Wu Z.J., Meng Q.B. and Zhang S.Y. (1997). Theoretical investigation of LaC_n^+ (n = 2-8) clusters. *Chem. Phys. Lett.*, **281**, 233–238.

[10235] Wu Z.J., Meng Q.B. and Zhang S.Y. (1998a). Theoretical investigation of LaC_3^{n+} (n = 0,1,2) clusters by density functional theory. *Int. J. Quantum Chem.*, **66**, 301–307.

[10236] Wu Z.J., Meng Q.B. and Zhang S.Y. (1998b). Theoretical investigation of LaC_2 and LaC_2^+ clusters. *J. Mol. Str. (Theochem)*, **431**, 165–172.

[10237] Würde K., Mazur A. and Pollmann J. (1994). Surface electronic structure of Pb(001), Pb(110), and Pb(111). *Phys. Rev. B*, **49**, 7679–7686.

[10238] Xantheas S.S., Fanourgakis G.S., Farantos S.C. and Velegrakis M. (1998). Spectroscopic constants of the $X^2\Sigma^+$ and $A^2\Pi$ states of Sr^+Ar from first principles: Comparison with experiment. *J. Chem. Phys.*, **108**, 46–49.

[10239] Xiao C.Y., Krüger S., Belling T., Mayer M. and Rösch N. (1999). Relativistic effects on geometry and electronic structure of small Pd_n species (n=1, 2, 4). *Int. J. Quantum Chem.*, **74**, 405–416.

[10240] Xie J.J., de Gironcoli S., Baroni S. and Scheffler M. (1999). First-principles calculation of the thermal properties of silver. *Phys. Rev. B*, **59**, 965–969.

[10241] Xu H. and Balasubramanian K. (1995). Spectroscopic constants and potential energy curves of GeF^+. *Chem. Phys. Lett.*, **237**, 7–13.

[10242] Xu J.H., Wang E.G., Ting C.S. and Su W.P. (1993). Tight-binding theory of the electronic structures for rhombohedral semimetals. *Phys. Rev. B*, **48**, 17271–17279.

[10243] Xu W.X., Schierbaum K.D. and Goepel W. (1997). Ab initio study of electronic structures of Pt_n clusters (n= 2-12). *Int. J. Quantum Chem.*, **62**, 427–436.

[10244] Xu X., Wu D.Y., Ren B., Xian H. and Tian Z.Q. (1999). On-top adsorption of hydrogen at platinum electrodes: a quantum-chemical study. *Chem. Phys. Lett.*, **311**, 193–201.

[10245] Xu X., Yuan Y.Z., Asakura K., Iwasawa Y., Wan H.L. and Tsai K.R. (1998). Structural properties of $[(AuPH_3)_6Pt(H_2)(PH_3)]^{2+}$: theoretical study of dihydrogen activation. *Chem. Phys. Lett.*, **286**, 163–170.

[10246] Xu Z.T., Bytheway I., Jia G.C. and Lin Z.Y. (1999). Theoretical studies of the acidity of the dihydrogen complexes *trans*-$[LM(H_2PCH_2CH_2PH_2)_2(\eta^2\text{-}H_2)]^{n+}$. *Organometallics*, **18**, 1761–1766.

[10247] Xu Z.T. and Lin Z.Y. (1998). Unusual five-center, four-electron bonding in a rhodium-bismuth complex with pentagonal-bipyramidal geometry. *Angew. Chem. Int. Ed. Engl.*, **37**, 1686–1689.

[10248] Xu Z.T. and Lin Z.Y. (1998). Unusual five-center, four-electron bonding in a rhodium-bismuth complex with pentagonal-bipyramidal geometry (in German). *Angew. Chem.*, **110**, 1815–1818.

[10249] Yabushita S., Zhang Z.Y. and Pitzer R.M. (1999). Spin-orbit configuration interaction using the graphical unitary group approach and relativistic core potential and spin-orbit operators. *J. Phys. Chem. A*, **103**, 5791–5800.

[10250] Yakhontov V.L. and Amusia M.Y. (1994). Hyperfine splitting computation in the $1s_{1/2}{}^{(e)}$ $2s_{1/2}{}^{(\mu)}$ state of the exotic $({}^4He^{2+}\text{-}\mu^-\text{-}e^-)^0$ and $({}^3He^{2+}\text{-}\mu^-\text{-}e^-)^0$ atoms. *J. Phys. B*, **27**, 3743–3765.

[10251] Yakhontov V.L. and Grant I.P. (1993). Parameter-free renormalization in the self-mass correction computation. *J. Phys. B*, **26**, L773–L781.

[10252] Yakovlev A.L., Neyman K.M., Zhidomirov G.M. and Rösch N. (1996). Interaction of CO molecules with electron-deficient Pt atoms in zeolites: A density functional model cluster study. *J. Phys. Chem.*, **100**, 3482–3487.

[10253] Yamagami H. (1998). All-electron spin-polarized relativistic linearized APW method: electronic and magnetic properties of BCC Fe, HCP Gd and uranium monochalcogenides. *J. Phys. Soc. Japan*, **67**, 3176–3190.

[10254] Yamaguchi K. and Miyagi H. (1998). Structural properties of molecular solid iodine under pressure: first-principles study of Raman-active A_g modes and hyperfine parameters. *Phys. Rev. B*, **57**, 11141–11148.

[10255] Yamanaka N. and Ichimura A. (1999). Transverse nuclear polarization in hydrogen-like ions. *Phys. Scripta*, **T80**, 488–490.

[10256] Yamanishi M., Hirao K. and Yamashita K. (1998). Theoretical study of the low-lying electronic states of XeO and XeS. *J. Chem. Phys.*, **108**, 1514–1521.

[10257] Yan Z.C. (1994). High-precision calculation of the isotope shift in the ratio $g_J({}^3\text{He}, 2\,{}^3S_1)/g_J({}^4\text{He}, 2\,{}^3S_1)$. *Phys. Rev. A*, **50**, 3809–3811.

[10258] Yan Z.C. and Drake G.W.F. (1994). High-precision calculations of the Zeeman effect in the $2\,{}^3P_J$, $2\,{}^1P_1$, $2\,{}^3S_1$, and $3\,{}^3P_J$ states of helium . *Phys. Rev. A*, **50**, R1980–R1983.

[10259] Yan Z.C. and Drake G.W.F. (1995). Eigenvalues and expectation values for the $1s^22s\,{}^2S$, $1s^22p\,{}^2P$, and $1s^23d\,{}^2D$ states of lithium. *Phys. Rev. A*, **52**, 3711–3717.

[10260] Yan Z.C. and Drake G.W.F. (1997). Lithium fine structure in the $1s^22p\,{}^2P_J$ states. *Phys. Rev. Lett.*, **79**, 1646–1649.

[10261] Yan Z.C. and Drake G.W.F. (1998). Relativistic and QED energies in lithium. *Phys. Rev. Lett.*, **81**, 774–777.

[10262] Yan Z.C. and Ho Y.K. (1999). Relativistic effects in positronium hydride. *Phys. Rev. A*, **60**, 5098–5100.

[10263] Yan Z.C., McKenzie D.K. and Drake G.W.F. (1996). Variational calculations of the Fermi contact term for the 2 ^2S, 2 ^2P, and 3 ^2S states of Li and the 2 ^2S state of Be^+. *Phys. Rev. A*, **54**, 1322–1327.

[10264] Yan Z.C., Tambasco M. and Drake G.W.F. (1998). Energies and oscillator strengths for lithiumlike ions. *Phys. Rev. A*, **57**, 1652–1661.

[10265] Yang D.S., Zgierski M.Z., Rayner D.M., Hackett P.A., Martinez A., Salahub D.R., Roy P.N. and Carrington Jr T. (1996). The stucture of Nb_3O and Nb_3O^+ determined by pulsed field ionization-zero electron kinetic energy photoelectron spectroscopy and density functional theory. *J. Chem. Phys.*, **103**, 5335–5342.

[10266] Yang H.Y. and Chung K.T. (1995). Energy, fine-structure, and hyperfine-structure studies of the core-excited states $1s2s2p^2(^5P)$ and $1s2p^3(^5S)$ for Be-like systems. *Phys. Rev. A*, **51**, 3621–3629.

[10267] Yang L. (1993). A CI study for K shell excitation and ionization energies in neutral sodium. *J. Phys. B*, **26**, 1813–1817.

[10268] Yang L., Heinemann D. and Kolb D. (1993). Fully numerical relativistic calculations for diatomic molecules using the finite-element method. *Phys. Rev. A*, **48**, 2700–2707.

[10269] Yang L.H., Smith A.P., Benedek R. and Koelling D.D. (1993). Effect of semicore banding on heavy-alkali-metal lattice constants: Corrections to the frozen-core approximation. *Phys. Rev. B*, **47**, 16101–16106.

[10270] Yang S.H., Drabold D.A., Adams J.B., Ordejón P. and Glassford K. (1997). Density functional studies of small platinum clusters. *J. Phys. CM*, **9**, L39–L45.

[10271] Yang X., Dagdigian P.J. and Alexander M.H. (1998). Experimental and theoretical study of the AlNe complex. *J. Chem. Phys.*, **108**, 3522–3530.

[10272] Yang Y.S., Hsu W.Y., Lee H.F., Huang Y.C., Yeh C.S. and Hu C.H. (1999). Experimental and theoretical studies of metal cation–pyridine complexes containing Cu and Ag. *J. Phys. Chem. A*, **103**, 11287–11292.

[10273] Yao G.H. and Chu S.I. (1993). Generalized pseudospectral methods with mappings for bound and resonance state problems. *Chem. Phys. Lett.*, **204**, 381–388.

[10274] Yarkony D.R. (1992). Spin-forbidden chemistry within the Breit-Pauli approximation. *Int. Rev. Phys. Chem.*, **11**, 195–242.

[10275] Yarkony D.R. (1996). Quenching of $CH(a^4\Sigma^-)$ by $CO(\Sigma^+)$: Surface of intersection, spin-orbit interactions, and the incorporation of Kramers' degeneracy. *J. Phys. Chem.*, **100**, 17439–17445.

[10276] Yatsimirskii K.B. (1995). Relativistic effects in chemistry (in Russian). *Teor. Eksp. Khim.*, **31(3)**, 181–199.

[10277] Yatsimirskii K.B. (1995). Relativistic effects in chemistry. *Theor. Exp. Chem.*, **31**, 153–168.

[10278] Yerokhin V.A., Artemyev A.N., Beier T., Plunien G., Shabaev V.M. and Soff G. (1999a). Two-electron self-energy corrections to the $2p_{1/2} - 2s$ transition energy in Li-like ions. *Phys. Rev. A*, **60**, 3522–3540.

[10279] Yerokhin V.A., Artemyev A.N., Beier T., Shabaev V.M. and Soff G. (1998). Direct evaluation of the two-electron self-energy corrections to the ground state energy of lithium-like ions. *J. Phys. B*, **31**, L691–L697.

[10280] Yerokhin V.A., Artemyev A.N., Beier T., Shabaev V.M. and Soff G. (1999b). Calculation of the screened self-energy and vacuum-polarization corrections in high-Z lithium-like ions. *Phys. Scripta*, **T80**, 495–497.

[10281] Yerokhin V.A., Artemyev A.N. and Shabaev V.M. (1997a). Two-electron self-energy contribution to the ground-state energy of helium-like ions. *Phys. Lett. A*, **234**, 361–366.

[10282] Yerokhin V.A. and Shabaev V.M. (1999). First-order self-energy correction in hydrogenlike systems. *Phys. Rev. A*, **60**, 800–811.

[10283] Yerokhin V.A., Shabaev V.M. and Artemyev A.N. (1997b). Self-energy correction to the hyperfine splitting of the $1s$ and $2s$ states in hydrogenlike ions. *JETP Letters*, **66**, 18–21.

[10284] Yildirim T., Harris A.B., Entin-Wohlman O. and Aharony A. (1994). Symmetry, spin-orbit interactions, and spin anisotropies. *Phys. Rev. Lett.*, **73**, 2919–2922.

[10285] Yndurain F.J. (1996). *Relativistic Quantum Mechanics with an Introduction to Field Theory.*. Springer-Verlag, Berlin.

[10286] Ynnerman A. and Froese Fischer C. (1995). Multi-configurational-Dirac-Fock calculation of the $2s^2\,{}^1S_0$-$2s2p\,{}^3P_1$ spin-forbidden transition for the Be-like isoelectronic sequence. *Phys. Rev. A*, **51**, 2020–2030.

[10287] Ynnerman A., James J., Lindgren I., Persson H. and Salomonson S. (1994). Many-body calculation of the $2p_{1/2,3/2}$-$2s_{1/2}$ transition energies in Li-like ^{238}U. *Phys. Rev. A*, **50**, 4671–4678.

[10288] Yokojima S., Komachiya M. and Fukuda R. (1993). Derivation of Bethe-Salpeter type N-body bound state equation. *Nucl. Phys. B*, **390**, 319–352.

[10289] Yoo C.S., Cynn H. and Söderlind P. (1998). Phase diagram of uranium at high pressures and temperatures. *Phys. Rev. B*, **57**, 10359–10362.

[10290] Yoo R.K., Ruscic B. and Berkowitz J. (1995). Photoionization of atomic bismuth. *J. Phys. B*, **28**, 1743–1759.

[10291] Young I.G. and Norrington P.H. (1994). Solution of the relativistic asymptotic equations in electron-ion scattering. *Comp. Phys. Comm.*, **83**, 215–226.

[10292] Young L., Hasegawa S., Kurtz C., Datta D. and Beck D.R. (1995). Hyperfine structure studies of Nb II: Experimental and relativistic configuration-interaction results. *Phys. Rev. A*, **51**, 3534–3540.

[10293] Young L., Kurtz C.A., Beck D.R. and Datta D. (1993). Hyperfine-structure studies of Zr II: Experimental and relativistic configuration-interaction results. *Phys. Rev. A*, **48**, 173–181.

[10294] Younk E.H. and Kunz A.B. (1997). An ab initio investigation of the electronic structure of lithium azide (LiN_3), sodium azide (NaN_3), and lead azide ($Pb(N_3)_2$). *Int. J. Quantum Chem.*, **63**, 615–621.

[10295] Yu A. and Takahashi H. (1998). Quantum processes in the field of a two-frequency circularly polarized plane electromagnetic wave. *Phys. Rev. E*, **57**, 2276–2282.

[10296] Yu M. and Dolg M. (1997). Covalent contributions to bonding in group 12 dimers M_2 (M = Zn, Cd, Hg). *Chem. Phys. Lett.*, **273**, 329–336.

[10297] Yuan J.M. (1995). Intra-atomic relativistic effects on the spin polarization in low-energy electron scattering from Ca, Sr, Ba, and Yb atoms. *Phys. Rev. A*, **52**, 4647–4655.

[10298] Yuan J.M. and Zhang Z.J. (1993a). Spin polarization of electrons elastically scattered from argon atoms in the Ramsauer-Townsend region. *Z. Phys. D*, **25**, 285–286.

[10299] Yuan J.M. and Zhang Z.J. (1993b). Enhanced spin polarization of elastic electron scattering from alkaline-earth-metal atoms in Ramsauer-Townsend and low-lying shape resonance regions. *Phys. Rev. A*, **48**, 2018–2023.

[10300] Yuan X., Dougherty R.W., Das T.P. and Andriessen J. (1995a). Relativistic calculations of the hyperfine interactions in the excited 7 $^2P_{3/2}$ and 7 $^2P_{1/2}$ states of the Ra^+ ion. *Phys. Rev. A*, **52**, 3563–3571.

[10301] Yuan X., Panigrahy S.N., Dougherty R.W., Das T.P. and Andriessen J. (1995b). Hyperfine structures of Ca^+ and Sr^+ ions: Summary of trends in hyperfine interactions in the alkaline-earth-metal ions and corresponding series with similar electronic structures. *Phys. Rev. A*, **52**, 197–207.

[10302] Yudanov I.V., Vent S., Neyman K., Pacchioni G. and Rösch N. (1997). Adsorption of Pd atoms and Pd_4 clusters on MgO(001) surface: a density functional study. *Chem. Phys. Lett.*, **275**, 245–252.

[10303] Zaitsevskii A., Teichteil C., Vigué J. and Bazalgette G. (1999). Quasirelativistic transition moment calculations using the multipartitioning perturbation theory: $B0^+(^3\Pi)\rightarrow X0^+(^1\Sigma^+)$ transitions in IF and ICl. *Chem. Phys. Lett.*, **307**, 277–282.

[10304] Zakharov O. and Cohen M.L. (1995). Theory of structural, electronic, vibrational, and superconducting properties of high-pressure phases of sulfur. *Phys. Rev. B*, **52**, 12572–12578.

[10305] Zakout I. and Sever R. (1994). Relativistic description of heavy $q\bar{q}$ bound states. *Phys. Rev. D*, **50**, 4611–4618.

[10306] Záliš S., Stoll H., Baerends E.J. and Kaim W. (1999). The d^0, d^1 and d^2 configurations in known and unknown tetrathiometal compounds MS_4^{n-} (M = Mo, Tc, Ru; W, Re, Os). A quantum chemical study. *Inorg. Chem.*, **38**, 6101–6105.

[10307] Zamick L. and Zheng D.C. (1994). Nuclear structure with Dirac phenomenology. *Phys. Rep.*, **242**, 233–251.

[10308] Zanasi R. and Lazzeretti P. (1998). On the stabilization of natural L-enantiomers of α-amino acids via parity-violating effects. *Chem. Phys. Lett.*, **286**, 240–242.

[10309] Zarić S., Couty M. and Hall M.B. (1997). Ab initio calculation of the geometry and vibrational frequencies of the triplet state of tungsten pentacarbonyl amine: A model for the unification of the preresonance Raman and the time-resolved infrared experiments. *J. Am. Chem. Soc.*, **119**, 2885–2888.

[10310] Zarzo A. and Martinez A. (1993). The quantum relativistic harmonic oscillator: spectrum of zeros of its wave functions. *J. Math. Phys.*, **34**, 2926–2935.

[10311] Zeman V., McEachran R.P. and Stauffer A.D. (1994a). Relativistic distorted-wave calculation of electron impact excitation of caesium. *J. Phys. B*, **27**, 3175–3188.

[10312] Zeman V., McEachran R.P. and Stauffer A.D. (1994b). Relativistic calculation of Stokes' parameters for intermediate energy electron impact excitation of caesium atoms. *Z. Phys. D*, **30**, 145–148.

[10313] Zeman V., McEachran R.P. and Stauffer A.D. (1995a). Relativistic distorted-wave calculation of electron impact excitation of caesium: II . *J. Phys. B*, **28**, 1835–1849.

[10314] Zeman V., McEachran R.P. and Stauffer A.D. (1995b). A test of the *LS* approximation for electron-impact excitation of caesium . *J. Phys. B*, **28**, 3063–3077.

[10315] Zeman V., McEachran R.P. and Stauffer A.D. (1996). Relativistic effects exhibited by the generalized Stokes parameters for electron impact excitation of quasi one-electron atoms. *J. Phys. B*, **29**, 5937–5946.

[10316] Zeman V., McEachran R.P. and Stauffer A.D. (1997). Relativistic calculation of superelastic electron - alkali atom scattering. *J. Phys. B*, **30**, 3475–3490.

[10317] Zeman V., McEachran R.P. and Stauffer A.D. (1998). Relativistic distorted-wave calculation of inelastic electron - alkali atom scattering. *Eur. Phys. J. D*, **1**, 117–128.

[10318] Zeng W.S., Heine V. and Jepsen O. (1997). The structure of barium in the hexagonal close-packed phase under high pressure. *J. Phys. CM*, **9**, 3489–3502.

[10319] Zhang H.L. (1998). Relativistic calculations of photoionization cross sections. *Phys. Rev. A*, **57**, 2640–2650.

[10320] Zhang H.L. and Pradhan A.K. (1994). Electron-impact excitation along an isoelectronic sequence: B-like ions. *Phys. Rev. A*, **50**, 3105–3116.

[10321] Zhang H.L. and Pradhan A.K. (1995a). Relativistic and radiation damping effects in electron-impact excitation of highly charged ions. *J. Phys. B*, **28**, L285–L292.

[10322] Zhang H.L. and Pradhan A.K. (1995b). Relativistic and electron-correlation effects in electron-impact excitation of Fe^{2+}. *J. Phys. B*, **28**, 3403–3414.

[10323] Zhang H.L. and Sampson D.H. (1993). Procedures for improving a relativistic distorted-wave approach for excitation of ions by electron impact. *Phys. Rev. A*, **47**, 208–214.

[10324] Zhang H.L. and Sampson D.H. (1994a). Relativistic distorted-wave collision strengths and oscillator strengths for the 105 $\Delta n = 0$ transitions with n=2 in the 85 B-like ions with $8 \leq Z \leq 92$. *At. Data Nucl. Data Tables*, **56**, 41–104.

[10325] Zhang H.L. and Sampson D.H. (1994b). Relativistic distorted-wave collision strengths and oscillator strengths for all possible $n = 2$ - $n = 3$ transitions in B-like ions. *At. Data Nucl. Data Tables*, **58**, 255–305.

[10326] Zhang H.L. and Sampson D.H. (1995). Relativistic collision strengths for optically allowed $\Delta n = 0$ transitions between magnetic sublevels of highly charged ions. *Phys. Rev. A*, **52**, 3827–3832.

[10327] Zhang H.L. and Sampson D.H. (1997). Relativistic distorted-wave collision strengths and oscillator strengths for all possible $n = 2$ - $n = 3$ transitions in C-like ions. *At. Data Nucl. Data Tables*, **65**, 183–271.

[10328] Zhang H.L. and Sampson D.H. (1999). Relativistic distorted-wave collision strengths and oscillator strengths for the 105 $\Delta n = 0$ transitions with n=2 in the 81 N-like ions with $12 \leq Z \leq 92$. *At. Data Nucl. Data Tables*, **72**, 153–216.

[10329] Zhang H.X. and Balasubramanian K. (1993). Spectroscopic constants and potential energy curves for 15 electronic states of Ag_2. *J. Chem. Phys.*, **98**, 7092–7097.

[10330] Zhang T. (1996a). QED corrections to O($\alpha^7 mc^2$) fine-structure splittings in helium. *Phys. Rev. A*, **53**, 3896–3914.

[10331] Zhang T. (1996b). Corrections to $O(\alpha^7(\ln\alpha)mc^2)$ fine-structure splittings and $O(\alpha^6(\ln\alpha)mc^2)$ energy levels in helium. *Phys. Rev. A*, **54**, 1252–1312.

[10332] Zhang T. (1997). Three-body corrections to $O(\alpha^6)$ fine structure in helium. *Phys. Rev. A*, **56**, 270–277.

[10333] Zhang T. and Drake G.W.F. (1994a). QED correction of O($\alpha^6 mc^2$) to the fine structure splittings of helium and positronium. *Phys. Rev. Lett.*, **72**, 4078–4081.

[10334] Zhang T. and Drake G.W.F. (1994b). A rigorous treatment of O($\alpha^6 mc^2$) QED corrections to the fine structure splittings of helium. *J. Phys. B*, **27**, L311–L316.

[10335] Zhang T. and Drake G.W.F. (1996). Corrections to O($\alpha^7 mc^2$) fine-structure splittings in helium. *Phys. Rev. A*, **54**, 4882–4922.

[10336] Zhang T. and Xiao L.X. (1994). Hyperfine correlations to order α^6 in a relativistic formalism for positronium. *Phys. Rev. A*, **49**, 2411–2414.

[10337] Zhang Y., Holzwarth A.W. and Williams R.T. (1998). Electronic band structures of the scheelite materials $CaMoO_4$, $CaWO_4$, $PbMoO_4$, and $PbWO_4$. *Phys. Rev. B*, **57**, 12738–12750.

[10338] Zhang Y., Zhao C.Y. and You X.Z. (1997). Systematic theoretical study of structures and bondings of the charge-transfer complexes of ammonia with HX, XY, and X_2 (X and Y are halogens). *J. Phys. Chem. A*, **101**, 2879–2885.

[10339] Zhang Y., Zhu Q.Y. and Pan S.F. (1992b). The energy levels and radiative transition probilities of Na-like niobium (in Chinese). *Acta Sci. Natur. Univ. Jilin*, **May(2)**, 60–63.

[10340] Zhang Z.Y. and Pitzer R.M. (1999). Application of relativistic quantum chemistry to the electronic energy levels of the uranyl ion. *J. Phys. Chem. A*, **103**, 6880–6886.

[10341] Zhao K. and Pitzer R.M. (1996). Electronic structure of C_{28}, Pa@C_{28}, and U@C_{28}. *J. Phys. Chem.*, **100**, 4798–4802.

[10342] Zheng Y.X. (1993). Regularity of weak solutions to a two-dimensional modified Dirac-Klein-Gordon system of equations. *Comm. Math. Phys.*, **151**, 67–87.

[10343] Zhorin V.V. and Liu G.K. (1998). Modelling crystal-field interactions for f-elements in $LaCl_3$. *J. Alloys Compounds*, **275-277**, 137–141.

[10344] Zhou F. and Spruch L. (1995). van der Waals and retardation (Casimir) interactions of an electron or an atom with multilayered walls. *Phys. Rev. A*, **52**, 297–310.

[10345] Zhou M.F. and Andrews L. (1998). Reactions of laser-ablated niobium, tantalum, and rhenium atoms with nitrogen atoms and molecules. Infrared spectra and density functional calculations of the metal nitride and dinitride molecules. *J. Phys. Chem. A*, **102**, 9061–9071.

[10346] Zhou M.F. and Andrews L. (1999a). Infrared spectra and density functional calculations of the CrO_2^-, Mo_2^- and WO_2^- molecular anions in solid neon. *J. Chem. Phys.*, **111**, 4230–4238.

[10347] Zhou M.F. and Andrews L. (1999b). Infrared spectra and density functional calculations of $RuCO^+$, $OsCO^+$, $Ru(CO)_x$, $Os(CO)_x$, $Ru(CO)_x^-$ ans $Os(CO)_x^-$ (x = 1-4) in solid neon. *J. Phys. Chem. A*, **103**, 6956–6968.

[10348] Zhou M.F. and Andrews L. (1999c). Infrared spectra of CNbO, CMO^-, OMCCO, $(C_2)MO_2$ and $M(CO)_x$ (x = 1-6)(M = Nb, Ta) in solid neon. *J. Phys. Chem. A*, **103**, 7785–7794.

[10349] Zhou M.F. and Andrews L. (1999d). Infrared spectra of $RhCO^+$, RhCO, and $RhCO^-$ in solid neon: A scale for charge in supported Rh(CO) catalyst systems. *J. Am. Chem. Soc.*, **121**, 9171–9175.

[10350] Zhou M.F. and Andrews L. (1999e). Infrared spectra and pseudopotential calculations for NUO^+, NUO, and NThO in solid neon. *J. Chem. Phys.*, **111**, 11044–11049.

[10351] Zhou M.F., Andrews L., Li J. and Bursten B.E. (1999a). Reaction of laser-ablated uranium atoms with CO: Infrared spectra of the CUO, CUO^-, OUCCO, $(\eta^2\text{-}C_2)UO_2$, and $U(CO)_x$ (x = 1-6) molecules in solid neon. *J. Am. Chem. Soc.*, **121**, 9712–9721.

[10352] Zhou M.F., Andrews L., Li J. and Bursten B.E. (1999b). Reactions of Th atoms with CO: The first thorium carbonyl complex and an unprecedented bent triplet insertion product. *J. Am. Chem. Soc.*, **121**, 12188–12189.

[10353] Zhu Y.F., Grant E.R. and Lefebvre-Brion H. (1993). Spin-orbit and rotational autoionization in HCl and DCl. *J. Chem. Phys.*, **99**, 2287–2299.

[10354] Zhu Z.H., Yin Y., Tan M.L., Fu Y.B. and Wang X.L. (1996). Energy levels for U^{++} and electronic states for UO_2^{++} (in Chinese). *Yuanzi Yu Fenzi Wuli Xuebao*, **13**, 468–470.

[10355] Zhuang P.F. and Heinz U. (1996). Relativistic kinetic equations for electromagnetic, scalar, and pseudoscalar interactions. *Phys. Rev. D*, **53**, 2096–2101.

[10356] Ziegler K. (1998). Delocalization of 2D Dirac fermions: the role of a broken supersymmetry. *Phys. Rev. Lett.*, **80**, 3113–3116.

[10357] Zilitis V.L. (1993). Theoretical determination of Rydberg energy levels for lithiumlike ions. *Opt. Spektr.*, **74**, 801–805.

[10358] Zilitis V.L. (1993). Theoretical determination of Rydberg energy levels for lithiumlike ions. *Opt. Spectr.*, **74**, 477–479.

[10359] Zilitis V.L. (1994). Theoretical study of the energy level structure for the isoelectronic sequence of aluminum-like ions. *Opt. Spektr.*, **76**, 891–895.

[10360] Zilitis V.L. (1994). Theoretical study of the energy level structure for the isoelectronic sequence of aluminum-like ions. *Opt. Spectr.*, **76**, 793–797.

[10361] Zilitis V.A. (1995). Theoretical study of the fine structure of $3\ ^2P$ and $3\ ^2D$ energy levels for ions of the isoelectronic series of aluminum (in Russian). *Opt. Spektr.*, **78**, 726–728.

[10362] Zilitis V.A. (1995). Theoretical study of the fine structure of $3\ ^2P$ and $3\ ^2D$ energy levels for ions of the isoelectronic series of aluminum. *Opt. Spectr.*, **78**, 652–654.

[10363] Znojil M. (1996). Harmonic oscillations in a quasi-relativistic regime. *J. Phys. A*, **29**, 2905–2917.

[10364] Zolotorev M. and Budker D. (1997). Parity nonconservation in relativistic hydrogenic ions. *Phys. Rev. Lett.*, **78**, 4717–4720.

[10365] Zolotorev M. and Budker D. (1999). Parity nonconservation in relativistic hydrogenic ions. *AIP Conf. Proc.*, **457**, 175–176.

[10366] Zouchoune B., Ogliaro F., Halet J.F., Saillard J.Y., Eveland J.R. and Whitmire K.H. (1998). Bonding analysis in inorganic transition-metal cubic clusters. 3. Metal-centered tetracapped $M_9(\mu_5\text{-}E)_4L_n$ species with a tetragonal distortion. *Inorg. Chem.*, **37**, 865–875.

[10367] Zumdahl S.S. (1995). *Chemical Principles, 2nd Ed.*. D. C. Heath, Lexington, MA, 1044 p. See pp. 548-549.

[10368] Zumdahl S.S. (1998). *Chemical Principles, 3rd Ed.*. Houghton Mifflin, Boston, MA, see pp. 548-549.

[10369] Zurita S., Rubio J., Illas F. and Barthelat J.C. (1996). Ab initio electronic structure of PtH^+, PtH, Pt_2, and Pt_2H from a one-electron pseudopotential approach. *J. Chem. Phys.*, **104**, 8500–8506.

Druck: Strauss Offsetdruck, Mörlenbach
Verarbeitung: Schäffer, Grünstadt

Lecture Notes in Chemistry

For information about Vols. 1–35
please contact your bookseller or Springer-Verlag

Editorial Policy

This series aims to report new developments in chemical research and teaching - quickly, informally and at a high level. The type of material considered for publication includes:

1. Preliminary drafts of original papers and monographs
2. Lectures on a new field, or presenting a new angle on a classical field
3. Seminar work-outs
4. Reports of meetings, provided they are
   a) of exceptional interest and
   b) devoted to a single topic.

Texts which are out of print but still in demand may also be considered if they fall within these categories.

The timeliness of a manuscript is more important than its form, which may be unfinished or tentative. Thus, in some instances, proofs may be merely outlined and results presented which have been or will later be published elsewhere. If possible, a subject index should be included. Publication of Lecture Notes is intended as a service to the international chemical community, in that a commercial publisher, Springer-Verlag, can offer a wider distribution to documents which would otherwise have a restricted readership. Once published and copyrighted, they can be documented in the scientific literature.

Manuscripts

Manuscripts should comprise not less than 100 and preferably not more than 500 pages. They are reproduced by a photographic process and therefore must be submitted in camera-ready form according to Springer-Verlag's specifications: technical instructions will be sent on request.

The text area should take care of the page length and width (12.2 x 19.3 cm when you use a 10 point font size, 15.3 x 24.2 cm for a 12 point font size).

Authors receive 50 free copies and are free to use the material in other publications.

Manuscripts should be sent to one of the editors or directly to Springer-Verlag, Heidelberg.